Birkhäuser

ISNM

International Series of Numerical Mathematics

Volume 171

More information about this series at https://link.springer.com/bookseries/4819

Peter Benner · Tobias Breiten ·
Heike Faßbender · Michael Hinze ·
Tatjana Stykel · Ralf Zimmermann
Editors

Model Reduction of Complex Dynamical Systems

Birkhäuser

Editors
Peter Benner
Dynamics of Complex Technical Systems
Max Planck Institute
Magdeburg, Sachsen-Anhalt, Germany

Tobias Breiten
Institute of Mathematics
Technical University of Berlin
Berlin, Germany

Heike Faßbender
Institute for Numerical Analysis
TU Braunschweig
Braunschweig, Germany

Michael Hinze
Mathematisches Institut
Universität Koblenz-Landau, Campus Koblenz
Koblenz, Germany

Tatjana Stykel
Institut für Mathematik
Universität Augsburg
Augsburg, Germany

Ralf Zimmermann
Department of Mathematics and Computer Science
University of Southern Denmark
Odense, Denmark

ISSN 0373-3149 ISSN 2296-6072 (electronic)
International Series of Numerical Mathematics
ISBN 978-3-030-72985-1 ISBN 978-3-030-72983-7 (eBook)
https://doi.org/10.1007/978-3-030-72983-7

Mathematics Subject Classification: 65-06, 93-06

This book is published under the imprint Birkhäuser, www.birkhauser-science.com by the registered company Springer Nature Switzerland AG
The registered company address is: Gewerbestrasse 11, 6330 Cham, Switzerland

Preface

The workshop series on *Model Reduction of Complex Dynamical Systems—MODRED* aims to bring together researchers and users of model order reduction techniques with focus on time-dependent problems. This includes, in particular,

- *system-theoretic methods* like, e.g., balanced truncation, Hankel norm approximation, rational interpolation (moment-matching, H_2-optimal reduction), proper orthogonal decomposition (POD) and generalizations, as well as reduced basis methods;
- *data-driven methods*, e.g., vector fitting, Loewner matrix and pencil based approaches, dynamic mode decomposition (DMD), and kernel-based methods;
- *surrogate modeling* for design and optimization, with special emphasis on control and data assimilation;
- *model reduction methods in applications*, e.g., control and network systems, computational electromagnetics, computational nanoelectronics, structural mechanics, fluid dynamics, and digital twins, in general.

The *MODRED* workshop series started in 2008 at Hamburg University, then under the name *Model Reduction for Circuit Simulation*. It was continued in Berlin 2010, Magdeburg 2013, and Odense 2017.[1] The fifth edition took place at Karl-Franzens-Universität Graz in Austria, August 28–30, 2019, with keynote contributions of

- Serkan Gugercin (Virginia Tech),
- Bernard Haasdonk (University of Stuttgart),
- Laura Iapichino (TU Eindhoven),
- J. Nathan Kutz (University of Washington), and
- Utz Wever (Siemens).

This volume contains papers related to presentations given there.

[1] This was supposed to be MODRED 2016, but due to certain constraints had to be moved to January 2017.

Methods and Techniques of Model Order Reduction

In *On Bilinear Time Domain Identification and Reduction in the Loewner Framework*, D. S. Karachalios, I. V. Gosea, and A. C. Antoulas propose a two-step procedure that learns a reduced bilinear control system based on time-domain measurements. In a first step, a linear surrogate model is fitted which subsequently is extended by a suitably fitted bilinear operator. The approach combines the Loewner framework with Volterra series representations.

Large-scale linear control systems are routinely tackled with the model reduction method of balanced truncation. Balanced truncation requires solving a pair of Lyapunov equations for two Gramians associated with the system. The work *Balanced Truncation for Parametric Linear Systems Using Interpolation of Gramians: a Comparison of Algebraic and Geometric Approaches* by N. T. Son, P.-Y. Gousenbourger, E. Massart, and T. Stykel addresses such linear control systems in the presence of additional parameter dependencies. Their approach to parametric MOR is to approximate the Gramians of the Lyapunov equations at a given parameter via interpolation. Two methods are proposed: direct matrix interpolation (called the algebraic approach) and interpolation of the Gramians on the matrix manifold of $n \times n$ positive semidefinite matrices of fixed rank (called the geometric approach). Both methods are juxtaposed and assessed by means of numerical examples.

Dynamic mode decomposition (DMD) is a data-driven method for learning the dynamics of complex nonlinear systems. This information, in turn, can be exploited to construct reduced-order models. In the contribution *Toward Fitting Structured Nonlinear Systems by Means of Dynamic Mode Decomposition* of I. V. Gosea and I. Pontes Duff, two specifications of the DMD method, namely, DMD with control and input-output DMD are extended to the case of fitting control problems with bilinear and quadratic-bilinear terms. The authors present the general procedure and detail the computation of the reduced-order matrices that are required for the task at hand. Then, the proposed approaches are demonstrated on the viscous Burgers' equation and the coupled van der Pol oscillators.

In *Clustering-based Model Order Reduction for Nonlinear Network Systems*, P. Benner, S. Grundel, and P. Mlinarić discuss model reduction of nonlinear multi-agent systems by using a combination of a projection-based model reduction method and a k-means clustering algorithm. An important property of this approach is that it preserves the network structure. Numerical examples including a nonlinear oscillator network are used to illustrate the performance of the approach.

In *Adaptive Interpolatory MOR by Learning the Error Estimator in the Parameter Domain*, S. Chellappa, L. Feng, V. de la Rubia, and P. Benner present an adaptive training technique for interpolatory model order reduction of parametric linear systems. Their approach is based on learning the error estimator over the parameter domain by evaluating on a coarse training parameter set only and interpolation using radial basis functions. The training parameter set is adaptively enlarged by new points identified by evaluating the interpolated error estimator on a fine training set. The efficiency of this approach is demonstrated by three numerical examples.

The contribution *A link between Gramian based model order reduction and moment matching* of C. Bertram and H. Faßbender addresses asymptotically stable, linear time-invariant single-input single-output dynamical system. In order to reduce such systems, a balancing-related approach is pursued that is derived from numerical integration. More precisely, an ordinary differential equation for the Gramians inherent to the LTI system at hand is considered and approximate solutions are obtained via Runge-Kutta methods. This corresponds to a quadrature framework for Lyapunov and Sylvester equations. Eventually, the work establishes a bridge between balanced POD and moment-matching techniques.

C. Himpe provides a detailed comparison of (empirical) Gramian-based model reduction methods in *Comparing (Empirical-Gramian-Based) Model Order Reduction Algorithms*. The performance of the individual MOR frameworks is compared to each other based on common benchmark problems and different system norms. Additionally, with MOR score, a new performance index is introduced and investigated in the empirical Gramian MOR context.

R. Ullmann, S. Sicklinger, and G. Müller discuss a parametric model order reduction approach for the frequency-domain analysis of complex industry models in their chapter *Optimization-based Parametric Model Order Reduction for the Application to the Frequency Domain Analysis of Complex Systems*. Here, the challenge arises from a high-dimensional input parameter space on the one hand, but the restriction to only a few full-order model evaluations due to budget constraints on the other hand. This is tackled using a global basis approach for model order reduction, in combination with an optimization-based greedy search strategy for the model training and an *a posteriori* statistical error evaluation based on Bayesian inference.

Control systems with time delay appear frequently in control engineering and design as well as in nano- and micro-electronics. A new model order reduction technique based on balanced truncation and relying on the solution of extended linear matrix inequalities is discussed in the chapter *On Extended Model Order Reduction for Linear Time Delay Systems* by S. Naderi Lordejani, B. Besselink, A. Chaillet, and N. van de Wouw. The new approach guarantees stability preservation and comes with an *a priori* error bound similar to the classical twice-the-tail-of-Hankel-singular-values bound known to hold for balanced truncation applied to stable linear time-invariant systems.

Applications of Model Order Reduction

A. Jungiewicz, C. Ludwig, S. Sun, U. Wever, and R. Wüchner discuss aspects of model order reduction in the design of digital twins of electric motors/generators or gas turbines. While thermal and mechanical parts are often available as mathematical models in matrix form, thus allowing classical projection-based model order reduction, coupling terms are often not accessible when using commercial software. Thus, they suggest an approach to infer a coupling model from data in their contribution *A Practical Method for the Reduction of Linear Thermo-mechanical Dynamic Equations*.

Reduced Order Methods in Medical Imaging are investigated by S. Chaturantabut, T. Freeze, E. S. Helou, and C. H. Lee. They study POD reduced-order modeling as a tool for image compression and reconstruction. They demonstrate how POD can be used very efficiently to compress large sets of medical tomography data into a much smaller set of representative modes that can be used to reconstruct any image in the set with a high degree of accuracy.

In their chapter *Efficient Krylov Subspace Techniques for Model Order Reduction of Automotive Structures in Vibroacoustic Applications*, H. K. Sreekumar1, R. Ullmann, S. Sicklinger, and S. C. Langer discuss the application of Krylov subspace methods for model order reduction of damped vibrating systems arising in acoustics. As they consider general damping models in contrast to many other papers that restrict themselves to particular damping structures, special emphasis is given to the fact that the system matrices can be complex rather than real for certain damping strategies. This requires special care when implementing Krylov subspace methods in order to achieve sufficient efficiency in an industrial context.

G. Pascarella and M. Fossati present an adaptive reduced-order model selection framework for the accurate reconstructions of vortex-dominated unsteady flows by means of the reduced basis method in the chapter *Model-based Adaptive MOR Framework for Unsteady Flows Around Lifting Bodies*. They illustrate the performance of the approach for two numerical examples, utilizing a variety of common reduced-order methods like POD, spectral POD, DMD, and recursive DMD.

Mathematical modeling of permanent magnet synchronous motors (PMSM) through nonlinear magnetostatics equations leads to quasilinear elliptic partial differential equations. To prepare for the model-based design of PMSM, M. Hinze and D. Korolev propose a certified reduced basis method for parameterized quasilinear elliptic problems in *Reduced Basis Methods for Quasilinear Elliptic PDEs with Applications to Permanent Magnet Synchronous Motors*. They apply the empirical interpolation method to reduce the non-polynomial nonlinearity, and thus to guarantee an efficient offline-online computational procedure.

Often, dynamical models incorporate certain invariants, which are bound to be preserved by the laws of physics, e.g., conservation of mass or energy. The classical methods of model reduction are not designed to preserve such invariants. The featured article *Structure-preserving Reduced Order Modeling of non-traditional Shallow Water Equation* by S. Yildiz, M. Uzunca, and B. Karasözen proposes an energy-preserving reduced-order model for the shallow water equation with full Coriolis force. First, a non-canonical Hamiltonian form for the full-order model is introduced that features a certain skew-gradient structure. Then, the method of proper orthogonal decomposition is adapted such that the skew-gradient structure also shows in the reduced-order model. The authors illustrate their approach by means of two numerical examples of increasing sophistication.

Benchmarks and Software for Model Order Reduction

S. Rave and J. Saak present *A Non-stationary Thermal-Block Benchmark Model for Parametric Model Order Reduction*. This benchmark of a parametric heat conduction problem is used in the following chapters of this volume to test implementations of different model reduction approaches.

P. Mlinarić, S. Rave, and J. Saak give an overview of the free software library pyMOR in *Parametric Model Order Reduction Using pyMOR*. pyMOR consists of several system-theoretic as well as reduced basis methods. In their contribution, they experimentally compare these approaches using the benchmark model presented in the previous chapter of this volume.

In *Matrix Equations, Sparse Solvers: M-M.E.S.S.-2.0.1—Philosophy, Features and Application for (Parametric) Model Order Reduction*, P. Benner, M. Köhler, and J. Saak describe the MATLAB toolbox M-M.E.S.S. in version 2.0.1 which provides solvers for large-scale sparse symmetric algebraic and differential Lyapunov and Riccati matrix equations. They also demonstrate the usage of this toolbox for balancing-related and interpolatory model reduction of different types of linear dynamical systems including first- and second-order systems, structured differential-algebraic equations, and parametric systems, whereas, in the latter case, again the parametric benchmark from the penultimate chapter is employed.

P. Benner and S. Werner present their model reduction toolbox in *MORLAB—The Model Order Reduction LABoratory*. MORLAB includes a variety of system-theoretic reduction techniques. The chapter includes a detailed introduction into the toolbox structure, function interfaces, and its documentation. The underlying mathematical principles (spectral splitting) are explained for first- and second-order linear time-invariant control systems.

Magdeburg, Germany	Peter Benner
Berlin, Germany	Tobias Breiten
Braunschweig, Germany	Heike Faßbender
Koblenz, Germany	Michael Hinze
Augsburg, Germany	Tatjana Stykel
Odense, Denmark	Ralf Zimmermann
November 2020	

Acknowledgements The editors thank the anonymous referees for their invaluable support in evaluating the contributions to this proceedings volume, and thus assuring the quality of the content of this book.

Contents

Methods and Techniques of Model Order Reduction

On Bilinear Time-Domain Identification and Reduction in the Loewner Framework

D. S. Karachalios, I. V. Gosea, and A. C. Antoulas

Abstract The *Loewner framework* (LF) in combination with *Volterra series* (VS) offers a non-intrusive approximation method that is capable of identifying bilinear models from time-domain measurements. This method uses harmonic inputs which establish a natural way for data acquisition. For the general class of nonlinear problems with VS representation, the growing exponential approach allows the derivation of the generalized kernels, namely, *symmetric generalized frequency response functions* (GFRFs). In addition, the homogeneity of the Volterra operator determines the accuracy in terms of how many kernels are considered. For the weakly nonlinear setup, only a few kernels are needed to obtain a good approximation. In this direction, the proposed adaptive scheme is able to improve the estimations of the computationally nonzero kernels. The Fourier transform associates these measurements with the derived GFRFs and the LF makes the connection with system theory. In the linear case, the LF associates the so-called S-parameters with the linear transfer function by interpolating in the frequency domain. The goal of the proposed method is to extend identification to the case of bilinear systems from time-domain measurements and to approximate other general nonlinear systems (by means of the Carleman bilinearizarion scheme). By identifying the linear contribution with the LF, a considerable reduction is achieved by means of the SVD. The fitted linear system has the same

D. S. Karachalios (✉) · I. V. Gosea · A. C. Antoulas
Max Planck Institute for Dynamics of Complex Technical Systems, Data-Driven System Reduction and Identification (DRI), Magdeburg, Germany
e-mail: karachalios@mpi-magdeburg.mpg.de

I. V. Gosea
e-mail: gosea@mpi-magdeburg.mpg.de

A. C. Antoulas
e-mail: aca@rice.edu

A. C. Antoulas
Electrical and Computer Engineering Department, Rice University, Houston, TX 77005, USA

Baylor College of Medicine, Houston, TX 77030, USA

P. Benner et al. (eds.), *Model Reduction of Complex Dynamical Systems*, International Series of Numerical Mathematics 171,
https://doi.org/10.1007/978-3-030-72983-7_1

McMillan degree as the original linear system. Then, the performance of the linear model is improved by augmenting a special nonlinear structure. In a nutshell, we learn reduced-dimension bilinear models directly from a potentially large-scale system that is simulated in the time domain. This is done by fitting first a linear model, and afterward, by fitting the corresponding bilinear operator.

1 Introduction

In natural sciences, evolutionary phenomena can be modeled as dynamical systems. An ever-increasing need for improving the approximation accuracy has motivated including more involved and detailed features in the modeling process, thus inevitably leading to large-scale dynamical systems [3]. To overcome this problem, efficient finite methods heavily rely on *model reduction*. Model reduction methods can be classified into two broad categories, namely, *SVD based* and *Krylov based* (moment matching).

The most prominent among the SVD-based methods is *balanced truncation* (BT). In general, balancing methods are based on the computation of controllability and observability *gramians* and lead to the elimination of state variables which are difficult to reach and to observe. Besides having high-computational cost of solving the associated matrix Lyapunov equations, the advantages of balancing methods include the preservation of stability and an a priori computable error bound. For more details on these topics as well as on other model reduction methods not treated here (e.g., proper orthogonal decomposition (POD)/reduced basis (RB)), we refer the reader to the book [3] and the surveys [9, 15].

One way to perform model reduction is by employing *tangential interpolation*. These methods are known as rational *Krylov methods* or *moment-matching* methods. Krylov-based methods are numerically efficient and have lower computational cost, but, in general, the preservation of other properties (e.g., stability or passivity) is not automatic. For an extensive study in *interpolatory model reduction*, we refer the reader to the recent book [4]. In what follows, we will consider exclusively *interpolatory model reduction methods* and, in particular, the LF. For recent surveys on the LF, see [2, 6, 29]. The sensitivity to noise in the LF was already discussed in [19, 30].

When *input-output* data are offered, *data-driven* methods, such as the *Loewner framework* (LF), *dynamic mode decomposition* (DMD) [37], *sparse identification of nonlinear systems (with control)* (SINDYc) in [27], *vector fitting* (VF) [23], *Hankel* [25] or *subspace* methods [8, 24, 26], *moment-matching* [36], and *operator inference* [13, 14, 32], remain the only feasible approaches for recovering the hidden information.

DMD-based methods represent viable alternatives that require state-derivative estimations.

While the underlying dynamical system acts as a black box, model identification tools are important for the reliability of the discovered models (i.e., stability,

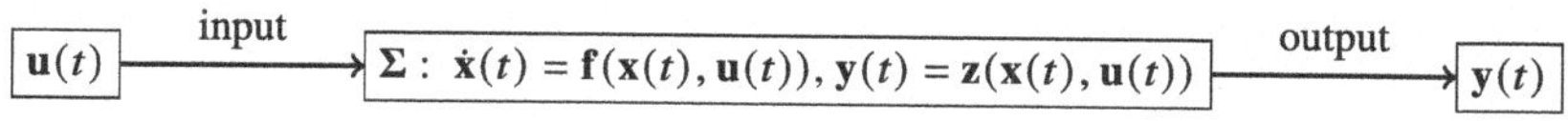

Fig. 1 Mathematical formalism for evolutionary phenomena

prediction). At the same time, these discovered models might have large dimension and hence are not suitable for fast numerical simulation and control. The LF is a direct data-driven interpolatory method able to identify and reduce models derived directly from measurements. For measured data in the frequency domain, the LF is well established for linear and nonlinear systems (e.g., bilinear or quadratic-bilinear systems) see [5, 22]. In the case of time-domain data, the LF was already applied for approximating linear models [21, 24, 31]. As the aim of this paper is to extend the identification and reduction procedure to the class of bilinear systems from time-domain data, we start our analysis by introducing the mathematical description of the *input*-$u(t)$ to *output*-$y(t)$ relation as depicted in Fig. 1. The differential and algebraic operators are denoted with $\mathbf{f}$ and, respectively, with $\mathbf{z}$. To achieve this goal, all the important steps from nonlinear system theory and interpolatory model reduction are summarized.

1.1 Outline of the Paper

The rest of the paper is organized as follows:

- Section 2 contains a brief description of system theory starting from the linear case followed by extensions to the nonlinear case by means of the Volterra series representation. The single-input and single-output case is addressed for both frequency- and time-domain representations.
- Section 3 introduces the Loewner framework as an interpolatory tool for model approximation; the results that are presented here actually set the foundation for identification and reduction of linear time-invariant systems.
- Section 4 introduces a special class of nonlinear systems, e.g., bilinear systems. The theoretical discussion for analyzing such systems starts with the growing exponential approach and the derivation of the generalized frequency response functions (GFRFs) up to the case where a double-tone input is assumed. In addition, the kernel separation strategy for improving the measurements and the linear identification/reduction part is presented. A concise algorithm that summarizes the method is presented.
- Section 5 presents the numerical experiments performed in order to illustrate the practical applicability of the newly proposed method. This section includes both a simple (low-dimensional) example and a large-scale example, compared to another state-of-the-art method.
- Section 6 presents the concluding remarks and also some potential future developments of the current method.

2 System Theory Preliminaries

In this section, we will briefly present some important material from system theory starting from the linear case.

2.1 Linear Systems

Consider SISO linear, time-invariant systems with n internal variables (called "states" whenever the matrix $\mathbf{E}$ is non-singular).

$$\Sigma_l : \begin{cases} \mathbf{E}\,\dot{\mathbf{x}}(t) = \mathbf{A}\mathbf{x}(t) + \mathbf{b}u(t), \\ \quad y(t) = \mathbf{c}\mathbf{x}(t),\ t \geq 0, \end{cases} \tag{1}$$

where $\mathbf{E},\ \mathbf{A} \in \mathbb{R}^{n\times n},\ \mathbf{b} \in \mathbb{R}^{n\times 1},\ \mathbf{c} \in \mathbb{R}^{1\times n}$. In the sequel, we will restrict our attention to invertible matrix $\mathbf{E}$ and with zero d-term ($d = 0$) in the state-output equation[1]. The explicit solution with the *convolution integral*[2] notation and the time-domain linear kernel $h(t)$ as the *impulse response* of the system can be written as

$$y(t) = \mathbf{c}e^{\mathbf{A}t}\mathbf{x}(0) + (h * u)(t),\ t \geq 0, \tag{2}$$

where multiplication with $\mathbf{E}^{-1}$ from the left has been performed in the differential part of Eq. (1). Also, we keep the same notation for the remaining matrix $\mathbf{A}$ and vector $\mathbf{b}$. By assuming zero initial conditions and performing a *Laplace transform*, we obtain the transfer function description:

$$H(s) = \frac{Y(s)}{U(s)} = \mathbf{c}(s\mathbf{I} - \mathbf{A})^{-1}\mathbf{b},\ s \in \mathbb{C}, \tag{3}$$

where $Y(s)$, $U(s)$ stand for the *input* and the *output* in the frequency domain.

2.2 Nonlinear Systems

A large class of nonlinear systems can be described by means of the Volterra-Wiener approach in [35]. Other relevant works on nonlinear systems and nonlinear modeling/identification include Schetzen (1980), Chen and Billings (1989), Boyd and Chua (1985) et. al.

The aim in this study is to identify and reduce special types of nonlinear systems (s.a., bilinear) from time-domain measurements. By knowing only the input and the

---

[1] The state-output equation often is represented as $y(t) = \mathbf{c}\mathbf{x}(t) + du(t)$.

[2] $(h * u)(t) = \int_{-\infty}^{\infty} h(\tau)u(t - \tau)d\tau$.

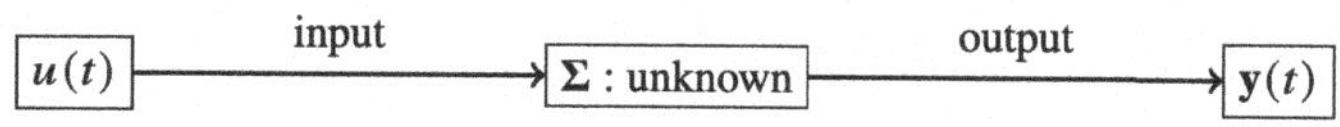

Fig. 2 The input-output mapping from the data-driven perspective with the unknown system Σ. Specific structures of the unknown system can be assumed/inspired by the physical problem. For instance, if the underlying physical phenomenon is fluid flow inside a control volume, quadratic models should be constructed, e.g., [22]

simulated or measured output in the time domain as in Fig. 2, we will identify the hidden model. In such situations where only snapshots are available, beyond the linear fit which is well established a nonlinear fit of a special type will be developed.

2.2.1 Approximation of Nonlinear Systems (Volterra Series)

The *input-output* relationship for a wide class of nonlinear systems [35] can be approximated by a Volterra series for sufficiently high N as

$$y(t) = \sum_{n=1}^{N} y_n(t), \quad y_n(t) = \int_{-\infty}^{\infty} \cdots \int_{-\infty}^{\infty} h_n(\tau_1, \ldots, \tau_n) \prod_{i=1}^{n} u(t - \tau_i) d\tau_i, \tag{4}$$

where $h_n(\tau_1, \ldots, \tau_n)$ is a *real-valued* function of $\tau_1, \ldots, \tau_n$ known as the nth-order Volterra kernel.

Definition 1 The nth-order generalized frequency response function (GFRF) is defined as

$$H_n(j\omega_1, \ldots, j\omega_n) = \int_{-\infty}^{\infty} \cdots \int_{-\infty}^{\infty} h_n(\tau_1, \ldots, \tau_n) e^{(-j \sum_{i=1}^{n} \omega_i \tau_i)} d\tau_1 \cdots d\tau_n, \tag{5}$$

which is the multidimensional Fourier[3] transform of $h_n(\tau_1, \ldots, \tau_n)$.

By applying the inverse Fourier transform of the nth-order GFRF, Eq. (5) can be written as

$$y_n(t) = \frac{1}{(2\pi)^n} \int_{-\infty}^{\infty} \cdots \int_{-\infty}^{\infty} H_n(j\omega_1, \ldots, j\omega_n) \prod_{i=1}^{n} U(j\omega_i) e^{j(\omega_1 + \cdots + \omega_n)t} d\omega_i. \tag{6}$$

The nth Volterra operator is defined as

$$V_n(u_1, u_2, ..., u_n) = \int_{-\infty}^{\infty} \cdots \int_{-\infty}^{\infty} h_n(\tau_1, ..., \tau_n) \prod_{i=1}^{n} u_i(t - \tau_i) d\tau_i, \tag{7}$$

[3] With ($j^2 = -1$), as the frequency $s = j\omega$ lies on the imaginary axis, the Laplace transform simplifies in most cases to Fourier transform (e.g., for square-integrable functions).

so that $y_n = V_n(u, u, ..., u)$ holds true.

> Homogeneity of the Volterra operator

The map $u(t) \rightarrow y_n(t)$ is homogeneous of degree n, that is, $\alpha u \rightarrow \alpha^n y_n$, $\alpha \in \mathbb{C}$. Each Volterra kernel $h_n(t)$ determines a symmetric multi-linear operator. Small amplitudes (e.g., $|\alpha| < \epsilon$) will allow ordering the nonlinear terms in such a way that terms with large powers of the amplitude (α^n) will be negligible. That is precisely the sense of approximating weakly nonlinear systems with Volterra series.

2.2.2 A Single-Tone Input

Consider the excitation of a system with an input consisting of two complex exponentials as in Eq. (8). Such inputs are typically used in chemical engineering applications as [33].

$$u(t) = A\cos(\omega t) = \left(\frac{A}{2}\right)e^{j\omega t} + \left(\frac{A}{2}\right)e^{-j\omega t}. \tag{8}$$

By using the above input in Eq. (4), we can derive the first Volterra term with $n = 1$ as

$$\begin{aligned} y_1(t) &= \int_{-\infty}^{\infty} h_1(\tau_1)[u(t-\tau_1)]d\tau_1 \\ &= \frac{A}{2}e^{j\omega t}\underbrace{\int_{-\infty}^{\infty} h_1(\tau_1)e^{-j\omega\tau_1}d\tau_1}_{H_1(j\omega)} + \frac{A}{2}e^{-j\omega t}\underbrace{\int_{-\infty}^{\infty} h_1(\tau_1)e^{j\omega\tau_1}d\tau_1}_{H_1(-j\omega)} \Rightarrow \\ y_1(t) &= \frac{A}{2}\left(e^{j\omega t}H_1(j\omega) + e^{-j\omega t}H_1(-j\omega)\right). \end{aligned} \tag{9}$$

Similarly, for the second term, we can derive

$$y_2(t) = \left(\frac{A}{2}\right)^2\left[e^{2j\omega}H_2(j\omega, j\omega) + 2e^0H_2(j\omega, -j\omega) + e^{-2j\omega}H_2(-j\omega, -j\omega)\right]. \tag{10}$$

Remark 1 (Conjugate symmetry): $H_2^*(j\omega, -j\omega) = H_2(-j\omega, j\omega),\ \forall\omega \in \mathbb{R}$.

The input amplitude is A, the angular frequency is ω, the imaginary unit is j, the first-order response function is $H_1(j\omega)$, and $H_n(j\omega, ..., j\omega)$, for $n \geq 2$, are the higher order FRFs or GFRFs. Then, the nth Volterra term can be written as

$$y_n(t) = \left(\frac{A}{2}\right)^n \sum_{p+q=n} {}^nC_q H_n^{p,q}(j\omega)e^{j\omega_{p,q}t},\ \omega_{p,q} = (p-q)\omega. \tag{11}$$

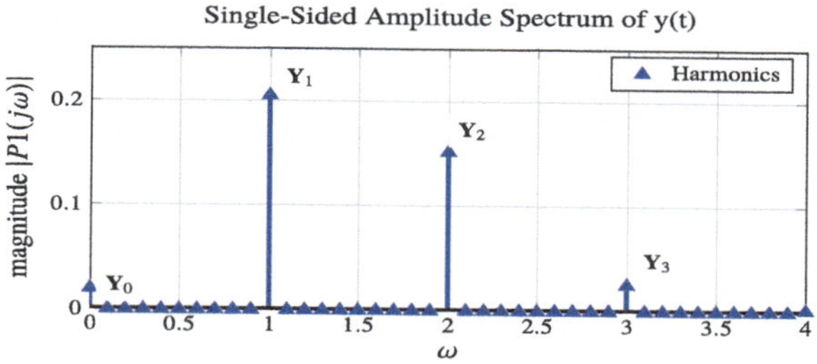

Fig. 3 An instance of the single-sided power spectrum with a singleton input with $\omega = 1$ is depicted. The underlying system is nonlinear and as a result higher harmonics appeared with a DC (direct current—non-periodic) term as well

where the following notations have been used:

$$H_n^{p,q}(j\omega) = H_n(\underbrace{j\omega, ..., j\omega}_{p-times}; \underbrace{-j\omega, ..., -j\omega}_{q-times}),\ \omega_{p,q} = (p-q)\omega,\ {}^nC_q = \frac{n!}{q!(n-q)!}. \tag{12}$$

2.2.3 Time-Domain Representation of Harmonics

The mth harmonic in the *time domain* can be computed by collecting the identical exponential power coefficients from Eq. (13) and by setting $p - q = m$, with $p = m + i - 1$ and $q = i - 1$ in Eq. (11). Hence, it follows that

$$y_{m^{th}}(t) = \sum_{i=1}^{\infty} \left(\frac{A}{2}\right)^{m+2i-2} {}^{m+2i-2}C_{i-1} H_{m+2i-2}^{m+i-1,i-1}(j\omega)e^{jm\omega t}. \tag{13}$$

2.2.4 Frequency-Domain Representation of Harmonics

The mth harmonic in the *frequency domain* by applying single-sided Fourier transform in Eq. (13) is the following:

$$Y_{m^{th}}(jm\omega) = \sum_{i=1}^{\infty} \left(\frac{A}{2}\right)^{m+2i-2} {}^{m+2i-2}C_{i-1} H_{m+2i-2}^{m+i-1,i-1}(j\omega)\delta(jm\omega), \tag{14}$$

where $\delta(\cdot)$ is the Dirac delta distribution. When a single-tone input excites a nonlinear dynamical system, the steady-state frequency response is characterized by a spectrum with higher harmonics (as can be seen, for example, in Fig. 3). This behavior is not observed in the linear case, where only one harmonic appears at the input frequency.

3 The Loewner Framework

We start with an account of the Loewner framework (LF) in the linear case [2, 6, 29]. The LF is an interpolatory method that seeks reduced models whose transfer function matches that of the original system at selected interpolation points. An important attribute is that it provides a trade-off between accuracy of fit and complexity of the model. It constructs models from given frequency data in a straightforward manner. In the case of SISO systems, we have the rational scalar interpolation problem to solve.

Consider a given set of complex data as

$$\{(s_k, f_k(s_k)) \in \mathbb{C} \times \mathbb{C} : k = 1, \ldots, 2n)\}.$$

We partition the data in two disjoint sets:

$$\mathbf{S} = [\underbrace{s_1, \ldots, s_n}_{\mu}, \underbrace{s_{n+1}, \ldots, s_{2n}}_{\lambda}], \ \mathbf{F} = [\underbrace{f_1, \ldots, f_n}_{\mathbb{V}}, \underbrace{f_{n+1}, \ldots, f_{2n}}_{\mathbb{W}}],$$

where $\mu_i = s_i$, $\lambda_i = s_{n+i}$, $v_i = f_i$, $w_i = f_{n+i}$ for $i = 1, \ldots, n$.
The objective is to find $H(s) \in \mathbb{C}$, such that

$$H(\mu_i) = v_i, \ i = 1, \ldots, n, \ \text{and} \ H(\lambda_j) = w_j, \ j = 1, \ldots, n. \tag{15}$$

The *left dataset* is denoted as

$$\mathbf{M} = [\mu_1, \cdots, \mu_n] \in \mathbb{C}^{1\times n}, \ \ \mathbb{V} = [v_1, \cdots, v_n]^T \in \mathbb{C}^{n\times 1}, \tag{16}$$

while the *right dataset* as

$$\mathbf{\Lambda} = [\lambda_1, \cdots, \lambda_n]^T \in \mathbb{C}^{n\times 1}, \ \ \mathbb{W} = [w_1, \cdots, w_n] \in \mathbb{C}^{1\times n}. \tag{17}$$

Interpolation points are determined by the problem or are selected to achieve given model reduction goals. For ways of choosing the interpolation grids and of partitioning the data into the left and right sets, we refer the reader to the recent survey [29].

3.1 The Loewner Matrix

Given a row array of complex numbers (μ_j, v_j), $j = 1, \ldots, n$, and a column array, (λ_i, w_i), $i = 1, \ldots, n$, (with λ_i and the μ_j mutually distinct) the associated *Loewner matrix* $\mathbb{L}$ and the shifted *Loewner matrix* $\mathbb{L}_s$ are defined as

$$\mathbb{L} = \begin{bmatrix} \frac{v_1-w_1}{\mu_1-\lambda_1} & \cdots & \frac{v_1-w_n}{\mu_1-\lambda_n} \\ \vdots & \ddots & \vdots \\ \frac{v_n-w_1}{\mu_n-\lambda_1} & \cdots & \frac{v_n-w_n}{\mu_n-\lambda_n} \end{bmatrix} \in \mathbb{C}^{n\times n},\ \mathbb{L}_s = \begin{bmatrix} \frac{\mu_1 v_1-\lambda_1 w_1}{\mu_1-\lambda_1} & \cdots & \frac{\mu_1 v_1-\lambda_n w_n}{\mu_1-\lambda_n} \\ \vdots & \ddots & \vdots \\ \frac{\mu_n v_n-\lambda_1 w_1}{\mu_n-\lambda_1} & \cdots & \frac{\mu_n v_n-\lambda_n w_n}{\mu_n-\lambda_n} \end{bmatrix} \in \mathbb{C}^{n\times n}.$$

Definition 2 If g is rational, i.e., $g(s) = \frac{p(s)}{q(s)}$, for appropriate polynomials p, q, the McMillan degree or the complexity of g is $\deg g = \max\{\deg(p), \deg(q)\}$.

Now, if $w_i = g(\lambda_i)$ and $v_j = g(\mu_j)$ are *samples* of a rational function g, the *main property* of Loewner matrices asserts the following.

Theorem 1 *[2] Let $\mathbb{L}$ be as above. If $k, q \geq \deg g$, then* $\operatorname{rank} \mathbb{L} = \deg g$.

In other words, the rank of $\mathbb{L}$ encodes the complexity of the underlying rational function g. Furthermore, the same result holds for matrix-valued functions g.

3.2 Construction of Interpolants

If the pencil $(\mathbb{L}_s, \mathbb{L})$ is regular, then $\mathbf{E} = -\mathbb{L}$, $\mathbf{A} = -\mathbb{L}_s$, $\mathbf{b} = \mathbb{V}$, $\mathbf{c} = \mathbb{W}$, is a minimal realization of an interpolant for the data, i.e., $H(s) = \mathbb{W}(\mathbb{L}_s - s\mathbb{L})^{-1}\mathbb{V}$. Otherwise, as shown in [2], the problem in Eq. (15) has a solution provided that

$$\operatorname{rank}\,[s\,\mathbb{L} - \mathbb{L}_s] = \operatorname{rank}\,[\mathbb{L},\ \mathbb{L}_s] = \operatorname{rank} \begin{bmatrix} \mathbb{L} \\ \mathbb{L}_s \end{bmatrix} = r,$$

for all $s \in \{\mu_i\} \cup \{\lambda_j\}$. Consider then the thin SVDs:

$$[\mathbb{L},\ \mathbb{L}_s] = \mathbf{Y}\widehat{\Sigma}_r\tilde{\mathbf{X}}^*, \quad \begin{bmatrix} \mathbb{L} \\ \mathbb{L}_s \end{bmatrix} = \tilde{\mathbf{Y}}\Sigma_r\mathbf{X}^*,$$

where $\widehat{\Sigma}_r$, $\Sigma_r \in \mathbb{R}^{r\times r}$, $\mathbf{Y} \in \mathbb{C}^{n\times r}$, $\mathbf{X} \in \mathbb{C}^{n\times r}$, $\tilde{\mathbf{Y}} \in \mathbb{C}^{2n\times r}$, $\tilde{\mathbf{X}} \in \mathbb{C}^{r\times 2n}$.

Remark 2 r can be chosen as the *numerical rank* (as opposed to the *exact rank*) of the Loewner pencil.

Theorem 2 *The quadruple* $(\tilde{\mathbf{A}}, \tilde{\mathbf{b}}, \tilde{\mathbf{c}}, \tilde{\mathbf{E}})$ *of size,* $r \times r$, $r \times 1$, $1 \times r$, $r \times r$, *given by*

$$\tilde{\mathbf{E}} = -\mathbf{Y}^T \mathbb{L}\mathbf{X}, \quad \tilde{\mathbf{A}} = -\mathbf{Y}^T \mathbb{L}_s \mathbf{X}, \quad \tilde{\mathbf{b}} = \mathbf{Y}^T \mathbb{V}, \quad \tilde{\mathbf{c}} = \mathbb{W}\mathbf{X},$$

is a descriptor realization of an (approximate) interpolant of the data with McMillan degree $r = rank(\mathbb{L})$, *where* $\tilde{H}(s) = \tilde{\mathbf{c}}(s\tilde{\mathbf{E}} - \tilde{\mathbf{A}})^{-1}\tilde{\mathbf{b}}$.

For more details on the construction/identification of linear systems with the LF, we refer the reader to [4, 6, 29] where both the SISO and MIMO cases are addressed together with other more technical aspects (e.g., how to impose the construction of real-valued models, etc.).

4 The Special Case of Bilinear Systems

In recent years, projection-based Krylov methods have extensively been applied for model reduction of bilinear systems. We mention the following contributions [1, 5, 7, 10–12, 17, 20, 34] and the references within.

Scalar bilinear systems are described by the set of matrices; $\Sigma_b = (\mathbf{A}, \mathbf{N}, \mathbf{b}, \mathbf{c}, \mathbf{E})$ and characterized by the following equations:

$$\Sigma_b : \begin{cases} \mathbf{E}\dot{\mathbf{x}}(t) = \mathbf{A}\mathbf{x}(t) + \mathbf{N}\mathbf{x}(t)u(t) + \mathbf{b}u(t), \\ y(t) = \mathbf{c}\mathbf{x}(t), \end{cases} \tag{18}$$

where $\mathbf{E}, \mathbf{A}, \mathbf{N} \in \mathbb{R}^{n\times n}$, $\mathbf{b} \in \mathbb{R}^{n\times 1}$, $\mathbf{c} \in \mathbb{R}^{1\times n}$, and $\mathbf{x} \in \mathbb{R}^{n\times 1}$, $u, y \in \mathbb{R}$. In what follows, we restrict our analysis to systems with non-singular $\mathbf{E}$ matrices (e.g., identity matrix).

4.1 The Growing Exponential Approach

The properties of the growing exponential approach can be adapted readily to the problem of *finding transfer functions* for constant-parameter (stationary) state equations. Let us consider the bilinear model in Eq. (18) with zero initial conditions. A single-tone input with amplitude $A < 1$ is considered as in Eq. (8).

$$u(t) = A\cos(\omega t) = \frac{A}{2}e^{j\omega t} + \frac{A}{2}e^{-j\omega t} = ae^{j\omega t} + ae^{-j\omega t}, \tag{19}$$

where $a = A/2$ and $a \in (0, \epsilon)$ with $0 < \epsilon < 1/2$ and for all $t \geq 0$. The steady-state solution for the differential equation in Eq. (18) can be written as follows:

$$\mathbf{x}(t) = \sum_{p,q\in\mathbb{N}}^{\infty} \mathbf{G}_n^{p,q}(\underbrace{j\omega,\ldots,j\omega}_{p-times},\underbrace{-j\omega,\ldots,-j\omega}_{q-times})a^{p+q}e^{j\omega(p-q)t}. \tag{20}$$

The symbol[4] $\mathbf{G}_n^{p,q}$ denotes the nth input to state frequency response containing p-times the frequency ω and q-times the frequency $-\omega$. By substituting in Eq. (18) and collecting the terms of the same exponential (as the $e^{j\omega_m t}$), we can derive the input to state frequency responses $\mathbf{G}_n$ for every n as follows:

$$\sum_{p,q\in\mathbb{N}}^{\infty} (j\omega(p-q)\mathbf{E}-\mathbf{A})\,\mathbf{G}_n^{p,q}a^{p+q}e^{j\omega(p-q)t} = \mathbf{b}(ae^{j\omega t}+ae^{-j\omega t})+$$
$$+\mathbf{N}\left(\sum_{p,q\in\mathbb{N}}^{\infty}\mathbf{G}_n^{p,q}a^{p+q+1}e^{j\omega(p+1-q)t}+\sum_{p,q\in\mathbb{N}}^{\infty}\mathbf{G}_n^{p,q}a^{p+q+1}e^{j\omega(p-q-1)t}\right).$$

For the first choices of p and q up to $p+q\leq 2$, $(1,0),(0,1),(2,0),(0,2),(1,1)$ and by denoting the resolvent $\boldsymbol{\Phi}(j\omega)=(j\omega\mathbf{E}-\mathbf{A})^{-1}\in\mathbb{C}^{n\times n}$, c.t. conjugate terms, we derive the first set of terms

$$\boldsymbol{\Phi}(j\omega)^{-1}\mathbf{G}_1^{1,0}ae^{j\omega t}+\boldsymbol{\Phi}(2j\omega)^{-1}\mathbf{G}_2^{2,0}a^2e^{2j\omega t}+\boldsymbol{\Phi}(0)^{-1}\mathbf{G}_2^{1,1}a^2+c.t.+\cdots=$$
$$\mathbf{N}\mathbf{G}_1^{1,0}a^2e^{2j\omega t}+\mathbf{N}\mathbf{G}_2^{2,0}a^3e^{3j\omega t}+\mathbf{N}\mathbf{G}_2^{1,1}a^3e^{j\omega t}+c.t.+\cdots+\mathbf{b}ae^{j\omega t}+c.t.$$

Collecting the same powers in both exponential and polynomial magnitudes, we compute the first and the second time/input-invariant GFRFs:

$$\begin{aligned}\mathbf{G}_1^{1,0}(j\omega)&=\boldsymbol{\Phi}(j\omega)\mathbf{b},\\ \mathbf{G}_2^{2,0}(j\omega)&=\boldsymbol{\Phi}(2j\omega)\mathbf{N}\mathbf{G}_1^{1,0}=\boldsymbol{\Phi}(2j\omega)\mathbf{N}\boldsymbol{\Phi}(j\omega)\mathbf{b}.\end{aligned} \tag{21}$$

Then, the following input to state transfer functions $\mathbf{G}_n$ using induction are

$$\begin{aligned}\mathbf{G}_n^{n,0}(j\omega)&=\boldsymbol{\Phi}(nj\omega)\mathbf{N}\boldsymbol{\Phi}((n-1)j\omega)\mathbf{N}\cdots\mathbf{N}\boldsymbol{\Phi}(j\omega)\mathbf{b},\\ \mathbf{G}_n^{0,n}(j\omega)&=\boldsymbol{\Phi}(-nj\omega)\mathbf{N}\boldsymbol{\Phi}(-(n-1)j\omega)\mathbf{N}\cdots\mathbf{N}\boldsymbol{\Phi}(-j\omega)\mathbf{b},\\ \mathbf{G}_n^{p,q}(j\omega)&=\boldsymbol{\Phi}((p-q)j\omega)\mathbf{N}\left[\mathbf{G}_{n-1}^{p,q-1}(j\omega)+\mathbf{G}_{n-1}^{p-1,q}(j\omega)\right],\ p,q\geq 1,\end{aligned} \tag{22}$$

for $n\geq 1$ and $p+q=n$. By multiplying with the output vector $\mathbf{c}$, we can further derive the input-output generalized frequency responses GFRFs as

[4] $\mathbf{G}_n^{p,q}=\mathbf{G}(\underbrace{j\omega,\ldots,j\omega}_{p-times};\underbrace{-j\omega,\ldots,-j\omega}_{q-times})$.

$$\begin{aligned} H_n^{n,0}(j\omega) &= \mathbf{c}\boldsymbol{\Phi}(nj\omega)\mathbf{N}\boldsymbol{\Phi}((n-1)j\omega)\mathbf{N}\cdots\mathbf{N}\boldsymbol{\Phi}(j\omega)\mathbf{b}, \\ H_n^{0,n}(j\omega) &= \mathbf{c}\boldsymbol{\Phi}(-nj\omega)\mathbf{N}\boldsymbol{\Phi}(-(n-1)j\omega)\mathbf{N}\cdots\mathbf{N}\boldsymbol{\Phi}(-j\omega)\mathbf{b}, \\ H_n^{p,q}(j\omega) &= \mathbf{c}\boldsymbol{\Phi}((p-q)j\omega)\mathbf{N}\left[\mathbf{G}_{n-1}^{p,q-1}(j\omega)+\mathbf{G}_{n-1}^{p-1,q}(j\omega)\right],\ p,q\geq 1. \end{aligned} \tag{23}$$

At this point, we can write the Volterra series by using the above specific structure of the GFRFs that were derived with the growing exponential approach for the bilinear case. An important property to notice is that the nth kernel is a multivariate function of order n. It is obvious that the identification of the nth-order FRF involves an n-dimensional frequency space. For that reason, next, we derive the general second symmetric kernel for the bilinear case with a double-tone input. Consider:

$$u(t) = A_1\cos(\omega_1 t) + A_2\cos(\omega_2 t) = \sum_{i=1}^{2}\alpha_i(e^{j\omega_i t}+e^{-j\omega_i t}), \tag{24}$$

where $\alpha_1 = \frac{A_1}{2}$ and $\alpha_2 = \frac{A_2}{2}$. In that case, with the growing exponential approach the state solution in steady state is

$$\mathbf{x}(t) = \sum_{m_1,\ldots,m_4\in\mathbb{N}}^{\infty} \mathbf{G}_n^{m_1,m_2,m_3,m_4}\alpha_1^{m_1+m_2}\alpha_2^{m_3+m_4}e^{j((m_1-m_2)\omega_1+(m_3-m_4)\omega_2)t}. \tag{25}$$

We are looking for the input to state frequency response $\mathbf{G}(j\omega_1, j\omega_2)$. By substituting to the bilinear model in Eq. (18) and collecting the appropriate terms while at the same time using the symmetry $\mathbf{G}(j\omega_1, j\omega_2) = \mathbf{G}(j\omega_2, j\omega_1)$, we conclude that

$$\mathbf{G}_2(j\omega_1, j\omega_2)=\frac{1}{2}\left[(j\omega_1+j\omega_2)\mathbf{E}-\mathbf{A}\right]^{-1}\mathbf{N}\left[(j\omega_1\mathbf{E}-\mathbf{A})^{-1}\mathbf{b}+(j\omega_2\mathbf{E}-\mathbf{A})^{-1}\mathbf{b}\right], \tag{26}$$

where by using the resolvent notation and multiplying with $\mathbf{c}$, we derive the *second-order symmetric generalized frequency response function as*

$$H_2(j\omega_1, j\omega_2) = \frac{1}{2}\mathbf{c}\boldsymbol{\Phi}(j\omega_1+j\omega_2)\mathbf{N}\left[\boldsymbol{\Phi}(j\omega_1)\mathbf{b}+\boldsymbol{\Phi}(j\omega_2)\mathbf{b}\right]. \tag{27}$$

4.2 The Kernel Separation Method

One way to deduce Volterra kernels is by means of interpolation. This problem is equivalent to that of estimating a polynomial with noisy coefficients. This interpolation scheme builds a linear system with a Vandermonde matrix which is invertible since the amplitudes are distinct and nonzero. The inverse of a Vandermonde matrix can be explicitly computed and there are stable ways to solve these equations [16]. The recently proposed method presented in [18] solves the exponentially ill-condition problem of the Vandermonde matrix with Arnoldi orthogonalization. The mth harmonic in the frequency domain is derived by applying a (single-sided) Fourier transform. More precisely, the explicit formulation is as follows:

$$\begin{aligned} Y_{m^{th}}(jm\omega) &= \sum_{i=1}^{\infty} \underbrace{\left(\frac{A}{2}\right)^{m+2i-2} {}_{m+2i-2}C_{i-1}}_{\alpha^{m+2i-2}} H_{m+2i-2}^{m+i-1,i-1}(j\omega)\delta(jm\omega) \\ &= \sum_{i=1}^{\infty} \alpha^{m+2i-2} H_{m+2(i-1)}^{m+i-1,i-1}(j\omega)\delta(jm\omega). \end{aligned} \tag{28}$$

We simplify the notation in order to reveal the adaptive method that will help us to estimate the GFRFs up to a specific order. Next, write the linear system of equations that connects the harmonic information with the higher Volterra kernels as follows:

$$\underbrace{\begin{bmatrix} Y_0(0j\omega) \\ Y_1(1j\omega) \\ Y_2(2j\omega) \\ Y_3(3j\omega) \\ \vdots \\ Y_m(mj\omega) \end{bmatrix}}_{\mathbf{Y}_{(\alpha,\omega)}} = \left\{ \underbrace{\begin{bmatrix} \alpha^0 & \alpha^2 & \alpha^4 & \cdots \\ \alpha^1 & \alpha^3 & \alpha^5 & \cdots \\ \alpha^2 & \alpha^4 & \alpha^6 & \cdots \\ \alpha^3 & \alpha^5 & \alpha^7 & \cdots \\ \vdots & \vdots & \vdots & \vdots \\ \alpha^m & \alpha^{m+2} & \alpha^{m+4} & \cdots \end{bmatrix}}_{\mathbf{M}_\alpha} \odot \underbrace{\begin{bmatrix} H_0^{0,0} & H_2^{1,1} & H_4^{2,2} & \cdots \\ H_1^{1,0} & H_3^{2,1} & H_5^{3,2} & \cdots \\ H_2^{2,0} & H_4^{3,1} & H_6^{4,2} & \cdots \\ H_3^{3,0} & H_5^{4,1} & H_7^{5,2} & \cdots \\ \vdots & \vdots & \vdots & \vdots \\ H_n^{n,0} & H_{n+2}^{n+1,1} & H_{n+4}^{n+2,2} & \cdots \end{bmatrix}}_{\mathbf{P}_\omega} \right\} \underbrace{\begin{bmatrix} 1 \\ 1 \\ 1 \\ 1 \\ \vdots \\ 1 \end{bmatrix}}_{\mathbf{e}_{n+1,1}}. \tag{29}$$

By introducing the Hadamard product notation[5] and by substituting the δ's with ones, we can compactly rewrite the above system in the following form:

$$\mathbf{Y}_{(\alpha,\omega)} = [\mathbf{M}_\alpha \odot \mathbf{P}_\omega] \cdot \mathbf{e}_{n+1,1}. \tag{30}$$

The above system offers the level of approximation we want to achieve. Note that the frequency response $\mathbf{Y}$ depends on both the amplitude and the frequency, while the right-hand side of Eq. (30) reveals the separation of the aforementioned quantities. As we neglect higher order Volterra kernels, the measurement set tends to be corrupted by noise.

> **Kernel separation and stage ℓ-approximation**

For a given system, the procedure consists in exciting it with a single-tone input. By varying the driving frequency, as well as the amplitude, we can approximate the GFRFs by minimizing the (2-norm) of the remaining systems.

---

[5] The Hadamard product is denoted with "$\odot$"; the matrix multiplication is performed element-wise.

$$\mathbf{Y}_{m+1,\ell}(jm\omega, \alpha_\ell) = \left[\mathbf{M}_{m+1,\ell}(\alpha_\ell) \odot \mathbf{P}_{m+1,\ell}(jm\omega)\right] \cdot \mathbf{e}_{n+1,1}. \tag{31}$$

The m-"direction" gives us the threshold up to the specific harmonic that we measure while the ℓ-"direction" gives us the level of the kernel separation that we want to achieve. For instance, for the second stage approximation, it holds $\ell = 2$ with $Y_m \approx 0,\ \forall m$ with $\ell = 2 < m = 3, 4, \ldots$.

4.3 Identification of the Matrix N

The difference between linear and bilinear models is the presence of the product between the input and the state that is scaled by the matrix **N**. As the LF is able to identify the linear part (**A**, **b**, **c**, **E**) of the bilinear model the only thing that remains is the identification of the matrix **N**. The matrix **N** enters linearly in the following kernels (as **E** has been considered invertible, for simplicity, it is assumed $\mathbf{E} = \mathbf{I}$):

- With a single-tone input the kernel $H_2^{1,1}$ can be written as

$$H_2(j\omega_1, -j\omega_1) = \frac{1}{2}\mathbf{c}\,(-\mathbf{A})^{-1}\,\mathbf{N}\left((j\omega_1\mathbf{I} - \mathbf{A})^{-1}\mathbf{b} + (-j\omega_1\mathbf{I} - \mathbf{A})^{-1}\mathbf{b}\right) \tag{32}$$

  and the kernel $H_2^{2,0}$ as

$$H_2(j\omega_1, j\omega_1) = \mathbf{c}\,(2j\omega_1\mathbf{I} - \mathbf{A})^{-1}\,\mathbf{N}(j\omega_1\mathbf{I} - \mathbf{A})^{-1}\mathbf{b}. \tag{33}$$

- While with a double-tone input the general kernel H_2 can be written as

$$H_2(j\omega_1, j\omega_2) = \frac{1}{2}\mathbf{c}\Big((j\omega_1 + j\omega_2)\mathbf{I} - \mathbf{A}\Big)^{-1}\mathbf{N}\Big((j\omega_1\mathbf{I} - \mathbf{A})^{-1}\mathbf{b} + (j\omega_2\mathbf{I} - \mathbf{A})^{-1}\mathbf{b}\Big). \tag{34}$$

We introduce the following notation:

$$\begin{aligned} \mathcal{O}(j\omega_1, j\omega_2) &= \frac{1}{2}\mathbf{c}\Big(j(\omega_1 + j\omega_2)\mathbf{I} - \mathbf{A}\Big)^{-1} \in \mathbb{C}^{1\times n}, \\ \mathcal{R}(j\omega_1, j\omega_2) &= \Big((j\omega_1\mathbf{I} - \mathbf{A})^{-1}\mathbf{b} + (j\omega_2\mathbf{I} - \mathbf{A})^{-1}\mathbf{b}\Big) \in \mathbb{C}^{n\times 1}. \end{aligned} \tag{35}$$

Then, Eq. (34) can be compactly rewritten as

$$H_2(j\omega_1, j\omega_2) = \mathcal{O}(j\omega_1, j\omega_2)\mathbf{N}\mathcal{R}(j\omega_1, j\omega_2). \tag{36}$$

Assume that k measurements of the function H_2 are available (measured) for k different pairs (ω_1, ω_2). By vectorizing in respect to the measurement set, we have for the kth measurement:

$$\underbrace{H_2(j\omega_1^{(k)}, j\omega_2^{(k)})}_{\mathbf{Y}^{(k)}} = \underbrace{O(j\omega_1^{(k)}, j\omega_2^{(k)})}_{O_{1,n}^{(k)}} \underbrace{\mathbf{N}}_{n\times n} \underbrace{\mathcal{R}(j\omega_1^{(k)}, j\omega_2^{(k)})}_{\mathcal{R}_{n,1}^{(k)}},$$

$$\text{For all } k \text{ measurements} \rightarrow \mathbf{Y}_{(1:k,1)} = \underbrace{\left(O_{(1,n)}^{(k)} \otimes \mathcal{R}_{(1,n)}^{T(k)}\right)}_{(1:k,n^2)} \underbrace{vec\,(\mathbf{N})}_{(1:n^2,1)}. \tag{37}$$

Note that Eqs. (32), (33), (34) can be equivalently rewritten as the one linear matrix equation given in Eq. (37). By filling out the above matrix $\left[O \otimes \mathcal{R}^T\right]$ with the information from $H_2(j\omega_1, -j\omega_1)$ and from $H_2(j\omega_1, j\omega_1)$ as well, the solution can be improved. Hence, we are able to solve Eq. (37) with full rank and identify the matrix $\mathbf{N}$. All the symmetry properties of the kernels are appropriately used, e.g., conjugate-real symmetry. For n denoting the dimension of the bilinear model and k the number of measurements, we have the following two cases[6]:

1. $k < n^2$ underdetermined $\rightarrow$ least-squares (LS) solution (minimizing the 2-norm) as in [28],
2. $k \geq n^2$ determined-rank completion $\rightarrow$ identification of $\mathbf{N}$,

Proposition 1 *Let $\Sigma_b = (\mathbf{A}, \mathbf{N}, \mathbf{b}, \mathbf{c}, \mathbf{E})$ be a bilinear system of dimension n for which the linear subsystem $\Sigma_l = (\mathbf{A}, \mathbf{b}, \mathbf{c}, \mathbf{E})$ is fully controllable and observable. Then, for $k \geq n^2$ measurements so that $(j\omega_1^{(k)}, j\omega_2^{(k)})$ are distinct complex pairs with $(\omega_1^{(k)}, \omega_2^{(k)}) \in \mathbb{R}_+^2$ and $\omega_1^{(k)} \neq \omega_2^{(k)}$, the following holds:*

$$rank\left(\underbrace{\begin{bmatrix} O^{(1)} \otimes \mathcal{R}^{T(1)} \\ O^{(2)} \otimes \mathcal{R}^{T(2)} \\ \vdots \\ O^{(k)} \otimes \mathcal{R}^{T(k)} \end{bmatrix}}_{(1:k\geq n^2,n^2)}\right) = n^2. \tag{38}$$

As the above result indicates, one would need at least n^2 measurements to identify the matrix $\mathbf{N}$ corresponding to bilinear system of dimension n.

4.4 A Separation Strategy for the second Kernel

To identify the nth Volterra kernel, we need an n-tone input signal. As we want to identify the second kernel, the input signal needs to be chosen as a double-tone Eq. (24). The propagating harmonics are $e^{(j(m_1-m_2)\omega_1+j(m_3-m_4)\omega_2)t}$ or more compactly

[6] The vectorization is row-wise, $vec(\mathbf{N}) = \left[\mathbf{N}(1, 1:n) \cdots \mathbf{N}(n, 1:n)\right]^T \in \mathbb{R}^{n^2\times 1}$.

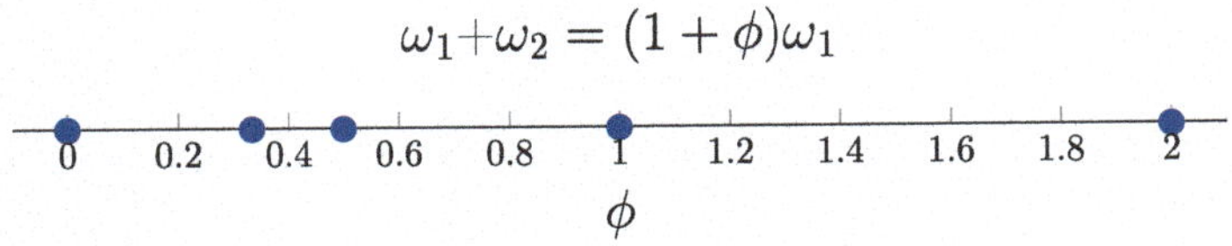

Fig. 4 This figure shows the constrains of ϕ (e.g., $\phi = 0, 1/3, 1/2, 1, 2, 3, \ldots$, etc.). By choosing ϕ's within the blue dots, we construct frequency bandwidths with a unique $(\omega_1 + \omega_2)$

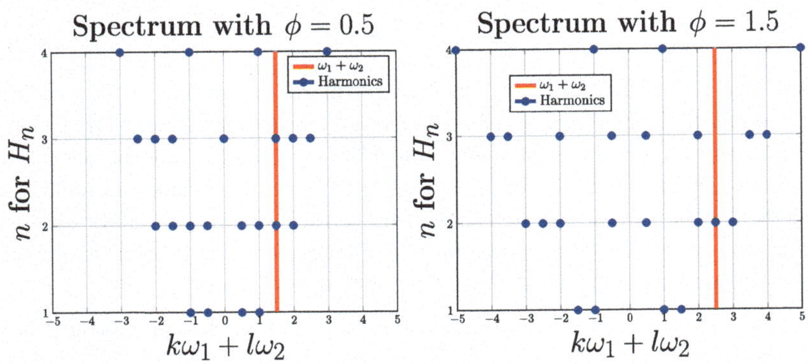

Fig. 5 Left pane: Overlapping kernels contributing to the same harmonic with invalid $\phi = 0.5$. Right pane: Uniquely defined harmonic at $(\omega_1 + \omega_2)$ with valid $\phi = 1.5$. Here, it holds $(n = k + l)$

$e^{(\pm kj\omega_1 \pm lj\omega_2)t}$, where $k, l \in \mathbb{N}$. The aim is to differentiate the $(\omega_1 + \omega_2)$ harmonic from the others harmonics. More precisely, we want the following result to hold:

$$\omega_1 + \omega_2 \neq k\omega_1 + l\omega_2, \ \forall (k, l) \in \mathbb{Z} \times \mathbb{Z} \setminus \{1, 1\}. \tag{39}$$

Suppose $\omega_2 = \phi\omega_1$, $\phi \in \mathbb{R}$. The suitable ϕ's where Eq. (39) holds are

$$\omega_1 + \phi\omega_1 = k\omega_1 + l\phi\omega_1 \Rightarrow 1 + \phi = k + l\phi \Rightarrow \phi = \frac{k-1}{1-l}, \ k, l \in \mathbb{Z} \setminus \{1\}. \tag{40}$$

By choosing ϕ so that the equality in Eq. (40) doesn't hold, with harmonic mixing index $m = k + l$, it makes the harmonic $(\omega_1 + \omega_2)$ uniquely defined in the frequency spectrum up to the mth kernel.

To visualize this feature, we choose $\omega_1 = 1$, and $\omega_2 = \omega_1\phi = \phi$, for harmonic mixing index $m = 4$. Then, the constraints of ϕ are depicted in Fig. 4 with blue dots.

Next, in Fig. 5 and on the left pane, one ϕ constraint that occurs commensurate harmonics is depicted with the second and the third kernel to contribute at the same harmonic. On the right pane, the harmonic is uniquely defined at $(\omega_1 + \omega_2)$ from the second kernel up to the mixing order $m = 4$.

The next result allows us to construct sweeping frequency schemes to get enough measurements for the $H_2(j\omega_1, j\omega_2)$. So, for every $\omega_1 > 0$ the following should hold:

$$\omega_2 \in (\phi_{i-1}\omega_1, \phi_i\omega_1)\,,\ i = 1, \ldots \tag{41}$$

where ϕ_i are the constraints (see Fig. 4 blue dots).

Remark 3 Note that in the proposed framework, the separation of the kernels that contribute at $(\omega_1 + \omega_2)$ harmonic is forced only under a specific mixing order m. We do not offer any general solution to this separation problem for multi-tone input, although techniques have been introduced such as in [16]. Therefore, it was also stated that the solution of the full separation of harmonics is, in general, not possible.

4.5 *The Loewner-Volterra Algorithm for Time-Domain Bilinear Identification and Reduction*

We start with a set of single-tone inputs $u(t) = \alpha_\ell \cos(\omega_1^{(i)} t),\ \ i = 1, ..., k$, with $\alpha_\ell < 1$. For those k measurements, we can estimate the linear kernel $H_1(j\omega_1^{(i)})$, the $H_2(j\omega_1^{(k)}, j\omega_1^{(k)})$ and the $H_2(j\omega_1^{(k)}, -j\omega_1^{(k)})$ by simply measuring the first harmonic as $\mathbf{Y}_1$, the second harmonic as $\mathbf{Y}_2$, and the DC term as $\mathbf{Y}_0$, from the frequency spectrum as shown in Fig. 3. To improve the accuracy of the estimations for the aforementioned kernels, we could further upgrade to an ℓ-stage approximation by varying the amplitude α_ℓ as explained in Sect. 4.2. This approach is necessary whenever higher harmonics are considered to be numerically nonzero, hence meaningful. The reason for this is that the first harmonic is hence corrupted by noise introduced by the term $H_3^{2,1}$ and the rest of the terms which appears on the second row of matrix $\mathbf{P}_\omega$ in Eq. (29).

Since the LF reveals the underlying order of the linear system denoted with r, the value of k should be at least equal to $2r$. Then, we can take the decision on what will be the order r of the reduced system by analyzing the singular value decay. Up to the previous step, we have identified the linear part with the LF, and we have filled the LS problem Eq. (37) with measurements from the diagonal of the second kernel and from the the perpendicular to the diagonal axis $(\omega_1, -\omega_1)$. Those measurements contribute to the problem, but with an underdetermined (rank deficient) LS problem.

We need more measurements of H_2 to reach the full rank (r^2) solution that will lead to the identification of $\mathbf{N}$. So, we proceed by measuring the H_2 out of the diagonal $(\omega_1 \neq \omega_2)$ with a double-tone input as $u(t) = \alpha_\ell \cos(\omega_1^{(k)} t) + \beta_\ell \cos(\omega_2^{(k)} t)$, for a set of frequency pairs (ω_1, ω_2) up to r^2. The kernel separation problem for the frequency (ω_1, ω_2) appears now. To deal with this problem, we follow the solution proposed in Sect. 4.4 (up to a mixing degree). Last, we solve the real[7] full-rank LS problem described in Eq. (37) by using all the symmetric properties of these kernels (i.e., real symmetry, conjugate symmetry, and the fact that $H_2(j\omega_1, j\omega_2) = H_2(j\omega_2, j\omega_1)$). An algorithm that summarizes the above procedure is presented below.

[7] Enforcing real-valued models has been discussed in [6, 29]; here, we follow the same approach.

(Algorithm) The Loewner-Volterra algorithm for bilinear identification and reduction from time-domain data.

Input/Data acquisition: Use as control input the signals: $u(t) = \alpha_\ell \cos(\omega_1^{(k)} t) + \beta_\ell \cos(\omega_2^{(k)} t),\ t \geq 0$, by sweeping the small amplitudes (< 1) and a particular range of frequencies.

Output: A bilinear system of dimension-r: $\Sigma_{b_r} : (\mathbf{A}_r, \mathbf{N}_r, \mathbf{b}_r, \mathbf{c}_r, \mathbf{E}_r)$

1. Apply one-tone input $u(t)$ with $\beta_\ell = 0,\ \omega_1^{(k)}$ for $k = 1, \ldots, n$, and collect the snapshots $y(t)$ in steady state.
2. Apply Fourier transform and collect the following measurements:
   - DC term: $Y_O(0 \cdot j\omega_1^{(k)})$,
   - 1st harmonic: $Y_I(1 \cdot j\omega_1^{(k)})$,
   - 2nd harmonic: $Y_{II}(2 \cdot j\omega_1^{(k)})$,

     $\vdots$
   - mth harmonic: $Y_{m^{th}}(m \cdot j\omega_1^{(k)})$ (last numerically nonzero harmonic).
3. If the second harmonic or higher harmonics are nonzero, the system is nonlinear. By sweeping the amplitude and using the adaptive scheme (stage ℓ-approximation) in Eq. (30), the estimations of the first and the second kernels can be improved. If the second and higher harmonics are equal to zero, the bilinear matrix $\mathbf{N}$ remains zero and the underlying system is linear.
4. Apply the linear LF, see Algorithm 1 in [29] by using the measurements (e.g., $H_1(j\omega_1^{(k)}) \approx 2Y_I(j\omega_1^{(k)})/\alpha_\ell$ for the second stage approximation $Y_m \approx 0$ for $m > 2$) and get the order r linear model.
5. If the system is nonlinear, by fitting a bilinear matrix $\mathbf{N}$ will improve the accuracy. Apply the two-tone input $u(t) = \alpha_\ell \cos(\omega_1^{(k)} t) + \beta_\ell \cos(\omega_2^{(k)} t)$ to get enough measurements ($\leq r^2$) to produce a full-rank LS problem. Measure the $(\omega_1 + \omega_2)$ harmonic as explained in Sect. 4.4 and get the estimations for the second kernel as: $H_2(j\omega_1^{(k)}, j\omega_2^{(k)}) \approx 2Y_{II}(j\omega_1^{(k)}, j\omega_2^{(k)})/(\alpha_\ell \beta_\ell)$.
6. Solve the full-rank least-squares problem as described in Eq. (37) and compute the real-valued bilinear matrix $\mathbf{N}$. When the inversion is not exact due to numerical issues, the least-squares solution is obtained with a thresholding SVD.

4.6 Computational Effort of the Proposed Method

In this section, we discuss the computational effort of the proposed method by analyzing each step. We comment on the applicability of large-scale problems and the relation with real-world scenarios.

Simulation of processes with harmonic inputs constitutes a classical technique which is applied in many engineering applications; data acquisition in the time domain is a common procedure. Nevertheless, using advanced electronic devices such as vector network analyzers (VNAs), frequency-domain data can also be obtained (directly). The Loewner framework applied in the case where frequency-domain data that are obtained from VNAs offers an excellent identification and reduction tool in the linear case (with many applications in electrical, mechanical, or civil engineering). In the context of the current paper, we deal with time-domain data for a special class of nonlinear problems.

For the purpose of identifying and reducing bilinear systems from time-domain measurements, the most expensive procedure is that of data collection. This is done by simulating time-domain models with Euler's method (bilinear models such as the ones approximating Burgers' equation). Nevertheless, the heavy computational cost of simulating large dimensional systems in time domain could be alleviated using parallel processing (e.g., for multiple computational clusters). The process of estimating transfer functions values by computing the Fourier transform hence remains robust. In addition, the LF can adaptively detect the decay of the singular values and hence the procedure can be terminated for a specific reduced order $r \ll n$.

In the beginning, a linear system of reduced dimension r is fitted using the LF. For the rest of the proposed algorithm, note that we will use the lower dimension r to our advantage, and hence the method remains robust. The next step is to compute the matrix $\mathbf{N}$ that characterizes the nonlinearity of bilinear systems. As the fitted linear system is of dimension r, we hence need to detect exactly r^2 unknowns (the entries of matrix $\mathbf{N}$). As presented in Sect. 4.3, this boils down to solving a full-rank LS problem that can be easily dealt with.

The aim of the newly proposed method is to accurately train bilinear models from time-domain data. We offer a first step approach toward complete identification of such systems within the Volterra series approximation approach. In many cases, large-scale systems are sparse (due to spatial domain semi-discretization) and hence reduction techniques can be applied. The new method deals with the inherent redundancies through the linear subsystem (compression by means of SVD). Afterward, it updates the nonlinear behavior by introducing an appropriate low-dimensional bilinear matrix that improves the overall approximation. Note also that the new method relies on the *controllability/observability* of the fitted linear system. Additionally, noise values up to a particular threshold can be handled as presented in Sect. 5; further analysis on noise-related issues is left for future research.

5 Numerical Examples

Example 1 (Identifying a low-order bilinear toy example) The aim of this experiment is to identify a simple bilinear model from time-domain measurements. Consider the following controllable/observable bilinear model Eq. (18) of dimension-2 with a *non-symmetric* matrix **N**, zero initial condition and matrices as

$$\mathbf{E} = \begin{bmatrix} 1 & 0 \\ 0 & 1 \end{bmatrix}, \ \mathbf{A} = \begin{bmatrix} -1 & -10 \\ 10 & -1 \end{bmatrix}, \ \mathbf{N} = \begin{bmatrix} 1 & -2 \\ 3 & -4 \end{bmatrix}, \ \mathbf{B} = \begin{bmatrix} 1 \\ 1 \end{bmatrix}, \ \mathbf{C} = \begin{bmatrix} 1 & 1 \end{bmatrix}. \quad (42)$$

We simulate the system in the time domain with an input as: $u(t) = A\cos(\omega t)$, magnitude $A = 0.01$, frequency $\omega \in \begin{bmatrix} 0.5 & 1 & 1.5 & 2 \end{bmatrix} 2\pi$, and time step $dt = 1e-4$. Next, the second-stage approximation results for the linear kernel $\tilde{H}_1$ in comparison with the theoretical values of H_1 are presented in Table 1.

With the estimations of the linear transfer function and by using the LF as the data-driven identification and reduction tool for linear systems, we identify the linear system ($\tilde{\mathbf{A}}, \tilde{\mathbf{b}}, \tilde{\mathbf{c}}, \tilde{\mathbf{E}}$). We stopped at the fourth measurement due to the fact that the underlying system is of second order (McMillan degree 2). Otherwise, more measurements will be needed to have a sufficient decay of the singular values as shown in Fig. 6. The singular value decay offers a choice of reduction. As long as the simulation of the system is done, with time step $dt = 1e-4$, the singular values with magnitude below that threshold are neglected.

Construction of the linear system with order $r = 2$, by using the theoretical noise-free measurements (subscript "t") appears next:

$$\tilde{\mathbf{A}}_t = \begin{bmatrix} -1.4513 & -8.8181 \\ 11.363 & -0.54868 \end{bmatrix}, \ \tilde{\mathbf{B}}_t = \begin{bmatrix} -0.92979 \\ 1.3967 \end{bmatrix}, \ \tilde{\mathbf{C}}_t = \begin{bmatrix} -0.76857 & 0.9203 \end{bmatrix}, \quad (43)$$

while by using the measured data with second-stage approximation results to the following:

$$\tilde{\mathbf{A}} = \begin{bmatrix} -1.458 & -8.8137 \\ 11.367 & -0.55162 \end{bmatrix}, \ \tilde{\mathbf{B}} = \begin{bmatrix} -0.9342 \\ 1.4 \end{bmatrix}, \ \tilde{\mathbf{C}} = \begin{bmatrix} -0.7675 & 0.91611 \end{bmatrix}. \quad (44)$$

Table 1 Measurements of the first (linear) kernel

Frequency ω	$\tilde{H}_1(j\omega)$-second stage	$H_1(j\omega)$-theoretical
$0.5 \cdot 2\pi$	$+0.026606 + 0.067106i$	$+0.026574 + 0.067115i$
$1.0 \cdot 2\pi$	$+0.071503 + 0.189600i$	$+0.071258 + 0.189700i$
$1.5 \cdot 2\pi$	$+0.752720 + 0.377300i$	$+0.754030 + 0.380870i$
$2.0 \cdot 2\pi$	$+0.134070 - 0.381970i$	$+0.133780 - 0.382520i$

[a]With 2nd-stage approximation $\tilde{H}_1(j\omega) \approx 2Y_1(j\omega)/A$

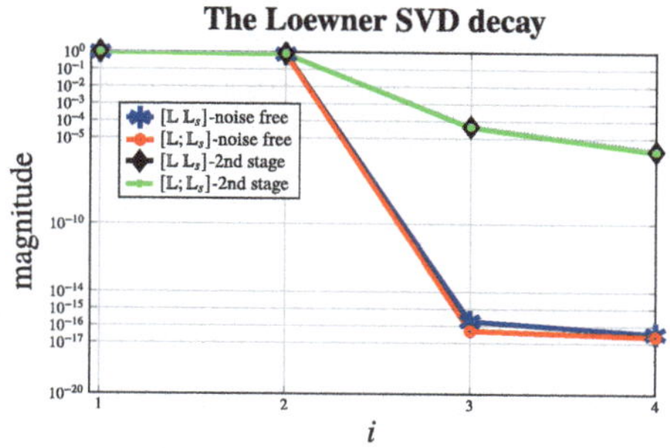

Fig. 6 The singular value decay of the LF as a fundamental characterization of the McMillan degree of the underlying linear system. Here, a truncation scheme of order $r = 2$ is recommended where the second stage approximation gave $\sigma_3/\sigma_1 = 4.721 \cdot 10^{-5}$, while for the noise-free case the third singular values have reached the machine precision

Table 2 Measurements of the H_2 on the diagonal and perpendicular to the diagonal

Freq. ω	$\tilde{H}_2(j\omega, j\omega)$	$H_2(j\omega, j\omega)$	$\tilde{H}_2(j\omega, -j\omega)$	$H_2(j\omega, -j\omega)$
$0.5 \cdot 2\pi$	+0.026440 − 0.124490i	+0.026570 − 0.124440i	+0.032190	+0.032177
$1.0 \cdot 2\pi$	−0.184590 + 0.298430i	−0.184510 + 0.298910i	+0.045648	+0.045641
$1.5 \cdot 2\pi$	+0.178080 + 0.305840i	+0.178160 + 0.307170i	+0.063936	+0.064350
$2.0 \cdot 2\pi$	+0.062642 − 0.054219i	+0.062588 − 0.054423i	−0.044927	−0.044998

[b]The estimations of the second kernel are given as: $\tilde{H}_2(j\omega, j\omega) \approx 4Y_2(j\omega, j\omega)/A^2$, on the diagonal, and $\tilde{H}_2(j\omega, -j\omega) \approx 2Y_2(j\omega, -j\omega)/A^2$, which is the DC term

> Identified linear dynamics

Even if the coordinate system is different, one crucial qualitative result is to compute the poles and zeros of the linear transfer function. For the identified system with the theoretical measurements (noise free), the poles and zeros are exactly as the original: $\tilde{p}_t = -1 \pm 10\text{i}$ and the zero is: $\tilde{z}_t = -1$ while for the second-stage approximation to the linear system, the corresponding results are: $\tilde{p} = -1.0048 \pm 9.9989\text{i}$, $\tilde{z} = -1.0042$.

At this point, we have recovered the linear part of the bilinear system up to an accuracy due to the truncation of Volterra series. The inexact simulations of the continuous system which are done with a finite time step $dt = 1e - 4$, and the Fourier accuracy led to quite accurate results with a perturbation of the order $\sim O(1e - 3)$ by comparing the theoretical poles and zeros. We proceed by collecting the measurements of the second kernel. Table 2, contains measurements of the second kernel with one-tone input.

We can get **N** by solving the least-squares problem by just minimizing the 2-norm as in [28]. This result was not toward the identification of the matrix **N** and here is the new approach working toward the identification of bilinear systems.

? Can we identify the matrix N?

The improvement relies on the rank deficiency problem that is produced by getting the least-squares solution without taking under consideration measurements out of the diagonal of the second kernel H_2. By filling in the least-squares problem in Eq. (37) with these extra equations, as Proposition 1 indicates, the problem solution upgrades to a full-rank inversion and the answer is affirmative.

Back to our introductory example, the rank of the least-squares problem is less than $r^2 = 4$. So, we need to increase the rank. We take measurements (≤ 4) out of the diagonal from the second kernel by using the input $u(t) = A_1 \cos(\omega_1) + B_1 \cos(\omega_2)$. Table 3 includes the theoretical and measured results.

The full-rank least-squares solution gave for the theoretical noise-free case and for the second-stage approximation the following results, respectively:

$$\tilde{\mathbf{N}}_t = \begin{bmatrix} -4.1542 & -2.0998 \\ 3.236 & 1.1542 \end{bmatrix}, \; \tilde{\mathbf{N}} = \begin{bmatrix} -4.1557 & -2.1084 \\ 3.2284 & 1.1513 \end{bmatrix} \tag{45}$$

> Coordinate transformation

By transforming all the matrices to the same coordinate system as in [26], we conclude to the

- **Noise-free case—exact identification**

$$\breve{\mathbf{A}}_t = \begin{bmatrix} -1.0 & -10.0 \\ 10.0 & -1.0 \end{bmatrix}, \; \breve{\mathbf{N}}_t = \begin{bmatrix} 1.0 & -2.0 \\ 3.0 & -4.0 \end{bmatrix}, \; \breve{\mathbf{B}}_t = \begin{bmatrix} 1.0 \\ 1.0 \end{bmatrix}, \; \breve{\mathbf{C}}_t = \begin{bmatrix} 1.0 \\ 1.0 \end{bmatrix}^T. \tag{46}$$

- **Simulated case—approximated identification**

$$\breve{\mathbf{A}} = \begin{bmatrix} -1.0037 & -9.9941 \\ 10.004 & -1.0059 \end{bmatrix}, \; \breve{\mathbf{N}} = \begin{bmatrix} 0.99525 & -1.997 \\ 3.006 & -3.9997 \end{bmatrix}, \; \breve{\mathbf{B}} = \begin{bmatrix} 0.99925 \\ 1.0003 \end{bmatrix}, \; \breve{\mathbf{C}} = \begin{bmatrix} 1.0 \\ 1.0 \end{bmatrix}^T. \tag{47}$$

Next, in Fig. 7, evaluation results for the linear and the second-order generalized transfer function are presented:

Finally, time-domain simulations for each system performed in Fig. 8 with a larger amplitude than the probing one.

Table 3 Measurements of the second kernel (out of the diagonal)

Frequencies (ω_1, ω_2)	$\tilde{H}_2(j\omega_1, j\omega_2)$	$H_2(j\omega_1, j\omega_2)$
$(0.2 \cdot 2\pi, 0.3 \cdot 2\pi)$	$+0.030440 - 0.039259i$	$+0.030429 - 0.039237i$
$(0.2 \cdot 2\pi, 0.6 \cdot 2\pi)$	$+0.031002 - 0.080364i$	$+0.031037 - 0.080315i$
$(0.4 \cdot 2\pi, 0.3 \cdot 2\pi)$	$+0.030948 - 0.062869i$	$+0.030961 - 0.062835i$
$(0.4 \cdot 2\pi, 0.6 \cdot 2\pi)$	$+0.026417 - 0.125320i$	$+0.026554 - 0.125260i$

[c]The estimation of the second kernel as $\tilde{H}_2(j\omega_1, j\omega_2) \approx 2Y_2(j\omega_1, j\omega_2)/(A_1 B_1)$. Here we use $\phi = 1.5$, to avoid the harmonic overlapping as explained in Sect. 4.4 and amplitudes as $A_1 = B_1 = 0.01$

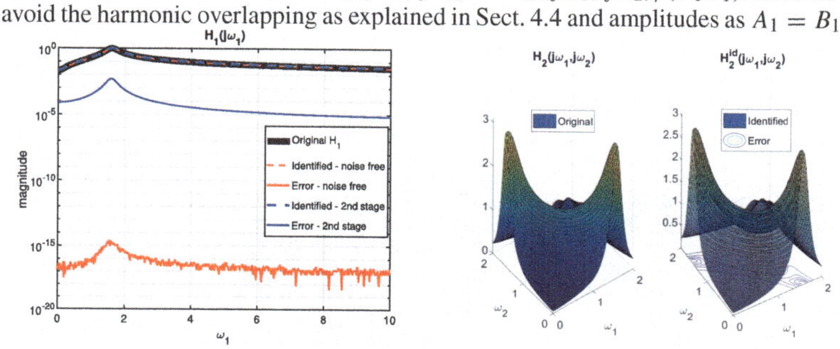

Fig. 7 The identified first and second kernel with second-stage approximation in comparison with the theoretical kernels

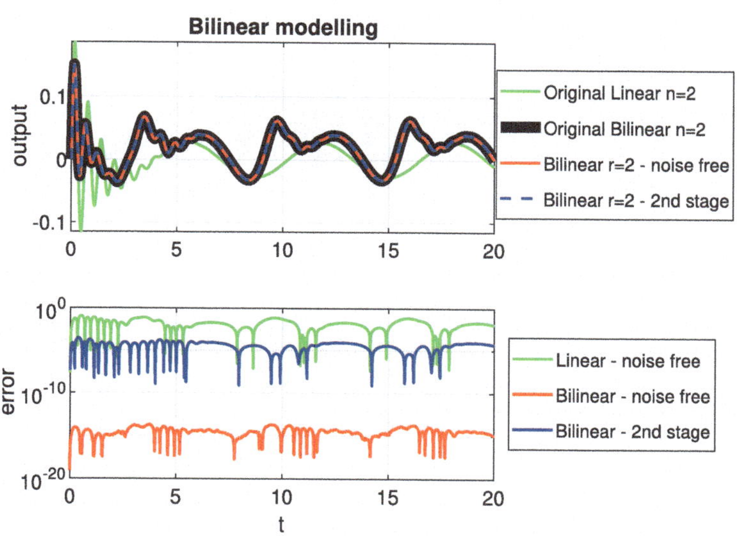

Fig. 8 The evaluation of the models with order $r = 2$ performed with input as $u(t) = \cos(t)$, $t \in [0, 20]$. The noise-free case has reached machine precision

Example 2 Time-domain reduction of the Burgers' Equation. This example illustrates the bilinear modeling and reduction concepts proposed in [5] for the viscous Burgers' equation from time-domain simulations. We simulate the system with 40 measurements as $\omega_k = j2\pi[0.1, 0.2, \ldots, 4]$. We present the corre-

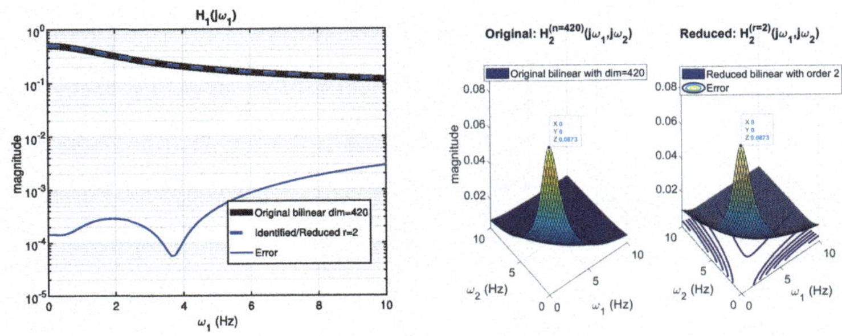

Fig. 9 The first and the second kernel evaluations in comparison with the originals

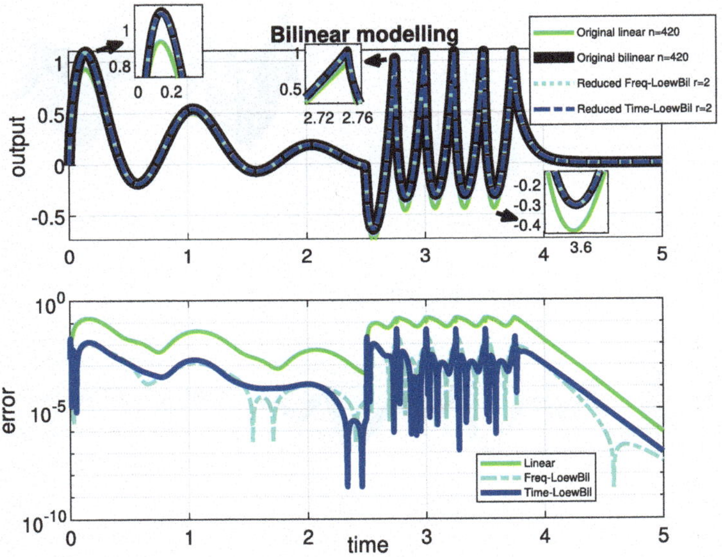

Fig. 10 Time-domain simulation for the Burgers' equation example; viscosity parameter ν is set as 1 and the dimension of the semi-discretized model is chosen to be 420. A comparison among the identified/reduced bilinear of order $r = 2$ with the linear and with the frequency-domain Loewner bilinear is depicted. The input is chosen as: $u(t) = (1 + 2\cos(2\pi t))e^{-t}, t \in [0, 2.5]$, $u(t) = 4\text{sawtooth}(8\pi t), t \in [2.5, 3.75]$, $u(t) = 0, t \in [3.75, 5]$

sponding results with initial system dimension $n = 420$ reduced by the proposed method to order $r = 2$ with the first normalized neglected singular value to be $\sigma_3/\sigma_1 = 4.6255 \cdot 10^{-4}$. As the order was chosen $r = 2$, the reduced bilinear matrix $\tilde{\mathbf{N}}$ was introduced by using the following measurements as $\omega_1 = j2\pi[0.2, 0.4]$ and $\omega_2 = j2\pi[0.3, 0.6]$. In Fig. 9, evaluation results are presented.

Lastly, in Fig. 10, a time-domain simulation reveals that the proposed method can improve the accuracy by fitting a nonlinear model. Table 4 contains approximation results both in the frequency and, also in the time-domain. For the example presented (dimension reduction from $n = 420$ to $r = 2$), we offer a comparison of the

Table 4 Summary of the results from the two examples with Time-LoewBil and comparison with [5] for Burgers' Example 2 of dimension $n = 420$

Error quantification	Time-LoewBil Example 1	Time-LoewBil Example 2	Freq-LoewBil Example 2
$\max_{\omega}\Vert H_1(j\omega) - \tilde{H}_1(j\omega)\Vert$	$5.077 \cdot 10^{-3}$	$2.937 \cdot 10^{-3}$	$4.430 \cdot 10^{-3}$
$\max_{t}\Vert y(t) - y_l(t)\Vert$	$1.213 \cdot 10^{-1}$	$1.699 \cdot 10^{-1}$	$1.699 \cdot 10^{-1}$
$\max_{(\omega_1,\omega_2)}\Vert H_2(j\omega_1, j\omega_2) - \tilde{H}_2(j\omega_1, j\omega_2)\Vert$	$2.794 \cdot 10^{-2}$	$3.077 \cdot 10^{-3}$	$2.991 \cdot 10^{-3}$
$\max_{t}\Vert y(t) - \tilde{y}_b(t)\Vert$	$\mathbf{2.739 \cdot 10^{-4}}$	$\mathbf{5.032 \cdot 10^{-2}}$	$\mathbf{5.278 \cdot 10^{-2}}$

[d]The evaluations of the kernels and the outputs (y_l: linear, $\tilde{y}_b$ reduced bilinear (r = 2)) took place over the domains depicted in Figs. 7, 8, 9, 10

newly proposed method (Time-LoewBil) with another method, i.e., the frequency-domain bilinear Loewner framework introduced in [5] (Freq-LoewBil). The common frequency grid was selected as described above while the sampling values of the tranfser functions (in the frequency-domain) were corrupted with white-noise. The noise magnitude of the latter was selected to match the noise values introduced by performing time-domain simulations with a time step of $dt = 1e - 4$.

Remark 4 (Computational cost for the discretized Burgers' model of dimension 420) The proposed time-domain Loewner bilinear method uses measurements corresponding to symmetric transfer functions. Such values can be directly inferred from time-domain data by processing the spectral domain, i.e., by computing the FFT of the observed output signals for oscillatory input signals. All experiments were performed on a computer with 12 GB RAM and an Intel(R) Core(TM) i7-10510U CPU running at 1.80 GHz, 2304 Mhz, 4 Cores, 8 Logical Processors. To simulate a system of dimension 420, each measurement took ~3 min. So, the data acquisition cost was reported in the range of 1 or 2 h where the identification/reduction part was almost direct. The proposed method seems to efficiently for moderate dimensions; for large-scale problems, the computational issues that appear belong to the class of "embarrassingly parallel" tasks; as the simulations are independent to each other, one can easily speed up the whole process by using instead parallel clusters.

Remark 5 (Discussion and comparison between the two methods) In what follows, we will state the pluses and minuses of the two methods applied for the second numerical example.

The frequency Loewner bilinear framework (Freq-LoewBil)

- Pluses: recovers the original bilinear system with high accuracy, incorporates linear and nonlinear transfer function measurements in a coupled way ("all at once"), can be easily extended to cope with higher order regular kernels, can also be viewed as a Petrov-Galerkin projection-based moment-matching approach.

- Minuses: It is not completely clear how to measure/obtain the frequency-domain data needed for this method; it uses measurements of regular transfer functions which cannot be (directly) inferred from time-domain simulations.

The time-Loewner bilinear framework (Time-LoewBil)

- Pluses: It uses measurements corresponding to symmetric transfer functions. Such values can be directly inferred from time-domain data by processing the spectral domain, i.e., by computing the FFT of the observed output signals for oscillatory input signals.
- Minuses: The fitted bilinear model is as good as the fitted linear model (it relies on the linear fit). As opposed to the first method, it fits the linear and nonlinear parts separately (not "all at once"). It introduces additional errors due to conversion from the time domain to the frequency domain. The latter disadvantage could also occur for the method in [5], provided that "regular transfer function" measurements could be successfully inferred from time-domain data.

6 Conclusion

The proposed method offers approximate bilinear system identification from time-domain measurements, since it is not possible to measure the corresponding kernels exactly. An adaptive scheme that improves the estimation of the kernels was presented. Our proposed method uses only *input-output* measurements without requiring state-space access. What makes this algorithm feasible is the combination of the data-driven Loewner framework with the nonlinear Volterra series framework.

We have shown that for the noise-free case, the proposed method achieves system identification from time-domain measurements through the symmetric kernels. Further study is required to quantify the effects of the noise introduced by the truncation of the Volterra series (in the ℓ-stage approximation). All the time-domain numerical simulations have been implemented by means of the backward Euler approximation scheme which certifies that this method can handle some level of numerical noise. Multi-stepping methods, e.g., Runge-Kutta can offer a significance improvement to the results and reduce the influence of numerical noise.

The variational approach is a theoretical method to identify regular kernels which are appropriate for system identification purposes [35]. However, these kernels do not have a physical meaning, i.e., cannot be directly measured from time-domain simulations. This is not an issue for the growing exponential approach. The derived transfer functions by means of this method can be measured from time-domain data. The difficulty in combining both derivations, i.e., symmetric and regular is also explained from the nth-dimensional integral that connects those through the triangular kernels. Extensions to the MIMO case and to other nonlinearity structures, e.g., quadratic or bilinear quadratic etc., are promising endeavors that will be the matter of future research.

References

1. Ahmad, M.I., Baur, U., Benner, P.: Implicit Volterra series interpolation for model reduction of bilinear systems. J. Comput. Appl. Math. 316, 15–28 (2017). https://doi.org/10.1016/j.cam.2016.09.048
2. Anderson, B.D.O., Antoulas, A.C.: Rational interpolation and state-variable realizations. Linear Algebra Appl. 137/138, 479–509 (1990)
3. Antoulas, A.C.: Approximation of large-scale dynamical systems. Advances in Design and Control, vol. 6. SIAM Publications, Philadelphia, PA (2005). https://doi.org/10.1137/1.9780898718713
4. Antoulas, A.C., Beattie, C.A., Güğercin, S.: Interpolatory Methods for Model Reduction. Society for Industrial and Applied Mathematics, Philadelphia, PA (2020). https://doi.org/10.1137/1.9781611976083
5. Antoulas, A.C., Gosea, I.V., Ionita, A.C.: Model reduction of bilinear systems in the Loewner framework. SIAM J. Sci. Comput. **38**(5), B889–B916 (2016). https://doi.org/10.1137/15M1041432
6. Antoulas, A.C., Lefteriu, S., Ionita, A.C.: Chapter 8: A Tutorial Introduction to the Loewner Framework for Model Reduction, pp. 335–376. https://doi.org/10.1137/1.9781611974829.ch8
7. Bai, Z., Skoogh, D.: A projection method for model reduction of bilinear dynamical systems. Linear Algebra Appl. 415(2–3), 406–425 (2006)
8. Bartee, J.F., Georgakis, C.: Bilinear identification of nonlinear processes. IFAC Proc. Vol. **27**(2), 47–52 (1994). https://doi.org/10.1016/S1474-6670(17)48128-6, http://www.sciencedirect.com/science/article/pii/S1474667017481286. IFAC Symposium on Advanced Control of Chemical Processes, Kyoto, Japan, 25–27 May 1994
9. Baur, U., Benner, P., Feng, L.: Model order reduction for linear and nonlinear systems: a system-theoretic perspective **21**(4), 331–358 (2014). https://doi.org/10.1007/s11831-014-9111-2
10. Benner, P., Breiten, T.: Interpolation-based $\mathcal{H}_2$-model reduction of bilinear control systems. SIAM J. Matrix Anal. Appl. 33(3), 859–885 (2012)
11. Benner, P., Breiten, T., Damm, T.: Generalized tangential interpolation for model reduction of discrete-time MIMO bilinear systems. Internat. J. Control 84(8), 1398–1407 (2011). DOI: 10.1080/00207179.2011.601761
12. Benner, P., Damm, T.: Lyapunov equations, energy functionals, and model order reduction of bilinear and stochastic systems. SIAM J. Control Optim. 49(2), 686–711 (2011). DOI: 10.1137/09075041X
13. Benner, P., Goyal, P., Heiland, J., Pontes Duff, I.: Operator inference and physics-informed learning of low-dimensional models for incompressible flows. e-prints 2010.06701, arXiv (2020). http://arxiv.org/abs/2010.06701. Math.DS
14. Benner, P., Goyal, P., Kramer, B., Peherstorfer, B., Willcox, K.: Operator inference for non-intrusive model reduction of systems with non-polynomial nonlinear terms. Comp. Meth. Appl. Mech. Eng. 372, 113433 (2020). DOI: 10.1016/j.cma.2020.113433
15. Benner, P., Gugercin, S., Willcox, K.: A survey of model reduction methods for parametric systems. SIAM Review **57**(4), 483–531 (2015). https://doi.org/10.1137/130932715
16. Boyd, S., Shing Tang, Y., Chua, L.O.: Measuring Volterra Kernels (1983)
17. Breiten, T.: Interpolatory methods for model reduction of large-scale dynamical systems. Dissertation, Department of Mathematics, Otto-von-Guericke University, Magdeburg, Germany (2013)
18. Brubeck, P.D., Nakatsukasa, Y., Trefethen, L.N.: Vandermonde with Arnoldi (2019)
19. Drmač, Z., Peherstorfer, B.: Learning low-dimensional dynamical-system models from noisy frequency-response data with Loewner rational interpolation (2019)
20. Flagg, G.M., Gugercin, S.: Multipoint Volterra series interpolation and $\mathcal{H}_2$ optimal model reduction of bilinear systems. SIAM J. Numer. Anal. 36(2), 549–579 (2015). https://doi.org/10.1137/130947830
21. Fosong, E., Schulze, P., Unger, B.: From time-domain data to low-dimensional structured models (2019)

22. Gosea, I.V., Antoulas, A.C.: Data-driven model order reduction of quadratic-bilinear systems. Numerical Linear Algebra Appl. **25**(6), e2200 (2018). https://doi.org/10.1002/nla.2200. E2200 nla.2200
23. Gustavsen, B., Semlyen, A.: Rational approximation of frequency domain responses by vector fitting. IEEE Transactions on Power Delivery **14**(3), 1052–1061 (1999). https://doi.org/10.1109/61.772353
24. Ionita, A.C.: Matrix pencils in time and frequency domain system identification, pp. 79–88 (2012). https://doi.org/10.1049/pbce076e_ch9
25. Isidori, A.: Direct construction of minimal bilinear realizations from nonlinear input-output maps. IEEE Transactions on Automatic Control **18**(6), 626–631 (1973)
26. Juang, J.N.: Continuous-time bilinear system identification. Nonlinear Dynamics **39**(1), 79–94 (2005). https://doi.org/10.1007/s11071-005-1915-z
27. Kaiser, E., Kutz, J.N., Brunton, S.L.: Sparse identification of nonlinear dynamics for model predictive control in the low-data limit. Proc. R. Soc. A: Math. Phys. Eng. Sci. **474**(2219), 20180335 (2018). https://doi.org/10.1098/rspa.2018.0335
28. Karachalios, D.S., Gosea, I.V., Antoulas, A.C.: A bilinear identification-modeling framework from time domain data. Proc. Appl. Math. Mech. 19(1), e201900246 (2019). DOI: 10.1002/pamm.201900246
29. Karachalios, D.S., Gosea, I.V., Antoulas, A.C.: The Loewner framework for system identification and reduction. In: Benner, P., Grivet-Talocia, S., Quarteroni, A., Rozza, G., Schilders, W.H.A., Silveira, L.M. (eds.), Handbook on Model Reduction, volume I of Methods and Algorithms (in press)
30. Lefteriu, S., Ionita, A.C., Antoulas, A.C.: Modeling Systems Based on Noisy Frequency and Time Domain Measurements, pp. 365–378. Springer, Berlin, Heidelberg (2010). https://doi.org/10.1007/978-3-540-93918-4_33
31. Peherstorfer, B., Gugercin, S., Willcox, K.: Data-driven reduced model construction with time-domain Loewner models. SIAM J. Sci. Comput. **39**(5), A2152–A2178 (2017). https://doi.org/10.1137/16M1094750
32. Peherstorfer, B., Willcox, K.: Data-driven operator inference for nonintrusive projection-based model reduction. Comput. Methods Appl. Mech. Eng. **306**, 196–215 (2016). https://doi.org/10.1016/j.cma.2016.03.025, http://www.sciencedirect.com/science/article/pii/S0045782516301104
33. Petkovska, M., Nikolić, D., Seidel-Morgenstern, A.: Nonlinear frequency response method for evaluating forced periodic operations of chemical reactors. Israel J. Chem. **58**(6-7), 663–681 (2018). https://doi.org/10.1002/ijch.201700132
34. Phillips, J.R.: Projection-based approaches for model reduction of weakly nonlinear, time-varying systems **22**(2), 171–187 (2003)
35. Rugh, W.J.: Nonlinear System Theory: The Volterra/Wiener Approach. The Johns Hopkins University Press, Baltimore (1981)
36. Scarciotti, G., Astolfi, A.: Data-driven model reduction by moment matching for linear and nonlinear systems. Automatica **79**, 340–351 (2017). https://doi.org/10.1016/j.automatica.2017.01.014, http://www.sciencedirect.com/science/article/pii/S0005109817300249
37. Schmid, P.J.: Dynamic mode decomposition of numerical and experimental data. Journal of Fluid Mechanics **656**, 5–28 (2010). https://doi.org/10.1017/S0022112010001217

Balanced Truncation for Parametric Linear Systems Using Interpolation of Gramians: A Comparison of Algebraic and Geometric Approaches

Nguyen Thanh Son, Pierre-Yves Gousenbourger, Estelle Massart, and Tatjana Stykel

Abstract When balanced truncation is used for model order reduction, one has to solve a pair of Lyapunov equations for two Gramians and uses them to construct a reduced-order model. Although advances in solving such equations have been made, it is still the most expensive step in this reduction method. Parametric model order reduction aims to determine reduced-order models for parameter-dependent systems. Popular techniques for parametric model order reduction rely on interpolation. Nevertheless, interpolation of Gramians is rarely mentioned which motivates our work. Here, we propose and compare two approaches for Gramian interpolation. In the first approach, the interpolated Gramian is computed as a linear combination of the data Gramians with positive coefficients. We show that, if the system depends affinely on the parameters, computation time can be saved by making part of the computations offline. The second approach aims at performing the interpolation on the manifold of fixed-rank positive semidefinite matrices. We resort then to interpolation algorithms

---

E. Massart—Most of this work was done when this author was with ICTEAM, UCLouvain.

N. T. Son (✉) · P.-Y. Gousenbourger
ICTEAM, UCLouvain, Avenue Georges Lemaître 4-6/L4.05.01, 1348 Louvain-la-Neuve, Belgium
e-mail: thanh.son.nguyen@uclouvain.be

P.-Y. Gousenbourger
e-mail: pierre-yves.gousenbourger@uclouvain.be

N. T. Son
Thai Nguyen University of Sciences, Thai Nguyen 25000, Vietnam

E. Massart
Mathematical Institute, University of Oxford, Radcliffe Observatory Quarter, Woodstock Road, Oxford OX26GG, UK
e-mail: estelle.massart@maths.ox.ac.uk

National Physical Laboratory, Hampton Road, Teddington, Middlesex TW11 0LW, UK

T. Stykel
Institute of Mathematics, University of Augsburg, Universitätsstr. 14, 86159 Augsburg, Germany
e-mail: stykel@math.uni-augsburg.de

P. Benner et al. (eds.), *Model Reduction of Complex Dynamical Systems*,
International Series of Numerical Mathematics 171,
https://doi.org/10.1007/978-3-030-72983-7_2

on Riemannian manifolds, and, more specifically, to curve and surface interpolation techniques assuming that the model depends on one or two parameters, respectively. The results of the interpolation step are then used to construct parametric reduced-order models, which are compared numerically on two benchmark problems.

1 Introduction

The need for increasingly accurate simulations in sciences and technology results in large-scale mathematical models. Simulation of those systems is usually time-consuming or even infeasible, especially with limited computer resources. Model order reduction (MOR) is a well-known tool to deal with such problems. Founded about half a century ago, this field is still getting attraction due to the fact that many complicated or large problems have not been considered and many advanced methods have not been invoked yet.

Often, the full-order model (FOM) depends on parameters. The reduced-order model (ROM), preferably parameter dependent as well, is therefore required to approximate the FOM on a given parameter domain. This problem, so-called parametric model order reduction (PMOR), has been addressed using various approaches such as Krylov subspace-based methods [1, 2], optimization [3], interpolation [4–7], and reduced basis techniques [8, 9], just to name a few. The reader is referred to the survey [10] and the contributed book [11] for more details. We focus here on interpolation-based methods to build a ROM for the linear parametric control system

$$\begin{aligned} E(\mu)\dot{x}(t,\mu) &= A(\mu)x(t,\mu) + B(\mu)u(t), \\ y(t,\mu) &= C(\mu)x(t,\mu), \end{aligned} \tag{1}$$

where $E(\mu),\ A(\mu) \in \mathbb{R}^{n\times n},\ B(\mu) \in \mathbb{R}^{n\times m},\ C(\mu) \in \mathbb{R}^{p\times n}$ with $p, m \ll n$, and $\mu \in \mathcal{D} \subset \mathbb{R}^{\ell}$. We assume that the matrix $E(\mu)$ is nonsingular and that all the eigenvalues of the pencil $\lambda E(\mu) - A(\mu)$ have negative real part for all $\mu \in \mathcal{D}$. This assumption allows us to avoid working with singular control systems and to restrict ourselves to the use of standard balanced truncation [12, 13]. The goal of PMOR is to approximate system (1) by a smaller parametric model

$$\begin{aligned} \tilde{E}(\mu)\dot{\tilde{x}}(t,\mu) &= \tilde{A}(\mu)\tilde{x}(t,\mu) + \tilde{B}(\mu)u(t), \\ \tilde{y}(t,\mu) &= \tilde{C}(\mu)\tilde{x}(t,\mu), \end{aligned} \tag{2}$$

where $\tilde{E}(\mu), \tilde{A}(\mu) \in \mathbb{R}^{r\times r}, \tilde{B}(\mu) \in \mathbb{R}^{r\times m}, \tilde{C}(\mu) \in \mathbb{R}^{p\times r}$ and $r \ll n$.

Interpolation-based methods work as follows. On a given sample grid μ_j, $j = 1, \ldots, q$, in the parameter domain $\mathcal{D}$, one computes a ROM associated with each μ_j. These ROMs can be obtained using any MOR method for non-parametric models [14] and are characterized by either their projection subspaces, coefficient matrices, or transfer functions. Then they are interpolated using standard methods such as Lagrange or spline interpolation. These approaches have been discussed intensively in many publications, see, e.g., [6, 15, 16] for interpolating local reduced system matrices, [4, 17] for interpolating projection subspaces, [5, 18] for interpolating reduced transfer functions, and [19] for a detailed discussion on the use of manifold interpolation for model reduction. Each of them has its own strengths and works well in some specific applications but fails to be superior to the others in a general setting.

When balanced truncation [20] is used, one has to solve a pair of Lyapunov equations for two Gramians. Although advances in solving such equations have been made, it is still the most expensive step in this reduction method. Therefore, any interpolation method that can circumvent this step is of interest. Unfortunately, to our knowledge, there has been no work addressing this issue. In this contribution, we propose to interpolate the solutions to the Lyapunov equations, i.e., the Gramians. It is noteworthy that in the large-scale setting, one should avoid working with full-rank solution matrices. Fortunately, in many practical cases, the solution of the Lyapunov equation can be well approximated by a symmetric positive semidefinite (SPSD) matrix of considerably smaller rank [21, 22]. Such approximations can be used in the square root balanced truncation method [23] to make the reduction procedure more computationally efficient.

To ensure that the SPSD property is preserved during the interpolation, we propose two approaches. A key feature of these two approaches is that they allow a direct manipulation of the low-rank factors of the SPSD matrices instead of the full Gramians, thereby reducing the computation cost (i.e., any $n \times n$ Gramian P, of rank k, is written as $P = XX^T$ for some factor matrix $X \in \mathbb{R}^{n \times k}$). In the first approach, which is the main content of Sect. 2, the target Gramians are written as a linear combination of the data Gramians with some given (positive) weights. An issue with this first method is that the "interpolated" low-rank factors have a considerably larger number of columns than the ones associated with the training data, and, as a consequence, they may contain redundant information. However, by applying the balanced truncation model reduction method, we can remove this redundancy. Moreover, assuming the affine dependence of the matrices E, A, B, C (see (1)) on the parameters, we can design an offline-online decomposition of the balanced truncation procedure, to reduce the computational cost of the operations that have to be done on-the-fly. We refer to this as the linear algebraic (or algebraic for short) approach.

The second approach, given in Sect. 3, consists of mapping beforehand all the matrices to the set of fixed-rank positive semidefinite matrices, and performing the interpolation directly in that set. This would ensure that the rank of the interpolated Gramians remains consistent with the ranks of the data Gramians. It was shown in [24, 25] that the set of SPSD matrices of fixed rank can be turned into a Riemannian manifold by equipping it with a differential structure. We can then resort to interpolation techniques specifically designed to work on Riemannian manifolds. Oldest techniques are based on subdivision schemes [26] or rolling procedures [27]. In the last decades, path fitting techniques rose up, such as least-squares smoothing [28] or more recently by means of Bézier splines [29, 30]. The latter will be employed here for interpolating the Gramians. The resulting PMOR method will be referred to as the geometric method in the sense that it strictly preserves the geometric structure of the data.

The rest of the chapter is organized as follows. In Sect. 2, we briefly recall balanced truncation for MOR, the square root balanced truncation procedure, and present the algebraic interpolation method. Section 3 is devoted to the geometric interpolation method. It first describes the geometry of the manifold of fixed-rank SPSD matrices, and then algorithms to perform interpolation on this manifold. The two proposed approaches are then compared numerically in Sect. 4, and the conclusion is given in Sect. 5.

2 Balanced Truncation for Parametric Linear Systems and Standard Interpolation

2.1 Balanced Truncation

Balanced truncation [14, 20] is a well-known method for model reduction. In this section, we briefly review the square root procedure proposed in [23] which is more numerically efficient than its original version. As in other projection-based methods, a balancing projection for system (1) must be constructed. This projection helps to balance the input and output energies on each state so that one can easily decide which state component should be truncated. To this end, one has to solve the pair of generalized Lyapunov equations

$$E(\mu)P(\mu)A^T(\mu) + A(\mu)P(\mu)E^T(\mu) = -B(\mu)B^T(\mu), \tag{3}$$

$$E^T(\mu)Q(\mu)A(\mu) + A^T(\mu)Q(\mu)E(\mu) = -C^T(\mu)C(\mu), \tag{4}$$

for the *controllability Gramian* $P(\mu)$ and the *observability Gramian* $Q(\mu)$. In practice, these Gramians are computed in the factorized form

$$P(\mu) = X(\mu)X^T(\mu), \qquad Q(\mu) = Y(\mu)Y^T(\mu),$$

with $X(\mu) \in \mathbb{R}^{n\times k_c}$ and $Y(\mu) \in \mathbb{R}^{n\times k_o}$. One can show that the eigenvalues of the matrix $P(\mu)E^T(\mu)Q(\mu)E(\mu)$ are real and non-negative [14]. The positive square roots of the eigenvalues of this matrix, $\sigma_1(\mu) \geq \cdots \geq \sigma_n(\mu) \geq 0$, are called the *Hankel singular values* of system (1). They can also be determined from the singular value decomposition (SVD)

$$Y^T(\mu)E(\mu)X(\mu) = [U_1(\mu)\;\; U_0(\mu)] \begin{bmatrix} \Sigma_1(\mu) & 0 \\ 0 & \Sigma_0(\mu) \end{bmatrix} [V_1(\mu)\;\; V_0(\mu)]^T, \tag{5}$$

where $[U_1(\mu)\;\; U_0(\mu)]$ and $[V_1(\mu)\;\; V_0(\mu)]$ have orthonormal columns, and

$$\Sigma_1(\mu) = \operatorname{diag}(\sigma_1(\mu), \ldots, \sigma_r(\mu)), \quad \Sigma_0(\mu) = \operatorname{diag}(\sigma_{r+1}(\mu), \ldots, \sigma_{k_{co}}(\mu))$$

with $k_{co} = \min(k_c, k_o)$. Then the ROM (2) is computed by projection

$$\begin{aligned} \tilde{E}(\mu) &= W^T(\mu)E(\mu)T(\mu), & \tilde{A}(\mu) &= W^T(\mu)A(\mu)T(\mu), \\ \tilde{B}(\mu) &= W^T(\mu)B(\mu), & \tilde{C}(\mu) &= C(\mu)T(\mu), \end{aligned} \tag{6}$$

where the projection matrices are given by

$$W(\mu) = Y(\mu)U_1(\mu)\Sigma_1^{-1/2}(\mu), \qquad T(\mu) = X(\mu)V_1(\mu)\Sigma_1^{-1/2}(\mu). \tag{7}$$

The $\mathcal{H}_\infty$-error of the approximation is shown to satisfy

$$\|H(\cdot, \mu) - \tilde{H}(\cdot, \mu)\|_{\mathcal{H}_\infty} \leq 2\big(\sigma_{r+1}(\mu) + \cdots + \sigma_{k_{co}}(\mu)\big),$$

where the $\mathcal{H}_\infty$-norm is defined as

$$\|H\|_{\mathcal{H}_\infty} = \sup_{\omega \in \mathbb{R}} \|H(\mathrm{i}\omega)\|_2,$$

where $\mathrm{i} = \sqrt{-1}$, and

$$\begin{aligned} H(s, \mu) &= C(\mu)(sE(\mu) - A(\mu))^{-1}B(\mu), \\ \tilde{H}(s, \mu) &= \tilde{C}(\mu)(s\tilde{E}(\mu) - \tilde{A}(\mu))^{-1}\tilde{B}(\mu) \end{aligned}$$

are the transfer functions of systems (1) and (2), respectively.

2.2 *Interpolation of Gramians for Parametric Model Order Reduction*

Since solving Lyapunov equations is the most expensive step of the balanced truncation procedure, we propose to compute the solution for only a few values of the parameter, and then interpolate those for other values of the parameter. To this end, on the chosen sample grid $\mu_1, \ldots, \mu_q \in \mathcal{D}$, we solve the Lyapunov equations (3) and (4) for $P(\mu_j) = P_j = X_j X_j^T$ and $Q(\mu_j) = Q_j = Y_j Y_j^T$, $j = 1, \ldots, q$. Note that the ranks of the local Gramians P_j and Q_j, $j = 1, \ldots, q$, do not need to be the same. Then we define the mappings

$$\begin{aligned} P : \mathcal{D} &\to \mathbb{R}^{n\times n}, & Q : \mathcal{D} &\to \mathbb{R}^{n\times n}, \\ \mu &\mapsto P(\mu), & \mu &\mapsto Q(\mu), \end{aligned}$$

interpolating the data points (μ_j, P_j) and (μ_j, Q_j), respectively, as

$$P(\mu) = \sum_{j=1}^{q} w_j(\mu) X_j X_j^T, \qquad Q(\mu) = \sum_{j=1}^{q} w_j(\mu) Y_j Y_j^T,$$

where $w_j(\mu)$ are some weights that will be detailed in Sect. 4. To preserve the positive semidefiniteness of the Gramians, we propose to use non-negative weights [31]. This methodology is compatible with the factorization structure since we can write

$$\begin{aligned} P(\mu) &= \sum_{j=1}^{q} \sqrt{w_j(\mu)} X_j \sqrt{w_j(\mu)} X_j^T &(8)\\ &= \left[\sqrt{w_1(\mu)} X_1 \cdots \sqrt{w_q(\mu)} X_q\right] \left[\sqrt{w_1(\mu)} X_1 \cdots \sqrt{w_q(\mu)} X_q\right]^T \\ &= X(\mu) X^T(\mu), &(9) \end{aligned}$$

and, similarly,

$$Q(\mu) = Y(\mu) Y^T(\mu) \text{ with } Y(\mu) = \left[\sqrt{w_1(\mu)} Y_1 \cdots \sqrt{w_q(\mu)} Y_q\right]. \qquad (10)$$

Note that the computation of the parametric Gramians is not the ultimate goal. After interpolation, we still have to proceed steps (5) and (6) to get the ROM. The computations required by these steps explicitly involve large matrices which may reduce the efficiency of the proposed method. To overcome this difficulty, we separate the computations into two stages. The first stage can be expensive but must be independent of μ so that it can be precomputed. The second step, where one has to compute the ROM at any new value $\mu \in \mathcal{D}$, must be fast. Ideally, its computational complexity should be independent of n, the dimension of the initial problem. Such a decomposition is often referred to as an *offline-online decomposition* and quite well known in the reduced basis community [32, 33]. Details are presented in the next subsection. Before that, we would like to drive the reader's attention to a related work [34], where we considered the problem of interpolating the solution of parametric Lyapunov equations using different interpolation techniques and compared the obtained results.

2.3 Offline-Online Decomposition

For the offline-online decomposition, we assume that the matrices of system (1) can be written as affine combinations of some parameter-independent matrices $\{E_i\}_{i=1,\ldots,q_E}$, $\{A_i\}_{i=1,\ldots,q_A}$, $\{B_i\}_{i=1,\ldots,q_B}$, and $\{C_i\}_{i=1,\ldots,q_C}$ as follows:

$$E(\mu) = \sum_{i=1}^{q_E} f_i^E(\mu) E_i, \qquad A(\mu) = \sum_{i=1}^{q_A} f_i^A(\mu) A_i,$$
$$B(\mu) = \sum_{i=1}^{q_B} f_i^B(\mu) B_i, \qquad C(\mu) = \sum_{i=1}^{q_C} f_i^C(\mu) C_i,$$

where q_E, q_A, q_B, q_C are small and the evaluations of $f_i^E, f_i^A, f_i^B, f_i^C$ are cheap. This assumption is common in PMOR and often fulfilled in practice, see, e.g., [1, 8, 17, 35] and examples in Sect. 4. Once the interpolated Gramians are available, we obtain

$$\begin{aligned} Y^T(\mu)E(\mu)X(\mu) &= \begin{bmatrix} \sqrt{w_1(\mu)}Y_1^T \\ \cdots \\ \sqrt{w_q(\mu)}Y_q^T \end{bmatrix} \sum_{i=1}^{q_E} f_i^E(\mu) E_i \left[\sqrt{w_1(\mu)}X_1 \ \cdots \ \sqrt{w_q(\mu)}X_q\right] \\ &= \sum_{i=1}^{q_E} f_i^E(\mu) \begin{bmatrix} w_{11}(\mu)Y_1^T E_i X_1 & \cdots & w_{1q}(\mu)Y_1^T E_i X_q \\ \vdots & \ddots & \vdots \\ w_{q1}(\mu)Y_q^T E_i X_1 & \cdots & w_{qq}(\mu)Y_q^T E_i X_q \end{bmatrix}, \end{aligned} \tag{11}$$

with $w_{lj}(\mu)=\sqrt{w_l(\mu)w_j(\mu)}$. Obviously, all $q_E q^2$ blocks $Y_l^T E_i X_j$ for $l, j=1,\ldots,q$ and $i = 1,\ldots,q_E$ can be precomputed and stored since they are independent of μ. After computing the SVD of (11), the projection matrices in (7) take the form

$$\begin{aligned} W(\mu) &= \left[\sqrt{w_1(\mu)}Y_1 \ \cdots \ \sqrt{w_q(\mu)}Y_q\right] U_1(\mu)\Sigma_1^{-1/2}(\mu), \\ T(\mu) &= \left[\sqrt{w_1(\mu)}X_1 \ \cdots \ \sqrt{w_q(\mu)}X_q\right] V_1(\mu)\Sigma_1^{-1/2}(\mu). \end{aligned}$$

The reduced matrices are then computed as in (6):

$$\begin{aligned} \tilde{E}(\mu) = W^T(\mu)E(\mu)T(\mu) &= \sum_{i=1}^{q_E} f_i^E(\mu)\Sigma_1^{-1/2}(\mu)U_1^T(\mu) \\ &\times \begin{bmatrix} w_{11}(\mu)Y_1^T E_i X_1 & \cdots & w_{1q}(\mu)Y_1^T E_i X_q \\ \vdots & \vdots & \vdots \\ w_{q1}(\mu)Y_q^T E_i X_1 & \cdots & w_{qq}(\mu)Y_q^T E_i X_q \end{bmatrix} V_1(\mu)\Sigma_1^{-1/2}(\mu), \end{aligned} \tag{12}$$

$$\begin{aligned} \tilde{A}(\mu) = W^T(\mu)A(\mu)T(\mu) &= \sum_{i=1}^{q_A} f_i^A(\mu)\Sigma_1^{-1/2}(\mu)U_1^T(\mu) \\ &\times \begin{bmatrix} w_{11}(\mu)Y_1^T A_i X_1 & \cdots & w_{1q}(\mu)Y_1^T A_i X_q \\ \vdots & \vdots & \vdots \\ w_{q1}(\mu)Y_q^T A_i X_1 & \cdots & w_{qq}(\mu)Y_q^T A_i X_q \end{bmatrix} V_1(\mu)\Sigma_1^{-1/2}(\mu), \end{aligned} \tag{13}$$

$$\tilde{B}(\mu) = W^T(\mu)B(\mu) = \sum_{i=1}^{q_B} f_i^B(\mu)\Sigma_1^{-1/2}(\mu)U_1^T(\mu) \begin{bmatrix} \sqrt{w_1(\mu)}Y_1^T B_i \\ \vdots \\ \sqrt{w_q(\mu)}Y_q^T B_i \end{bmatrix}, \tag{14}$$

$$\begin{aligned}\tilde{C}(\mu) &= C(\mu)T(\mu) \\ &= \sum_{i=1}^{q_C} f_i^C(\mu)\left[\sqrt{w_1(\mu)}C_i X_1 \cdots \sqrt{w_q(\mu)}C_i X_q\right] V_1(\mu)\Sigma_1^{-1/2}(\mu).\end{aligned} \tag{15}$$

Again, all matrix blocks, that are independent of μ, can be computed and stored beforehand. The offline-online procedure can thus be summarized as follows:

Offline For $\mu_1, \ldots, \mu_q \in \mathcal{D}$,

- solve the Lyapunov equations (3) and (4) for $P_j \approx X_j X_j^T$ and $Q_j \approx Y_j Y_j^T$, $j = 1, \ldots, q$;
- compute and store all the parameter-independent matrix blocks mentioned in (11)–(15).

Online Given $\mu \in \mathcal{D}$,

- assemble precomputed matrix blocks and compute the SVD of (11);
- assemble precomputed matrix blocks and compute the reduced matrices (12)–(15).

3 Interpolation on the Manifold $\mathcal{S}_+(k, n)$

As we have seen above, the interpolated Gramians obtained by the algebraic approach in (9) and (10), in general, have a considerably higher rank than the approximated local Gramians P_j and Q_j, $j = 1, \ldots, q$ obtained by directly solving the Lyapunov equations to a reasonable accuracy. This fact somewhat puts more computational burden on the last steps of the model reduction procedure, especially when the number of grid points is large. If the Gramians can be well approximated by symmetric positive semidefinite matrices of low rank on the whole parameter domain and this rank does not vary significantly we can assume that all approximated local Gramians P_j, $j = 1, \ldots, q$, have a fixed rank. Numerically, this can be almost always achieved by first setting a (very) small tolerance for the low-rank solver when solving the Lyapunov equations at the training points and then truncating all the Gramians to the smallest rank obtained. Hence, with some relaxation, we can assume that $P_j \in \mathcal{S}_+(k_P, n)$ for $j = 1, \ldots, q$ and $Q_j \in \mathcal{S}_+(k_Q, n)$ for $j = 1, \ldots, q$, where $\mathcal{S}_+(k, n)$ is the set of $n \times n$ positive semidefinite matrices of rank k. This set admits a manifold structure [25, 36], and therefore our second interpolation method relies on this geometric property.

Informally speaking, a d-dimensional manifold is a set $\mathcal{M}$ that can be mapped locally through a set of bijections, called *charts*, to (an open subset of) the Euclidean space $\mathbb{R}^d$. Under some additional compatibility assumptions, the collection of charts forms a differentiable structure and the set $\mathcal{M}$ endowed with this structure is called a d-dimensional manifold. The set of charts allows rewriting locally any problem defined on $\mathcal{M}$ into a problem defined on a subset of $\mathbb{R}^d$. We will see that $\mathcal{S}_+(k, n)$ is a *matrix manifold*, i.e., a manifold whose points can be represented by matrices.

Many matrix manifolds are either *embedded submanifolds* of $\mathbb{R}^{m\times n}$, the manifold is then seen as a subset of the Euclidean space $\mathbb{R}^{m\times n}$, or *quotient manifolds* of $\mathbb{R}^{m\times n}$, each point is representing then a set of equivalent points of $\mathbb{R}^{m\times n}$, for a given equivalence relationship. In each case, the differentiable structure of the manifold is inherited from the differentiable structure on $\mathbb{R}^{m\times n}$.

As charts are defined locally, they are not very practical for numerical computations. Their use can be avoided by resorting to other tools specific for working on manifolds. The most important for this work are *tangent spaces*, *exponential*, and *logarithmic maps*. The tangent space $T_x\mathcal{M}$ is the first-order approximation to the manifold $\mathcal{M}$ around $x \in \mathcal{M}$, where the point x is called the *foot* of the tangent space. When the tangent spaces are endowed with a *Riemannian metric* (an inner product $g_x : T_x\mathcal{M} \times T_x\mathcal{M} \to \mathbb{R}$ smoothly varying with x), the manifold is called a *Riemannian manifold.*

The Riemannian metric allows defining geodesics (curves with zero acceleration) on the manifold. This in turn leads to the exponential map which allows mapping tangent vectors to the manifold by following the geodesic starting at the foot of the tangent vector, and whose initial velocity is given by the tangent vector itself. Its reciprocal map is the logarithmic map mapping points from the manifold to a given tangent space. For further details on Riemannian manifolds, we refer to [37, 38].

3.1 A Quotient Geometry of $\mathcal{S}_+(k, n)$

The manifold $\mathcal{S}_+(k, n)$ is here seen as a quotient manifold $\mathbb{R}_*^{n\times k}/O_k$, where $\mathbb{R}_*^{n\times k}$ is the set of full-rank $n \times k$ matrices endowed with the Euclidean metric and O_k is the orthogonal group in dimension k. This geometry has been developed in [25, 36, 39] and has already been used in, e.g., [40–42] for solving different fitting problems. It relies on the fact that any matrix $A \in \mathcal{S}_+(k, n)$ can be factorized as $A = YY^T$ with $Y \in \mathbb{R}_*^{n\times k}$. As the factorization is not unique, this leads to the equivalence relationship:

$$Y_1 \sim Y_2 \quad \text{if and only if} \quad Y_1 = Y_2 Q \text{ with } Q \in O_k.$$

For any $Y \in \mathbb{R}_*^{n\times k}$, the set

$$[Y] := \{YQ \ : \ Q \in O_k\}$$

of points equivalent to Y is called the *equivalence class* of Y. The quotient manifold $\mathbb{R}_*^{n\times k}/O_k$ is the set of all equivalence classes.

The fact that on the manifold $\mathbb{R}_*^{n\times k}/O_k$, any point is a set of points in $\mathbb{R}_*^{n\times k}$ makes it difficult to perform computations directly on elements of $\mathbb{R}_*^{n\times k}/O_k$. Instead of manipulating sets of points, most algorithms on quotient manifolds are only manipulating representatives of the equivalence classes.

The tangent space $T_Y\mathbb{R}_*^{n\times k}$ to the manifold $\mathbb{R}_*^{n\times k}$ at some point Y is the direct sum of two subspaces: the vertical space $\mathcal{V}_Y$ which is, by definition, tangent to $[Y]$, and the horizontal space $\mathcal{H}_Y$ which is its orthogonal complement with respect to the Euclidean metric in $\mathbb{R}_*^{n\times k}$. Horizontal vectors will allow representing tangent vectors to the quotient manifold $\mathbb{R}_*^{n\times k}/O_k$ in a "tangible way", i.e., in a way suitable for numerical computations. Indeed, given a point $Y \in \mathbb{R}_*^{n\times k}$, any tangent vector $\xi_{[Y]} \in T_{[Y]}\mathbb{R}_*^{n\times k}/O_k$ can be identified with a unique horizontal vector $\bar{\xi}_Y \in \mathcal{H}_Y$ (the identification means here that the two vectors act identically as differential operators), see [43, §3.5.8]. This vector is called the *horizontal lift* of $\xi_{[Y]}$ at Y.

The Riemannian metric is naturally inherited from the Euclidean metric in $\mathbb{R}_*^{n\times k}$ (see [25]). When defined, the associated exponential map can be written as

$$\mathrm{Exp}_{[Y]}(\xi_{[Y]}) = [Y + \bar{\xi}_Y], \tag{16}$$

where Y is an arbitrary element of the equivalence class $[Y]$ and $\bar{\xi}_Y$ is the unique horizontal lift of the tangent vector $\xi_{[Y]}$ at Y. Accordingly, for $[Y_1], [Y_2] \in \mathbb{R}_*^{n\times k}/O_k$, the logarithm of $[Y_2]$ at $[Y_1]$, denoted by $\mathrm{Log}_{[Y_1]}([Y_2])$, is a vector in $T_{[Y_1]}\mathbb{R}_*^{n\times k}/O_k$ whose horizontal lift at Y_1 is given by

$$\overline{\mathrm{Log}_{[Y_1]}([Y_2])}_{Y_1} = Y_2 Q^T - Y_1, \tag{17}$$

where Q is the orthogonal factor of the polar decomposition of $Y_1^T Y_2$, when unique. We refer the interested reader to [25] for more information on the domain of definition of these mappings. In all the datasets considered here, we have never faced issues related to ill-definitions of these tools.

3.2 *Curve and Surface Interpolation on Manifolds*

To interpolate the matrices P_i and Q_i on $\mathcal{S}_+(k_P, n)$ and $\mathcal{S}_+(k_Q, n)$, respectively, we consider an intrinsic interpolation technique on Riemannian manifolds. Here, we briefly review it in the specific framework of curve and surface interpolation on Riemannian manifolds, and refer to [30] for the related problem of curve fitting (i.e., relaxing the interpolation constraint).

Curves. Consider a Riemannian manifold $\mathcal{M}$ (here, the set of $n \times n$ positive semidefinite matrices of rank k), and a set of data points $d_1 \dots, d_q \in \mathcal{M}$ (e.g., the matrices P_i) associated with parameter values $t_1 < \dots < t_q \in \mathbb{R}$ (here, the values μ_i). Curve interpolation on $\mathcal{M}$ is often done by encapsulating the interpolation into an optimization problem, e.g., one seeks the curve $\mathbf{B} : [t_1, t_q] \to \mathcal{M}$ minimizing

$$\min_{\mathbf{B}} \int_{t_1}^{t_q} \left\| \frac{\mathrm{D}^2\mathbf{B}(t)}{\mathrm{d}t^2} \right\|_{\mathbf{B}(t)}^2 \mathrm{d}t \text{ such that } \mathbf{B}(t_i) = d_i, \quad i = 1, \ldots, q, \tag{18}$$

where the operator $\mathrm{D}^2/\mathrm{d}t^2$ is the Levi-Civita second covariant derivative (also named acceleration vector field [43, p. 102]) of the manifold-valued function $\mathbf{B}$ and $\|\cdot\|_{\mathbf{B}(t)}$ is the norm (inherited from the Riemannian metric) on the tangent space at $\mathbf{B}(t)$ [43]. Different techniques exist to solve this problem, but nearly none of them tackle (18) directly on $\mathcal{M}$, as the computational effort would be so high that it would not bring any advantages to most of the applications.

An efficient way to approximate the optimal solution is to transfer the interpolation problem to a carefully chosen tangent space $T_x\mathcal{M}$ at a point $x \in \mathcal{M}$, such that $T_x\mathcal{M}$ approximates $\mathcal{M}$ in the area where the data points are defined. The transfer to $T_x\mathcal{M}$ is usually done by mapping the data points to $T_x\mathcal{M}$ via the logarithmic map or an accurate approximation of it. As the tangent space is a Euclidean space, solving the Euclidean version of (18) is easy and computationally tractable since the Levi-Civita second covariant derivative reduces to a classical second derivative. Actually, the solution can even be written in closed form as it is the interpolating natural cubic spline when (18) is minimized over the Sobolev space $H^2(t_1, t_q)$ [45]. When an approximated curve is computed on $T_x\mathcal{M}$, it is mapped back to $\mathcal{M}$ via the exponential map or an appropriate retraction, see [37, 43] for a detailed exposition on Riemannian geometry. Curves obtained in this way are noted $\mathbf{B}^{TS}(t)$, where the superscript TS comes from **T**angent **S**pace.

It should, however, be noted that the tangent space $T_x\mathcal{M}$ is a good approximation of $\mathcal{M}$ only in a close neighborhood of x, and in most of the cases, the data points cannot all lie in this neighborhood. This is why the so-called *blended curve* exploits multiple tangent spaces [30]. It is built as a C^1-composite curve

$$\begin{aligned} \mathbf{B} : [t_1, t_q] &\to \mathcal{M} \\ t &\mapsto f_i(t - t_i), \quad \text{when } t \in [t_i, t_{i+1}],\ i = 1, \ldots, q-1. \end{aligned}$$

Here, $f_i(t)$ is the weighted mean of two curves $\mathbf{B}^{TS}(t)$ computed, respectively, on the tangent spaces based at d_i and d_{i+1}. This weighted mean is what gives its name (blended) to the technique.

Surfaces. Interpolation via surfaces is a little bit more intricate. Assume that we have a set of data points $d_{ij} \in \mathcal{M}$ (e.g., the Gramian matrices) associated with some parameter values $(t_i^{(1)}, t_j^{(2)}) \in \mathbb{R}^2$, with $i = 1, \ldots, q_1$ and $j = 1, \ldots, q_2$, i.e., the points are located on a grid defined by $t_1^{(1)} < \cdots < t_{q_1}^{(1)}$ and $t_1^{(2)} < \cdots < t_{q_2}^{(2)}$. To remain consistent with Sect. 2, we assume that $q_1 q_2 = q$, the total number of training points (i.e., the training points are located on a grid). We rely on Bézier surfaces presented in [44] as a generalization of Euclidean Bézier surfaces [46], inspired from the generalization of curves to manifolds already presented by Popiel et al. [47].

Let us first recall the definition of Bézier curves and Bézier splines in the Euclidean setting. Consider a Euclidean space $\mathbb{R}^r$. A Euclidean Bézier curve and Bézier surface of degree $K \in \mathbb{N}$ are functions $\beta_{K,\ell}$, $\ell = 1, 2$, defined as

$$\begin{aligned} \beta_{K,1}(\cdot; b_0, \ldots, b_K) : [t_1, t_q] &\to \mathbb{R}^r \\ t &\mapsto \sum_{i=0}^{K} b_i B_{iK}(t), \end{aligned} \tag{19}$$

$$\begin{aligned} \beta_{K,2}(\cdot, \cdot; (b_{ij})_{i,j=0,\ldots,K}) : [t_1^{(1)}, t_{q_1}^{(1)}] \times [t_1^{(2)}, t_{q_2}^{(2)}] &\to \mathbb{R}^r \\ (t^{(1)}, t^{(2)}) &\mapsto \sum_{i,j=0}^{K} b_{ij} B_{iK}(t^{(1)}) B_{jK}(t^{(2)}), \end{aligned} \tag{20}$$

where $B_{jK}(t) = \binom{K}{j} t^j (1-t)^{K-j}$ are the Bernstein polynomials, and $b_i \in \mathbb{R}^r$ (resp. $b_{ij} \in \mathbb{R}^r$) are the *control points* of the curve (resp. surface). Since the Bernstein polynomials form a partition of unity, the surface can be seen as a convex combination of the control points. Hence, Eq. (20) is equivalent to computing two Bézier curves (19): the first one in the $t^{(1)}$ direction and the second one in the $t^{(2)}$ direction, i.e.,

$$\begin{aligned} \beta_{K,2}(t^{(1)}, t^{(2)}; (b_{ij})_{i,j=0,\ldots,K}) &= \sum_{j=0}^{K} \left(\sum_{i=0}^{K} b_{ij} B_{iK}(t^{(1)}) \right) B_{jK}(t^{(2)}) \\ &= \beta_{K,1}\left(t^{(2)}; (\beta_{K,1}(t^{(1)}; (b_{ij})_{i=0,\ldots,K}))_{j=0,\ldots,K}\right). \end{aligned}$$

This equivalence allows us to easily generalize Bézier surfaces to a manifold $\mathcal{M}$ by using the generalization of Bézier *curves* based on the De Casteljau algorithm, see [47] for details.

To interpolate data points $d_{ij} \in \mathcal{M}$ associated with parameter values $(t_i^{(1)}, t_j^{(2)})$, with $i = 1, \ldots, q_1$ and $j = 1, \ldots, q_2$, one seeks the C^1-composite surface

$$\begin{aligned} \mathbf{B} : [t_1^{(1)}, t_{q_1}^{(1)}] \times [t_1^{(2)}, t_{q_2}^{(2)}] &\to \mathcal{M} \\ (t^{(1)}, t^{(2)}) &\mapsto \beta_{K,2}(t^{(1)} - t_k^{(1)}, t^{(2)} - t_l^{(2)}; (b_{ij}^{kl})_{i,j=0,\ldots,K}) \end{aligned}$$

when $t^{(1)} \in [t_k^{(1)}, t_{k+1}^{(1)}]$ and $t^{(2)} \in [t_l^{(2)}, t_{l+1}^{(2)}]$. Here, $\beta_{K,2}(\cdot, \cdot; (b_{ij}^{kl})_{i,j=0,\ldots,K})$ denotes a Bézier surface on $\mathcal{M}$, and $b_{ij}^{kl} \in \mathcal{M}$ are the control points to be determined such that interpolation is guaranteed and that the mean squared second derivative of the piecewise surface is minimized. This is done with a technique close to the one used for curves, i.e., transferring the optimization problem on carefully chosen tangent spaces. The only difference here is that the curve itself is not computed on the tangent space; instead, the optimality conditions obtained on a Euclidean space are generalized to manifolds. We refer to [44] for a detailed presentation of the optimization of the control points, and to [29] for a complete discussion on the C^1-conditions to patch several Bézier surfaces together.

In the following, we briefly summarize the steps for the resulting PMOR procedure:

- solve the Lyapunov equations (3) and (4) for $P_j \approx X_j X_j^T$ and $Q_j \approx Y_j Y_j^T$, $j = 1, \ldots, q$;
- interpolate the above data to get $P(\mu) = X(\mu)X^T(\mu)$ and $Q(\mu) = Y(\mu)Y^T(\mu)$ at the test points using either curve or surface interpolation depending on the dimension of the parameter domain;
- perform squaring balanced truncation at each test point by computing the SVD (5) and the reduced system matrices (6), (7).

4 Numerical Examples

In this section, we consider two numerical examples. We first describe the general setting of our experiments. Regarding the choice of the positive weights used in the algebraic approach, we have initially considered two options: weights defined based on the distance (in the parameter domain) from the test point to the training points, and the ones used in classical linear splines. Observe that the last choice results in a local interpolation, as instead of q matrix blocks in each factor of (9), we have only two (resp., four) of them for models with one (resp., two) parameter(s), regardless of the number of training points. This feature has two main advantages. The first one is that it allows a computational cost reduction, as many columns of the matrices $X(\mu)$ and $Y(\mu)$ defined in (9) and (10) are then zero. The second is the fact that, if we want to improve the accuracy by increasing the number of training points, more computation will be required in the offline stage but this makes no changes in the online stage. In other words, the online computation cost to reconstruct the ROM at a given parameter value does not depend on the number of training points. This local interpolation is thus much less affected by the so-called curse of dimensionality when the number of parameters increases compared to the conventional approach. As our tests, moreover, revealed that the latter delivers a smaller error, we only use linear splines here. For the geometric approach, as described in the previous section, and based on the numerical comparisons performed in [34], we choose the blended curves interpolation technique for the case of one parameter. When the model has two parameters, we use piecewise Bézier surface interpolation.

To verify the accuracy of ROMs, we compute an approximate $\mathcal{H}_\infty$-norm of the absolute errors in the frequency response defined as

$$\begin{aligned} \|H(\cdot,\mu) - \tilde{H}(\cdot,\mu)\|_{\mathcal{H}_\infty} &= \sup_{\omega\in\mathbb{R}} \|H(\mathrm{i}\omega,\mu) - \tilde{H}(\mathrm{i}\omega,\mu)\|_2 \\ &\approx \sup_{\omega_j\in[\omega_{\min},\omega_{\max}]} \|H(\mathrm{i}\omega_j,\mu) - \tilde{H}(\mathrm{i}\omega_j,\mu)\|_2, \end{aligned} \tag{21}$$

where $H(s,\mu)$ and $\tilde{H}(s,\mu)$ are the transfer functions of the FOM (1) and the ROM (2).

Regarding the efficiency measure, all computations are performed with MATLAB R2018a on a standard desktop using 64-bit OS Windows 10, equipped with 3.20 GHz 16 GB Intel Core i7-8700U CPU.

4.1 A model for heat conduction in solid material

This model is adapted from the one used in [48]. Consider the heat equation

$$\begin{aligned} \frac{\partial \vartheta}{\partial t} - \nabla \cdot (\sigma \nabla \vartheta) = f \quad &\text{in} \quad \Omega \times (0, T), \\ \vartheta = 0 \quad &\text{on } \partial\Omega \times (0, T), \end{aligned} \tag{22}$$

with the heat conductivity coefficient

$$\sigma(\xi) = \begin{cases} 1 + \mu^{(i)} & \text{for } \xi \in D_i, \ i = 1, 2, \\ 1 & \text{for } \xi \in \Omega \backslash (D_1 \cup D_2), \end{cases} \tag{23}$$

where the subdomains $D_i \subset \Omega = (0, 4)^2$, $i = 1, 2$, are two disks of radius 0.5 centered at (1, 1) and (3, 3), respectively, and the parameter $\mu = (\mu^{(1)}, \mu^{(2)})$ varies in $\mathcal{D} = [1, 10] \times [4, 10]$. Equation (22) with the source term $f \equiv 1$ is discretized using the finite element method with piecewise linear basis functions resulting in a system (1) of dimension $n = 1580$ with the symmetric positive definite mass matrix $E(\mu) \equiv E$ and the stiffness matrix

$$A(\mu) = \mu^{(1)} A_1 + \mu^{(2)} A_2 + A_3, \tag{24}$$

where A_1 and A_2 are symmetric negative semidefinite, and A_3 is symmetric negative definite. The input matrix $B(\mu) \equiv B \in \mathbb{R}^n$ originates from the source function f, and the output matrix is given by $C(\mu) \equiv C = [1/n \ \dots \ 1/n] \in \mathbb{R}^{1\times n}$.

First, we fix a uniform grid $\mu_1, \dots, \mu_q \in \mathcal{D}$, which will be specified in the caption of the error figures. At those points, we solve (3) and (4) using the low-rank ADI method [49] with a prescribed tolerance 10^{-10}. We end up with local approximate solutions whose rank varies from 25 to 27. In order to apply the geometric interpolation method, we truncate them to make all the Gramians of rank 25, and we work on the manifold $\mathcal{S}_+(25, 1580)$. Note that for the algebraic method presented in Sect. 2, the local solutions at training points do not necessarily have the same rank. However, as we will compare the two methods, we also use the truncated Gramians.

The interpolated Gramians are then computed from the local Gramians, and are used to construct the ROMs at the training points. We choose the reduced order r in (5) by requiring $\sigma_r(\mu)/\sigma_1(\mu) < 10^{-8}$, which in our case gives r between 12 and 15 for the algebraic approach and 11 and 12 for the geometric approach (depending on the test point considered). In Fig. 1, we plot the approximate $\mathcal{H}_\infty$-norm of the

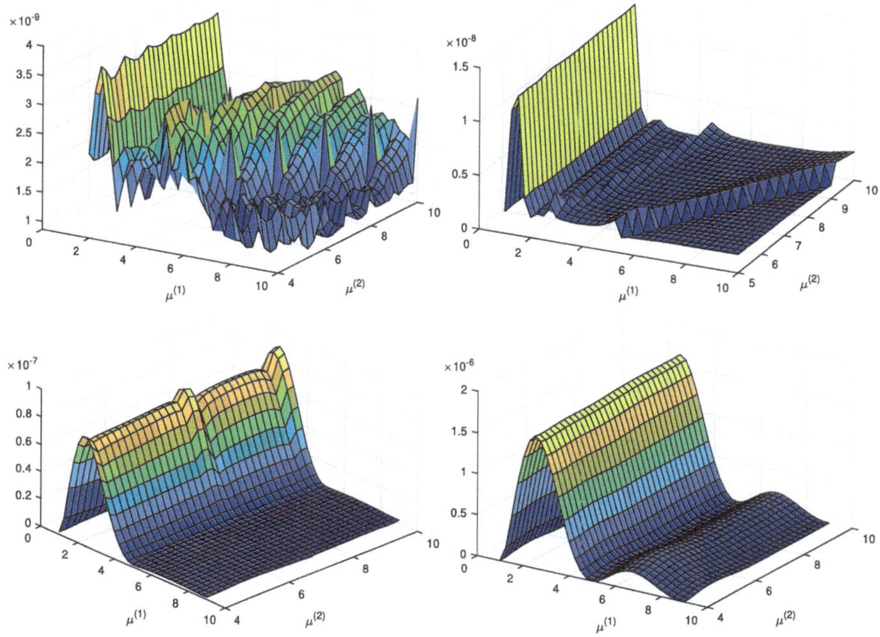

Fig. 1 The heat conduction model: absolute errors $\|H(\cdot,\mu) - \tilde{H}(\cdot,\mu)\|_{\mathcal{H}_\infty}$ at test points. Top figures: training grid $[1:1:10] \times [5:1:10]$ and test grid $[1:0.25:10] \times [5:0.2:10]$; bottom figures: training grid $[1:4:9] \times [4:3:10]$ and test grid $[1:0.25:9] \times [4:0.2:10]$ (using MATLAB notation). The left figures present the errors for the ROMs obtained by the algebraic method and the right figures present that computed by the geometric method

absolute errors, as defined in (21). For a better readability of the plots, we simply choose the set of test points as regularly spaced between the training points, as specified in the caption of the figures. It can be observed that, in the same setting, the algebraic method delivers a slightly smaller error than the geometric one. Moreover, the figures show that the error corresponding to a small μ_1 tends to be larger. This suggests that we should use more interpolation data in this area. To this end, we try an adaptively finer grid for the algebraic method and obtain the result shown in Fig. 2 (left). Furthermore, to give the reader a view on the relative errors of the method, we plot the $\mathcal{H}_\infty$-norm of the full-order transfer function in Fig. 2 (right).

We now report the time consumed by the two proposed methods in the second setting, i.e., corresponding to Fig. 1 (bottom). First, solving two Lyapunov equations at 9 training points needs 2.15 s. Then, the interpolation of the low-rank solutions of these two equations using the geometric approach at 1023 test points costs 26.88 s. Regarding computation time, clearly this method can be a good candidate to estimate quickly the solutions of parametric Lyapunov equations. For model reduction, once the interpolated Gramians are available, evaluating the ROM at the prescribed test

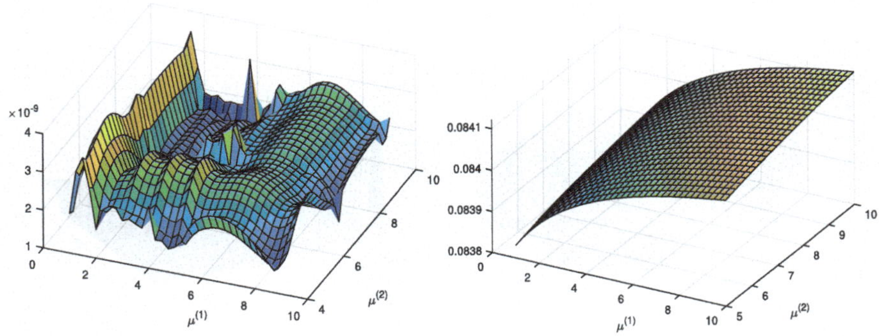

Fig. 2 The heat conduction model: (left) the absolute error $\|H(\cdot,\mu) - \tilde{H}(\cdot,\mu)\|_{\mathcal{H}_\infty}$ with adaptive grid $[1\ 2\ 3\ 4\ 5\ 9] \times [4:3:10]$; (right) the $\mathcal{H}_\infty$-norm of the full-order transfer function on the parameter domain

Table 1 The heat conduction model: time consumed by the different tasks (s)

		Geometric approach	Algebraic approach
Offline:	Solving the Lyapunov equations at training parameters	2.15	2.15
	Preparation for interpolation	–	0.2
Online:	Interpolation	26.88	–
	Computation of the ROMs	1.47	1.33

points takes 1.47 s. In comparison, for the algebraic approach, the offline stage lasts 0.2 s and the online one costs 1.33 s. We summarize these details in Table 1.

4.2 An Anemometer Model

In the second example, we verify the numerical behavior of the proposed methods when applied to fairly large problems. To this end, we consider a model for a thermal-based flow sensor, see [50] and references therein. Simulation of this device requires solving a convection-diffusion partial differential equation of the form

$$\rho c \frac{\partial \vartheta}{\partial t} = \nabla \cdot (\kappa \nabla \vartheta) - \rho c \mu \nabla \vartheta + \dot{q}, \tag{25}$$

where ρ denotes the mass density, c is the specific heat, κ is the thermal conductivity, μ is the fluid velocity, ϑ is the temperature, and $\dot{q}$ is the heat flow into the system caused by the heater. The considered model is restricted to the case $\rho = 1, c = 1$, $\kappa = 1$ and $\mu \in [0, 1]$ which corresponds to the one-parameter model. The finite ele-

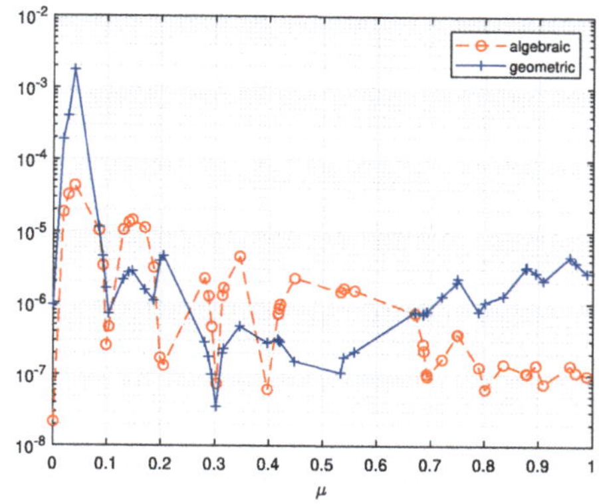

Fig. 3 The anemometer model: absolute errors $\|H(\cdot,\mu) - \tilde{H}(\cdot,\mu)\|_{\mathcal{H}_\infty}$ at test points

ment discretization of (25) leads to system (1) of order $n = 29008$ with the symmetric positive definite mass matrix $E(\mu) \equiv E$ and the stiffness matrix $A(\mu) = A_1 + \mu A_2$, where A_1 is symmetric negative definite, A_2 is non-symmetric negative semidefinite. The input matrix $B \in \mathbb{R}^n$ and the output matrix $C \in \mathbb{R}^{1\times n}$ are parameter independent. The reader is referred to as [51] and references therein for more detailed descriptions and numerical data.

For this model, we use the training grid $[0 : 0.1 : 1]$ while the test grid is made of 50 points randomly generated within the range of the parameter domain. The tolerance for the low-rank ADI solver (used to solve the Lyapunov equations) is 10^{-9}, which results in local Gramians of rank ranging from 25 to 39. For balanced truncation, we take the tolerance 10^{-7}. The resulting ROMs have different reduced orders at test points: the ROMs produced by the algebraic approach have orders between 16 and 17 while those obtained by the geometric approach are between 9 and 17. The absolute errors are represented in Fig. 3. One can see that at the middle of the parameter domain, the geometric approach provides a better approximation than the one computed by the algebraic method, while near the two ends, we observe the reverse result. We would like to note that cubic spline interpolation might have a larger error near the ends because at those two ends, we have to impose endpoint conditions that may not match with the nature of the data. This phenomenon is also observed in, e.g., [7, Sect. 4.1] where a cubic spline is used to interpolate the transfer function. As our geometric approach uses cubic spline interpolation on tangent spaces, this would be a possible explanation for the behavior observed in Fig. 3.

The time consumed by different tasks is summarized in Table 2.

5 Conclusion

We presented two methods for interpolating the Gramians of parameter-dependent linear dynamical systems in the framework of parametric balanced truncation model reduction. The first method is merely based on linear algebra which takes no geo-

Table 2 The heat anemometer model: time consumed by different tasks (s)

		Geometric approach	Algebraic approach
Offline:	Solving the Lyapunov equations at training parameters	199.18	199.18
	Preparation for interpolation	–	25.55
Online:	Interpolation	18.40	–
	Computation of the ROMs	0.81	0.06

metric structure of the data into account. When the matrices of the system depend affinely on the parameters, it can be combined with the reduction process which enables an offline-online decomposition, reducing the amount of online computations. The second method exploits the positive semidefiniteness of the data and recent developments in matrix manifold theory. It reformulates the problem as an interpolation problem on the underlying manifold and relies on recent interpolation techniques blending interpolating curves computed on different tangent spaces. This method is expected to work well if the rank of the training data does not change much from one training point to another. While the error obtained using the geometric approach is seemingly a bit larger, it results in lower reduced orders than the algebraic approach.

Acknowledgements This work was supported by the Fonds de la Recherche Scientifique—FNRS and the Fonds Wetenschappelijk Onderzoek—Vlaanderen under EOS Project no. 30468160. The third author is supported by the National Physical Laboratory. The authors would like to thank the authors of [48] for sharing their data with them.

References

1. Feng, L.H., Rudnyi, E.B., Korvink, J.G.: Preserving the film coefficient as a parameter in the compact thermal model for fast electrothermal simulation. IEEE Trans. Computer-Aided Design Integr. Circuits Syst. **24**(12), 1838–1847 (2005). https://doi.org/10.1109/TCAD.2005.852660
2. Li, Y.-T., Bai, Z., Su, Y.: A two-directional Arnoldi process and its application to parametric model order reduction. J. Comput. Appl. Math. **226**(1), 10–21 (2009). https://doi.org/10.1016/j.cam.2008.05.059
3. Baur, U., Beattie, C., Benner, P., Gugercin, S.: Interpolatory projection methods for parameterized model reduction. SIAM J. Sci. Comput. **33**(5), 2489–2518 (2011). https://doi.org/10.1137/090776925
4. Amsallem, D., Farhat, C.: Interpolation method for adapting reduced-order models and application to aeroelasticity. AIAAJ **46**(7), 1803–1813 (2008). https://doi.org/10.2514/1.35374
5. Baur, U., Benner, P.: Modellreduktion für parametrisierte Systeme durch balanciertes Abschneiden und Interpolation. at-Automatisierungstechnik **57**(8), 411–422 (2009). https://doi.org/10.1524/auto.2009.0787
6. Panzer, H., Mohring, J., Eid, R., Lohmann, B.: Parametric model order reduction by matrix interpolation. at – Automatisierungstechnik **58**(8), 475–484 (2010). https://doi.org/10.1524/auto.2010.0863

7. Son, N.T.: Interpolation based parametric model order reduction. Ph.D. thesis, Universität Bremen, Germany (2012)
8. Haasdonk, B., Ohlberger, M.: Efficient reduced models and a posteriori error estimation for parametrized dynamical systems by offline/online decomposition. Math. Comput. Model. Dyn. Syst. **17**, 145–161 (2011). https://doi.org/10.1080/13873954.2010.514703
9. Son, N.T., Stykel, T.: Solving parameter-dependent Lyapunov equations using the reduced basis method with application to parametric model order reduction. SIAM J. Matrix Anal. Appl. **38**(2), 478–504 (2017). https://doi.org/10.1137/15M1027097
10. Benner, P., Gugercin, S., Willcox, K.: A survey of projection-based model reduction methods for parametric dynamical systems. SIAM Rev. **57**(4), 483–531 (2015). https://doi.org/10.1137/130932715
11. Benner, P., Ohlberger, M., Patera, A., Rozza, G., Urban, K. (eds.): Model Reduction of Parametrized Systems, vol. 17. Springer (2019). https://doi.org/10.1007/978-3-319-58786-8
12. Dai, L.: Singular Control Systems. Lecture Notes in Control and Information Sciences, vol. 118. Springer, Berlin, Heidelberg (1989)
13. Stykel, T.: Gramian-based model reduction for descriptor systems. Math. Control Signals Syst. **16**, 297–319 (2004). https://doi.org/10.1007/s00498-004-0141-4
14. Antoulas, A.: Approximation of Large-Scale Dynamical Systems. SIAM, Philadelphia, PA (2005). https://doi.org/10.1137/1.9780898718713
15. Degroote, J., Vierendeels, J., Willcox, K.: Interpolation among reduced-order matrices to obtain parameterized models for design, optimization and probabilistic analysis. Int. J. Numer. Meth. Fl. **63**(2), 207–230 (2010). https://doi.org/10.1002/fld.2089
16. Amsallem, D., Farhat, C.: An online method for interpolating linear reduced-order models. SIAM J. Sci. Comput. **33**(5), 2169–2198 (2011). https://doi.org/10.1137/100813051
17. Son, N.T.: A real time procedure for affinely dependent parametric model order reduction using interpolation on Grassmann manifolds. Int. J. Numer. Methods Eng. **93**(8), 818–833 (2013). https://doi.org/10.1002/nme.4408
18. Son, N.T., Stykel, T.: Model order reduction of parameterized circuit equations based on interpolation. Adv. Comput. Math. **41**(5), 1321–1342 (2015). https://doi.org/10.1007/s10444-015-9418-z
19. Zimmermann, R.: Manifold interpolation and model reduction. arXiv:1902.06502v2 (2019)
20. Moore, B.: Principal component analysis in linear systems: controllability, observability, and model reduction. IEEE Trans. Automat. Control **26**(1), 17–32 (1981). https://doi.org/10.1109/TAC.1981.1102568
21. Penzl, T.: Eigenvalue decay bounds for solutions of Lyapunov equations: the symmetric case. Syst. Control Lett. **40**(2), 139–144 (2000). https://doi.org/10.1016/S0167-6911(00)00010-4
22. Antoulas, A., Sorensen, D., Zhou, Y.: On the decay rate of the Hankel singular values and related issues. Syst. Control Lett. **46**(5), 323–342 (2002). https://doi.org/10.1016/S0167-6911(02)00147-0
23. Tombs, M.S., Postlethweite, I.: Truncated balanced realization of a stable non-minimal state-space system. Internat. J. Control **46**(4), 1319–1330 (1987). https://doi.org/10.1080/00207178708933971
24. Vandereycken, B., Absil, P.A., Vandewalle, S.: Embedded geometry of the set of symmetric positive semidefinite matrices of fixed rank. In: Proceedings of the IEEE 15th Workshop on Statistical Signal Processing (Washington, DC), pp. 389–392. IEEE (2009). https://doi.org/10.1109/SSP.2009.5278558
25. Massart, E., Absil, P.A.: Quotient geometry with simple geodesics for the manifold of fixed-rank positive-semidefinite matrices. SIAM J. Matrix Anal. Appl. **41**(1), 171–198 (2020)
26. Dyn, N.: Linear and nonlinear subdivision schemes in geometric modeling. In: Cucker, F., Pinkus, A., Todd, M.J., Foundations of Computational Mathematics, Hong Kong 2008, vol. 363, pp. 68–92 (2009). https://doi.org/10.1017/CBO9781139107068.004
27. Hüper, K., Silva Leite, F.: On the geometry of rolling and Interpolation curves on S^n, SO_n, and Grassmann manifolds. J. Dyn. Control Syst. **13**(4), 467–502 (2007). https://doi.org/10.1007/s10883-007-9027-3

28. Machado, L., Silva Leite, F., Krakowski, K.: Higher-order smoothing splines versus least squares problems on Riemannian manifolds. J. Dyn. Control Syst. **16**(1), 121–148 (2010). https://doi.org/10.1007/s10883-010-9080-1
29. Absil, P.-A., Gousenbourger, P.-Y., Striewski, P., Wirth, B.: Differentiable Piecewise-Bézier Surfaces on Riemannian Manifolds. SIAM J. Imaging Sci. **9**(4), 1788–1828 (2016). https://doi.org/10.1137/16M1057978
30. Gousenbourger, P.-Y., Massart, E., Absil, P.-A.: Data fitting on manifolds with composite Bézier-like curves and blended cubic splines. J. Math. Imaging Vis. **61**(5), 645–671 (2019). https://doi.org/10.1007/s10851-018-0865-2
31. Allasia, G.: Simultaneous interpolation and approximation by a class of multivariate positive operators. Numer. Alg. **34**, 147–158 (2003). https://doi.org/10.1023/B:NUMA.0000005359.72118.b6
32. Patera, A.T., Rozza, G.: Reduced Basis Approximation and A Posteriori Error Estimation for Parametrized Partial Differential Equations. MIT Pappalardo Graduate Monographs in Mechanical Engineering. MIT, MA (2007)
33. Hesthaven, J., Rozza, G., Stamm, B.: Certified Reduced Basis Methods for Parametrized Partial Differential Equations. SpringerBriefs in Mathematics. Springer, Cham (2016). https://doi.org/10.1007/978-3-319-22470-1
34. Massart, E., Gousenbourger, P.-Y., Son, N.T., Stykel, T., Absil, P.-A.: Interpolation on the manifold of fixed-rank positive-semidefinite matrices for parametric model order reduction: preliminary results. ESANN **2019**, 281–286 (2019)
35. Daniel, L., Siong, O.C., Chay, L. S., Lee, K.H., White, J.: A multiparameter moment-matching model-reduction approach for generating geometrically parameterized interconnect performance models. IEEE Trans. Computer-Aided Design Integr. Circuits Syst. **23**(5), 678–693 (2004). https://doi.org/10.1109/TCAD.2004.826583
36. Journée, M., Bach, F., Absil, P.-A., Sepulchre, R.: Low-rank optimization on the cone of positive semidefinite matrices. SIAM J. Optim. **20**(5), 2327–2351 (2010). https://doi.org/10.1137/080731359
37. do Carmo, M.P.: Riemannian Geometry. Birkhäuser, Boston (1992)
38. Lee, J.M.: Introduction to Riemannian Manifolds. Graduate Texts in Mathematics, Springer (2018)
39. Massart, E., Absil, P.-A., Hendrickx, J. M.: Curvature of the manifold of fixed-rank positive-semidefinite matrices endowed with the Bures-Wasserstein metric. GSI2019, 739–748 (2019)
40. Gousenbourger, P.-Y., Massart, E., Musolas, A., Absil, P.-A., Jacques, L., Hendrickx, J. M., Marzouk, Y.: Piecewise-Bézier C^1 smoothing on manifolds with application to wind field estimation. ESANN2017, 305–310 (2017)
41. Sabbagh, D., Ablin, P., Varoquaux, G., Gramfort, A., Engemann, D.A.: Manifold-regression to predict from MEG/EEG brain signals without source modeling, arXiv:1906.02687 (2019)
42. Szczapa, B., Daoudi M., Berretti S., Del Bimbo A., Pala P., Massart E.: Fitting, Comparison, and Alignment of Trajectories on Positive Semi-Definite Matrices with Application to Action Recognition, arxiv:1908.00646 (2019)
43. Absil, P.-A., Mahony, R., Sepulchre, R.: Optimization Algorithms on Matrix Manifolds. Princeton University Press (2008)
44. Absil, P.-A., Gousenbourger, P.-Y., Striewski, P., Wirth, B.: Differentiable piecewise-Bézier interpolation on Riemannian manifolds. ESANN2016, 95–100 (2016)
45. Green, P.J., Silverman, B.W.: Nonparametric Regression and Generalized Linear Models: A Roughness Penalty Approach. CRC Press (1993)
46. Farin, G.E.: Curves and Surfaces for CAGD, 5th edn. Academic Press (2002)
47. Popiel, T., Noakes, L.: Bézier curves and C^2 interpolation in Riemannian manifolds. J. Approx. Theory **148**(2), 111–127 (2007). https://doi.org/10.1016/j.jat.2007.03.002
48. Kressner, D., Plešinger, M., Tobler, C.: A preconditioned low-rank CG method for parameter-dependent Lyapunov matrix equations. Numer. Linear Algebra Appl. **21**(5), 666–684 (2014). https://doi.org/10.1002/nla.1919

49. Li, J.R., White, J.: Low rank solution of Lyapunov equations. SIAM J. Matrix Anal. Appl. **24**(1), 260–280 (2002). https://doi.org/10.1137/S0895479801384937
50. Moosmann, C., Rudnyi, E.B., Greiner, A., Korvink, J.G., Hornung, M.: Parameter preserving model order reduction of a flow meter. In: Technical Proceedings of the 2005 Nanotechnology Conference and Trade Show (Nanotech 2005, Anaheim, California, USA), vol. 3, pp. 684–687. NSTINanotech (2005)
51. The MORwiki Community: Anemometer. MORwiki – Model Order Reduction Wiki (2018). http://modelreduction.org/index.php/Anemometer

Toward Fitting Structured Nonlinear Systems by Means of Dynamic Mode Decomposition

Ion Victor Gosea and Igor Pontes Duff

Abstract The dynamic mode decomposition (DMD) is a data-driven method used for identifying the dynamics of complex nonlinear systems. It extracts important characteristics of the underlying dynamics using measured time-domain data produced either by means of experiments or by numerical simulations. In the original methodology, the measurements are assumed to be approximately related by a linear operator. Hence, a linear discrete-time system is fitted to the given data. However, often, nonlinear systems modeling physical phenomena have a particular known structure. In this contribution, we propose an identification and reduction method based on the classical DMD approach allowing to fit a structured nonlinear system to the measured data. We mainly focus on two types of nonlinearities: bilinear and quadratic bilinear. By enforcing this additional structure, more insight into extracting the nonlinear behavior of the original process is gained. Finally, we demonstrate the proposed methodology for different examples, such as Burgers' equation and the coupled van der Pol oscillators.

1 Introduction

Mathematical models are commonly used to simulate, optimize, and control the behavior of real dynamical processes. A common way to derive those models is to use the first principles, generally leading to a set of ordinary or partial differential equations. For high complex dynamics, fine discretization leads to high fidelity models, which require numerous equations and variables. In some situations, the high

---

I. V. Gosea (✉) · I. Pontes Duff
Max Planck Institute, Magdeburg, Germany
e-mail: gosea@mpi-magdeburg.mpg.de

I. Pontes Duff
e-mail: pontes@mpi-magdeburg.mpg.de

P. Benner et al. (eds.), *Model Reduction of Complex Dynamical Systems*,
International Series of Numerical Mathematics 171,
https://doi.org/10.1007/978-3-030-72983-7_3

model is given as a black box setup, i.e., by solvers that allow the computation of the full-model states for a given set of initial conditions and inputs, but does not provide the dynamical system realization. In order to better understand such dynamics processes, it is often beneficial to construct surrogate models using simulated data. This justifies the development of identification or data-driven model reduction methods. Indeed, with the ever-increasing availability of measured/simulated data in different scientific disciplines, the need for incorporating this information in the identification and reduction process has steadily grown. The data-driven model reduction problem consists of determining low-order models from the provided data obtained either by experimentation or numerical simulations. Methods such as Dynamic Mode Decomposition (DMD) have drawn considerable research endeavors.

DMD is a data-driven method for analyzing complex systems. The purpose is to learn/extract the important dynamic characteristics (such as unstable growth modes, resonance, and spectral properties) of the underlying dynamical system by means of measured time-domain data. These can be acquired through experiments in a practical setup or artificially through numerical simulations (by exciting the system). It was initially proposed in [29] in the context of analyzing numerical and experimental fluid mechanics problems. Additionally, it is intrinsically related to the Koopman operator analysis, see [22, 27]. Since its introduction, several extensions have been proposed in the literature, e.g., the exact DMD [31], the extended DMD [11], and the higher order DMD [20]. Also, in order to address control problems, DMD with control inputs was proposed in [25], and then extended to the case where outputs are also considered in [1, 9]. The reader is referred to [19] for a comprehensive monograph on the topic.

Often, nonlinear systems modeling physical phenomena have a particular known structure, such as bilinear and quadratic terms. In the present work, our primary goal is to embed nonlinear structures in the DMD framework. To this aim, we propose an identification and data-driven reduction method based on the classical DMD approach allowing to fit a bilinear and quadratic-bilinear structures to the measured data. The choice to fit such terms is due to the fact most systems with analytical nonlinearities (e.g., rational, trigonometrical, polynomial) can be exactly reformulated as quadratic-bilinear systems [15]. Our work is rooted in the two variants, DMD with control and input-output DMD, and can be considered as an extension of those methodologies.

There exist vast literature on learning nonlinear dynamics from data, and we review the most relevant literature for our work. One approach is the so-called Loewner framework, which enables to construct low-order models from frequency-domain data. It was initially proposed in [21], and later extended to bilinear [3] and quadratic-bilinear case [14]. Another approach is the operator inference, proposed [24]. This approach infers polynomial low-order models as a solution of a least-squares problem based on the initial conditions, inputs, and trajectories of the states. This approach was recently extended to systems with non-polynomials [8]. Also, the authors in [26] show how the use of lifting transformations can be beneficial to identify the system. Finally, the approach proposed in [23] introduces a method based on operator inference enabling to learn exactly the reduced models that are traditionally constructed with model reduction. It is worth mentioning that the operator inference approach [24] can be seen as an extension to DMD for nonlinear systems. Indeed, in this framework, the reduced-order model is allowed to have polynomial terms on the state and its matrices are obtained by solving a least-squares problems. The main difference is that this optimization problem is set using the reduced trajectories as the data (see the introduction of [24] for more details).

In this work, we aim at fitting nonlinear model structures using the DMD setup, i.e., by using the full-model trajectories, which is the main difference from [24]. Additionally, besides the quadratic structure on the state, we also consider reduced-order models having bilinear structure on the state and input.

The rest of the paper is organized as follows. Section 2 recalls some results on the classical DMD, DMD with control, and the input-output DMD. In Sect. 3, we present the main contribution of the paper, which is the incorporation of bilinear and quadratic-bilinear terms in the DMD setup. Finally, in Sect. 4, we demonstrate the proposed methodology for different examples, such as Burgers' equation and the coupled van der Pol oscillators.

2 Dynamic Mode Decomposition

In this section, we briefly recall the classical DMD framework [29]. To this aim, we analyze time-invariant systems of ordinary differential equations (ODEs) written compactly in a continuous-time setting as follows:

$$\dot{\mathbf{x}}(t) = f(\mathbf{x}(t)), \tag{1}$$

where $\mathbf{x}(t) \in \mathbb{R}^n$ is the state vector and $f : \mathbb{R}^n \to \mathbb{R}^n$ is the system nonlinearity.

By means of sampling the variable $\mathbf{x}$ in (1) at uniform intervals of time, we collect a series of vectors $\mathbf{x}(t_k)$ for sampling times $t_0, t_1, \ldots, t_m$. For simplicity, denote $\mathbf{x}_k := \mathbf{x}(t_k)$.

DMD aims at analyzing the relationship between pairs of measurements from a dynamical system. The measurements $\mathbf{x}_k$ and $\mathbf{x}_{k+1}$, as previously introduced, are assumed to be approximately related by a linear operator for all $k \in \{0, 1, \ldots, m-1\}$.

$$\mathbf{x}_{k+1} \approx \mathbf{A}\mathbf{x}_k, \tag{2}$$

where $\mathbf{A} \in \mathbb{R}^{n \times n}$. This approximation is assumed to hold for all pairs of measurements. Next, group together the sequence of collected snapshots of the discretized state $\mathbf{x}(t)$ and use the following notations:

$$\mathbf{X} = \begin{bmatrix} \mathbf{x}_0 \ \mathbf{x}_1 \ \ldots \ \mathbf{x}_{m-1} \end{bmatrix} \in \mathbb{R}^{n \times m}, \ \ \mathbf{X}_s = \begin{bmatrix} \mathbf{x}_1 \ \mathbf{x}_2 \ \ldots \ \mathbf{x}_m \end{bmatrix} \in \mathbb{R}^{n \times m}. \tag{3}$$

The DMD method is based on finding a best-fit solution of an operator A so that the following relation is (approximately) satisfied

$$\mathbf{X}_s = \mathbf{A}\mathbf{X}, \tag{4}$$

which represents the block version of Eq. (2). Moreover, the above relation does not need to hold exactly. Previous work has theoretically justified using this approximating operator on data generated by nonlinear dynamical systems. For more details, see [30]. A best-fit solution is explicitly given as follows:

$$\mathbf{A} = \mathbf{X}_s\mathbf{X}^\dagger, \tag{5}$$

where $\mathbf{X}^\dagger \in \mathbb{R}^{m\times n}$ is the Moore-Penrose inverse of matrix $\mathbf{X} \in \mathbb{R}^{n\times m}$. In the above statement, by "best-fit" it is meant the solution that minimizes the least-squares error in the Frobenius norm (see [9]). More precisely, the matrix $\mathbf{A}$ in (5) is the solution of the following optimization problem:

$$\underset{\hat{\mathbf{A}}\in\mathbb{R}^{n\times n}}{\arg\min}\left(\|\mathbf{X}_s - \hat{\mathbf{A}}\mathbf{X}\|_{\mathrm{F}}\right). \tag{6}$$

The so-called DMD modes are given by the eigenvectors of matrix $\mathbf{A}$ in (5), collected in matrix $\mathbf{T}$ with $\mathbf{A} = \mathbf{T}\mathbf{\Lambda}\mathbf{T}^{-1}$. These spatial modes of system (1) are computed at a single frequency and are connected to the Koopman operator, see [22].

In this work, we will mainly focus on the construction of the reduced-order model rather than the evaluation of the DMD modes.

2.1 *Dynamic Mode Decomposition with Control (DMDc)*

Dynamic mode decomposition with control (DMDc) was introduced in [25] and it modifies the basic framework characterizing DMD. The novelty is given by including measurements of a control input $u(t) \in \mathbb{R}$. It is hence assumed that the dynamics of the original system of ODEs includes an input dependence, i.e.,

$$\dot{\mathbf{x}}(t) = f(\mathbf{x}(t), u(t)), \tag{7}$$

which represents a directs extension of (1). In (7), it is assumed that $f : \mathbb{R}^n \times \mathbb{R} \to \mathbb{R}^n$. Then, continue as in the classical DMD case without control to collect a discretized solution $\mathbf{x}$ at particular time instances.

In this setup, a trio of measurements are now assumed to be connected. The goal of DMDc is to analyze the relationship between a future state measurement $\mathbf{x}_{k+1}$ with the current measurement $\mathbf{x}_k$ and the current control u_k.

The motivation for this method is that, understanding the dynamic characteristics of systems that have both internal dynamics and applied external control is of great use for many applications, such as for controller design and sensor placement.

The DMDc method is used to discover the underlying dynamics subject to a driving control input by quantifying its effect to the time-domain measurements corresponding to the underlying dynamical system.

A pair of linear operators represented by matrices $\mathbf{A} \in \mathbb{R}^{n \times n}$ and $\mathbf{B} \in \mathbb{R}^n$ provides the following dependence for each trio of measurement data snapshots $(\mathbf{x}_{k+1}, \mathbf{x}_k, \mathbf{u}_k)$

$$\mathbf{x}_{k+1} = \mathbf{A}\mathbf{x}_k + \mathbf{B}u_k, \quad 0 \leq k \leq m-1. \tag{8}$$

Next, denote the sequence of control input snapshots with

$$\mathbf{U} = \begin{bmatrix} \mathbf{u}_0 \ \mathbf{u}_1 \ \dots \ \mathbf{u}_{m-1} \end{bmatrix} \in \mathbb{R}^{1 \times m}. \tag{9}$$

The first step is to augment the matrix $\mathbf{X}$ with the row vector $\mathbf{U}$ and similarly group together the $\mathbf{A}$ and $\mathbf{B}$ matrices by using the notations:

$$\mathbf{G} = [\mathbf{A} \ \ \mathbf{B}] \in \mathbb{R}^{n \times (n+1)}, \quad \boldsymbol{\Omega} = \begin{bmatrix} \mathbf{X} \\ \mathbf{U} \end{bmatrix} \in \mathbb{R}^{(n+1) \times m}. \tag{10}$$

The matrix $\mathbf{G}$ introduced above will be referred to as the system matrix since it incorporates the matrices corresponding to the system to be fitted.

By letting the index k vary in the range $\{0, 1, \dots, m-1\}$, one can compactly rewrite the m equations in the following matrix format:

$$\mathbf{X}_s = \mathbf{A}\mathbf{X} + \mathbf{B}\mathbf{U} = [\mathbf{A} \ \ \mathbf{B}] \begin{bmatrix} \mathbf{X} \\ \mathbf{U} \end{bmatrix} := \mathbf{G}\boldsymbol{\Omega}. \tag{11}$$

Thus, similar to standard DMD, compute a pseudo-inverse and solve for $\mathbf{G}$ as

$$\mathbf{G} = \mathbf{X}_s \boldsymbol{\Omega}^{\dagger} \ \Rightarrow \ [\mathbf{A} \ \ \mathbf{B}] = \mathbf{X}_s \begin{bmatrix} \mathbf{X} \\ \mathbf{U} \end{bmatrix}^{\dagger}. \tag{12}$$

The matrix $\mathbf{G} \in \mathbb{R}^{n \times (n+1)}$ in (12) is actually the solution of the following optimization problem:

$$\underset{\hat{\mathbf{G}} \in \mathbb{R}^{n \times (n+1)}}{\arg\min} \left(\left\| \mathbf{X}_s - \hat{\mathbf{G}} \begin{bmatrix} \mathbf{X} \\ \mathbf{U} \end{bmatrix} \right\|_F \right). \tag{13}$$

To explicitly compute the matrix in (12), we first find the singular value decomposition (SVD) of the augmented data matrix $\boldsymbol{\Omega}$ as follows

$$\boldsymbol{\Omega} = \mathbf{V}\boldsymbol{\Sigma}\mathbf{W}^T \approx \tilde{\mathbf{V}}\tilde{\boldsymbol{\Sigma}}\tilde{\mathbf{W}}^T, \tag{14}$$

where the full-scale and reduced-order matrices have the following dimensions:

$$\begin{cases} \mathbf{V} \in \mathbb{R}^{(n+1)\times(n+1)}, & \mathbf{\Sigma} \in \mathbb{R}^{(n+1)\times m}, \quad \mathbf{V} \in \mathbb{R}^{m\times m}, \\ \tilde{\mathbf{V}} \in \mathbb{R}^{(n+1)\times p}, & \tilde{\mathbf{\Sigma}} \in \mathbb{R}^{p\times p}, \quad \tilde{\mathbf{V}} \in \mathbb{R}^{m\times r}. \end{cases}$$

The truncation index is denoted with p, where $p \leqslant n$. The pseudo-inverse $\mathbf{\Omega}^{\dagger}$ is computed using the matrices from the SVD in (14), i.e., as $\mathbf{\Omega}^{\dagger} \approx \tilde{\mathbf{W}}\tilde{\mathbf{\Sigma}}^{-1}\tilde{\mathbf{V}}^T$.

By splitting up the matrix $\mathbf{V}^T$ as $\tilde{\mathbf{V}}^T = [\tilde{\mathbf{V}}_1^T \;\; \tilde{\mathbf{V}}_2^T]$, recover the system matrices as

$$\overline{\mathbf{A}} = \mathbf{X}_s\tilde{\mathbf{W}}\tilde{\mathbf{\Sigma}}^{-1}\tilde{\mathbf{V}}_1^T, \;\; \overline{\mathbf{B}} = \mathbf{X}_s\tilde{\mathbf{W}}\tilde{\mathbf{\Sigma}}^{-1}\tilde{\mathbf{V}}_2^T. \tag{15}$$

As mentioned in, there is one additional step. By performing another (short) SVD of the matrix $\mathbf{X}_s$, write

$$\mathbf{X}_s \approx \hat{\mathbf{V}}\hat{\mathbf{\Sigma}}\hat{\mathbf{W}}^T, \tag{16}$$

where $\hat{\mathbf{V}} \in \mathbb{R}^{(n+1)\times r}, \;\; \hat{\mathbf{\Sigma}} \in \mathbb{R}^{r\times r}, \;\; \hat{\mathbf{V}} \in \mathbb{R}^{m\times r}$. Note that the two SVDs will likely have different truncation values. The following reduced-order approximations of $\mathbf{A}$ and $\mathbf{B}$ are hence computed as

$$\tilde{\mathbf{A}} = \hat{\mathbf{V}}^T\overline{\mathbf{A}}\hat{\mathbf{V}} = \hat{\mathbf{V}}^T\mathbf{X}_s\tilde{\mathbf{W}}\tilde{\mathbf{\Sigma}}^{-1}\tilde{\mathbf{V}}_1^T\hat{\mathbf{V}} \in \mathbb{R}^{r\times r}, \; \tilde{\mathbf{B}} = \hat{\mathbf{V}}^T\overline{\mathbf{B}} = \hat{\mathbf{V}}^T\mathbf{X}_s\tilde{\mathbf{W}}\tilde{\mathbf{\Sigma}}^{-1}\tilde{\mathbf{V}}_2^T \in \mathbb{R}^r. \tag{17}$$

2.2 *Input-Output Dynamic Mode Decomposition*

In this section, we discuss the technique proposed in [1] known as input-output dynamic mode decomposition (ioDMD). This method constructs an input-output reduced-order model and can be viewed as an extension of DMDc for the case with observed outputs. As stated in the original work [1], this method represents a combination of POD and system identification techniques. The proposed method discussed here is similar in a sense to the algorithms for subspace state-space system identification (N4SID) introduced in [32] and can be also applied to large-scale systems.

We consider as given a system of ODEs whose dynamics is described by the same equations as in (7). Additionally, assume that observations are collected in the variable $y(t) \in \mathbb{R}$, as function of the state variable $\mathbf{x}$ and of the control u, written as

$$y(t) = g(\mathbf{x}(t), u(t)), \tag{18}$$

where $g : \mathbb{R}^n \times \mathbb{R} \to \mathbb{R}$.

As before, the next step is to collect snapshots of both variable $\mathbf{x}(t)$ and of the output $y(t)$ sampled at some positive time instances $t_0, t_1, \ldots t_{m-1}$. Again, for simplicity of the exposition, denote with $y_k := y(t_k)$.

We enforce the following dependence for each trio of measurement data snapshots given by $(\mathbf{y}_k, \mathbf{x}_k, \mathbf{u}_k)$

$$\mathbf{y}_k = \mathbf{C}\mathbf{x}_k + \mathbf{D}u_k, \quad 0 \leq k \leq m-1. \tag{19}$$

Afterward, collect the output values in a row vector as follows:

$$\mathbf{Y} = \begin{bmatrix} \mathbf{y}_0 \; \mathbf{y}_1 \; \ldots \; \mathbf{y}_{m-1} \end{bmatrix} \in \mathbb{R}^{1\times m}. \tag{20}$$

The ioDMD method aims at fitting the given set of snapshot measurements collected in matrices $\mathbf{X}_s$, $\mathbf{X}$ and vectors $\mathbf{U}$ and $\mathbf{Y}$ to a linear discrete-time system characterized by the following equations:

$$\begin{aligned} \mathbf{X}_s &= \mathbf{A}\mathbf{X} + \mathbf{B}\mathbf{U}, \\ \mathbf{Y} &= \mathbf{C}\mathbf{X} + \mathbf{D}\mathbf{U}, \end{aligned} \tag{21}$$

where, as before, $\mathbf{A} \in \mathbb{R}^{n\times n}$ and $\mathbf{B} \in \mathbb{R}^n$, and also $\mathbf{C}^T \in \mathbb{R}^n$, $\mathbf{D} \in \mathbb{R}$. Note that the first equation in (21) exactly corresponds to the driving matrix equation of DMDc presented in (12). Moreover, write the system of equations in (21) compactly as

$$\begin{bmatrix} \mathbf{X}_s \\ \mathbf{Y} \end{bmatrix} = \begin{bmatrix} \mathbf{A} \; \mathbf{B} \\ \mathbf{C} \; \mathbf{D} \end{bmatrix} \begin{bmatrix} \mathbf{X} \\ \mathbf{U} \end{bmatrix}. \tag{22}$$

Next, we adapt the definition of the system matrix $\mathbf{G}$ from (10) by incorporating an extra line as follows:

$$\mathbf{G} = \begin{bmatrix} \mathbf{A} \; \mathbf{B} \\ \mathbf{C} \; \mathbf{D} \end{bmatrix} \in \mathbb{R}^{(n+1)\times(n+1)}, \tag{23}$$

while $\mathbf{\Omega} = \begin{bmatrix} \mathbf{X} \\ \mathbf{U} \end{bmatrix} \in \mathbb{R}^{(n+1)\times m}$ is as before. Introduce a new notation that will become useful also in the next sections. It represents an augmentation of the shifted state matrix $\mathbf{X}_s$ with the output observation vector $\mathbf{Y}$, i.e.,

$$\mathbf{\Gamma} = \begin{bmatrix} \mathbf{X}_s \\ \mathbf{Y} \end{bmatrix} \in \mathbb{R}^{(n+1)\times m}. \tag{24}$$

Again, the solution of Eq. (22) will be computed as a best-fit type of approach. Hence, similarly to the DMDc case, recover the matrices $\mathbf{A}$, $\mathbf{B}$, $\mathbf{C}$, and $\mathbf{D}$ by computing the pseudo-inverse of matrix $\mathbf{\Omega}$ and writing

$$\mathbf{G} = \mathbf{\Gamma}\mathbf{\Omega}^\dagger \;\Rightarrow\; \begin{bmatrix} \mathbf{A} \; \mathbf{B} \\ \mathbf{C} \; \mathbf{D} \end{bmatrix} = \begin{bmatrix} \mathbf{X}_s \\ \mathbf{Y} \end{bmatrix} \begin{bmatrix} \mathbf{X} \\ \mathbf{U} \end{bmatrix}^\dagger. \tag{25}$$

The matrix $\mathbf{G} \in \mathbb{R}^{(n+1)\times(n+1)}$ in (12) is actually the solution of the following optimization problem:

$$\underset{\hat{\mathbf{G}}\in\mathbb{R}^{(n+1)\times(n+1)}}{\arg\min} \left(\left\| \begin{bmatrix} \mathbf{X}_s \\ \mathbf{Y} \end{bmatrix} - \hat{\mathbf{G}} \begin{bmatrix} \mathbf{X} \\ \mathbf{U} \end{bmatrix} \right\|_F \right). \tag{26}$$

Similar to the procedure covered in Sect. 2.1, one could further lower the dimension of the recovered system matrices by employing an additional SVD of the matrix $\mathbf{\Gamma}$, as was done in (16).

3 The Proposed Extensions

In this section, we present the main contribution of the paper. We propose extensions of the methods previously introduced in Sects. 2.1 and 2.2, e.g., DMDc and, respectively, ioDMD to fit nonlinear structured systems. More specifically, the discrete-time models that are fitted using these procedures will no longer be linear as in (21); the new models will contain nonlinear (bilinear or quadratic) terms.

3.1 Bilinear Systems

Bilinear systems are a class of mildly nonlinear systems for which the nonlinearity is given by the product between the state variable and the control input. More exactly, the characterizing system of ODEs is written as in (7) but for a specific choice of mapping f, i.e., $f(\mathbf{x}, \mathbf{u}) = \mathbf{A}\mathbf{x} + \mathbf{N}\mathbf{x}u + \mathbf{B}u$. Additionally, assume that the observed output y depends linearly on the state $\mathbf{x}$. Hence, in what follows, we will make use of the following description of bilinear systems (with a single input and a single output):

$$\begin{aligned} \dot{\mathbf{x}}(t) &= \mathbf{A}\mathbf{x}(t) + \mathbf{N}\mathbf{x}(t)u(t) + \mathbf{B}u(t), \\ y(t) &= \mathbf{C}\mathbf{x}(t), \end{aligned} \tag{27}$$

where the matrix $\mathbf{N} \in \mathbb{R}^{n\times n}$ scales the product of the state variable $\mathbf{x}$ with the control input u. In practice, bilinear control systems are used to approximate nonlinear systems with more general, analytic nonlinearities. This procedure is known as Carleman's linearization; for more details see [28].

Bilinear systems are a class of nonlinear systems that received considerable attention in the last four or five decades. Contributions that range from realization theory in [16], classical system identification in [12], or to subspace identification in [13]. In more recent years (last two decades), model order reduction of bilinear systems (in both continuous- and discrete-time domains) was extensively studied with contributions covering balanced truncation in [33], Krylov subspace methods in [10], interpolation-based $\mathcal{H}_2$ method in [4, 6], or data-driven Loewner approach in [3, 17].

3.1.1 The General Procedure

We start by collecting snapshots of the state $\mathbf{x}$ for multiple time instances t_k. We enforce that the snapshot $\mathbf{x}_{k+1}$ at time t_{k+1} depends on the snapshot $\mathbf{x}_k$ in the following way:

$$\mathbf{x}_{k+1} = \mathbf{A}\mathbf{x}_k + \mathbf{N}\mathbf{x}_k u_k + \mathbf{B}u_k, \quad \text{for } 0 \leq k \leq m-1. \tag{28}$$

We denote the sequence of state and input snapshots as in (3) and in (9). Again, by varying the index k in the interval $\{1, 2, \ldots, m-1\}$, one can compactly rewrite the $m-1$ equations in the following matrix format:

$$\mathbf{X}_s = \mathbf{A}\mathbf{X} + \mathbf{N}\mathbf{X}\mathbf{U}_\mathbf{D} + \mathbf{B}\mathbf{U}, \tag{29}$$

where $\mathbf{U}_\mathbf{D} = \text{diag}(u_0, u_1, \ldots, u_{m-1}) \in \mathbb{R}^{m\times m}$. One can hence write $\mathbf{U} = \mathbf{L}\mathbf{U}_\mathbf{D}$, with $\mathbf{L} = [1\ 1\ \ldots\ 1] \in \mathbb{R}^{1\times m}$ and then introduce the matrix $\mathbf{Z} \in \mathbb{R}^{(n+1)\times m}$ as

$$\mathbf{Z} = \begin{bmatrix} \mathbf{L}\mathbf{U}_\mathbf{D} \\ \mathbf{X}\mathbf{U}_\mathbf{D} \end{bmatrix} = \begin{bmatrix} \mathbf{L} \\ \mathbf{X} \end{bmatrix} \mathbf{U}_\mathbf{D}. \tag{30}$$

The next step is to augment the matrix $\mathbf{X}$ with matrix $\mathbf{Z}$ and denote this new matrix with $\boldsymbol{\Omega} \in \mathbb{R}^{(2n+1)\times m}$ as an extension of the matrix previously introduced in (10), i.e.,

$$\boldsymbol{\Omega} = \begin{bmatrix} \mathbf{X} \\ \mathbf{Z} \end{bmatrix}. \tag{31}$$

For the case in which we extend the DMDc method in Sect. 2.1 to fitting bilinear dynamics (no output observations), we propose a slightly different definition for the matrix $\mathbf{G}$. We hence append the matrix $\mathbf{N}$ to the originally introduced system matrix in (10). Then, Eq. (29) can be written in a factorized way as $\boldsymbol{\Gamma} = \mathbf{G}\boldsymbol{\Omega}$, where the matrices for this particular setup are as follows:

$$\mathbf{G} = \begin{bmatrix} \mathbf{A}\ \mathbf{B}\ \mathbf{N} \end{bmatrix} \in \mathbb{R}^{n\times(2n+1)}, \quad \boldsymbol{\Gamma} = \mathbf{X}_s. \tag{32}$$

Alternatively, for the case where output observations y_k are also available, we enforce a special bilinear dependence for each trio of measurement data snapshots as

$$\mathbf{y}_k = \mathbf{C}\mathbf{x}_k + \mathbf{F}\mathbf{x}_k u_k + \mathbf{D}u_k, \quad 0 \leq k \leq m-1, \tag{33}$$

where $\mathbf{F}^T \in \mathbb{R}^n$. Note that (33) represents a natural extension of the relation imposed in (19). Therefore, fitting a linear structure is instead enforced.

Afterward, we collect the equations in (33) for each index k and hence write

$$\mathbf{Y} = \mathbf{C}\mathbf{X} + \mathbf{F}\mathbf{X}\mathbf{U}_\mathbf{D} + \mathbf{D}\mathbf{U}, \tag{34}$$

with the same notations as in (30). Then, by combining (29) and (34), we can write all snapshot matrix quantities in the following structured equalities:

$$\begin{aligned} \mathbf{X}_s &= \mathbf{A}\mathbf{X} + \mathbf{N}\mathbf{X}\mathbf{U}_\mathbf{D} + \mathbf{B}\mathbf{U}, \\ \mathbf{Y} &= \mathbf{C}\mathbf{X} + \mathbf{F}\mathbf{X}\mathbf{U}_\mathbf{D} + \mathbf{D}\mathbf{U}. \end{aligned} \tag{35}$$

This system of equations can be then written in a factorized way as before, i.e., $\mathbf{\Gamma} = \mathbf{G}\mathbf{\Omega}$, where the matrices for this particular setup are given below:

$$\mathbf{G} = \begin{bmatrix} \mathbf{A} & \mathbf{B} & \mathbf{N} \\ \mathbf{C} & \mathbf{D} & \mathbf{F} \end{bmatrix} \in \mathbb{R}^{(n+1)\times(2n+1)}, \quad \mathbf{\Gamma} = \begin{bmatrix} \mathbf{X}_s \\ \mathbf{Y} \end{bmatrix}. \tag{36}$$

Finally, the last step is to recover the matrix $\mathbf{G}$ and split it block-wise in order to put together a system realization. Consequently, this all boils down to solving the equation $\mathbf{\Gamma} = \mathbf{G}\mathbf{\Omega}$ (in either of the two cases, with or without output observations included). More precisely, the objective matrix $\mathbf{G} \in \mathbb{R}^{(n+1)\times(2n+1)}$ in (36) is the solution of the following optimization problem:

$$\underset{\hat{\mathbf{G}}\in\mathbb{R}^{(n+1)\times(2n+1)}}{\arg\min} \left(\left\| \begin{bmatrix} \mathbf{X}_s \\ \mathbf{Y} \end{bmatrix} - \hat{\mathbf{G}} \begin{bmatrix} \mathbf{X} \\ \mathbf{Z} \end{bmatrix} \right\|_F \right) \Leftrightarrow \underset{\hat{\mathbf{G}}\in\mathbb{R}^{(n+1)\times(2n+1)}}{\arg\min} \left(\left\| \mathbf{\Gamma} - \hat{\mathbf{G}}\mathbf{\Omega} \right\|_F \right). \tag{37}$$

As shown in the previous sections, solving for $\mathbf{G}$ in (37) involves computing the pseudo-inverse of matrix $\mathbf{\Omega} \in \mathbb{R}^{(2n+1)\times m}$ from (31). More precisely, we write the solution as

$$\mathbf{G} = \mathbf{\Gamma}\mathbf{\Omega}^{\dagger}. \tag{38}$$

Remark 1 Note that the observation map g corresponding to the original dynamical system, as introduced in (18), need not have a bilinear structure as in (33). It could include more complex nonlinearities or could even be linear. In the later case, the recovered matrix $\mathbf{F}$ will typically have a low norm.

3.1.2 Computation of the Reduced-Order Matrices

In this section, we present specific/practical details for retrieving the system matrices in the case of the proposed procedure in Sect. 3.1.1. We solve the equation $\mathbf{\Gamma} = \mathbf{G}\mathbf{\Omega}$ for which the matrices are given as in (36), i.e., the case containing output observations. We compute an SVD of the augmented data matrix $\mathbf{\Omega}$ giving

$$\mathbf{\Omega} = \mathbf{V}\mathbf{\Sigma}\mathbf{W}^T \approx \tilde{\mathbf{V}}\tilde{\mathbf{\Sigma}}\tilde{\mathbf{W}}^T, \tag{39}$$

where the full-scale and reduced-scale matrices derived from SVD are as follows:

$$\begin{cases} \mathbf{V} \in \mathbb{R}^{(2n+1)\times(2n+1)}, & \boldsymbol{\Sigma} \in \mathbb{R}^{(2n+1)\times(m-1)}, \quad \mathbf{V} \in \mathbb{R}^{(m-1)\times(m-1)}, \\ \tilde{\mathbf{V}} \in \mathbb{R}^{(2n+1)\times p}, & \tilde{\boldsymbol{\Sigma}} \in \mathbb{R}^{p\times p}, \quad \tilde{\mathbf{V}} \in \mathbb{R}^{(m-1)\times p}. \end{cases}$$

The truncation index is denoted with p, where $p \leqslant n$. The computation of the pseudo-inverse $\boldsymbol{\Omega}^{\dagger}$ is done by the SVD approach, i.e., $\boldsymbol{\Omega}^{\dagger} \approx \tilde{\mathbf{W}}\tilde{\boldsymbol{\Sigma}}^{-1}\tilde{\mathbf{V}}^T$. By splitting up the $\mathbf{V}^T$ matrix as $\tilde{\mathbf{V}}^T = [\tilde{\mathbf{V}}_1^T \;\; \tilde{\mathbf{V}}_2^T \;\; \tilde{\mathbf{V}}_3^T]$, one can recover the system matrices as

$$\begin{aligned} \overline{\mathbf{A}} &= \mathbf{X}_s\tilde{\mathbf{W}}\tilde{\boldsymbol{\Sigma}}^{-1}\tilde{\mathbf{V}}_1^T, \quad \overline{\mathbf{B}} = \mathbf{X}_s\tilde{\mathbf{W}}\tilde{\boldsymbol{\Sigma}}^{-1}\tilde{\mathbf{V}}_2^T, \quad \overline{\mathbf{N}} = \mathbf{X}_s\tilde{\mathbf{W}}\tilde{\boldsymbol{\Sigma}}^{-1}\tilde{\mathbf{V}}_3^T, \\ \overline{\mathbf{C}} &= \mathbf{Y}\tilde{\mathbf{W}}\tilde{\boldsymbol{\Sigma}}^{-1}\tilde{\mathbf{V}}_1^T, \quad \overline{\mathbf{D}} = \mathbf{Y}\tilde{\mathbf{W}}\tilde{\boldsymbol{\Sigma}}^{-1}\tilde{\mathbf{V}}_2^T, \quad \overline{\mathbf{F}} = \mathbf{Y}\tilde{\mathbf{W}}\tilde{\boldsymbol{\Sigma}}^{-1}\tilde{\mathbf{V}}_3^T. \end{aligned} \tag{40}$$

By performing another (short) SVD for the matrix $\mathbf{X}_s$, we can write

$$\mathbf{X}_s \approx \hat{\mathbf{V}}\tilde{\boldsymbol{\Sigma}}\hat{\mathbf{W}}^T, \tag{41}$$

where $\hat{\mathbf{V}} \in \mathbb{R}^{(n+1)\times r}$, $\hat{\boldsymbol{\Sigma}} \in \mathbb{R}^{r\times r}$, $\hat{\mathbf{W}} \in \mathbb{R}^{(m-1)\times p}$. Note that the two SVDs could have different truncation values denoted with p and r. Using the transformation $\mathbf{x} = \hat{\mathbf{V}}\tilde{\mathbf{x}}$, the following reduced-order matrices can be computed:

$$\begin{aligned} \tilde{\mathbf{A}} &= \hat{\mathbf{V}}^T\overline{\mathbf{A}}\hat{\mathbf{V}} \in \mathbb{R}^{r\times r}, \quad \tilde{\mathbf{B}} = \hat{\mathbf{V}}^T\overline{\mathbf{B}} \in \mathbb{R}^{r}, \quad \tilde{\mathbf{N}} = \hat{\mathbf{V}}^T\overline{\mathbf{N}}\hat{\mathbf{V}} \in \mathbb{R}^{r\times r}, \\ \tilde{\mathbf{C}} &= \overline{\mathbf{C}}\hat{\mathbf{V}} \in \mathbb{R}^{1\times r}, \quad \tilde{\mathbf{D}} = \overline{\mathbf{D}} \in \mathbb{R}, \quad \tilde{\mathbf{F}} = \hat{\mathbf{V}}^T\overline{\mathbf{F}}\hat{\mathbf{V}} \in \mathbb{R}^{1\times r}. \end{aligned} \tag{42}$$

3.1.3 Conversions Between Discrete-Time and Continuous-Time Representations

The DMD-type approaches available in the literature identify continuous-time systems by means of linear discrete-time models. In this contribution, we make use of the same philosophy, in the sense that the models fitted are discrete time. We extend the DMDc and ioDMD approaches by allowing bilinear or quadratic terms to appear in these models as well.

As also mentioned in [9], one can compute a continuous-time model that represents a first-order approximation of the discrete-time model obtained by DMD-type approaches.

Assume that we are in the bilinear setting presented in Sect. 3.1 and that we already have computed a reduced-order discrete-time model given by matrices $\{\tilde{\mathbf{A}}, \tilde{\mathbf{B}}, \tilde{\mathbf{N}}, \tilde{\mathbf{C}}, \tilde{\mathbf{D}}, \tilde{\mathbf{F}}\}$, i.e., following the explicit derivations in (42). Then, a continuous-time model $\{\hat{\mathbf{A}}, \hat{\mathbf{B}}, \hat{\mathbf{N}}, \hat{\mathbf{C}}, \hat{\mathbf{D}}, \hat{\mathbf{F}}\}$ can also be derived. By assuming that the standard first-order Euler method was used for simulating the original system (with a small enough time step size $0 < \Delta_t \ll 1$), we can write that

$$\begin{aligned} \mathbf{x}_{k+1} &= \mathbf{x}_k + \Delta_t(\hat{\mathbf{A}}\mathbf{x}_k + \hat{\mathbf{B}}\mathbf{u}_k + \hat{\mathbf{N}}\mathbf{x}_k\mathbf{u}_k) \Rightarrow \\ \tilde{\mathbf{A}}\mathbf{x}_k + \tilde{\mathbf{B}}\mathbf{u}_k + \tilde{\mathbf{N}}\mathbf{x}_k\mathbf{u}_k &= \mathbf{x}_k + \Delta_t\hat{\mathbf{A}}\mathbf{x}_k + \Delta_t\hat{\mathbf{B}}\mathbf{u}_k + \Delta_t\hat{\mathbf{N}}\mathbf{x}_k\mathbf{u}_k \Rightarrow \\ &\begin{cases} \hat{\mathbf{A}} = \Delta_t^{-1}(\tilde{\mathbf{A}} - \mathbf{I}), \quad \hat{\mathbf{B}} = \Delta_t^{-1}\hat{\mathbf{B}}, \quad \hat{\mathbf{N}} = \Delta_t^{-1}\hat{\mathbf{N}}, \\ \hat{\mathbf{C}} = \tilde{\mathbf{C}}, \quad \hat{\mathbf{D}} = \tilde{\mathbf{D}}, \quad \hat{\mathbf{F}} = \tilde{\mathbf{F}}. \end{cases} \end{aligned} \tag{43}$$

Observe that for the ioDMD type of approaches, the feed-through terms that appear in the output-state equation are the same in both discrete and continuous representations.

3.2 Quadratic-Bilinear Systems

Next, we extend the method in Sect. 2.1 for fitting another class of nonlinear systems, i.e., quadratic-bilinear (QB) systems. Additional to the bilinear terms that enter the differential equations, we assume that quadratic terms are also present. More precisely, the system of the ODEs is written as in (7) but for a specific choice of nonlinear mapping f, i.e.,

$$f(\mathbf{x}, \mathbf{u}) = \mathbf{A}\mathbf{x} + \mathbf{Q}(\mathbf{x} \otimes \mathbf{x}) + \mathbf{N}\mathbf{x}u + \mathbf{B}u,$$

where "$\otimes$" denotes the Kronecker product, the matrix $\mathbf{Q} \in \mathbb{R}^{n \times n^2}$ scales the product of the state $\mathbf{x}$ with itself, and $\mathbf{N} \in \mathbb{R}^{n \times n}$ is as shown in Sect. 3.1.

Quadratic-bilinear systems appear in many applications for which the original system of ODEs inherently has the required quadratic structure. For example, after semi-discretizing Burgers' or Navier-Stokes equations in the spatial domain, one obtains a system of differential equations with quadratic nonlinearities (and also with bilinear terms). Moreover, many smooth analytic nonlinear systems that contain combinations of nonlinearities such as exponential, trigonometric, polynomial functions, etc. can be equivalently rewritten as QB systems. This is performed by employing so-called lifting techniques. More exactly, one needs to introduce new state variables in order to simplify the nonlinearities and hence derive new differential equations corresponding to these variables. Model order reduction of QB systems was a topic of interest in the last years with contributions ranging from projection-based approaches in [5, 15] to optimal $\mathcal{H}_2$-based approximation in [7], or data-driven approaches in the Loewner framework in [2, 14].

Similar to the procedure described in Sect. 3.1, we enforce that the snapshot $\mathbf{x}_{k+1}$ at time t_{k+1} depends on the snapshot $\mathbf{x}_k$ in the following way:

$$\mathbf{x}_{k+1} = \mathbf{A}\mathbf{x}_k + \mathbf{Q}(\mathbf{x}_k \otimes \mathbf{x}_k) + \mathbf{N}\mathbf{x}_k u_k + \mathbf{B}u_k, \quad \text{for } 0 \leq k \leq m-1. \tag{44}$$

Next, by varying the k in the range $\{1, 2, \ldots, m-1\}$, compactly rewrite the m equations in (44) in the following matrix format:

$$\mathbf{X}_s = \mathbf{A}\mathbf{X} + \mathbf{Q}(\mathbf{X} \otimes \mathbf{X})\mathbf{H} + \mathbf{N}\mathbf{X}\mathbf{U}_\mathbf{D} + \mathbf{B}\mathbf{U}, \tag{45}$$

with $\mathbf{U_D} = \text{diag}(u_0, u_1, \ldots, u_{m-1}) \in \mathbb{R}^{m\times m}$ and $\mathbf{H} = \left[\mathbf{e}_1 \otimes \mathbf{e}_1\ \mathbf{e}_2 \otimes \mathbf{e}_2\ \ldots\ \mathbf{e}_m \otimes \mathbf{e}_m\right] \in \mathbb{R}^{m^2\times m}$. Here, $\mathbf{e}_k$ is the unit vector of length n that contains the 1 on position k. Additionally, we introduce the matrix $\mathbf{T}$ that depends on the state matrix $\mathbf{X}$ as

$$\mathbf{T} = \left[\mathbf{X}_1 \otimes \mathbf{X}_1\ \mathbf{X}_2 \otimes \mathbf{X}_2\ \ldots\ \mathbf{X}_m \otimes \mathbf{X}_m\right] \in \mathbb{R}^{n^2\times m}.$$

Note that the equality holds as follows $\mathbf{T} = (\mathbf{X} \otimes \mathbf{X})\mathbf{H}$. Next, we augment the matrix $\mathbf{X}$ with both matrices $\mathbf{Z}$ and $\mathbf{T}$ and group together the matrices $\mathbf{A}$, $\mathbf{Q}$, $\mathbf{N}$ and $\mathbf{B}$ by using the notations:

$$\mathbf{G} = [\mathbf{A}\ \mathbf{B}\ \mathbf{N}\ \mathbf{Q}] \in \mathbb{R}^{n\times(n^2+2n+1)},\quad \mathbf{\Omega} = \begin{bmatrix}\mathbf{X}\\ \mathbf{Z}\\ \mathbf{T}\end{bmatrix} \in \mathbb{R}^{(n^2+2n+1)\times m},\quad \mathbf{\Gamma} = \mathbf{X}_s. \tag{46}$$

Hence, by using the above notations, rewrite Eq. (45) as follows: $\mathbf{\Gamma} = \mathbf{G}\mathbf{\Omega}$.

More precisely, the objective matrix $\mathbf{G} \in \mathbb{R}^{n\times(n^2+2n+1)}$ in (46) is the solution of the following optimization problem:

$$\underset{\hat{\mathbf{G}}\in\mathbb{R}^{n\times(n^2+2n+1)}}{\arg\min} \left(\left\|\mathbf{\Gamma} - \hat{\mathbf{G}}\mathbf{\Omega}\right\|_F\right). \tag{47}$$

Thus, one can recover the matrix $\mathbf{G}$ by solving an optimization problem, e.g., the one given in (47). This is explicitly done by computing the Moore-Pseudo pseudo-inverse of matrix $\mathbf{\Omega} \in \mathbb{R}^{(n^2+2n+1)\times m}$, and then writing $\mathbf{G} = \mathbf{\Gamma}\mathbf{\Omega}^\dagger$.

As previously shown in Sect. 3.1, we can again adapt the procedure for fitting QB systems in the ioDMD format by involving output observation measurements y_k. The procedure for quadratic-bilinear systems is similar to that for bilinear systems and we prefer to skip the exact description to avoid duplication. For more details, see the derivation in Sect. 6.1.

Remark 2 Note that the Kronecker product of the vector $\mathbf{x} \in \mathbf{R}^n$ with itself, i.e., $\mathbf{x}^{(2)} = \mathbf{x} \otimes \mathbf{x}$ has indeed duplicate components. For $n = 2$, one can write

$$\mathbf{x}^{(2)} = \left[x_1^2\ x_1x_2\ x_2x_1\ x_2^2\right]^T.$$

Thus, since matrix $\mathbf{G}$ is explicitly written in terms of $\mathbf{Q}$ as in (46), the linear system of equations $\mathbf{\Gamma} = \mathbf{G}\mathbf{\Omega}$ does not have an unique solution. By using the Moore-Penrose inverse, one implicitly regularizes the least-squares problem in (47). Additionally, note that using a different least-squares solver (with or without regularization) could indeed produce a different result.

Remark 3 It is to be noted that the operator inference procedure avoids the non-uniqueness by accounting for duplicates in the vector $\mathbf{x} \otimes \mathbf{x}$. This is done by introducing a special Kronecker product for which the duplicate terms are removed. For more details, we refer the reader to Sect. 2.3 from [8].

4 Numerical Experiments

4.1 The Viscous Burgers' Equation

Consider the partial differential viscous Burgers' equation:

$$\frac{\partial v(x,t)}{\partial t} + v(x,t)\frac{\partial v(x,t)}{\partial x} = \nu \frac{\partial^2 v(x,t)}{\partial x^2}, \quad (x,t) \in (0,L) \times (0,T),$$

with i.c. $v(x,0) = 0,\ x \in [0,L],\ v(0,t) = u(t),\ v(L,t) = 0,\ t \geqslant 0$. The viscosity parameter is denoted with ν.

Burgers' equation has a convective term, an unsteady term and a viscous term; it can be viewed as a simplification of the Navier-Stokes equations.

By means of semi-discretization in the space domain, one can obtain the following nonlinear (quadratic) model (see [5]) described by the following system of ODEs:

$$\dot{v}_k = \begin{cases} -\frac{1}{2h} v_1 v_2 + \frac{\nu}{h^2}(v_2 - 2v_1) + (\frac{v_1}{2h} + \frac{\nu}{h^2})u, & k = 1, \\ -\frac{v_k}{2h}(v_{k+1} - v_{k-1}) + \frac{\nu}{h^2}(v_{k+1} - 2v_k + v_{k-1}), & 2 \leqslant k \leqslant n_0 - 1, \\ -\frac{1}{2h} v_n v_{n-1} + \frac{\nu}{h^2}(-2v_n + 2v_{n-1}), & k = n_0. \end{cases} \tag{48}$$

Next, by means of the Carleman linearization procedure in [28], one can approximate the above nonlinear system of order n_0 with a bilinear system of order $n = n_0^2 + n_0$. The procedure is as follows: let $\mathbf{v} = [v_1\ v_2\ \ldots\ v_n]^T$ be the original state variable in (48). Then, introduce the augmented state variable $\mathbf{x} = \begin{bmatrix} \mathbf{v} \\ \mathbf{v} \otimes \mathbf{v} \end{bmatrix} \in \mathbb{R}^{n_0^2+n_0}$ corresponding to the system described by the following equations:

$$\begin{aligned} \dot{\mathbf{x}} &= \mathbf{A}\mathbf{x} + \mathbf{N}\mathbf{x}u + \mathbf{B}u, \\ \mathbf{y} &= \mathbf{C}\mathbf{x}. \end{aligned} \tag{49}$$

The continuous-time bilinear model in (49) is going to be used in following the numerical experiments.

Start by choosing the viscosity parameter to be $\nu = 0.01$. Then, choose $n_0 = 40$ as the dimension of the original discretization, and hence the bilinear system in (49) is of order $n = 1640$. Perform a time-domain simulation of this system by approximating the derivative as follows $\dot{\mathbf{x}}(t_k) \approx \frac{\mathbf{x}(t_{k+1}) - \mathbf{x}(t_k)}{t_{k+1} - t_k} = \frac{\mathbf{x}_{k+1} - \mathbf{x}_k}{\Delta_t}$. We use as time step $\delta_t = 10^{-3}$ and the time horizon to be $[0, 10]s$. The control input is chosen to be $u(t) = 0.5\cos(10t)e^{-0.3t}$.

Hence, collect 10^4 snapshots of the trio $(\mathbf{x}_k, u_k, y_k)$ that are arranged in the required matrix format as presented in the previous sections. The first step is to perform an SVD for the matrix $\boldsymbol{\Omega} \in \mathbb{R}^{3281 \times 10^4}$. The first 200 normalized singular values are presented in Fig. 1. Choose the tolerance value $\tau_p = 10^{-10}$ which corresponds to a truncation order of $p = 86$ (for computing the pseudo-inverse of matrix $\boldsymbol{\Omega}$). On the same plot

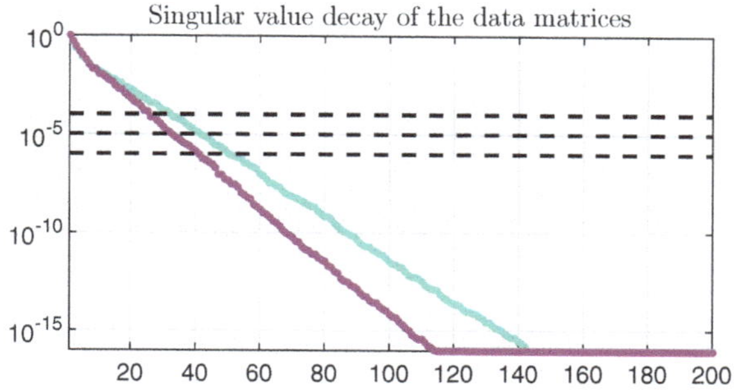

Fig. 1 The normalized first 200 singular values of matrices $\mathbf{\Omega}$ (with cyan) and $\mathbf{\Gamma}$ (with magenta). The three dotted black lines correspond to the three tolerance levels chosen for τ_r

in Fig. 1, we also display the normalized singular values of matrix $\mathbf{\Gamma} \in \mathbb{R}^{1641 \times 10^4}$. Note that machine precision is reached at the 112th singular value. We select three tolerance values $\tau_r \in \{10^{-4}, 10^{-5}, 10^{-6}\}$ for truncating matrices obtained from the SVD of $\mathbf{\Gamma}$.

In what follows, we compute reduced-order discrete-time models that have dimension r, as in (42). Next, these models are converted using (43) into a continuous-time model.

4.1.1 Experiment 1—Validating the Trained Models

In this first setup, we perform time-domain simulations of the reduced-order models for the same conditions as in the training stage, i.e., in the time horizon $[0, 10]s$ and by using the control input $u(t) = 0.5\cos(10t)e^{-0.3t}$. Hence, we are validating the trained models on the training data.

Start by choosing the first tolerance value, e.g., $\tau_r = 10^{-4}$. This corresponds to a truncation value of $r = 25$. We compute term $\hat{\mathbf{D}} = 1.1744\text{e} - 14$ and a also bilinear feed-through term with $\|\hat{\mathbf{F}}\|_2 = 6.7734\text{e} - 04$. We simulate both the original large-scale bilinear system and the reduced-order system. The results are presented in Fig. 2. Note that the observed output curves deviate substantially. One way to improve this behavior is to decrease the tolerance value.

For the next experiment, choose the tolerance value to be $\tau_r = 10^{-5}$. This corresponds to a truncation value of $r = 32$. After computing the required matrices, notice that the D term is again numerically 0, while the norm of the matrix $\hat{\mathbf{F}}$ slightly decreases to the value $6.9597e - 04$. Perform numerical simulations and depict the two outputs and the approximation error in Fig. 3. Observe that the approximation quality significantly improved, but there is still room for improvement.

Finally, the tolerance value is chosen as $\tau_r = 10^{-6}$. For this particular choice, it follows that the truncation value is $r = 40$. In this case, the output of the reduced-order model faithfully reproduces the original output, as it can be observed in Fig. 4. Note also that the approximation error stabilizes within the range $(10^{-4}, 10^{-5})$.

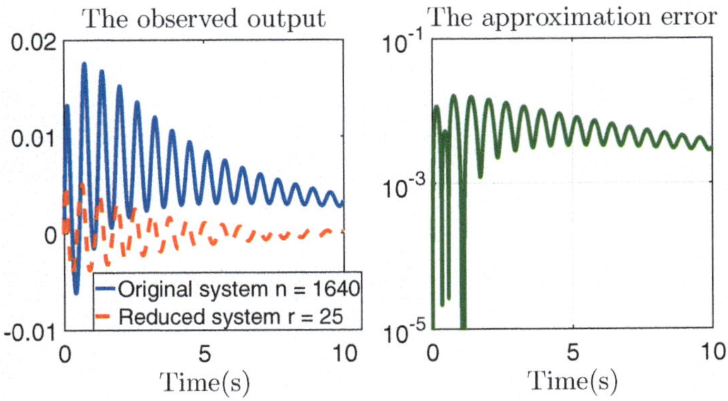

Fig. 2 Left plot: the observed outputs; right plot: the corresponding approximation error

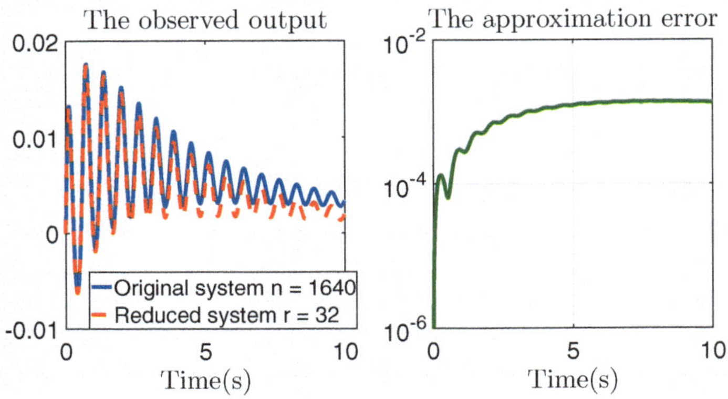

Fig. 3 Left plot: the observed outputs; right plot: the corresponding approximation error

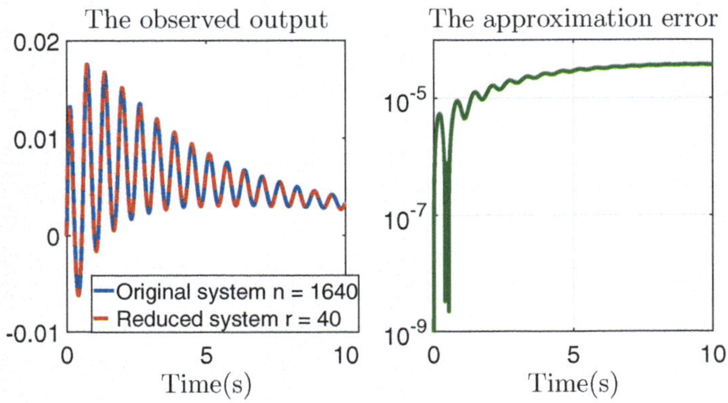

Fig. 4 Left plot: the observed outputs; right plot: the corresponding approximation error

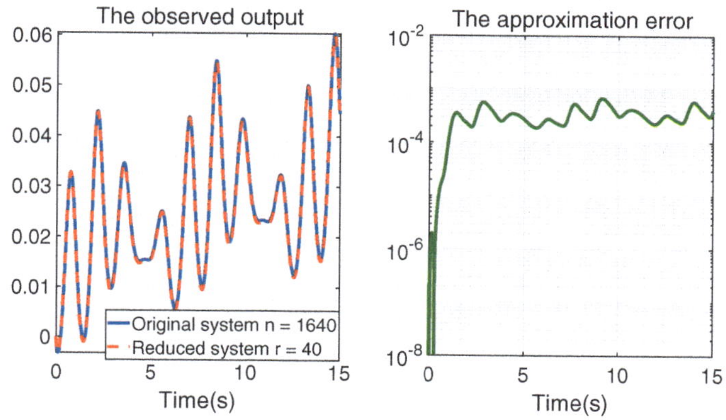

Fig. 5 Left plot: the observed outputs; right plot: the corresponding approximation error

4.1.2 Experiment 2—Testing the Trained Models

In the second setup, we perform time-domain simulations of the reduced-order models for different conditions than those used in the training stage, i.e., the time horizon is extended to $[0, 15]s$ and two other control inputs are used. Moreover, we keep the truncation value to be $r = 40$ (corresponding to tolerance $\tau_r = 10^{-6}$).

First, choose the testing control input to be $u_1(t) = \sin(4t)/4 - \cos(5t)/5$. The time-domain simulations showing the observed outputs are depicted in Fig. 5. Moreover, on the same figure, the magnitude of the approximation is presented. We observe that the output of the learned reduced model accurately approximates the output of the original system.

Afterward, choose the testing control input to be $u_2(t) = \frac{\text{square}(2t)}{5(t+1)}$. Note that square$(2t)$ is a square wave with period π. The time-domain simulations showing the observed outputs are depicted in Fig. 6. Moreover, on the same figure, the magnitude of the approximation is presented. We observe that the output of the learned reduced model does not approximate the output of the original system as well as in the previous experiments.

4.2 *Coupled van der Pol Oscillators*

Consider the coupled van der Pol oscillators along a limit cycle example given in [18]. The dynamics are characterized by the following six differential equations with linear and nonlinear (cubic) terms:

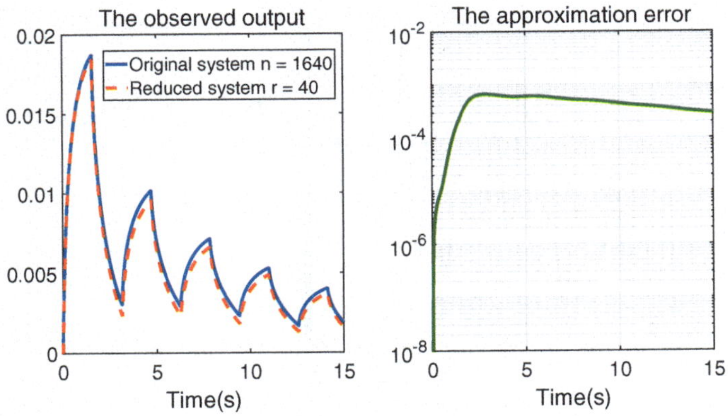

Fig. 6 Left plot: the observed outputs; right plot: the corresponding approximation error

$$\begin{aligned}
\dot{x}_1 &= x_2, \\
\dot{x}_2 &= -x_1 - \mu(x_1^2 - 1)x_2 + a(x_3 - x_1) + b(x_4 - x_2), \\
\dot{x}_3 &= x_4, \\
\dot{x}_4 &= -x_3 - \mu(x_3^2 - 1)x_4 + a(x_1 - x_3) + b(x_2 - x_4), \\
&\quad + a(x_5 - x_3) + b(x_6 - x_4) + u, \\
\dot{x}_5 &= x_6, \\
\dot{x}_6 &= -x_5 - \mu(x_5^2 - 1)x_6 + a(x_3 - x_5) + b(x_4 - x_6).
\end{aligned} \tag{50}$$

Choose the output to be $y = x_3$. Hence, the state-output equation is written as $y = \mathbf{C}\mathbf{x}$ with $\mathbf{C} = \begin{bmatrix} 0\ 0\ 1\ 0\ 0\ 0 \end{bmatrix}$ and $\mathbf{x} = \begin{bmatrix} x_1\ x_2\ x_3\ x_4\ x_5\ x_6 \end{bmatrix}^T$. Choose the parameters in (50) as follows: $\mu = 0.5$, $a = 0.5$ and $b = 0.2$.

Note that by introducing three additional surrogate states, e.g., $x_7 = x_1^2$, $x_8 = x_2^2$, and $x_9 = x_3^2$, one can rewrite the cubic nonlinear system in (50) of order $n = 6$ as an order $n_q = 9$ quadratic-bilinear system.

Perform time-domain simulations of the cubic system of order $n = 6$ and collect data from 500 snapshots using the explicit Euler method with step size $\Delta_t = 0.01$. The chosen time horizon is hence [0,5]s. The control input is a square wave with period $\pi/5$ and amplitude 30, i.e., $u(t) = 30$ square(10t).

Compute the pseudo-inverse of matrix $\mathbf{\Omega} \in \mathbb{R}^{49\times 500}$ and select as truncation value $p = 19$ (the 20th normalized singular value drops below machine precision).

We compute a reduced-order quadratic-bilinear model of order $r = 5$. We made this choice since the fifth normalized singular value of matrix $\mathbf{\Gamma} \in \mathbb{R}^{7\times 500}$ is 5.8651e-04 while the sixth is numerically 0, i.e., 3.5574e-16. We hence fit an order $r = 5$ quadratic-bilinear system that approximates the original order $n = 6$ cubic polynomial system. Note that the only nonzero feed-through quantity in the recovered state-output equation is given by $\hat{\mathbf{C}} = \begin{bmatrix} -0.1067\ 0.5580\ 0.0797\ -0.4145\ -0.7065 \end{bmatrix}$.

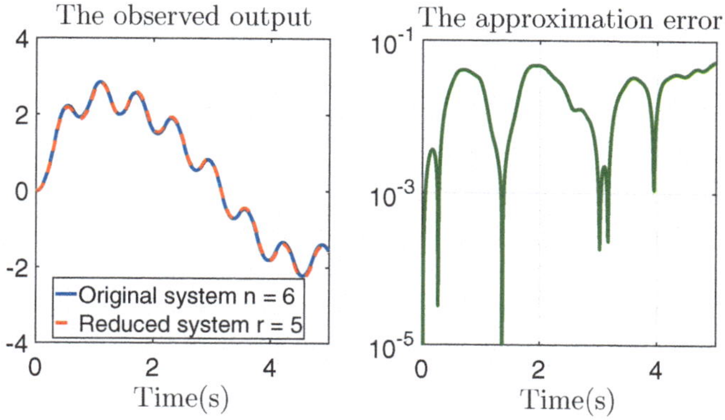

Fig. 7 Left plot: the observed outputs; Right plot: the corresponding approximation error

Next, we perform time-domain simulations in the same manner as in Sect. 4.1.1, i.e., by validating the reduced models on the training data. The results are depicted in Fig. 5. One can observe that the two outputs match well. In this particular setup, it follows that the response of the sixth-order cubic system (that can be equivalently written as a ninth-order QB system) can be accurately approximated with the response of a fifth-order QB system. The approximation error is presented in Fig. 7.

5 Conclusion

In this paper, we have proposed extensions of the DMDc and ioDMD recently proposed methods. The philosophy is similar to that of the original methods, but instead of fitting discrete-time linear systems, we impose a more complex structure to the fitted models. More precisely, we fit bilinear or quadratic terms to augment the existing linear quantities (both in the differential and in the output equations). The numerical results presented were promising, and they have shown the strength of the method. Indeed, there is a clear trade-off to be made between approximation quality and the dimension of the fitted model.

Nevertheless, this represents a first step toward extending DMD-type methods, and a more involved analysis of the method's advantages and disadvantages could represent an appealing future endeavor. Moreover, another contribution could be made by comparing the proposed methods in this work with the recently introduced operator inference-type methods. For the quadratic-bilinear case, additional challenges arise when storing the large-scale matrices involved and also when computing the classical SVD for such big non-sparse matrices.

6 Appendix

6.1 Computation of the Reduced-Order Matrices for the Quadratic-Bilinear Case

In this section, we present practical details for retrieving the system matrices in the case of the proposed procedure in Sect. 3.2. We solve the equation $\mathbf{\Gamma} = \mathbf{G}\mathbf{\Omega}$ for which the matrices are given as in (46), i.e., the case without output observations. We again utilize an SVD, now performed on the matrix $\mathbf{\Omega}$, i.e.,

$$\mathbf{\Omega} = \mathbf{V}\mathbf{\Sigma}\mathbf{W}^T \approx \tilde{\mathbf{V}}\tilde{\mathbf{\Sigma}}\tilde{\mathbf{W}}^T, \tag{51}$$

where the full-scale and reduced-order SVD matrices have the following dimensions:

$$\begin{cases} \mathbf{V} \in \mathbb{R}^{(n^2+2n+1)\times(n^2+2n+1)}, & \mathbf{\Sigma} \in \mathbb{R}^{(n^2+2n+1)\times(m-1)}, \quad \mathbf{W} \in \mathbb{R}^{(m-1)\times(m-1)}, \\ \tilde{\mathbf{V}} \in \mathbb{R}^{(n^2+2n+1)\times p}, & \tilde{\mathbf{\Sigma}} \in \mathbb{R}^{p\times p}, \quad \tilde{\mathbf{W}} \in \mathbb{R}^{(m-1)\times p}. \end{cases}$$

The truncation index is denoted with r, and written as before $\mathbf{\Omega}^\dagger \approx \tilde{\mathbf{W}}\tilde{\mathbf{\Sigma}}^{-1}\tilde{\mathbf{V}}^T$.

By splitting up the matrix $\mathbf{V}^T$ as $\tilde{\mathbf{V}}^T = [\tilde{\mathbf{V}}_1^T \ \tilde{\mathbf{V}}_2^T \ \tilde{\mathbf{V}}_3^T \ \tilde{\mathbf{V}}_4^T]$, with

$$\tilde{\mathbf{V}}_1, \tilde{\mathbf{V}}_3 \in \mathbb{R}^{n\times r}, \quad \tilde{\mathbf{V}}_2 \in \mathbb{R}^{1\times r}, \quad \tilde{\mathbf{V}}_4 \in \mathbb{R}^{n^2\times r},$$

recover the matrices

$$\overline{\mathbf{A}} = \mathbf{X}_s\tilde{\mathbf{W}}\tilde{\mathbf{\Sigma}}^{-1}\tilde{\mathbf{V}}_1^T, \quad \overline{\mathbf{B}} = \mathbf{X}_s\tilde{\mathbf{W}}\tilde{\mathbf{\Sigma}}^{-1}\tilde{\mathbf{V}}_2^T, \quad \overline{\mathbf{N}} = \mathbf{X}_s\tilde{\mathbf{W}}\tilde{\mathbf{\Sigma}}^{-1}\tilde{\mathbf{V}}_3^T, \quad \overline{\mathbf{Q}} = \mathbf{X}_s\tilde{\mathbf{W}}\tilde{\mathbf{\Sigma}}^{-1}\tilde{\mathbf{V}}_4^T. \tag{52}$$

Again, perform an additional SVD, e.g., $\mathbf{X}_s \approx \hat{\mathbf{V}}\hat{\mathbf{\Sigma}}\hat{\mathbf{W}}^T$, where $\hat{\mathbf{V}} \in \mathbb{R}^{(n+1)\times r}$, $\hat{\mathbf{\Sigma}} \in \mathbb{R}^{r\times r}$, $\hat{\mathbf{V}} \in \mathbb{R}^{(m-1)\times r}$. Using the transformation $\mathbf{x} = \hat{\mathbf{V}}\tilde{\mathbf{x}}$, the following reduced-order approximations are computed:

$$\begin{aligned}
\tilde{\mathbf{A}} &= \hat{\mathbf{V}}^T\overline{\mathbf{A}}\hat{\mathbf{V}} = \hat{\mathbf{V}}^T\mathbf{X}_s\tilde{\mathbf{W}}\tilde{\mathbf{\Sigma}}^{-1}\tilde{\mathbf{V}}_1^T\hat{\mathbf{V}} \in \mathbb{R}^{r\times r}, \\
\tilde{\mathbf{B}} &= \hat{\mathbf{V}}^T\overline{\mathbf{B}} = \hat{\mathbf{V}}^T\mathbf{X}_s\tilde{\mathbf{W}}\tilde{\mathbf{\Sigma}}^{-1}\tilde{\mathbf{V}}_2^T \in \mathbb{R}^{r}, \\
\tilde{\mathbf{N}} &= \hat{\mathbf{V}}^T\overline{\mathbf{A}}\hat{\mathbf{V}} = \hat{\mathbf{V}}^T\mathbf{X}_s\tilde{\mathbf{W}}\tilde{\mathbf{\Sigma}}^{-1}\tilde{\mathbf{V}}_3^T\hat{\mathbf{V}} \in \mathbb{R}^{r\times r}, \\
\tilde{\mathbf{Q}} &= \hat{\mathbf{V}}^T\overline{\mathbf{Q}}(\hat{\mathbf{V}} \otimes \hat{\mathbf{V}}) = \hat{\mathbf{V}}^T\mathbf{X}_s\tilde{\mathbf{W}}\tilde{\mathbf{\Sigma}}^{-1}\tilde{\mathbf{V}}_2^T(\hat{\mathbf{V}} \otimes \hat{\mathbf{V}}) \in \mathbb{R}^{r\times r^2}.
\end{aligned}$$

References

1. Annoni, J., Gebraad, P., Seiler, P.: Wind farm flow modeling using an input-output reduced-order model. In: 2016 American Control Conference (ACC), pp. 506–512. IEEE (2016)
2. Antoulas, A.C., Gosea, I.V., Heinkenschloss, M.: On the Loewner framework for model reduction of Burgers' equation. In: King, R. (ed.) Active Flow and Combustion Control, pp. 255–270. Springer (2018)
3. Antoulas, A.C., Gosea, I.V., Ionita, A.C.: Model reduction of bilinear systems in the Loewner framework. SIAM J. Sci. Comput. **38**(5), B889–B916 (2016)
4. Benner, P., Breiten, T.: Interpolation-based $\mathcal{H}_2$-model reduction of bilinear control systems. SIAM J. Matrix Anal. Appl. **33**(3), 859–885 (2012)
5. Benner, P., Breiten, T.: Two-sided projection methods for nonlinear model order reduction. SIAM J. Sci. Comput. **37**(2), B239–B260 (2015)
6. Benner, P., Breiten, T., Damm, T.: Generalised tangential interpolation for model reduction of discrete-time mimo bilinear systems. Int. J. Control **84**(8), 1398–1407 (2011)
7. Benner, P., Goyal, P., Gugercin, S.: $\mathcal{H}_2$-quasi-optimal model order reduction for quadratic-bilinear control systems. SIAM J. Matrix Anal. Appl. **39**(2), 983–1032 (2018)
8. Benner, P., Goyal, P., Kramer, B., Peherstorfer, B., Willcox, K.: Operator inference for non-intrusive model reduction of systems with non-polynomial nonlinear terms. arXiv preprint arXiv:2002.09726 (2020)
9. Benner, P., Himpe, C., Mitchell, T.: On reduced input-output dynamic mode decomposition. Adv. Comput. Math. **44**, 1751–1768 (2018)
10. Breiten, T., Damm, T.: Krylov subspace methods for model order reduction of bilinear control systems. Syst. Control Lett. **59**(8), 443–450 (2010)
11. Chen, K.K., Tu, J.H., Rowley, C.W.: Variants of dynamic mode decomposition: boundary condition, koopman, and fourier analyses. J. Nonlinear Sci. **22**(6), 887–915 (2012)
12. Dorissen, H.T.: A method for bilinear system identification, pp. 143–148 (1990)
13. Favoreel, W., de Moor, B., Van Overschee, P.: Subspace identification of bilinear systems subject to white inputs. IEE Trans. Autom. Control **44**(6), 1157–1165 (1999)
14. Gosea, I.V., Antoulas, A.C.: Data-driven model order reduction of quadratic-bilinear systems. Numer. Linear Algebra Appl. **25**(6), e2200 (2018)
15. Gu, C.: A projection-based nonlinear model order reduction approach using quadratic-linear representation of nonlinear systems. IEEE Trans. Comput.-Aided Design Integr. Circuits Syst. **30**, 1307–1320 (2011)
16. Isidori, A., Ruberti, A.: Realization theory of bilinear systems. In: Mayne, D.Q., Brockett, R.W. (eds.) Geometric Methods in System Theory. Springer, Dordrecht (1973)
17. Karachalios, D.S., Gosea, I.V., Antoulas, A.C.: On bilinear time domain identification and reduction in the Loewner framework. In: Model Reduction of Complex Dynamical Systems, International Series of Numerical Mathematics. Springer (2020). Accepted September 2020
18. Kawano, Y., Scherpen, J.: Empirical differential balancing for nonlinear systems, pp. 6326–6331 (2017)
19. Kutz, J.N., Brunton, S.L., Brunton, B.W., Proctor, J.L.: Dynamic Mode Decomposition: Data-driven Modeling of Complex Systems. SIAM (2016)
20. Le Clainche, S., Vega, J.M.: Higher order dynamic mode decomposition. SIAM J. Appl. Dyn. Syst. **16**(2), 882–925 (2017)
21. Mayo, A.J., Antoulas, A.C.: A framework for the solution of the generalized realization problem. Linear Algebra Appl. **425**(2–3), 634–662 (2007)
22. Mezić, I.: Analysis of fluid flows via spectral properties of the koopman operator. Annu. Rev. Fluid Mech. **45**, 357–378 (2013)
23. Peherstorfer, B.: Sampling low-dimensional markovian dynamics for pre-asymptotically recovering reduced models from data with operator inference. arXiv preprint arXiv:1908.11233 (2019)
24. Peherstorfer, B., Willcox, K.: Data-driven operator inference for nonintrusive projection-based model reduction. Comput. Methods Appl. Mech. Eng. **306**, 196–215 (2016)

25. Proctor, J.L., Brunton, S.L., Kutz, J.N.: Dynamic mode decomposition with control. SIAM J. Appl. Dyn. Syst. **15**(1), 142–161 (2016)
26. Qian, E., Kramer, B., Marques, A.N., Willcox, K.: Transform & learn: a data-driven approach to nonlinear model reduction. In: AIAA Aviation 2019 Forum, p. 3707 (2019)
27. Rowley, C.W., Mezić, I., Bagheri, S., Schlatter, P., Henningson, D.S.: Spectral analysis of nonlinear flows. J. Fluid Mech. **641**, 115–127 (2009)
28. Rugh, W.J.: Nonlinear System Theory - The Volterra/Wiener Approach. University Press (1981)
29. Schmid, P.J.: Dynamic mode decomposition of numerical and experimental data. J. Fluid Mech. **656**, 5–28 (2010)
30. Tu, J.H., Luchtenburg, D.M., Rowley, C.W., Brunton, S.L., Kutz, J.N.: On dynamic mode decomposition: theory and applications. J. Comput. Dyn. **1**, 391–421 (2014)
31. Tu, J.H., Rowley, C.W., Luchtenburg, D.M., Brunton, S.L., Kutz, J.N.: On dynamic mode decomposition: theory and applications. J. Comput. Dyn. **1**, 391–421 (2014)
32. Van Overschee, P., de Moor, B.: N4SID: subspace algorithms for the identification of combined deterministic-stochastic systems. Automatica **30**(1), 75–93 (1994)
33. Zhang, L., Lam, J., Huang, B., Yang, G.H.: On gramians and balanced truncation of discrete-time bilinear systems. Int. J. Control **76**(2003), 414–427 (1999)

Clustering-Based Model Order Reduction for Nonlinear Network Systems

Peter Benner, Sara Grundel, and Petar Mlinarić

Abstract Clustering by projection has been proposed as a way to preserve network structure in linear multi-agent systems. Here, we extend this approach to a class of nonlinear network systems. Additionally, we generalize our clustering method which restores the network structure in an arbitrary reduced-order model obtained by projection. We demonstrate this method on a number of examples.

1 Introduction

Nonlinear network systems appear in various application areas, including energy distribution networks, water networks, multi-robot networks, and chemical reaction networks. Model order reduction (MOR) enables faster simulation, optimization, and control of large-scale network systems. However, standard methods generally do not preserve the network structure. Preserving the network structure is necessary, e.g., if an optimization method assumes this structure.

Clustering was proposed in the literature as a way to preserve the multi-agent structure. Methods based on equitable partitions were described in [4, 16, 24] with an extension to almost equitable partitions in [17]. Based on this, a priori error expressions were developed in [23] with generalizations in [14]. Ishizaki et al. [12] developed a clustering-based $\mathcal{H}_\infty$-MOR method based on positive tridiagonalization and reducible clusters, applicable to linear time-invariant systems with asymptotically stable and symmetric dynamics matrices. In [11], they presented an efficient clustering-based method also based on reducible clusters for $\mathcal{H}_2$-MOR of linear

P. Benner · S. Grundel · P. Mlinarić (✉)
Max Planck Institute for Dynamics of Complex Technical Systems, Sandtorstr. 1, 39106 Magdeburg, Germany
e-mail: mlinaric@mpi-magdeburg.mpg.de

P. Benner
e-mail: benner@mpi-magdeburg.mpg.de

S. Grundel
e-mail: grundel@mpi-magdeburg.mpg.de

P. Benner et al. (eds.), *Model Reduction of Complex Dynamical Systems*, International Series of Numerical Mathematics 171,
https://doi.org/10.1007/978-3-030-72983-7_4

positive networks, which include systems with Laplacian-based dynamics. Cheng et al. [5, 6] developed a method based on agent dissimilarity. Besselink et al. [3] studied networks of identical passive systems over weighted and directed graphs with tree structures.

In this work, we extend the clustering-based approach for linear time-invariant multi-agent systems from [20, 21]. There we proposed a method combining the iterative rational Krylov algorithm (IRKA) [1] and QR decomposition-based clustering [26]. We generalize this approach to be able to combine any projection-based MOR method and clustering algorithm. Extending to arbitrary projection-based MOR methods allows applying the method to nonlinear network systems. For the clustering algorithm, we motivate the use of the k-means algorithm [9]. We show that for a class of nonlinear multi-agent systems, clustering by Galerkin projection preserves network structure, which additionally avoids the need for hyper-reduction to simplify the nonlinear part.

The outline of this paper is as follows. First, we provide some background information on linear multi-agent systems in Sect. 2. In Sect. 3, we recall our clustering-based MOR method for linear multi-agent systems and generalize it to a framework which allows combining any projection-based MOR method and clustering algorithm. In Sect. 4, we extend clustering by projection to a class of nonlinear multi-agent systems, which also permits the applicability of our framework. We demonstrate the approach numerically in Sect. 5 and conclude with Sect. 6.

We use i to denote the imaginary unit ($\mathrm{i}^2 = -1$), $\mathbb{C}_-$ as the open left complex half-plane, and $\mathbb{C}_+$ as the right. Furthermore, we use $\operatorname{diag}(v)$ to denote the diagonal matrix with the vector v as its diagonal and $\operatorname{col}(v_1, v_2, \ldots, v_k)$ as the vector obtained by concatenating $v_1, v_2, \ldots, v_k$. We call a square matrix A Hurwitz if all its eigenvalues have negative real parts. Similarly, for square matrices A and B, with B invertible, we call the matrix pair (A, B) Hurwitz if $B^{-1}A$ is Hurwitz. For a rectangular matrix A, $\operatorname{im}(A)$ denotes the subspace generated by the columns of A. For a rational matrix function $H\colon \mathbb{C} \to \mathbb{C}^{p\times m}$, i.e., a matrix-valued function whose components are rational functions, the $\mathcal{H}_2$ and $\mathcal{H}_\infty$ norms are

$$\|H\|_{\mathcal{H}_2} = \left(\int_{-\infty}^{\infty} \|H(\mathrm{i}\omega)\|_{\mathrm{F}}^2 \,\mathrm{d}\omega\right)^{1/2},$$
$$\|H\|_{\mathcal{H}_\infty} = \sup_{\omega\in\mathbb{R}} \|H(\mathrm{i}\omega)\|_2,$$

if all the poles of H have negative real parts and undefined otherwise.

2 Preliminaries

We present some basic concepts from graph theory in Sect. 2.1, graph partitions in Sect. 2.2, before moving on to linear multi-agent systems in Sect. 2.3, and clustering-based MOR in Sect. 2.4. Additionally, we give remarks on MOR for non-asymptotically stable linear multi-agent systems in Sect. 2.5.

2.1 Graph Theory

The notation in this section is based on [7, 19].

A *graph* G consists of a *vertex set* V and an *edge set* E encoding the relation between vertices. *Undirected* graphs are those for which the edge set is a subset of the set of all unordered pairs of vertices, i.e., $\mathsf{E} \subseteq \{\{i, j\} : i, j \in \mathsf{V},\ i \neq j\}$. On the other hand, a graph is *directed* if $\mathsf{E} \subseteq \{(i, j) : i, j \in \mathsf{V},\ i \neq j\}$. We think of an edge (i, j) as an arrow starting from vertex i and ending at j. We only consider *simple* graphs, i.e., graphs without self-loops or multiple copies of the same edge. Additionally, we only consider *finite graphs*, i.e., graphs with a finite number of vertices $\mathsf{n} := |\mathsf{V}|$. Without loss of generality, let $\mathsf{V} = \{1, 2, \ldots, \mathsf{n}\}$.

For an undirected graph, a *path* of length ℓ is a sequence of distinct vertices $i_0, i_1, \ldots, i_\ell$ such that $\{i_k, i_{k+1}\} \in \mathsf{E}$ for $k = 0, 1, \ldots, \ell - 1$. For a directed graph, a *directed path* of length ℓ is a sequence of distinct vertices $i_0, i_1, \ldots, i_\ell$ such that $(i_k, i_{k+1}) \in \mathsf{E}$ for $k = 0, 1, \ldots, \ell - 1$. An undirected graph is *connected* if there is a path between any two distinct vertices $i, j \in \mathsf{V}$. A directed graph is *strongly connected* if there is a directed path between any two distinct vertices $i, j \in \mathsf{V}$.

We can associate weights to edges of a graph by a *weight function* $\mathsf{w}\colon \mathsf{E} \to \mathbb{R}$. If $\mathsf{w}(\mathsf{e}) > 0$ for all $\mathsf{e} \in \mathsf{E}$, the tuple $\mathsf{G} = (\mathsf{V}, \mathsf{E}, \mathsf{w})$ is called a *weighted graph*. In the following, we will focus on weighted graphs. In particular, we will directly generalize concepts for unweighted graphs from [7, 19], as was done in [23].

The *adjacency matrix* $\mathsf{A} = [\mathsf{a}_{ij}]_{i,j \in \mathsf{V}} \in \mathbb{R}^{\mathsf{n} \times \mathsf{n}}$ of an undirected weighted graph is defined component-wise by

$$\mathsf{a}_{ij} := \begin{cases} \mathsf{w}(\{i, j\}), & \text{if } \{i, j\} \in \mathsf{E}, \\ 0, & \text{otherwise}, \end{cases}$$

and for a directed weighted graph as

$$\mathsf{a}_{ij} := \begin{cases} \mathsf{w}((j, i)), & \text{if } (j, i) \in \mathsf{E}, \\ 0, & \text{otherwise}. \end{cases}$$

For every vertex $i \in \mathsf{V}$, its *in-degree* is $d_i := \sum_{j=1}^{\mathsf{n}} \mathsf{a}_{ij}$. The diagonal matrix $\mathsf{D} := \operatorname{diag}(d_1, d_2, \ldots, d_n)$ is called the *in-degree matrix*. Notice that $\mathsf{D} = \operatorname{diag}(\mathsf{A}\mathbb{1})$, where $\mathbb{1}$ is the vector of all ones.

Let $\mathsf{e}_1, \mathsf{e}_2, \ldots, \mathsf{e}_{|\mathsf{E}|}$ be all the edges of G in some order. The *incidence matrix* $\mathsf{R} \in \mathbb{R}^{\mathsf{n} \times |\mathsf{E}|}$ of a directed graph G is defined component-wise

$$[\mathsf{R}]_{ik} := \begin{cases} -1, & \text{if } \mathsf{e}_k = (i, j) \text{ for some } j \in \mathsf{V}, \\ 1, & \text{if } \mathsf{e}_k = (j, i) \text{ for some } j \in \mathsf{V}, \\ 0, & \text{otherwise.} \end{cases}$$

If G is undirected, we assign some orientation to every edge to define a directed graph G^o, and define the incidence matrix of G to be the incidence matrix of G^o. The *weight matrix* is defined as $\mathsf{W} := \operatorname{diag}(\mathsf{w}(\mathsf{e}_1), \mathsf{w}(\mathsf{e}_2), \ldots, \mathsf{w}(\mathsf{e}_{|\mathsf{E}|}))$.

The *(in-degree) Laplacian matrix* L is defined by $\mathsf{L} := \mathsf{D} - \mathsf{A}$. For undirected graphs, it can be checked that $\mathsf{L} = \mathsf{RWR}^{\mathrm{T}}$, using

$$\mathsf{RWR}^{\mathrm{T}} = \sum_{\{i,j\} \in \mathsf{E}} \mathsf{a}_{ij}(e_i - e_j)(e_i - e_j)^{\mathrm{T}},$$

which is independent of the order of edges defining R and W or the orientation of edges in G^o. From the definition of L, it directly follows that the sum of each row in L is zero, i.e., $\mathsf{L}\mathbb{1} = 0$. From $\mathsf{L} = \mathsf{RWR}^{\mathrm{T}}$, we immediately see that, for undirected weighted graphs, the Laplacian matrix L is symmetric positive semidefinite.

The following theorem, based on Theorem 2.8 in [19], states how connectedness of a graph is related to the spectral properties of L.

Theorem 1 *Let* $\mathsf{G} = (\mathsf{V}, \mathsf{E}, \mathsf{w})$ *be an undirected weighted graph,* L *its Laplacian matrix, and* $0 = \lambda_1 \leq \lambda_2 \leq \cdots \leq \lambda_{\mathsf{n}}$ *the eigenvalues of* L. *Then the following statements are equivalent:*

1. G *is connected,*
2. $\lambda_2 > 0$,
3. $\ker(\mathsf{L}) = \operatorname{im}(\mathbb{1})$.

2.2 Graph Partitions

A nonempty subset $\mathsf{C} \subseteq \mathsf{V}$ is called a *cluster* of V. A *graph partition* π of the graph G is a partition of its vertex set V. The *characteristic vector* of a cluster $\mathsf{C} \subseteq \mathsf{V}$ is the vector $p(\mathsf{C}) \in \mathbb{R}^{\mathsf{n}}$ defined with

$$[p(\mathsf{C})]_i := \begin{cases} 1 & \text{if } i \in \mathsf{C}, \\ 0 & \text{otherwise.} \end{cases}$$

The *characteristic matrix of a partition* $\pi = \{\mathsf{C}_1, \mathsf{C}_2, \ldots, \mathsf{C}_{\mathsf{r}}\}$ is the matrix $\mathsf{P} \in \mathbb{R}^{\mathsf{n} \times \mathsf{r}}$ defined by

$$\mathsf{P} := \begin{bmatrix} p(\mathsf{C}_1) & p(\mathsf{C}_2) & \cdots & p(\mathsf{C}_{\mathsf{r}}) \end{bmatrix}.$$

Note that $\mathsf{P}^{\mathsf{T}}\mathsf{P} = \operatorname{diag}(|\mathsf{C}_1|, |\mathsf{C}_2|, \ldots, |\mathsf{C}_{\mathsf{r}}|)$.

2.3 Linear Multi-agent Systems

Here, we focus on linear time-invariant multi-agent systems (cf. [3, 5, 6, 11–13, 22, 23]). Additionally, we restrict ourselves to multi-agent systems defined over an undirected, weighted, and connected graph $\mathsf{G} = (\mathsf{V}, \mathsf{E}, \mathsf{w})$.

The dynamics of the ith agent, for $i \in \mathsf{V} = \{1, 2, \ldots, \mathsf{n}\}$, is

$$\begin{aligned} E\dot{x}_i(t) &= Ax_i(t) + Bv_i(t), \\ z_i(t) &= Cx_i(t), \end{aligned}$$

with system matrices $E, A \in \mathbb{R}^{n \times n}$, input matrix $B \in \mathbb{R}^{n \times m}$, output matrix $C \in \mathbb{R}^{p \times n}$, state $x_i(t) \in \mathbb{R}^n$, input $v_i(t) \in \mathbb{R}^m$, and output $z_i(t) \in \mathbb{R}^p$. We assume the matrix E to be invertible. The interconnections are

$$\mathsf{m}_i v_i(t) = \sum_{j=1}^{\mathsf{n}} \mathsf{a}_{ij} K \left(z_j(t) - z_i(t)\right) + \sum_{k=1}^{\mathsf{m}} \mathsf{b}_{ik} u_k(t),$$

for $i = 1, 2, \ldots, \mathsf{n}$, with inertias $\mathsf{m}_i > 0$, coupling matrix $K \in \mathbb{R}^{m \times p}$, external inputs $u_k(t) \in \mathbb{R}^m$, $k = 1, 2, \ldots, \mathsf{m}$, where $\mathsf{A} = [\mathsf{a}_{ij}]$ is the adjacency matrix of the graph G. The outputs are

$$y_\ell(t) = \sum_{j=1}^{\mathsf{n}} \mathsf{c}_{\ell j} z_j(t)$$

for $\ell = 1, 2, \ldots, \mathsf{p}$. Define

$$\begin{gathered} \mathsf{M} := \operatorname{diag}(\mathsf{m}_i) \in \mathbb{R}^{\mathsf{n} \times \mathsf{n}}, \ \mathsf{B} := [\mathsf{b}_{ik}] \in \mathbb{R}^{\mathsf{n} \times \mathsf{m}}, \ \mathsf{C} := [\mathsf{c}_{\ell j}] \in \mathbb{R}^{\mathsf{p} \times \mathsf{n}}, \\ x(t) := \operatorname{col}(x_i(t)) \in \mathbb{R}^{\mathsf{n}n}, \ v(t) := \operatorname{col}(v_i(t)) \in \mathbb{R}^{\mathsf{n}m}, \ z(t) := \operatorname{col}(z_i(t)) \in \mathbb{R}^{\mathsf{n}p}, \\ u(t) := \operatorname{col}(u_k(t)) \in \mathbb{R}^{\mathsf{m}m}, \text{ and } y(t) := \operatorname{col}(y_\ell(t)) \in \mathbb{R}^{\mathsf{p}p}. \end{gathered}$$

Then the agent dynamics can be rewritten as

$$\begin{aligned} (I_\mathsf{n} \otimes E)\dot{x}(t) &= (I_\mathsf{n} \otimes A)x(t) + (I_\mathsf{n} \otimes B)v(t), \\ z(t) &= (I_\mathsf{n} \otimes C)x(t), \end{aligned}$$

interconnection as

$$(\mathsf{M} \otimes I_n)v(t) = (-\mathsf{L} \otimes K)z(t) + (\mathsf{B} \otimes I_m)u(t),$$

and output as

$$y(t) = (\mathsf{C} \otimes I_p)z(t).$$

Therefore, we have

$$\begin{aligned}(\mathsf{M} \otimes E)\dot{x}(t) &= (\mathsf{M} \otimes A - \mathsf{L} \otimes BKC)x(t) + (\mathsf{B} \otimes B)u(t),\\ y(t) &= (\mathsf{C} \otimes C)x(t).\end{aligned} \tag{1}$$

Of particular interest are *leader-follower multi-agent systems* where only some agents (*leaders*) receive external input, while other agents (*followers*) receive no inputs. Let $\mathsf{m} \in \{1, 2, \ldots, \mathsf{n}\}$ be the number of leaders, $\mathsf{V}_\mathrm{L} = \{\mathsf{v}_1, \mathsf{v}_2, \ldots, \mathsf{v}_\mathsf{m}\} \subseteq \mathsf{V}$ the set of leaders, and $\mathsf{V}_\mathrm{F} = \mathsf{V} \setminus \mathsf{V}_\mathrm{L}$ the set of followers. Then, with B defined by

$$\mathsf{b}_{ik} := \begin{cases} 1, & \text{if } i = \mathsf{v}_k,\\ 0, & \text{otherwise,}\end{cases}$$

the system (1) becomes a leader-follower multi-agent system. One important class is multi-agent systems with *single-integrator agents*, i.e., with $n = 1$, $A = 0$, and $B = C = K = E = 1$. Thus, system (1) becomes

$$\begin{aligned}\mathsf{M}\dot{x}(t) &= -\mathsf{L}x(t) + \mathsf{B}u(t),\\ y(t) &= \mathsf{C}x(t).\end{aligned} \tag{2}$$

The property of interest for multi-agent systems is *synchronization*.

Definition 1 The system (1) is *synchronized* if

$$\lim_{t \to \infty} (x_i(t) - x_j(t)) = 0,$$

for all $i, j \in \mathsf{V}$ and all initial conditions $x(0) = x_0$ and $u \equiv 0$.

In words, this means that the agents' states converge to the same trajectory for zero input and arbitrary initial condition. The following results give a characterization ([15, Theorem 1], [22, Lemma 4.2]).

Proposition 1 *Let a system* (1) *be given, where* L *is the Laplacian matrix of an undirected, weighted, and connected graph. Then the system* (1) *is synchronized if and only if* $(A - \lambda BKC, E)$ *is Hurwitz for all nonzero eigenvalues* λ *of* (L, M).

Note that linear multi-agent systems with single-integrator agents, as in (2), are always synchronized since $(A - \lambda BKC, E) = (-\lambda, 1)$.

2.4 Clustering-Based Model Order Reduction

By choosing some matrices $V, W \in \mathbb{R}^{\mathsf{n}n \times \mathsf{r}n}$, we get the reduced model for (2)

$$\begin{aligned} W^{\mathrm{T}}\mathsf{M}V\dot{\hat{x}}(t) &= -W^{\mathrm{T}}\mathsf{L}V\hat{x}(t) + W^{\mathrm{T}}\mathsf{B}u(t), \\ \hat{y}(t) &= \mathsf{C}V\hat{x}(t), \end{aligned} \tag{3}$$

or, for (1),

$$\begin{aligned} W^{\mathrm{T}}(\mathsf{M} \otimes E)V\dot{\hat{x}}(t) &= W^{\mathrm{T}}(\mathsf{M} \otimes A - \mathsf{L} \otimes BKC)V\hat{x}(t) + W^{\mathrm{T}}(\mathsf{B} \otimes B)u(t), \\ \hat{y}(t) &= (\mathsf{C} \otimes C)V\hat{x}(t), \end{aligned} \tag{4}$$

which is not necessarily a multi-agent system. As suggested in [5] (similar to [12, 23]), using

$$V = W = \mathsf{P}, \tag{5}$$

in (3), or in general

$$V = W = \mathsf{P} \otimes I_n, \tag{6}$$

in (4), preserves the structure, where P is a characteristic matrix of a partition π of the vertex set V. In particular, $\mathsf{P}^{\mathrm{T}}\mathsf{M}\mathsf{P}$ is a positive definite diagonal matrix and $\mathsf{P}^{\mathrm{T}}\mathsf{L}\mathsf{P}$ is the Laplacian matrix of the reduced graph.

2.5 Model Reduction for Non-asymptotically Stable Systems

Note that the system (2) is not (internally) asymptotically stable since L has a zero eigenvalue. Similarly, the system (1) is not asymptotically stable if A is not Hurwitz. First, we discuss a decomposition into the asymptotically and the non-asymptotically stable part. This allows an extension of MOR methods and the computation of system norms. Next, we analyze stability of clustering-based reduced models.

For an arbitrary linear time-invariant system

$$\begin{aligned} \mathcal{E}\dot{x}(t) &= \mathcal{A}x(t) + \mathcal{B}u(t), \\ y(t) &= \mathcal{C}x(t), \end{aligned}$$

with invertible $\mathcal{E}$, let $\mathcal{T} = \begin{bmatrix}\mathcal{T}_- & \mathcal{T}_+\end{bmatrix}$ and $\mathcal{S} = \begin{bmatrix}\mathcal{S}_- & \mathcal{S}_+\end{bmatrix}$ be invertible matrices such that

$$\mathcal{S}^{\mathrm{T}}\mathcal{E}\mathcal{T} = \begin{bmatrix}\mathcal{E}_- & 0 \\ 0 & \mathcal{E}_+\end{bmatrix}, \quad \mathcal{S}^{\mathrm{T}}\mathcal{A}\mathcal{T} = \begin{bmatrix}\mathcal{A}_- & 0 \\ 0 & \mathcal{A}_+\end{bmatrix},$$

with $\sigma(\mathcal{A}_-, \mathcal{E}_-) \subset \mathbb{C}_-$ and $\sigma(\mathcal{A}_+, \mathcal{E}_+) \subset \overline{\mathbb{C}_+}$. In particular, this means that $\operatorname{im}(\mathcal{T}_-)$ is a direct sum of generalized (right) eigenspaces corresponding to the eigenvalues of $(\mathcal{A}, \mathcal{E})$ with negative real parts and analogously for $\operatorname{im}(\mathcal{T}_+)$, $\operatorname{im}(\mathcal{S}_-)$, $\operatorname{im}(\mathcal{S}_+)$. If we denote

$$\mathcal{S}^{\mathrm{T}}\mathcal{B} = \begin{bmatrix} \mathcal{B}_- \\ \mathcal{B}_+ \end{bmatrix}, \quad \mathcal{C}\mathcal{T} = \begin{bmatrix} \mathcal{C}_- & \mathcal{C}_+ \end{bmatrix},$$

this gives us that $\mathcal{H} = \mathcal{H}_- + \mathcal{H}_+$, where

$$\begin{aligned} \mathcal{H}(s) &= \mathcal{C}\left(s\mathcal{E} - \mathcal{A}\right)^{-1}\mathcal{B}, \\ \mathcal{H}_-(s) &= \mathcal{C}_-\left(s\mathcal{E}_- - \mathcal{A}_-\right)^{-1}\mathcal{B}_-, \\ \mathcal{H}_+(s) &= \mathcal{C}_+\left(s\mathcal{E}_+ - \mathcal{A}_+\right)^{-1}\mathcal{B}_+. \end{aligned}$$

Note that $\mathcal{H}_-$ and $\mathcal{H}_+$ have poles in $\mathbb{C}_-$ and $\overline{\mathbb{C}_+}$, respectively. For the transfer function $\hat{\mathcal{H}}$ of a reduced model to be such that $\|\mathcal{H} - \hat{\mathcal{H}}\|_{\mathcal{H}_2}$ and $\|\mathcal{H} - \hat{\mathcal{H}}\|_{\mathcal{H}_\infty}$ are defined, it is necessary that $\mathcal{H}$ and $\hat{\mathcal{H}}$ have the same non-asymptotically stable part. This means that $\hat{\mathcal{H}} = \hat{\mathcal{H}}_- + \mathcal{H}_+$ for some $\hat{\mathcal{H}}_-$, $\hat{\mathcal{H}}_-(s) = \hat{\mathcal{C}}_-\left(s\hat{\mathcal{E}}_- - \hat{\mathcal{A}}_-\right)^{-1}\hat{\mathcal{B}}_-$, with poles in $\mathbb{C}_-$, i.e., $\hat{\mathcal{H}}_-$ is a reduced model for $\mathcal{H}_-$. If we use a projection-based MOR method with matrices $\mathcal{V}_-$, $\mathcal{W}_-$ to get

$$\hat{\mathcal{E}}_- = \mathcal{W}_-^{\mathrm{T}}\mathcal{E}_-\mathcal{V}_-, \quad \hat{\mathcal{A}}_- = \mathcal{W}_-^{\mathrm{T}}\mathcal{A}_-\mathcal{V}_-, \quad \hat{\mathcal{B}}_- = \mathcal{W}_-^{\mathrm{T}}\mathcal{B}_-, \quad \hat{\mathcal{C}}_- = \mathcal{C}_-\mathcal{V}_-,$$

then the overall basis matrices are

$$\mathcal{V} = \begin{bmatrix} \mathcal{T}_-\mathcal{V}_- & \mathcal{T}_+ \end{bmatrix} \quad \text{and} \quad \mathcal{W} = \begin{bmatrix} \mathcal{S}_-\mathcal{W}_- & \mathcal{S}_+ \end{bmatrix}. \tag{7}$$

Then we can compute the norm of $\mathcal{H} - \hat{\mathcal{H}}$ by computing the norm of $\mathcal{H}_- - \hat{\mathcal{H}}_-$.

To analyze linear multi-agent systems, we want to find an invertible matrix T such that

$$\mathsf{T}^{\mathrm{T}}\mathsf{M}\mathsf{T} = \begin{bmatrix} \mathsf{M}_- & 0 \\ 0 & \mathsf{m}_+ \end{bmatrix} \text{ and } \mathsf{T}^{\mathrm{T}}\mathsf{L}\mathsf{T} = \begin{bmatrix} \mathsf{L}_- & 0 \\ 0 & 0 \end{bmatrix},$$

where $\sigma(-\mathsf{L}_-, \mathsf{M}_-) \subset \mathbb{C}_-$. We see that if

$$\mathsf{T} = \begin{bmatrix} \mathsf{T}_- & \mathbb{1}_{\mathsf{n}} \end{bmatrix},$$

then

$$\mathsf{T}^{\mathrm{T}}\mathsf{M}\mathsf{T} = \begin{bmatrix} \mathsf{T}_-^{\mathrm{T}}\mathsf{M}\mathsf{T}_- & \mathsf{T}_-^{\mathrm{T}}\mathsf{M}\mathbb{1}_{\mathsf{n}} \\ \mathbb{1}_{\mathsf{n}}^{\mathrm{T}}\mathsf{M}\mathsf{T}_- & \mathbb{1}_{\mathsf{n}}^{\mathrm{T}}\mathsf{M}\mathbb{1}_{\mathsf{n}} \end{bmatrix} \text{ and } \mathsf{T}^{\mathrm{T}}\mathsf{L}\mathsf{T} = \begin{bmatrix} \mathsf{T}_-^{\mathrm{T}}\mathsf{L}\mathsf{T}_- & 0 \\ 0 & 0 \end{bmatrix}.$$

To have $\mathsf{T}_-^{\mathrm{T}}\mathsf{M}\mathbb{1}_{\mathsf{n}} = 0$, we need for the columns of T_- to be orthogonal to $\mathsf{M}\mathbb{1}_{\mathsf{n}}$. This will also ensure that $\sigma(-\mathsf{L}_-, \mathsf{M}_-) = \sigma(-\mathsf{T}_-^{\mathrm{T}}\mathsf{L}\mathsf{T}_-, \mathsf{T}_-^{\mathrm{T}}\mathsf{M}\mathsf{T}_-) \subset \mathbb{C}_-$. Additionally, T_- should be such that both $\mathsf{T}_-^{\mathrm{T}}\mathsf{M}\mathsf{T}_-$ and $\mathsf{T}_-^{\mathrm{T}}\mathsf{L}\mathsf{T}_-$ are sparse. We chose the form

$$\mathsf{T}_- = \begin{bmatrix} \alpha_1 & & & \\ -\beta_1 & \alpha_2 & & \\ & -\beta_2 & \ddots & \\ & & \ddots & \alpha_{\mathsf{n}-1} \\ & & & -\beta_{\mathsf{n}-1} \end{bmatrix}$$

with some $\alpha_i, \beta_i > 0, i = 1, 2, \ldots, \mathsf{n}-1$, which we determine next. From $e_i^{\mathrm{T}} \mathsf{T}_-^{\mathrm{T}} \mathsf{M} \mathbb{1}_{\mathsf{n}} = 0$, we find $\alpha_i \mathsf{m}_i = \beta_i \mathsf{m}_{i+1}$. If we additionally set $\alpha_i^2 + \beta_i^2 = 1$, we get

$$\alpha_i = \frac{\mathsf{m}_{i+1}}{\sqrt{\mathsf{m}_i^2 + \mathsf{m}_{i+1}^2}} \quad \text{and} \quad \beta_i = \frac{\mathsf{m}_i}{\sqrt{\mathsf{m}_i^2 + \mathsf{m}_{i+1}^2}}.$$

Therefore, the decomposition of a multi-agent system (2) is

$$H(s) = \mathsf{C}\mathsf{T}_-\left(s\mathsf{M}_- - \mathsf{L}_-\right)^{-1} \mathsf{T}_-^{\mathrm{T}} \mathsf{B} + \frac{1}{s\mathsf{m}_+} \mathsf{C} \mathbb{1}_{\mathsf{n}} \mathbb{1}_{\mathsf{n}}.^{\mathrm{T}} \mathsf{B}$$

Similarly, for the reduced model (3) with (5), we get that the non-asymptotically stable part is

$$\frac{1}{s \mathbb{1}_{\mathsf{r}}^{\mathrm{T}} \mathsf{P}^{\mathrm{T}} \mathsf{M} \mathsf{P} \mathbb{1}_{\mathsf{r}}} \mathsf{C} \mathsf{P} \mathbb{1}_{\mathsf{r}} \mathbb{1}_{\mathsf{r}}^{\mathrm{T}} \mathsf{P}^{\mathrm{T}} \mathsf{B},$$

which is equal to the non-asymptotically stable part of the original model since $\mathsf{P}\mathbb{1}_{\mathsf{r}} = \mathbb{1}_{\mathsf{n}}$. Therefore, the transfer function of the error system has only poles with negative real parts.

Next, to analyze system (1), let T and S be invertible matrices such that

$$S^{\mathrm{T}} E T = \begin{bmatrix} E_- & 0 \\ 0 & E_+ \end{bmatrix} \text{ and } S^{\mathrm{T}} A T = \begin{bmatrix} A_- & 0 \\ 0 & A_+ \end{bmatrix},$$

where $\sigma(A_-, E_-) \subset \mathbb{C}_-$ and $\sigma(A_+, E_+) \subset \overline{\mathbb{C}_+}$. Then

$$\begin{aligned} & \begin{bmatrix} I_{(\mathsf{n}-1)n} & 0 \\ 0 & S \end{bmatrix}^{\mathrm{T}} (\mathsf{T} \otimes I_n)^{\mathrm{T}} (\mathsf{M} \otimes E)(\mathsf{T} \otimes I_n) \begin{bmatrix} I_{(\mathsf{n}-1)n} & 0 \\ 0 & T \end{bmatrix} \\ &= \begin{bmatrix} I_{(\mathsf{n}-1)n} & 0 \\ 0 & S \end{bmatrix}^{\mathrm{T}} \begin{bmatrix} \mathsf{M}_- \otimes E & 0 \\ 0 & \mathsf{m}_+ E \end{bmatrix} \begin{bmatrix} I_{(\mathsf{n}-1)n} & 0 \\ 0 & T \end{bmatrix} \\ &= \begin{bmatrix} \mathsf{M}_- \otimes E & 0 & 0 \\ 0 & \mathsf{m}_+ E_- & 0 \\ 0 & 0 & \mathsf{m}_+ E_+ \end{bmatrix} \end{aligned}$$

and

$$\begin{bmatrix} I_{(\mathsf{n}-1)n} & 0 \\ 0 & S \end{bmatrix}^{\mathrm{T}} (\mathsf{T} \otimes I_n)^{\mathrm{T}} (\mathsf{M} \otimes A - \mathsf{L} \otimes BKC)(\mathsf{T} \otimes I_n) \begin{bmatrix} I_{(\mathsf{n}-1)n} & 0 \\ 0 & T \end{bmatrix}$$

$$= \begin{bmatrix} I_{(\mathsf{n}-1)n} & 0 \\ 0 & S \end{bmatrix}^{\mathrm{T}} \begin{bmatrix} \mathsf{M}_- \otimes A - \mathsf{L}_- \otimes BKC & 0 \\ 0 & \mathsf{m}_+ A \end{bmatrix} \begin{bmatrix} I_{(\mathsf{n}-1)n} & 0 \\ 0 & T \end{bmatrix}$$

$$= \begin{bmatrix} \mathsf{M}_- \otimes A - \mathsf{L}_- \otimes BKC & 0 & 0 \\ 0 & \mathsf{m}_+ A_- & 0 \\ 0 & 0 & \mathsf{m}_+ A_+ \end{bmatrix}.$$

Since the original system is assumed to be synchronized, we have that $\sigma(\mathsf{M}_- \otimes A - \mathsf{L}_- \otimes BKC, \mathsf{M}_- \otimes E) \subset \mathbb{C}_-$. Note that the overall transformation matrices are

$$\begin{aligned} \mathcal{T} &= (\mathsf{T} \otimes I_n) \begin{bmatrix} I_{(\mathsf{n}-1)n} & 0 \\ 0 & T \end{bmatrix} = \left(\begin{bmatrix} \mathsf{T}_- & \mathbb{1}_\mathsf{n} \end{bmatrix} \otimes I_n \right) \begin{bmatrix} I_{(\mathsf{n}-1)n} & 0 \\ 0 & T \end{bmatrix} \\ &= \begin{bmatrix} \mathsf{T}_- \otimes I_n & \mathbb{1}_\mathsf{n} \otimes I_n \end{bmatrix} \begin{bmatrix} I_{(\mathsf{n}-1)n} & 0 \\ 0 & T \end{bmatrix} = \begin{bmatrix} \mathsf{T}_- \otimes I_n & (\mathbb{1}_\mathsf{n} \otimes I_n) T \end{bmatrix} \\ &= \begin{bmatrix} \mathsf{T}_- \otimes I_n & \mathbb{1}_\mathsf{n} \otimes T \end{bmatrix} = \begin{bmatrix} \mathsf{T}_- \otimes I_n & \mathbb{1}_\mathsf{n} \otimes T_- & \mathbb{1}_\mathsf{n} \otimes T_+ \end{bmatrix}, \\ \mathcal{S} &= \begin{bmatrix} \mathsf{T}_- \otimes I_n & \mathbb{1}_\mathsf{n} \otimes S_- & \mathbb{1}_\mathsf{n} \otimes S_+ \end{bmatrix}. \end{aligned}$$

Therefore, the non-asymptotically stable part is

$$\begin{aligned} &(\mathsf{C} \otimes C)(\mathbb{1}_\mathsf{n} \otimes T_+)\left(s\mathsf{m}_+ E_+ - \mathsf{m}_+ A_+\right)^{-1} (\mathbb{1}_\mathsf{n} \otimes S_+)^{\mathrm{T}} (\mathsf{B} \otimes B) \\ &= \left(\mathsf{C}\mathbb{1}_\mathsf{n} \otimes C T_+\right)\left(s\mathsf{m}_+ E_+ - \mathsf{m}_+ A_+\right)^{-1} \left(\mathbb{1}_\mathsf{n}^{\mathrm{T}} \mathsf{B} \otimes S_+^{\mathrm{T}} B\right). \end{aligned}$$

Assuming that a clustering-based reduced model is synchronized, we see that it has the same non-asymptotically stable part as the original model.

It remains to consider synchronization preservation. In the single-integrator case, clustering using any partition preserves synchronization. In the general case, using Theorem 1, we need that $(A - \hat{\lambda} BKC, E)$ is Hurwitz for all nonzero eigenvalues $\hat{\lambda}$ of $(\hat{\mathsf{L}}, \hat{\mathsf{M}})$. Since, in general, $\sigma(\hat{\mathsf{L}}, \hat{\mathsf{M}})$ is not a subset of $\sigma(\mathsf{L}, \mathsf{M})$, we need an additional assumption. Based on the interlacing property [8], we know that all nonzero eigenvalues $(\hat{\mathsf{L}}, \hat{\mathsf{M}})$ are in $[\lambda_2, \lambda_\mathsf{n}]$, where $0 = \lambda_1 < \lambda_2 \leq \cdots \leq \lambda_\mathsf{n}$ are the eigenvalues of (L, M). Therefore, if $(A - \lambda BKC, E)$ is Hurwitz for all $\lambda \in [\lambda_2, \lambda_\mathsf{n}]$, we get that every partition preserves synchronization.

3 Clustering for Linear Multi-agent Systems

In this section, we motivate our general approach for clustering-based linear multi-agent systems. Since clustering is generally a difficult combinatorial problem (see, e.g., [25]), we propose a heuristic approach for finding suboptimal partitions.

For simplicity, we first consider multi-agent systems with single-integrator agents as in (2). Let

$$H(s) = \mathsf{C}(s\mathsf{M} + \mathsf{L})^{-1}\mathsf{B},$$
$$\hat{H}(s) = \mathsf{C}V\left(sW^{\mathrm{T}}\mathsf{M}V + W^{\mathrm{T}}\mathsf{L}V\right)^{-1}W^{\mathrm{T}}\mathsf{B}$$

be the transfer functions of systems (2) and (3), respectively, where $V, W \in \mathbb{R}^{\mathsf{n} \times r_{\mathsf{P}}}$ are obtained using a projection-based method such as balanced truncation or IRKA and the construction in (7).

In [20], motivated by (5) and the properties of clustering using QR decomposition with column pivoting (see [26, Sect. 3], [20, Lemma 1]), we proposed applying it to the set of rows of V or W to recover the partition. Here, we want to emphasize that the approach is not restricted to this choice of clustering algorithm. In particular, the following result on the forward error in the Petrov-Galerkin projection [2, Theorem 3.3] motivates using the k-means clustering [9].

Theorem 2 *Let $V_1, V_2, W_1, W_2 \in \mathbb{R}^{\mathsf{n} \times r_{\mathsf{P}}}$ be full-rank matrices and*

$$\mathcal{V}_i = \operatorname{im}(V_i), \quad \mathcal{W}_i = \operatorname{im}(W_i), \quad \hat{H}_i(s) = CV_i\left(sW_i^{\mathrm{T}}EV_i - W_i^{\mathrm{T}}AV_i\right)^{-1}W_i^{\mathrm{T}}B,$$

for $i = 1, 2$. Then

$$\frac{\|\hat{H}_1 - \hat{H}_2\|_{\mathcal{H}_\infty}}{\frac{1}{2}\left(\|\hat{H}_1\|_{\mathcal{H}_\infty} + \|\hat{H}_2\|_{\mathcal{H}_\infty}\right)} \le M \max(\sin\Theta(\mathcal{V}_1, \mathcal{V}_2), \sin\Theta(\mathcal{W}_1, \mathcal{W}_2)),$$

where

$$M = 2\max(M_1, M_2),$$
$$M_1 = \frac{\max_{\omega\in\mathbb{R}} \|C\|_2 \left\| V_1(\mathrm{i}\omega W_1^{\mathrm{T}}EV_1 - W_1^{\mathrm{T}}AV_1)^{-1}W_1^{\mathrm{T}}B\right\|_2 \|\hat{H}_1(\mathrm{i}\omega)\|_2^{-1}}{\min_{\omega\in\mathbb{R}} \cos\Theta\left(\ker\left(W_2^{\mathrm{T}}(\mathrm{i}\omega E - A)^{-1}\right)^{\perp}, \mathcal{V}_2\right)},$$
$$M_2 = \frac{\max_{\omega\in\mathbb{R}} \left\| CV_2(\mathrm{i}\omega W_2^{\mathrm{T}}EV_2 - W_2^{\mathrm{T}}AV_2)^{-1}W_2^{\mathrm{T}}\right\|_2 \|B\|_2 \|\hat{H}_2(\mathrm{i}\omega)\|_2^{-1}}{\min_{\omega\in\mathbb{R}} \cos\Theta\left(\operatorname{im}\left((\mathrm{i}\omega E - A)^{-1}V_1\right), \mathcal{W}_1\right)},$$

and $\Theta(\mathcal{M}, \mathcal{N})$ is the largest principal angle between subspaces $\mathcal{M}, \mathcal{N} \subseteq \mathbb{R}^n$.

The motivation for looking at this bound is, if we take $\hat{H}_1$ to be a projection-based reduced-order model that is very close to the original model, i.e., $\|H - \hat{H}_1\|_{\mathcal{H}_\infty}$ is small, then we could look for a clustering-based reduced-order model $\hat{H}_2$ by finding a characteristic matrix of a partition P such that $\operatorname{im}(\mathsf{P})$ is close to $\mathcal{V}_1$ and $\mathcal{W}_1$.

Note that, to use Theorem 2, we need to use the asymptotically stable parts of $\hat{H}_1$ and $\hat{H}_2$ such that $\|\hat{H}_1\|_{\mathcal{H}_\infty}$ and $\|\hat{H}_2\|_{\mathcal{H}_\infty}$ are defined. As discussed in Sect. 2.5, $\hat{H}_1$

and $\hat{H}_2$ need to have the same non-asymptotically stable part as H for the $\mathcal{H}_2$ and $\mathcal{H}_\infty$ errors to be defined.

Next, we show how the bounds motivate the use of k-means clustering. The sine of the largest principal angle between two subspaces $\mathcal{V}_1, \mathcal{V}_2 \subseteq \mathbb{R}^n$ is defined by (see [2, Sect. 3.1])

$$\sin\Theta(\mathcal{V}_1, \mathcal{V}_2) := \sup_{v_1 \in \mathcal{V}_1} \inf_{v_2 \in \mathcal{V}_2} \frac{\|v_2 - v_1\|_2}{\|v_1\|_2}.$$

Furthermore, if Π_1 and Π_2 are orthogonal projectors onto $\mathcal{V}_1$ and $\mathcal{V}_2$, then $\sin\Theta(\mathcal{V}_1, \mathcal{V}_2) = \|(I - \Pi_2)\Pi_1\|_2$. Therefore, we have

$$\sin\Theta(\mathcal{V}_1, \mathcal{V}_2) = \left\|\left(I - V_2 V_2^{\mathrm{T}}\right) V_1\right\|_2,$$

for any V_1 and V_2 with orthonormal columns such that $\mathcal{V}_1 = \operatorname{im}(V_1)$, $\mathcal{V}_2 = \operatorname{im}(V_2)$. If additionally V_1 is the $V \in \mathbb{R}^{\mathsf{n} \times r_P}$ from the projection-based method and $V_2 = \mathsf{P}(\mathsf{P}^{\mathrm{T}}\mathsf{P})^{-1/2} \in \mathbb{R}^{\mathsf{n} \times \mathsf{r}}$, then

$$\begin{aligned}
(\sin\Theta(\mathcal{V}_1, \mathcal{V}_2))^2 &\le \left\|\left(I - \mathsf{P}(\mathsf{P}^{\mathrm{T}}\mathsf{P})^{-1}\mathsf{P}^{\mathrm{T}}\right) V\right\|_{\mathrm{F}}^2 \\
&= \left\|\left(I - \begin{bmatrix} p(\mathsf{C}_1) & \cdots & p(\mathsf{C}_\mathsf{r}) \end{bmatrix} \begin{bmatrix} |\mathsf{C}_1|^{-1} & & \\ & \ddots & \\ & & |\mathsf{C}_\mathsf{r}|^{-1} \end{bmatrix} \begin{bmatrix} p(\mathsf{C}_1)^{\mathrm{T}} \\ \vdots \\ p(\mathsf{C}_\mathsf{r})^{\mathrm{T}} \end{bmatrix}\right) V\right\|_{\mathrm{F}}^2 \\
&= \left\|\left(I - \sum_{i=1}^{\mathsf{r}} \frac{1}{|\mathsf{C}_i|} p(\mathsf{C}_i) p(\mathsf{C}_i)^{\mathrm{T}}\right) V\right\|_{\mathrm{F}}^2 \\
&= \left\|\sum_{i=1}^{\mathsf{r}} \sum_{p \in \mathsf{C}_i} \left(e_p e_p^{\mathrm{T}} - \frac{1}{|\mathsf{C}_i|} e_p p(\mathsf{C}_i)^{\mathrm{T}}\right) V\right\|_{\mathrm{F}}^2 \\
&= \sum_{i=1}^{\mathsf{r}} \sum_{p \in \mathsf{C}_i} \left\|V_{p,:} - \frac{1}{|\mathsf{C}_i|} \sum_{q \in \mathsf{C}_i} V_{q,:}\right\|_2^2,
\end{aligned}$$

which is equal to the k-means cost functional for the set of rows of V, where $V_{p,:}$ is the pth row of V (and similarly for $V_{q,:}$). Therefore, applying the k-means algorithm to the rows of V will minimize an upper bound on the largest principal angle between $\operatorname{im}(V)$ and $\operatorname{im}(\mathsf{P})$.

The advantage of using k-means compared to QR decomposition-based clustering is in that the latter can only, given $V \in \mathbb{R}^{\mathsf{n} \times r_P}$, return a partition with r_P clusters. On the other hand, k-means clustering can return a partition with any number of clusters r. This makes it more efficient when $r_P \ll \mathsf{r}$ and projection-based MOR method already generate a good subspace $\operatorname{im}(V)$.

For multi-agent systems (1) with agents of order n, we have the matrices V and W as in (6). QR decomposition-based clustering can then be extended as in Algorithm 2 from [21] by clustering the block columns of V^{T} (or W^{T}). For the k-means algorithm, we can show in a similar way as in the single-integrator case that clustering the block rows leads to minimizing an upper bound of the largest principal angle. Therefore, k-means can be directly applied to the set of block rows of V or W.

Note that the approach is not limited to the two clustering algorithms mentioned here. Any clustering algorithm over the set of (block-)rows of V or W can be used. In particular, if only partitions with certain properties are wanted (e.g., those that only cluster neighboring agents), then special clustering algorithms could be used (e.g., agglomerative clustering taking into account the connectivity of the graph).

4 Clustering for Nonlinear Multi-agent Systems

In this section, we extend the approach from the previous section to a class of nonlinear multi-agent systems. We describe the class of multi-agent systems in Sect. 4.1. Next, in Sect. 4.2, we show that clustering by projection preserves structure for this class of systems.

4.1 Nonlinear Multi-agent Systems

Here, we consider a class of nonlinear multi-agent systems. In particular, let the dynamics of the ith agent, for $i = 1, 2, \ldots, \mathsf{n}$, be defined by the control-affine system

$$\dot{x}_i(t) = A(x_i(t)) + B(x_i(t))v_i(t), \tag{8a}$$

$$z_i(t) = C(x_i(t)), \tag{8b}$$

with functions $A\colon \mathbb{R}^n \to \mathbb{R}^n$, $B\colon \mathbb{R}^n \to \mathbb{R}^{n\times m}$, $C\colon \mathbb{R}^n \to \mathbb{R}^p$, state $x_i(t) \in \mathbb{R}^n$, input $v_i(t) \in \mathbb{R}^m$, and output $z_i(t) \in \mathbb{R}^p$. Furthermore, let the interconnections be

$$\mathsf{m}_i v_i(t) = \sum_{j=1}^{\mathsf{n}} \mathsf{a}_{ij} K(z_i(t), z_j(t)) + \sum_{k=1}^{\mathsf{m}} \mathsf{b}_{ik} u_k(t), \tag{8c}$$

for $i=1, 2, \ldots, \mathsf{n}$, with inertias $\mathsf{m}_i>0$ and $\mathsf{M}=\operatorname{diag}(\mathsf{m}_i)$, coupling $K\colon \mathbb{R}^p \times \mathbb{R}^p \to \mathbb{R}^m$, external input $u_k(t) \in \mathbb{R}^m$, $k = 1, 2, \ldots, \mathsf{m}$, where $\mathsf{A} = [\mathsf{a}_{ij}]$ is the adjacency matrix of the graph G, and $\mathsf{B} = [\mathsf{b}_{ik}]$. Additionally, let the external output be

$$y_\ell(t) = \sum_{j=1}^{\mathsf{n}} \mathsf{c}_{\ell j} z_j(t), \tag{8d}$$

with $\mathsf{C} = [\mathsf{c}_{\ell j}]$. We assume functions A, B, C, K are continuous and that there is a unique global solution $x(t) = \mathrm{col}(x_1(t), x_2(t), \ldots, x_\mathsf{n}(t))$ for any admissible $u(t)$.

4.2 Clustering by Projection

We want to find the form of the reduced-order model obtained from Galerkin projection with $V = \mathsf{P} \otimes I_n$. We can rewrite (8) to

$$
\begin{aligned}
(\mathsf{M} \otimes I_n)\dot{x}(t) &= f(x(t), u(t)), \\
y(t) &= g(x(t)),
\end{aligned}
$$

for some functions f and g. The reduced model is

$$
\begin{aligned}
(\mathsf{P}^{\mathrm{T}}\mathsf{M}\mathsf{P} \otimes I_n)\dot{\hat{x}}(t) &= (\mathsf{P}^{\mathrm{T}} \otimes I_n) f\big((\mathsf{P} \otimes I_n)\hat{x}(t), u(t)\big), \\
\hat{y}(t) &= g\big((\mathsf{P} \otimes I_n)\hat{x}(t)\big),
\end{aligned}
\tag{9}
$$

with $\hat{x}(t) = \mathrm{col}(\hat{x}_1(t), \hat{x}_2(t), \ldots, \hat{x}_\mathsf{r}(t))$ and $\hat{x}_i(t) \in \mathbb{R}^n$. Let $\pi(j) \in \{1, 2, \ldots, \mathsf{r}\}$ be such that $j \in \mathsf{C}_{\pi(j)}$, for $j \in \{1, 2, \ldots, \mathsf{n}\}$. Premultiplying (9) with $e_\imath^{\mathrm{T}} \otimes I_n$ for some $\imath \in \{1, 2, \ldots, \mathsf{r}\}$, we find

$$
\begin{aligned}
&\hat{\mathsf{m}}_\imath \dot{\hat{x}}_\imath(t) \\
&= \sum_{i \in \mathsf{C}_\imath} \left(\mathsf{m}_i A(\hat{x}_\imath(t)) + B(\hat{x}_\imath(t)) \left(\sum_{j=1}^{\mathsf{n}} \mathsf{a}_{ij} K\big(C(\hat{x}_\imath(t)), C(\hat{x}_{\pi(j)}(t))\big) + \sum_{k=1}^{\mathsf{m}} \mathsf{b}_{ik} u_k(t) \right) \right) \\
&= \hat{\mathsf{m}}_\imath A(\hat{x}_\imath(t)) \\
&\quad + B(\hat{x}_\imath(t)) \left(\sum_{i \in \mathsf{C}_\imath} \sum_{j=1}^{\mathsf{n}} \mathsf{a}_{ij} K\big(C(\hat{x}_\imath(t)), C(\hat{x}_{\pi(j)}(t))\big) + \sum_{i \in \mathsf{C}_\imath} \sum_{k=1}^{\mathsf{m}} \mathsf{b}_{ik} u_k(t) \right) \\
&= \hat{\mathsf{m}}_\imath A(\hat{x}_\imath(t)) \\
&\quad + B(\hat{x}_\imath(t)) \left(\sum_{\jmath=1}^{\mathsf{r}} \sum_{i \in \mathsf{C}_\imath} \sum_{j \in \mathsf{C}_\jmath} \mathsf{a}_{ij} K\big(C(\hat{x}_\imath(t)), C(\hat{x}_\jmath(t))\big) + \sum_{k=1}^{\mathsf{m}} \sum_{i \in \mathsf{C}_\imath} \mathsf{b}_{ik} u_k(t) \right) \\
&= \hat{\mathsf{m}}_\imath A(\hat{x}_\imath(t)) + B(\hat{x}_\imath(t)) \left(\sum_{\jmath=1}^{\mathsf{r}} \hat{\mathsf{a}}_{\imath\jmath} K\big(C(\hat{x}_\imath(t)), C(\hat{x}_\jmath(t))\big) + \sum_{k=1}^{\mathsf{m}} \hat{\mathsf{b}}_{\imath k} u_k(t) \right),
\end{aligned}
$$

for

$$
\hat{\mathsf{m}}_\imath = \sum_{i \in \mathsf{C}_\imath} \mathsf{m}_i, \quad \hat{\mathsf{a}}_{\imath\jmath} = \sum_{i \in \mathsf{C}_\imath} \sum_{j \in \mathsf{C}_\jmath} \mathsf{a}_{ij}, \quad \hat{\mathsf{b}}_{\imath k} = \sum_{i \in \mathsf{C}_\imath} \mathsf{b}_{ik}.
$$

Defining $\hat{\mathsf{M}} := \operatorname{diag}(\hat{\mathsf{m}}_\imath)$, $\hat{\mathsf{A}} := [\hat{\mathsf{a}}_{\imath\jmath}]$, and $\hat{\mathsf{B}} := [\hat{\mathsf{b}}_{\imath k}]$, we see that $\hat{\mathsf{M}} = \mathsf{P}^{\mathrm{T}}\mathsf{M}\mathsf{P}$, $\hat{\mathsf{A}} = \mathsf{P}^{\mathrm{T}}\mathsf{A}\mathsf{P}$, and $\hat{\mathsf{B}} = \mathsf{P}^{\mathrm{T}}\mathsf{B}$. For the output, we have

$$\hat{y}_\ell(t) = \sum_{j=1}^{\mathsf{n}} \mathsf{c}_{\ell j} C\big(\hat{x}_{\pi(j)}(t)\big) = \sum_{\jmath=1}^{\mathsf{r}} \sum_{j \in \mathsf{C}_\jmath} \mathsf{c}_{\ell j} C\big(\hat{x}_\jmath(t)\big) = \sum_{\jmath=1}^{\mathsf{r}} \hat{\mathsf{c}}_{\ell\jmath} C\big(\hat{x}_\jmath(t)\big),$$

where

$$\hat{\mathsf{c}}_{\ell\jmath} = \sum_{j \in \mathsf{C}_\jmath} \mathsf{c}_{\ell j}.$$

Thus, for $\hat{\mathsf{C}} := [\hat{\mathsf{c}}_{\ell\jmath}]$, we have $\hat{\mathsf{C}} = \mathsf{C}\mathsf{P}$. Therefore, we showed how to construct a reduced model of the same structure as the original multi-agent system. Based on this, to find a good partition, we can apply any projection-based MOR method for nonlinear systems (e.g., proper orthogonal decomposition [10]) and cluster the block rows of the matrix used to project the system.

5 Numerical Examples

Here, we demonstrate our approach for different network examples, beginning with a small linear multi-agent system in Sect. 5.1. Next, in Sect. 5.2, we use the van der Pol oscillator network.

The source code of the implementations used to compute the presented results can be obtained from

https://doi.org/10.5281/zenodo.3924653

and is authored by Petar Mlinarić.

5.1 Small Network Example

To illustrate distance to optimality, we use the leader-follower multi-agents system example from [23] with 10 single-integrator agents shown in Fig. 1, where we can compute the $\mathcal{H}_2$ and $\mathcal{H}_\infty$ errors for all possible partitions. The Laplacian and input matrices are

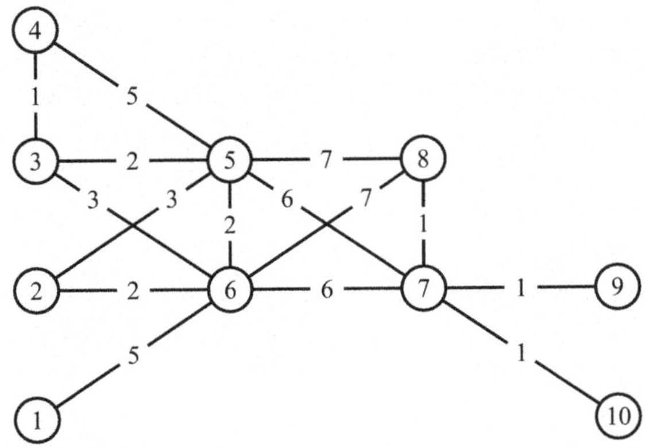

Fig. 1 Undirected weighted graph with 10 vertices from [23]

$$
\mathsf{L} = \begin{bmatrix}
5 & 0 & 0 & 0 & 0 & -5 & 0 & 0 & 0 & 0 \\
0 & 5 & 0 & 0 & -3 & -2 & 0 & 0 & 0 & 0 \\
0 & 0 & 6 & -1 & -2 & -3 & 0 & 0 & 0 & 0 \\
0 & 0 & -1 & 6 & -5 & 0 & 0 & 0 & 0 & 0 \\
0 & -3 & -2 & -5 & 25 & -2 & -6 & -7 & 0 & 0 \\
-5 & -2 & -3 & 0 & -2 & 25 & -6 & -7 & 0 & 0 \\
0 & 0 & 0 & 0 & -6 & -6 & 15 & -1 & -1 & -1 \\
0 & 0 & 0 & 0 & -7 & -7 & -1 & 15 & 0 & 0 \\
0 & 0 & 0 & 0 & 0 & 0 & -1 & 0 & 1 & 0 \\
0 & 0 & 0 & 0 & 0 & 0 & -1 & 0 & 0 & 1
\end{bmatrix}, \quad
\mathsf{B} = \begin{bmatrix}
0 & 0 \\ 0 & 0 \\ 0 & 0 \\ 0 & 0 \\ 0 & 0 \\ 1 & 0 \\ 0 & 1 \\ 0 & 0 \\ 0 & 0 \\ 0 & 0
\end{bmatrix},
$$

and we chose the edge ordering and orientation such that the incidence and edge-weights matrices are

$$
\mathsf{R} = \begin{bmatrix}
-1 & 0 & 0 & 0 & 0 & 0 & 0 & 0 & 0 & 0 & 0 & 0 & 0 & 0 & 0 \\
0 & -1 & -1 & 0 & 0 & 0 & 0 & 0 & 0 & 0 & 0 & 0 & 0 & 0 & 0 \\
0 & 0 & 0 & -1 & -1 & -1 & 0 & 0 & 0 & 0 & 0 & 0 & 0 & 0 & 0 \\
0 & 0 & 0 & 1 & 0 & 0 & -1 & 0 & 0 & 0 & 0 & 0 & 0 & 0 & 0 \\
0 & 1 & 0 & 0 & 1 & 0 & 1 & -1 & -1 & -1 & 0 & 0 & 0 & 0 & 0 \\
1 & 0 & 1 & 0 & 0 & 1 & 0 & 1 & 0 & 0 & -1 & -1 & 0 & 0 & 0 \\
0 & 0 & 0 & 0 & 0 & 0 & 0 & 0 & 1 & 0 & 1 & 0 & -1 & -1 & -1 \\
0 & 0 & 0 & 0 & 0 & 0 & 0 & 0 & 0 & 1 & 0 & 1 & 1 & 0 & 0 \\
0 & 0 & 0 & 0 & 0 & 0 & 0 & 0 & 0 & 0 & 0 & 0 & 0 & 1 & 0 \\
0 & 0 & 0 & 0 & 0 & 0 & 0 & 0 & 0 & 0 & 0 & 0 & 0 & 0 & 1
\end{bmatrix}
$$

and $\mathsf{W} = \operatorname{diag}(5, 3, 2, 1, 2, 3, 5, 2, 6, 7, 6, 7, 1, 1, 1)$, respectively. The output matrix is $\mathsf{C} = \mathsf{W}^{1/2}\mathsf{R}^{\mathsf{T}}$.

For this example, we focus on partitions with five clusters. There are in total 42 525 such partitions. Table 1 shows the 15 best partitions with respect to the $\mathcal{H}_2$ and $\mathcal{H}_\infty$ errors.

First, we used IRKA to find a reduced model of order $\mathsf{r} = 5$. It found a reduced model with relative $\mathcal{H}_2$ error of 3.30412×10^{-2}, which is 3.88 times better than the best partition. The partition resulting from QR decomposition-based clustering applied to IRKA's V matrix is

$$
\{\{1, 3\}, \{2, 4, 9, 10\}, \{5, 8\}, \{6\}, \{7\}\},
$$

with the associated relative $\mathcal{H}_2$ error of 0.150654. It is more than four times worse than using IRKA, but note that this partition is the 14th best partition and that the best partition produces

Table 1 Top 15 partitions with 5 clusters by $\mathcal{H}_2$ error and $\mathcal{H}_\infty$ error for reducing the multi-agent system in Sect. 5.1

Rank	Relative $\mathcal{H}_2$ error	Partition
1	0.128053	$\{\{1,8\},\{2,3,4,9,10\},\{5\},\{6\},\{7\}\}$
2	0.131311	$\{\{1,2,3,4\},\{5,8\},\{6\},\{7\},\{9,10\}\}$
3	0.137466	$\{\{1,2,3,4,9,10\},\{5\},\{6\},\{7\},\{8\}\}$
4	0.137473	$\{\{1,3,8\},\{2,4,9,10\},\{5\},\{6\},\{7\}\}$
5	0.143700	$\{\{1,5,8\},\{2,3,4\},\{6\},\{7\},\{9,10\}\}$
6	0.145900	$\{\{1,2,3\},\{4,9,10\},\{5,8\},\{6\},\{7\}\}$
7	0.146196	$\{\{1,8\},\{2,3,4,9\},\{5,10\},\{6\},\{7\}\}$
8	0.146196	$\{\{1,8\},\{2,3,4,10\},\{5,9\},\{6\},\{7\}\}$
9	0.147022	$\{\{1,2,3,8\},\{4,9,10\},\{5\},\{6\},\{7\}\}$
10	0.149240	$\{\{1,8,10\},\{2,3,4,9\},\{5\},\{6\},\{7\}\}$
11	0.149240	$\{\{1,8,9\},\{2,3,4,10\},\{5\},\{6\},\{7\}\}$
12	0.149654	$\{\{1,8\},\{2,4,9,10\},\{3,5\},\{6\},\{7\}\}$
13	0.150440	$\{\{1,5\},\{2,3,4,9,10\},\{6\},\{7\},\{8\}\}$
14	0.150654	$\{\{1,3\},\{2,4,9,10\},\{5,8\},\{6\},\{7\}\}$
15	0.151684	$\{\{1,2,8\},\{3,4,9,10\},\{5\},\{6\},\{7\}\}$

Rank	Relative $\mathcal{H}_\infty$ error	Partition
1	0.253975	$\{\{1,3,5,8\},\{2,4\},\{6\},\{7\},\{9,10\}\}$
2	0.254376	$\{\{1,2,5,8\},\{3,4\},\{6\},\{7\},\{9,10\}\}$
3	0.254818	$\{\{1,5,8\},\{2,3,4\},\{6\},\{7\},\{9,10\}\}$
4	0.259483	$\{\{1,2,3,5,8\},\{4\},\{6\},\{7\},\{9,10\}\}$
5	0.260859	$\{\{1,2,4\},\{3,5,8\},\{6\},\{7\},\{9,10\}\}$
6	0.262244	$\{\{1,2,3,4\},\{5,8\},\{6\},\{7\},\{9,10\}\}$
7	0.266387	$\{\{1,3,4\},\{2,5,8\},\{6\},\{7\},\{9,10\}\}$
8	0.273663	$\{\{1,4\},\{2,3,5,8\},\{6\},\{7\},\{9,10\}\}$
9	0.276919	$\{\{1,4,5,8\},\{2,3\},\{6\},\{7\},\{9,10\}\}$
10	0.286961	$\{\{1,3,4,5,8\},\{2\},\{6\},\{7\},\{9,10\}\}$
11	0.288414	$\{\{1,2,3\},\{4,5,8\},\{6\},\{7\},\{9,10\}\}$
12	0.293773	$\{\{1,5\},\{2,3,4,8\},\{6\},\{7\},\{9,10\}\}$
13	0.294028	$\{\{1,2,3,4,8\},\{5\},\{6\},\{7\},\{9,10\}\}$
14	0.299845	$\{\{1,2\},\{3,4,5,8\},\{6\},\{7\},\{9,10\}\}$
15	0.305583	$\{\{1,2,4,8\},\{3,5\},\{6\},\{7\},\{9,10\}\}$

about 1.18 times better error. Using k-means clustering gives

$$\{\{1,2,3\},\{4,9,10\},\{5,8\},\{6\},\{7\}\},$$

with relative $\mathcal{H}_2$ error of 0.1459 and taking the sixth place.

We notice by (5) that W can also be used to find a good partition. In this example, QR decomposition-based clustering returns the partition

$$\{\{1,2,3,9,10\},\{4,8\},\{5\},\{6\},\{7\}\},$$

with the relative $\mathcal{H}_2$ error 0.179746, which is worse than using only V from IRKA. Using k-means clustering returns

$$\{\{1, 2, 3, 4, 8\}, \{5\}, \{6\}, \{7\}, \{9, 10\}\}$$

with the relative $\mathcal{H}_2$ error 0.156788.

Using the first five left singular vectors of $\begin{bmatrix} V & W \end{bmatrix}$ to take into account both V and W, using QR decomposition-based clustering produces

$$\{\{1, 2, 3, 4, 8\}, \{5\}, \{6\}, \{7\}, \{9, 10\}\}$$

with the relative $\mathcal{H}_2$ error 0.189487, which further increases the error. On the other hand, k-means clustering gives us

$$\{\{1, 2, 3, 4\}, \{5, 8\}, \{6\}, \{7\}, \{9, 10\}\},$$

which is the second best partition in terms of the $\mathcal{H}_2$ error and sixth best in terms of the $\mathcal{H}_\infty$ error. Furthermore, using balanced truncation instead of IRKA produces the same partition, using either of the two clustering algorithms and the three choices of matrices.

Therefore, at least in this example, clustering the rows of V and/or W gives close to optimal partitions. Additionally, k-means clustering performs better than QR decomposition-based clustering.

5.2 van der Pol Oscillators

Here, we use the van der Pol oscillator network example from [18], where the agents are given by

$$\dot{x}_{i,1}(t) = x_{i,2}(t) + \sigma v_i(t), \tag{10a}$$

$$\dot{x}_{i,2}(t) = \mu\big(1 - x_{i,1}(t)^2\big)x_{i,2}(t) - x_{i,1}(t) - c v_i(t), \tag{10b}$$

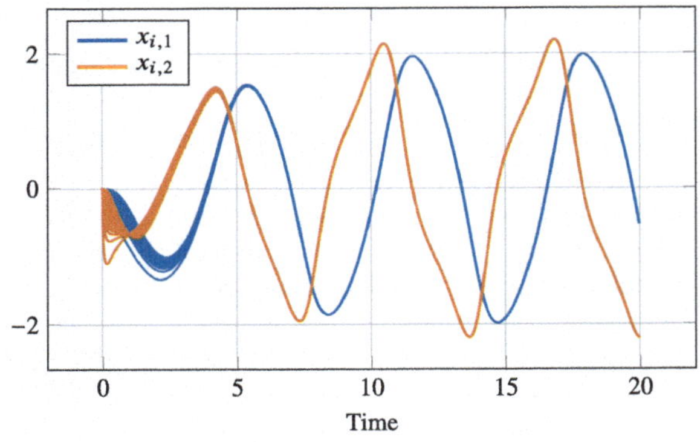

Fig. 2 State trajectory of the van der Pol oscillator network (10) for zero initial condition and input $u(t) = e^{-t}$

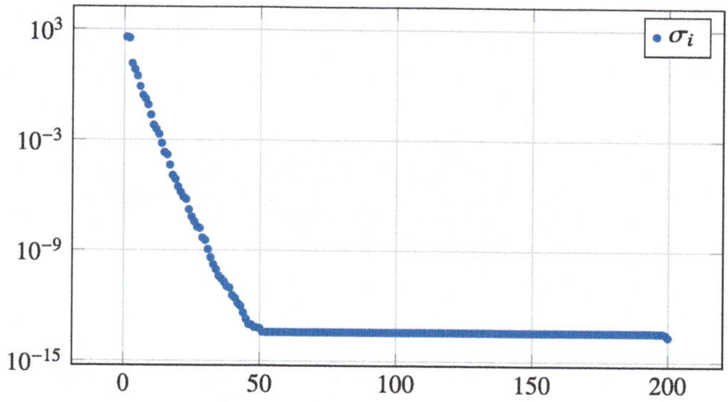

Fig. 3 POD singular values based on snapshots from Fig. 2

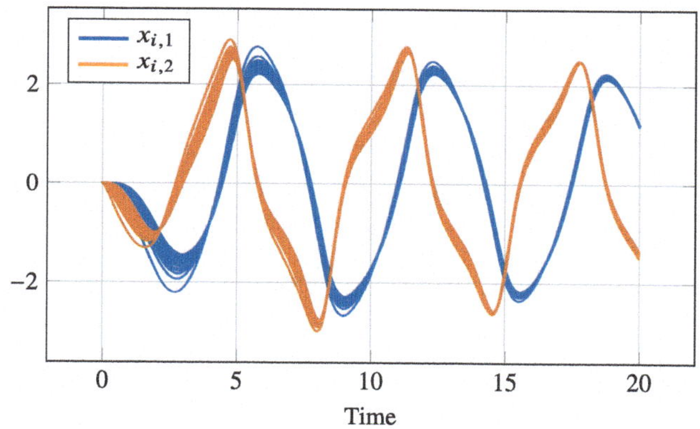

Fig. 4 van der Pol oscillator state trajectory for zero initial condition and input $u(t) = e^{-t/10} \sin t$

and interconnections by

$$v_i(t) = \sum_{j=1}^{\mathsf{n}} \mathsf{a}_{ij}\big((x_{i,1}(t) - x_{j,1}(t)) + (x_{i,2}(t) - x_{j,2}(t))\big) + \sum_{k=1}^{\mathsf{m}} \mathsf{b}_{ik} u_k(t), \tag{10c}$$

with $\mu = 0.5$ and $\sigma = 0.1$. Additionally, we chose a larger 10×10 grid graph ($\mathsf{n} = 100$), set the input matrix to be $\mathsf{B} = e_1$ (i.e., one of the corner agents receives external input), and used $c = 100$ to have synchronization.

Figure 2 shows the state trajectory of the system for zero initial condition and input $u(t) = e^{-t}$, using an adaptive BDF integrator producing 987 snapshots. We used these snapshots to find the POD modes, with associated singular values shown in Fig. 3.

Changing the input to $u(t) = e^{-t/10} \sin t$ gives the trajectory in Fig. 4. Applying k-means clustering to the first two POD modes to generate 10 clusters produces a reduced model with the error trajectory in Fig. 5. We computed the relative $\mathcal{L}_2$ error for k-means clustering using the first two POD modes with different number of clustering, which can be seen in Fig. 6. For this example, we see that the error decays exponentially with the order of the reduced model.

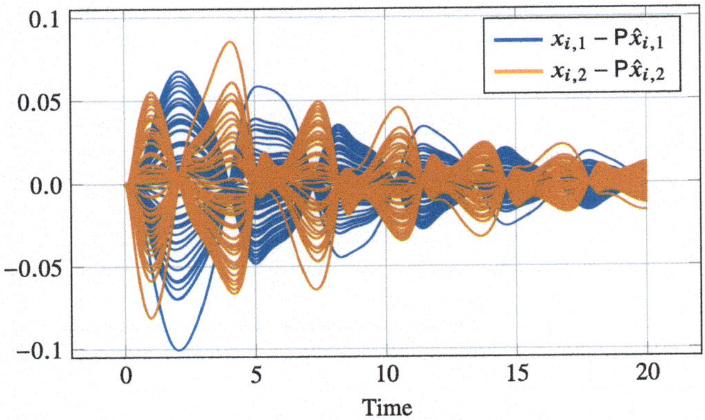

Fig. 5 van der Pol oscillator error when using k-means with the first two POD modes for zero initial condition and input $u(t) = e^{-t/10}\sin t$

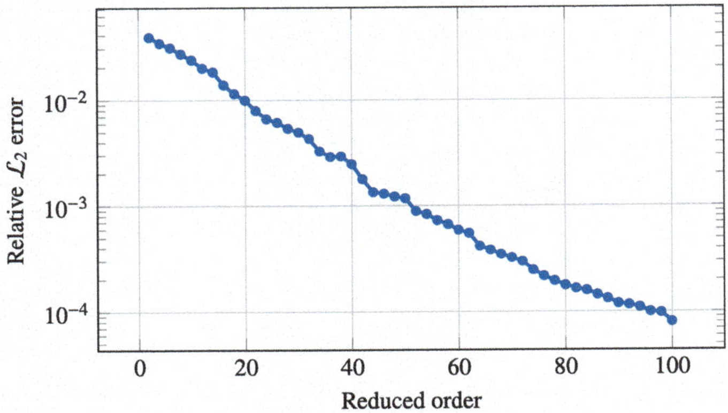

Fig. 6 Relative $\mathcal{L}_2$ error for zero initial condition and test input $u(t) = e^{-t/10}\sin t$ for k-means clustering using the first two POD modes

6 Conclusions

We extended clustering by projection to a class of nonlinear multi-agent systems and presented our clustering-based MOR method, combining any projection-based MOR method and a clustering algorithm, for reduction of multi-agent systems using graph partitions. In particular, we motivated the use of the k-means algorithm.

Our numerical test for a small network shows that our algorithm finds close to optimal partitions. We also illustrated our method for a larger nonlinear oscillator network.

Acknowledgements This work was partially funded by the Deutsche Forschungsgemeinschaft (DFG, German Research Foundation) under the project SA 3477/1-1.

References

1. Antoulas, A.C., Beattie, C.A., Gugercin, S.: Interpolatory model reduction of large-scale dynamical systems. In: Mohammadpour, J., Grigoriadis, K.M. (eds.) Efficient Modeling and Control of Large-Scale Systems, pp. 3–58. Springer, US (2010). https://doi.org/10.1007/978-1-4419-5757-3_1
2. Beattie, C.A., Gugercin, S., Wyatt, S.: Inexact solves in interpolatory model reduction. Linear Algebra Appl. **436**(8), 2916–2943 (2012). https://doi.org/10.1016/j.laa.2011.07.015
3. Besselink, B., Sandberg, H., Johansson, K.H.: Clustering-based model reduction of networked passive systems. IEEE Trans. Autom. Control **61**(10), 2958–2973 (2016). https://doi.org/10.1109/TAC.2015.2505418
4. Chapman, A., Mesbahi, M.: UAV flocking with wind gusts: adaptive topology and model reduction. In: American Control Conference (ACC), pp. 1045–1050 (2011). https://doi.org/10.1109/ACC.2011.5990799
5. Cheng, X., Kawano, Y., Scherpen, J.M.A.: Graph structure-preserving model reduction of linear network systems. In: European Control Conference (ECC), pp. 1970–1975 (2016). https://doi.org/10.1109/ECC.2016.7810580
6. Cheng, X., Kawano, Y., Scherpen, J.M.A.: Model reduction of multi-agent systems using dissimilarity-based clustering. IEEE Trans. Autom. Control (2018). https://doi.org/10.1109/TAC.2018.2853578
7. Godsil, C., Royle, G.: Algebraic graph theory. Graduate Texts in Mathematics, vol. 207. Springer, New York (2001). https://doi.org/10.1007/978-1-4613-0163-9
8. Golub, G.H., Van Loan, C.F.: Matrix Computations, 4th edn. Johns Hopkins Studies in the Mathematical Sciences. Johns Hopkins University Press, Baltimore (2013)
9. Hartigan, J.A., Wong, M.A.: Algorithm AS 136: a k-means clustering algorithm. J. R. Stat. Soc. Ser. C (Appl. Stat.) **28**(1), 100–108 (1979). https://doi.org/10.2307/2346830
10. Hinze, M., Volkwein, S.: Proper orthogonal decomposition surrogate models for nonlinear dynamical systems: error estimates and suboptimal control. In: Benner, P., Mehrmann, V., Sorensen, D. (eds.) Dimension Reduction of Large-Scale Systems, Lect. Notes Comput. Sci. Eng., vol. 45, pp. 261–306. Springer, Berlin/Heidelberg, Germany (2005)
11. Ishizaki, T., Kashima, K., Girard, A., Imura, J., Chen, L., Aihara, K.: Clustered model reduction of positive directed networks. Automatica J. IFAC **59**, 238–247 (2015). https://doi.org/10.1016/j.automatica.2015.06.027
12. Ishizaki, T., Kashima, K., Imura, J., Aihara, K.: Model reduction and clusterization of large-scale bidirectional networks. IEEE Trans. Autom. Control **59**(1), 48–63 (2014). https://doi.org/10.1109/TAC.2013.2275891
13. Ishizaki, T., Ku, R., Imura, J.: Clustered model reduction of networked dissipative systems. In: American Control Conference (ACC), pp. 3662–3667 (2016). https://doi.org/10.1109/ACC.2016.7525482
14. Jongsma, H.J., Mlinarić, P., Grundel, S., Benner, P., Trentelman, H.L.: Model reduction of linear multi-agent systems by clustering with $\mathcal{H}_2$ and $\mathcal{H}_\infty$ error bounds. Math. Control Signals Syst. **30**(6) (2018). https://doi.org/10.1007/s00498-018-0212-6
15. Li, Z., Duan, Z., Chen, G., Huang, L.: Consensus of multiagent systems and synchronization of complex networks: a unified viewpoint. IEEE Trans. Circuits Syst. I, Regular Papers **57**(1), 213–224 (2010). https://doi.org/10.1109/TCSI.2009.2023937
16. Martini, S., Egerstedt, M., Bicchi, A.: Controllability decompositions of networked systems through quotient graphs. In: 47th IEEE Conference on Decision and Control (CDC), pp. 5244–5249 (2008). https://doi.org/10.1109/CDC.2008.4739213
17. Martini, S., Egerstedt, M., Bicchi, A.: Controllability analysis of multi-agent systems using relaxed equitable partitions. Int. J. Syst., Control Commun. **2**(1/2/3), 100–121 (2010). https://doi.org/10.1504/IJSCC.2010.031160
18. Massioni, P., Scorletti, G.: Consensus analysis of large-scale nonlinear homogeneous multi-agent formations with polynomial dynamics. Internat. J. Robust Nonlinear Control **28**(17), 5605–5617 (2018). https://doi.org/10.1002/rnc.4334

19. Mesbahi, M., Egerstedt, M.: Graph Theoretic Methods in Multiagent Networks. Princeton Series in Applied Mathematics. Princeton University Press, Princeton, NJ (2010). https://doi.org/10.1515/9781400835355
20. Mlinarić, P., Grundel, S., Benner, P.: Efficient model order reduction for multi-agent systems using QR decomposition-based clustering. In: 54th IEEE Conference on Decision and Control (CDC), pp. 4794–4799 (2015). https://doi.org/10.1109/CDC.2015.7402967
21. Mlinarić, P., Grundel, S., Benner, P.: Clustering-based model order reduction for multi-agent systems with general linear time-invariant agents. In: 22nd International Symposium on Mathematical Theory of Networks and Systems (MTNS), pp. 230–235. Minneapolis, MN, USA (2016). http://hdl.handle.net/11299/181518
22. Monshizadeh, N., Trentelman, H.L., Camlibel, M.K.: Stability and synchronization preserving model reduction of multi-agent systems. Syst. Control Lett. **62**(1), 1–10 (2013). https://doi.org/10.1016/j.sysconle.2012.10.011
23. Monshizadeh, N., Trentelman, H.L., Camlibel, M.K.: Projection-based model reduction of multi-agent systems using graph partitions. IEEE Trans. Control Netw. Syst. **1**(2), 145–154 (2014). https://doi.org/10.1109/TCNS.2014.2311883
24. Rahmani, A., Ji, M., Mesbahi, M., Egerstedt, M.: Controllability of multi-agent systems from a graph-theoretic perspective. SIAM J. Control Optim. **48**(1), 162–186 (2009). https://doi.org/10.1137/060674909
25. Schaeffer, S.E.: Graph clustering. Comput. Sci. Rev. **1**(1), 27–64 (2007). https://doi.org/10.1016/j.cosrev.2007.05.001
26. Zha, H., He, X., Ding, C., Simon, H., Gu, M.: Spectral relaxation for k-means clustering. In: Proceedings of the 14th International Conference on Neural Information Processing Systems: Natural and Synthetic, pp. 1057–1064 (2001). https://papers.nips.cc/paper/1992-spectral-relaxation-for-k-means-clustering.pdf

Adaptive Interpolatory MOR by Learning the Error Estimator in the Parameter Domain

Sridhar Chellappa, Lihong Feng, Valentín de la Rubia, and Peter Benner

Abstract Interpolatory methods offer a powerful framework for generating reduced-order models (ROMs) for non-parametric or parametric systems with time-varying inputs. Choosing the interpolation points adaptively remains an area of active interest. A greedy framework has been introduced in [12, 14] to choose interpolation points automatically using a posteriori error estimators. Nevertheless, when the parameter range is large or if the parameter space dimension is larger than two, the greedy algorithm may take considerable time, since the training set needs to include a considerable number of parameters. As a remedy, we introduce an adaptive training technique by learning an efficient a posteriori error estimator over the parameter domain. A fast learning process is created by interpolating the error estimator using radial basis functions (RBF) over a fine parameter training set, representing the whole parameter domain. The error estimator is evaluated only on a coarse training set including a few parameter samples. The algorithm is an extension of the work in [9] to interpolatory model order reduction (MOR) in frequency domain. Beyond the work in [9], we use a newly proposed inf-sup-constant-free error estimator in the frequency domain [14], which is often much tighter than the error estimator using the inf-sup constant. Three numerical examples demonstrate the efficiency and validity of the proposed approach.

S. Chellappa (✉) · L. Feng · P. Benner
Max Planck Institute for Dynamics of Complex Technical Systems, Sandtorstraße 1, 39106 Magdeburg, Germany
e-mail: chellappa@mpi-magdeburg.mpg.de

L. Feng
e-mail: feng@mpi-magdeburg.mpg.de

P. Benner
e-mail: benner@mpi-magdeburg.mpg.de

V. de la Rubia
Departamento de Matemática Aplicada a las TIC, ETSI de Telecomunicación, Universidad Politécnica de Madrid, 28040 Madrid, Spain
e-mail: valentin.delarubia@upm.es

P. Benner et al. (eds.), *Model Reduction of Complex Dynamical Systems*, International Series of Numerical Mathematics 171,
https://doi.org/10.1007/978-3-030-72983-7_5

1 Introduction

MOR based on system theory and interpolation [3, 12, 13, 16, 17] has been developed as a class of efficient MOR methods among others. Detailed summary of those methods and comparison of them with other classes of MOR methods can be found in some survey papers and books [1, 2, 4, 7, 8, 18, 30]. A major advantage of the interpolatory methods is their flexibility in reducing systems with time- or parameter-varying inputs, since they are based on the transfer function or the input-output relation of the systems, which is independent of the input signal. On the contrary, the snapshot MOR methods, such as proper orthogonal decomposition (POD) and the reduced basis methods (RBM) are input dependent, and are often less efficient in reducing systems with varying inputs as compared with the interpolatory MOR methods [7].

A major topic of interest in interpolatory MOR methods is how to determine the interpolation points, so as to adaptively construct the ROM. Many methods have appeared in the last 10 years, some are heuristic [5, 15, 21, 23], some entail high computational complexity [12, 33], and some are inefficient for systems with more than one parameter [17, 22]. Random interpolation points are used in [6].

Recently, a new error estimator for the reduced transfer function error and an algorithm for iteratively choosing the interpolation points are proposed in [14], which overcomes many difficulties being faced by the above-mentioned interpolatory methods. It is neither heuristic nor needs a high computational cost. Moreover, it is a parametric MOR method and applicable to systems with more than two parameters. One shortcoming of the method is that the interpolation points are selected from a given training set, which must be decided a priori and becomes larger and larger with the increase of the parameter range or the parameter space dimension. Such a technique is standard also for the RBM, where a training set must be given before a greedy algorithm starts. This makes the greedy algorithm slow down when there is a large number of samples in the training set due to the large dimension or large range of the parameter domain. This is due to the fact that at each iteration of the greedy algorithm, an error estimator needs to be repeatedly computed for all the samples in the training set. Many adaptive training techniques have been proposed recently for RBM [9, 19, 20, 24]. In contrast, no efficient training techniques are proposed for the interpolatory MOR methods, though similar greedy algorithms using fixed training sets are proposed in [12, 14]. In this work, we extend the adaptive training technique in [9] for RBM to an adaptive training technique for the interpolatory MOR methods in [12, 14].

The main contribution of this work is an efficient algorithm to adaptively choose interpolation points for parametric, linear time-invariant (LTI) systems having a wide range of parameter values or with a large parameter space dimension. Compared with the greedy algorithms proposed in [12, 14], we have added two new ingredients to the greedy algorithms: (i) a surrogate for the error estimator, which can be cheaply computed and (ii) an adaptive sampling approach using the surrogate estimator.

The aim is that instead of computing the error estimator over the whole parameter domain for the ROM construction, a surrogate estimator is computed. In this way, the error estimator is computed only on a coarse training set at each iteration of the greedy algorithm, and for parameters outside the coarse training, the surrogate estimator is computed. Finally, the training set needs to be initialized by including only a few parameter samples, and can be iteratively updated using the surrogate estimator instead of the error estimator itself. As a consequence, a significant amount of computational cost can be saved for such systems.

The idea is similar to the one in [9] for the RBM. However, in [9], an error estimator in time domain is used, where the inf-sup constant needs to be computed for each parameter in the training set, which is computationally inefficient for large-scale systems. In this work, we use an inf-sup-constant-free error estimator newly proposed in [14]. It is suitable for interpolatory MOR methods, since it estimates the transfer function error in the frequency domain.

The paper is organized as follows. In Sect. 2, we briefly review interpolatory MOR methods based on projection. The greedy interpolatory methods [12, 14] for parametric systems are reviewed in Sect. 3. In Sect. 4, we introduce the basic idea of RBF interpolation and elaborate on the process of learning the error estimator using a surrogate estimator constructed by RBF interpolation. Based on this surrogate estimator, we propose the greedy algorithm IPSUE with adaptive training technique for adaptively choosing the interpolation points in a more efficient and fully adaptive way. We present numerical results on three real-word examples in Sect. 5 to show the robustness of IPSUE and conclude the work in the end.

2 Interpolatory MOR

In this work, we are interested in MOR of parametric LTI systems in the *state-space representation* given by

$$\Sigma(\boldsymbol{\mu}) : \begin{cases} \mathbf{E}\dot{\mathbf{x}}(t, \boldsymbol{\mu}) = \mathbf{A}(\boldsymbol{\mu})\mathbf{x}(t, \boldsymbol{\mu}) + \mathbf{B}(\boldsymbol{\mu})\mathbf{u}(t), \\ \mathbf{y}(t, \boldsymbol{\mu}) = \mathbf{C}(\boldsymbol{\mu})\mathbf{x}(t, \boldsymbol{\mu}), \;\; \mathbf{x}(0, \boldsymbol{\mu}) = 0. \end{cases} \tag{1}$$

Here, $\boldsymbol{\mu} := \left[\mu^1, \mu^2, \ldots, \mu^d\right]^\mathsf{T} \in \mathbb{R}^d$ is the vector of parameters (geometric or physical). $\mathbf{x}(t, \boldsymbol{\mu}) \in \mathbb{R}^n$ is the state vector and n is typically very large. $\mathbf{u}(t) \in \mathbb{R}^m$ is the input vector and $\mathbf{y}(t, \boldsymbol{\mu}) \in \mathbb{R}^p$ is the output vector. $\mathbf{A}(\boldsymbol{\mu}) \in \mathbb{R}^{n\times n}$ is the state matrix, $\mathbf{B}(\boldsymbol{\mu}) \in \mathbb{R}^{n\times m}$ is the input matrix, and $\mathbf{C}(\boldsymbol{\mu}) \in \mathbb{R}^{p\times n}$ is the output matrix. For the case when $m = p = 1$, Eq. (1) is referred to as a single-input, single-output (SISO) system. Otherwise, it is known as multi-input, multi-output (MIMO) system.

The ROM we seek should preserve the same structure of the FOM but have a much smaller dimension. We assume that the state vector lies (approximately) in the span of a low-dimensional linear subspace $\mathcal{V} \subset \mathbb{R}^{n\times r}$, $r \ll n$, such that $\mathbf{x}(t, \boldsymbol{\mu}) \approx$

$\mathbf{V}\hat{\mathbf{x}}(t, \boldsymbol{\mu})$. Column vectors in the matrix $\mathbf{V} \in \mathbb{R}^{n\times r}$ constitute an orthogonal basis of $\mathcal{V}$. Replacing $\mathbf{x}(t, \boldsymbol{\mu})$ in Eq. (1) with its approximation $\mathbf{V}\hat{\mathbf{x}}(t, \boldsymbol{\mu})$ and further imposing Petrov-Galerkin projection on the residual introduced by the approximation in a test subspace $\mathcal{W} \subset \mathbb{R}^{n\times r}$ leads to

$$\mathbf{W}^{\mathsf{T}}\left(\mathbf{V}\dot{\hat{\mathbf{x}}}(t, \boldsymbol{\mu}) - \mathbf{A}(\boldsymbol{\mu})\mathbf{V}\hat{\mathbf{x}}(t, \boldsymbol{\mu}) - \mathbf{B}(\boldsymbol{\mu})\mathbf{u}(t)\right) \equiv \mathbf{0},$$

where the column vectors in the matrix $\mathbf{W} \in \mathbb{R}^{n\times r}$ correspond to an orthogonal basis of $\mathcal{W}$. The resulting ROM is given as

$$\hat{\Sigma}(\boldsymbol{\mu}) : \begin{cases} \hat{\mathbf{E}}\dot{\hat{\mathbf{x}}}(t, \boldsymbol{\mu}) = \hat{\mathbf{A}}(\boldsymbol{\mu})\hat{\mathbf{x}}(t, \boldsymbol{\mu}) + \hat{\mathbf{B}}(\boldsymbol{\mu})\mathbf{u}(t), \\ \hat{\mathbf{y}}(t, \boldsymbol{\mu}) = \hat{\mathbf{C}}(\boldsymbol{\mu})\hat{\mathbf{x}}(t, \boldsymbol{\mu}), \;\; \hat{\mathbf{x}}(0, \boldsymbol{\mu}) = 0. \end{cases} \tag{2}$$

Here, $\hat{\mathbf{x}}(t, \boldsymbol{\mu}) \in \mathbb{R}^r$ is the reduced state vector, $\hat{\mathbf{E}}(\boldsymbol{\mu}) = \mathbf{W}^{\mathsf{T}}\mathbf{E}(\boldsymbol{\mu})\mathbf{V} \in \mathbb{R}^{r\times r}$, $\hat{\mathbf{A}}(\boldsymbol{\mu}) = \mathbf{W}^{\mathsf{T}}\mathbf{A}(\boldsymbol{\mu})\mathbf{V} \in \mathbb{R}^{r\times r}$, $\hat{\mathbf{B}}(\boldsymbol{\mu}) = \mathbf{W}^{\mathsf{T}}\mathbf{B}(\boldsymbol{\mu}) \in \mathbb{R}^{r\times m}$, $\hat{\mathbf{C}}(\boldsymbol{\mu}) = \mathbf{C}(\boldsymbol{\mu})\mathbf{V} \in \mathbb{R}^{p\times r}$ are the reduced system matrices, and $\hat{\mathbf{y}}(t, \boldsymbol{\mu})$ is the reduced output vector. The goal of MOR is to find the two subspaces $\mathcal{V}, \mathcal{W} \in \mathbb{R}^{n\times r}$. Different MOR methods vary in how they generate the matrices $\mathbf{W}, \mathbf{V}$.

Interpolatory MOR methods construct $\mathbf{V}, \mathbf{W} \in \mathbb{R}^{n\times r}$ based on the transfer function of the system, which is independent of the input signal. The transfer function of the system described in Eq. (1) is given by

$$\mathbf{H}(\tilde{\boldsymbol{\mu}}) := \mathbf{C}(\boldsymbol{\mu})\big(\overbrace{s\mathbf{E} - \mathbf{A}(\boldsymbol{\mu})}^{=:\mathscr{A}(\tilde{\boldsymbol{\mu}})}\big)^{-1}\mathbf{B}(\boldsymbol{\mu}). \tag{3}$$

Here, $\tilde{\boldsymbol{\mu}} := \left[s, \mu^1, \mu^2, \ldots, \mu^d\right]^{\mathsf{T}} \in \mathbb{R}^{d+1}$ is the vector of parameters with the additional Laplace variable $s \in \jmath\mathbb{R}$, where $\jmath$ is the imaginary unit. The corresponding ROM of Eq. (3) is of the form

$$\hat{\mathbf{H}}(\tilde{\boldsymbol{\mu}}) := \hat{\mathbf{C}}(\boldsymbol{\mu})\hat{\mathscr{A}}(\tilde{\boldsymbol{\mu}})^{-1}\hat{\mathbf{B}}(\boldsymbol{\mu}), \tag{4}$$

with $\hat{\mathscr{A}}(\tilde{\boldsymbol{\mu}}) := s\hat{\mathbf{E}} - \hat{\mathbf{A}}(\boldsymbol{\mu})$.

Many interpolatory methods have been proposed for linear systems, especially for linear non-parametric systems. The most representative methods are the moment-matching methods [16, 17], where the $\mathcal{H}_2$-optimal method IRKA [17] constructs a ROM satisfying the necessary conditions of local optimality. All these methods are known to be applicable to non-parametric systems. Later, IRKA is extended to MOR for parametric systems [3], where some pairs of projection matrices are constructed for given samples of parameters, then they are combined together to get the final pair of projection matrices. No rule is used for selecting the samples. In [22], a method for parametric systems is proposed based on $\mathcal{H}_2 \otimes \mathcal{L}_2$-optimality, but is only applicable to systems with one parameter and is facing high computational complexity for systems with $n \geq 1000$.

Choosing interpolation points using a greedy algorithm guided by an a posteriori error bound is proposed in [12]. However, computing the error estimator needs to compute the smallest singular values of the large matrix $\mathscr{A}(\tilde{\mu})$, the inf-sup constant. An inf-sup-constant-free error estimator is newly proposed in [14], which can be efficiently computed, and is also much tighter than the error bound [12] for many systems with small inf-sup constants. A similar greedy algorithm is proposed in [14] for choosing the interpolation points using the new error estimator. The adaptive training approach proposed in this work is based on the greedy algorithm and the new error estimator in [14]. In the next section, we briefly review the error estimator and the corresponding greedy algorithm.

3 Greedy Method for Choosing Interpolation Points

The transfer function can be seen as a mapping from the space of inputs $\mathbb{R}^m$ to the space of outputs $\mathbb{R}^p$ passing through a high-dimensional intermediate state in $\mathbb{R}^n$. If we look at the matrix product $\mathscr{A}^{-1}(\tilde{\mu})\mathbf{B}(\mu)$ in $\mathbf{H}(\mu)$, we may consider the primal system

$$\mathscr{A}(\tilde{\mu})\mathbf{X}_{\text{pr}}(\tilde{\mu}) = \mathbf{B}(\mu). \tag{5}$$

Here, $\mathbf{X}_{\text{pr}}(\tilde{\mu}) \in \mathbb{R}^n$ is the primal state vector. The reduced primal system is defined as

$$\hat{\mathscr{A}}(\tilde{\mu})\hat{\mathbf{X}}_{\text{pr}}(\tilde{\mu}) = \hat{\mathbf{B}}(\mu). \tag{6}$$

The approximate primal solution is given by $\tilde{\mathbf{X}}_{\text{pr}}(\tilde{\mu}) := \mathbf{V}\hat{\mathbf{X}}_{\text{pr}}(\tilde{\mu})$ and the corresponding residual is

$$\mathbf{r}_{\text{pr}}(\tilde{\mu}) = \mathbf{B}(\mu) - \mathscr{A}(\tilde{\mu})\tilde{\mathbf{X}}_{\text{pr}}(\tilde{\mu}). \tag{7}$$

Additionally, by considering the matrix product $\mathbf{C}(\mu)\mathscr{A}^{-1}(\tilde{\mu})$ in $\mathbf{H}(\mu)$, we have the following dual system:

$$\mathscr{A}^{\mathsf{T}}(\tilde{\mu})\mathbf{X}_{\text{du}}(\tilde{\mu}) = \mathbf{C}^{\mathsf{T}}(\mu). \tag{8}$$

$\mathbf{X}_{\text{du}}(\tilde{\mu}) \in \mathbb{R}^n$ is the dual state vector. The reduced dual system is given as

$$\hat{\mathscr{A}}^{\mathsf{T}}(\tilde{\mu})\hat{\mathbf{X}}_{\text{du}}(\tilde{\mu}) = \hat{\mathbf{C}}^{\mathsf{T}}(\mu). \tag{9}$$

The approximate dual solution is given by $\tilde{\mathbf{X}}_{\text{du}}(\tilde{\mu}) := \mathbf{V}_{\text{du}}\hat{\mathbf{X}}_{\text{du}}(\tilde{\mu})$ and the corresponding residual is

$$\mathbf{r}_{\text{du}}(\tilde{\mu}) = \mathbf{C}^{\mathsf{T}}(\mu) - \mathscr{A}^{\mathsf{T}}(\tilde{\mu})\tilde{\mathbf{X}}_{\text{du}}(\tilde{\mu}). \tag{10}$$

For parametric LTI systems, adopting the spirit of the RBM, [12] introduced a method to automatically generate a ROM through a greedy algorithm. The authors introduce a primal-dual residual-based a posteriori error estimator for the transfer function approximation error $\|\mathbf{H}(\tilde{\boldsymbol{\mu}}) - \hat{\mathbf{H}}(\tilde{\boldsymbol{\mu}})\|$, for both SISO and MIMO systems. For SISO systems, it reads

$$|\mathbf{H}(\tilde{\boldsymbol{\mu}}) - \hat{\mathbf{H}}(\tilde{\boldsymbol{\mu}})| \leq \frac{\|\mathbf{r}_{\mathrm{pr}}(\tilde{\boldsymbol{\mu}})\|_2 \|\mathbf{r}_{\mathrm{du}}(\tilde{\boldsymbol{\mu}})\|_2}{\sigma_{\min}(\tilde{\boldsymbol{\mu}})}. \tag{11}$$

Here, $\sigma_{\min}(\tilde{\boldsymbol{\mu}})$, called the inf-sup constant, is the smallest singular value of the matrix $\mathscr{A}(\tilde{\boldsymbol{\mu}})$ as defined in Eq. (3). The primal and dual residuals are given in Eq. (7) and Eq. (10), respectively. The work [14] improves the method in [12] by avoiding the calculation of the inf-sup constant required for the error estimator. This is achieved by introducing a dual-residual system

$$\mathscr{A}^{\mathsf{T}}(\tilde{\boldsymbol{\mu}})\mathbf{e}_{\mathrm{du}}(\tilde{\boldsymbol{\mu}}) = \mathbf{r}_{\mathrm{du}}(\tilde{\boldsymbol{\mu}}). \tag{12}$$

The following proposition from [14] gives the a posteriori error bound.

Proposition 1 *The transfer function approximation error can be bounded as*

$$|\mathbf{H}(\tilde{\boldsymbol{\mu}}) - \hat{\mathbf{H}}(\tilde{\boldsymbol{\mu}})| \leq \left|\tilde{\mathbf{X}}_{du}^T(\tilde{\boldsymbol{\mu}})\mathbf{r}_{pr}(\tilde{\boldsymbol{\mu}})\right| + \left|\mathbf{e}_{du}^T(\tilde{\boldsymbol{\mu}})\mathbf{r}_{pr}(\tilde{\boldsymbol{\mu}})\right|.$$

For a proof of Proposition 1, we refer to [14]. In this form, the error bound is not computationally efficient since one needs to solve the full-order dual-residual system (12) to obtain $\mathbf{e}_{\mathrm{du}}(\tilde{\boldsymbol{\mu}})$. Instead, system (12) is reduced by an orthogonal matrix $\mathbf{V}_e \in \mathbb{R}^{n\times\ell}$ to obtain

$$\hat{\mathscr{A}}_e^{\mathsf{T}}(\tilde{\boldsymbol{\mu}})\hat{\mathbf{e}}_{\mathrm{du}}(\tilde{\boldsymbol{\mu}}) = \hat{\mathbf{r}}_{\mathrm{du,e}}(\tilde{\boldsymbol{\mu}}), \tag{13}$$

where $\hat{\mathscr{A}}_e(\tilde{\boldsymbol{\mu}}) := \mathbf{V}_e^{\mathsf{T}} \mathscr{A}(\tilde{\boldsymbol{\mu}}) \mathbf{V}_e$ and $\hat{\mathbf{r}}_{\mathrm{du,e}}(\tilde{\boldsymbol{\mu}}) := \mathbf{V}_e^{\mathsf{T}} \mathbf{r}_{\mathrm{du}}(\tilde{\boldsymbol{\mu}})$. The projection matrices $\mathbf{V}$, $\mathbf{V}_{\mathrm{du}}$ and $\mathbf{V}_e$ corresponding to the primal, dual, and the dual-residual system are generated offline.

By using the approximate solution to the dual-residual system, an efficiently computable error estimator is obtained as

$$|\mathbf{H}(\tilde{\boldsymbol{\mu}}) - \hat{\mathbf{H}}(\tilde{\boldsymbol{\mu}})| \lesssim \left|\tilde{\mathbf{X}}_{\mathrm{du}}^{\mathsf{T}}(\tilde{\boldsymbol{\mu}})\mathbf{r}_{\mathrm{pr}}(\tilde{\boldsymbol{\mu}})\right| + \left|\tilde{\mathbf{e}}_{\mathrm{du}}^{\mathsf{T}}(\tilde{\boldsymbol{\mu}})\mathbf{r}_{\mathrm{pr}}(\tilde{\boldsymbol{\mu}})\right| =: \Delta(\tilde{\boldsymbol{\mu}}), \tag{14}$$

where $\tilde{\mathbf{e}}_{\mathrm{du}}(\tilde{\boldsymbol{\mu}}) := \mathbf{V}_e \hat{\mathbf{e}}_{\mathrm{du}}(\tilde{\boldsymbol{\mu}})$. For ease of comparison, we first present the greedy algorithm for parametric systems introduced in [14] as Algorithm 1.

Algorithm 1 Greedy ROM Construction for Parametric Systems [14]

Input: System matrices $\mathbf{A}(\boldsymbol{\mu}), \mathbf{B}(\boldsymbol{\mu}), \mathbf{C}(\boldsymbol{\mu})$, training set Ξ of cardinality N_μ covering the interesting parameter ranges, tolerance ϵ_{tol}.
Output: Projection matrix $\mathbf{V}$.

```
Initialize V = [ ], V_du = [ ], V_e = [ ], ε = 1 + ε_tol, fix η, the number of moments to be matched.
Initial interpolation point μ̃^1: the first sample in Ξ. μ̃_α^1: the last sample in Ξ. Set i = 1.
while ε > ε_tol do
    Solve Eq. (5) at interpolation point μ̃ = μ̃^i and update projection matrix
        V = orth([V mmm(𝒜(μ̃^i), B(μ̃^i), η, μ̃^i)]).
    Solve Eq. (8) at interpolation point μ̃ = μ̃^i and update projection matrix
        V_du = orth([V_du mmm(𝒜^T(μ̃^i), C^T(μ̃^i), η, μ̃^i)]).
    Solve Eq. (12) at interpolation point μ̃ = μ̃_α^i and update projection matrix
        V_e = orth([V_e V_du mmm(𝒜^T(μ̃_α^i), C^T(μ̃_α^i), η, μ̃_α^i)]).
    i = i + 1.
    μ̃^i = arg max_{μ̃∈Ξ} Δ(μ̃).
    μ̃_α^i = arg max_{μ̃∈Ξ} |ẽ_du^T(μ̃) r_pr(μ̃)|.
    ε = Δ(μ̃^i).
end while
```

It is automatic apart from the need for determining, a priori, a suitable training set Ξ. The method proceeds by picking points from Ξ that maximize the error estimator at every iteration and updating the three projection matrices $\mathbf{V}, \mathbf{V}_{\mathrm{du}}, \mathbf{V}_e$. However, there is no principled way to select the training set a priori. If not adequately sampled, the training set may result in a ROM whose error is not uniformly below the tolerance. When the parameters involved can take on a wide range of values, or if many parameters are involved, then the number of parameter samples in Ξ becomes large and the offline computation costs rise. We propose to solve this issue by constructing a surrogate model for $\Delta(\tilde{\boldsymbol{\mu}})$ in Eq. (14) and assure that computing the surrogate is much cheaper than computing the error estimator itself. The next section discusses this idea.

Remark 1 The algorithm is also applicable to MIMO systems. In this case, the transfer function is matrix-valued. The key is how to compute the error estimator $\Delta(\tilde{\boldsymbol{\mu}})$. We first estimate the error of the reduced transfer function entry-wise, i.e.,

$$|\mathbf{H}_{ij}(\tilde{\boldsymbol{\mu}}) - \hat{\mathbf{H}}_{ij}(\tilde{\boldsymbol{\mu}})| \lesssim \left|\tilde{\mathbf{X}}_{\mathrm{du}}^{\mathsf{T}}(\tilde{\boldsymbol{\mu}})\mathbf{r}_{\mathrm{pr}}(\tilde{\boldsymbol{\mu}})\right| + \left|\tilde{\mathbf{e}}_{\mathrm{du}}^{\mathsf{T}}(\tilde{\boldsymbol{\mu}})\mathbf{r}_{\mathrm{pr}}(\tilde{\boldsymbol{\mu}})\right| =: \Delta_{ij}(\tilde{\boldsymbol{\mu}}). \tag{15}$$

Note that the ij-th entry of the transfer function corresponds to the input signal at the j-th input port and the signal at the i-th output port. Then, $\tilde{\mathbf{X}}_{\mathrm{du}}(\tilde{\boldsymbol{\mu}})$ is the solution to the dual system by considering the right-hand side as the i-th row vector of $\mathbf{C}(\boldsymbol{\mu})$, namely, $\mathbf{C}^{\mathsf{T}}(:,i)$ in Eq. (9). Correspondingly, the residual $\mathbf{r}_{\mathrm{pr}}(\tilde{\boldsymbol{\mu}})$ is obtained by solving Eq. (6) with the right-hand side being $\mathbf{B}(:,j)$, the j-th column of $\mathbf{B}(\boldsymbol{\mu})$. Then, $\Delta(\tilde{\boldsymbol{\mu}}) = \arg\max_{i,j} \Delta_{ij}(\tilde{\boldsymbol{\mu}})$.

Remark 2 In order to build the projection matrices $(\mathbf{V}, \mathbf{V}_{\mathrm{du}}, \mathbf{V}_e)$, [14] makes use of the multi-moment matching (MMM) algorithm from [13]. The algorithm provides an orthogonal basis for the solution at a given interpolation point, obtained through multivariate power series expansion of the state vector. To be focused on our main contribution, we refer to [13, 14] for detailed computations. For use in the proposed algorithm, we give below the call to the algorithm in MATLAB®notation as

$$\mathbf{V}_{\mathsf{mmm}} = \mathtt{mmm}(\mathcal{A}(\tilde{\mu}_0), \mathcal{B}(\tilde{\mu}_0), \eta, \tilde{\mu}_0).$$

Here, $\mathcal{A}(\tilde{\mu}_0)$ denotes an arbitrary matrix evaluated at a given interpolation point $\tilde{\mu}_0$, $\mathcal{B}(\tilde{\mu}_0)$ corresponds to the right-hand side matrix in Eqs. (5), (8) and (12), respectively. η is the number of moments to be matched in the power series. When $\eta = 0$, the MMM algorithm is equivalent to RBM, see [14] for more explanations.

4 Adaptive Training by Learning the Error Estimator in the Parameter Domain

In this section, we propose an adaptive training technique, so that the greedy algorithm starts with a training set with small cardinality, which is then updated iteratively by using a surrogate error estimator. Different works have considered surrogate models of error estimators/indicators [9, 10, 25]. All of these consider a surrogate in the context of the RBM. In this work, we deal with the frequency-domain interpolatory MOR methods and focus on a surrogate model of an error estimator for the transfer function approximation error. The method we propose here is essentially an extension of the RBF-based error surrogate in [9] to the frequency domain. Beyond the work in [9], we introduce a learning process in Sect. 4.2 to show in detail how a surrogate estimator is constructed for any parameter in the whole parameter domain. We begin by introducing the method of RBF interpolation.

4.1 Radial Basis Functions

Radial basis functions belong to the family of *kernel methods* and are a popular technique to generate surrogate models of multivariate functions $f : \mathbb{R}^{d+1} \to \mathbb{R}$, defined in a domain $\Omega \subset \mathbb{R}^{d+1}$. It may be the case that the function f itself is unknown and one only knows a set of inputs $M = \{\tilde{\mu}_1, \tilde{\mu}_2, \ldots, \tilde{\mu}_\ell\} \subset \Omega$ and the corresponding function evaluations $F = \{f_1, f_2, \ldots, f_\ell\} \subset \mathbb{R}$. Or, it may be the case that f is known, but very expensive to evaluate repeatedly. For either case, RBF serves to generate an interpolant $g : \mathbb{R}^{d+1} \to \mathbb{R}$ of f given by

$$g(\tilde{\mu}) = \sum_{i=1}^{\ell} c_i \Phi(\|\tilde{\mu} - \tilde{\mu}_i\|), \; \forall \tilde{\mu} \in \Omega, \tag{16}$$

such that it interpolates the original function at the set of input points (or centers) in M, i.e., $f(\tilde{\mu}_i) = g(\tilde{\mu}_i)$, $i = 1, \ldots, \ell$. Moreover, $|f(\tilde{\mu}) - g(\tilde{\mu})| \ll \texttt{tol}$ for all $\tilde{\mu} \in \Omega$. The functions $\Phi(\cdot)$ are the kernels defined as $\Psi(\tilde{\mu}_1, \tilde{\mu}_2) := \Phi(\|\tilde{\mu}_1 - \tilde{\mu}_2\|)$ for all $\tilde{\mu}_1, \tilde{\mu}_2 \in \Omega$. They are called radial basis functions owing to their radial dependence on $\tilde{\mu}$. The coefficients $\{c_i\}_{i=1}^{\ell}$ are determined by solving the linear system of equations

$$\underbrace{\begin{bmatrix} \Psi(\tilde{\mu}_1, \tilde{\mu}_1) & \cdots & \Psi(\tilde{\mu}_1, \tilde{\mu}_\ell) \\ \vdots & \ddots & \vdots \\ \Psi(\tilde{\mu}_\ell, \tilde{\mu}_1) & \cdots & \Psi(\tilde{\mu}_\ell, \tilde{\mu}_\ell) \end{bmatrix}}_{\mathbf{R}} \underbrace{\begin{bmatrix} c_1 \\ \vdots \\ c_\ell \end{bmatrix}}_{c} = \underbrace{\begin{bmatrix} f(\tilde{\mu}_1) \\ \vdots \\ f(\tilde{\mu}_\ell) \end{bmatrix}}_{b}. \tag{17}$$

We need $\mathbf{R}$ to be invertible. Assuming that the centers $\tilde{\mu}_i$ are pairwise distinct, it can be shown that $\mathbf{R}$ is positive definite for a suitable choice of the RBF $\Phi(\cdot)$ and thus Eq. (17) has a unique solution. The class of RBF giving rise to positive definite $\mathbf{R}$ is limited. As a workaround, some additional constraints are imposed in practice, i.e.,

$$\sum_{i=1}^{\ell} c_i p_j(\tilde{\mu}_i) = 0, \quad j = 1, 2, \ldots, D,$$

so that a larger class of $\Phi(\cdot)$ can be admitted. Moreover, the addition of these constraints help in the exact recovery of polynomial functions. We refer to [11] for more details. The functions $p_1, p_2, \ldots, p_D$ are a basis of the polynomial space with suitable degree. In practice, we choose D to be equal to the dimension of the parameter space plus one: $(d + 1) + 1$. With the new conditions imposed, the radial basis interpolant now becomes

$$g(\tilde{\mu}) := \sum_{i=1}^{\ell} c_i \Phi(\|\tilde{\mu} - \tilde{\mu}_i\|) + \sum_{j=1}^{D} \lambda_j p_j(\tilde{\mu}). \tag{18}$$

This results in a saddle-point system of dimension $N_{\text{RBF}} := (D + \ell) \times (D + \ell)$:

$$\begin{bmatrix} \mathbf{R} & \mathbf{P} \\ \mathbf{P}^{\mathsf{T}} & 0 \end{bmatrix} \begin{bmatrix} c \\ \lambda \end{bmatrix} = \begin{bmatrix} b \\ 0 \end{bmatrix}. \tag{19}$$

With a proper choice of $p_1, p_2, \ldots, p_D$, the augmented coefficient matrix is positive definite for a wider choice of kernel functions $\Phi(\cdot)$. Following the common approach in the RBF literature [11], in our numerical experiments, we consider degree-1 polynomials in $(d+1)$ variables and the matrix $\mathbf{P}$ takes the form

$$\mathbf{P} = \begin{bmatrix} 1 & s_1 & \cdots & \mu_1^d \\ \vdots & \vdots & \ddots & \vdots \\ 1 & s_\ell & \cdots & \mu_\ell^d \end{bmatrix},$$

where $s_j, \mu_j^1, \ldots, \mu_j^d$ with $j = 1, 2, \ldots, \ell$ are entries of the j-th parameter sample $\tilde{\boldsymbol{\mu}}_j$ in the training set, i.e., $\tilde{\boldsymbol{\mu}}_j := [s_j, \mu_j^1, \ldots, \mu_j^d]^\mathsf{T}$. We refer to [32] for an exhaustive theoretical analysis of RBFs and the recent review paper [29] that analyzes RBFs in the larger context of kernel-based surrogate models.

4.2 Learning the Error Estimator over the Parameter Domain

As highlighted in the Introduction, one of the main bottlenecks of the standard greedy algorithm is that the error estimator $\Delta(\tilde{\boldsymbol{\mu}})$ needs to be determined at every parameter in the training set. In order to evaluate it cheaply, we first construct a surrogate model of the error estimator by learning the error estimator in the whole parameter domain using RBF interpolation. We have the multivariate function $f(\tilde{\boldsymbol{\mu}}) := \Delta(\tilde{\boldsymbol{\mu}})$ and the learning step involves determining the coefficients c in Eq. (19). First, we evaluate the error estimator at a small number of parameters in a coarse training set $\Xi_c : \{\tilde{\boldsymbol{\mu}}_1, \ldots, \tilde{\boldsymbol{\mu}}_{N_c}\}$. These points shall serve as the centers $\tilde{\boldsymbol{\mu}}$ of the RBF interpolation with N_c as the number of centers. Note that, with regard to the discussion in Sect. 4.1, we have $\ell = N_c$.

Next, we define the kernel function $\Phi(\cdot)$ and set up the linear system defined in Eq. (19). Many choices of the kernel function exist, and in the numerical experiments we have used the inverse multiquadric and the thin-plate spline kernel functions. For a deeper discussion, we refer to [32]. We note here that the assembling of the kernel matrix $\mathbf{R}$ can be done efficiently and software implementations exist to achieve this [29]. The right-hand side is defined by $b := [\Delta(\tilde{\boldsymbol{\mu}}_1), \ldots, \Delta(\tilde{\boldsymbol{\mu}}_{N_c})]^\mathsf{T}$.

Equation 19 constitutes a small, dense system of linear equations. The computational cost of its solution scales as $O((N_c + D)^3)$. However, since N_c, D are small, the cost remains under control. Once knowing c after solving Eq. (19), the interpolant $g(\tilde{\boldsymbol{\mu}})$ of the error estimator is obtained over the whole parameter domain employing only function evaluations in Eq. (18). Thus, the learned surrogate of the error estimator is $g(\tilde{\boldsymbol{\mu}})$. It is not difficult to see that computing the error estimator over the parameter domain is more expensive than using the surrogate $g(\tilde{\boldsymbol{\mu}})$.

- The cost of computing the surrogate $g(\tilde{\boldsymbol{\mu}})$ for *all the* parameter samples $\tilde{\boldsymbol{\mu}}$ in a certain parameter set with cardinality N_f is

  - Solving small, dense ROMs: $O(r^3) \times N_c$.
  - Matrix-vector products to evaluate residuals: $O(nr) \times N_c$.
  - Vector-vector inner products to evaluate Eq. (14) : $O(n) \times N_c$.
  - Identifying the coefficients c by solving Eq. (19): $O((N_c + D)^3)$.
  - Evaluating the interpolants through function evaluation Eq. (18) over a parameter set with cardinality N_f: $O(N_c + D) \times N_f$.

- The cost of evaluating the error estimator $\Delta(\tilde{\mu})$ for *all the* parameter samples $\tilde{\mu}$ in a certain parameter set with cardinality N_f are

  - Solving small, dense ROMs: $O(r^3) \times N_f$.
  - Matrix-vector products to evaluate residuals: $O(nr) \times N_f$.
  - Vector-vector inner products to evaluate Eq. (14) : $O(n) \times N_f$.

Here, n is the full-order dimension of the system; the reduced size r is as small as $N_c + D$, i.e., $r \approx N_c + D$. For $N_f \gg N_c$, it is clear that computing the error estimator is more expensive than computing the surrogate.

4.3 Adaptive Choice of Interpolation Points with Surrogate Error Estimator

In Algorithm 2, we present the proposed adaptive method to choose interpolation points using a surrogate error estimator. We call the algorithm IPSUE—Interpolation Points using SUrrogate error Estimator. To follow the learning process in Sect. 4.2, in practice, we do not consider the entire domain $\mathbb{R}^{d+1}$, but a fine representation of it given by $\Xi_f := \{\tilde{\mu}_1, \ldots, \tilde{\mu}_{N_f}\}$, containing $N_f \gg N_c$ parameters. Therefore, we consider two training sets: a coarse training set Ξ_c and a fine training set Ξ_f. In Step 4 of Algorithm 2, we perform Steps 4–6 from Algorithm 1. In Step 5, the error estimator is evaluated only over the coarse training set, an important distinction from Algorithm 1. In Step 6, the argument of the maximum is chosen as the next interpolation point for $\mathbf{V}$. Step 7 selects the parameter that maximizes the second summand of the error estimator and uses it as the interpolation point $(\tilde{\mu}^i_\alpha)$ for enriching $\mathbf{V}_e$ in the next iteration. As noted in [14], it is important that the interpolation points $\tilde{\mu}^i$ and $\tilde{\mu}^i_\alpha$ are distinct, in order to ensure that $\mathbf{V}_{\text{du}} \neq \mathbf{V}_e$. Then, in Step 8, using $\Delta(\tilde{\mu})$ for all $\tilde{\mu} \in \Xi_c$ we learn the error estimator over the parameter domain (represented by Ξ_f) by determining $g(\tilde{\mu})$ for all $\tilde{\mu} \in \Xi_f$. In Step 9, $n_a^{(1)}$ new parameters are identified from Ξ_f such that they have the largest errors measured by $g(\tilde{\mu})$. The coarse training set is then updated with the newly identified points.

Algorithm 2 Interpolation Points using SUrrogate error Estimator (IPSUE) algorithm

Input: System matrices $\mathbf{A}(\boldsymbol{\mu})$, $\mathbf{B}(\boldsymbol{\mu})$, $\mathbf{C}(\boldsymbol{\mu})$, coarse training set Ξ_c of cardinality N_c, fine training set Ξ_f of cardinality N_f covering the interesting parameter ranges, tolerance ϵ_{tol}.

Output: Projection matrix $\mathbf{V}$.

Initialize $\mathbf{V} = [\,]$, $\mathbf{V}_{\text{du}} = [\,]$, $\mathbf{V}_e = [\,]$, $\epsilon = 1 + \epsilon_{tol}$, fix η, the number of moments to be matched. Set $i = 1$.

Initial interpolation point $\tilde{\boldsymbol{\mu}}^1$: a random sample from Ξ_c selected using `rand()`, $\tilde{\boldsymbol{\mu}}_\alpha^1$: another random sample from Ξ_c selected using `rand()`. Here, `rand()` is the intrinsic MATLAB®function.

while $\epsilon > \epsilon_{tol}$ **do**

  Perform Steps 4–6 from Algorithm 1.

  Use Eq. (14) and obtain $\Delta(\tilde{\boldsymbol{\mu}})$, $\forall \tilde{\boldsymbol{\mu}} \in \Xi_c$.

  $\tilde{\boldsymbol{\mu}}^{i+1} = \arg\max\limits_{\tilde{\boldsymbol{\mu}} \in \Xi_c} \Delta(\tilde{\boldsymbol{\mu}})$.

  $\tilde{\boldsymbol{\mu}}_\alpha^{i+1} = \arg\max\limits_{\tilde{\boldsymbol{\mu}} \in \Xi_c} \left|\tilde{\mathbf{e}}_{\text{du}}^{\mathsf{T}}(\tilde{\boldsymbol{\mu}})\mathbf{r}_{\text{pr}}(\tilde{\boldsymbol{\mu}})\right|$.

  Form the RBF interpolant $g(\tilde{\boldsymbol{\mu}})$ of the error estimator $\Delta(\tilde{\boldsymbol{\mu}})$ over Ξ_f.

  Select $n_a^{(1)}$. Identify $\{\tilde{\boldsymbol{\mu}}_1^{(1)}, \ldots, \tilde{\boldsymbol{\mu}}_{n_a}^{(1)}\}$ from Ξ_f with the largest errors for $g(\tilde{\boldsymbol{\mu}})$. Usually, $n_a^{(1)} = 1$.

  Update the coarse training set with the newly identified parameters, $\Xi_c := \Xi_c \cup \left\{\tilde{\boldsymbol{\mu}}_1^{(1)}, \ldots, \tilde{\boldsymbol{\mu}}_{n_a}^{(1)}\right\}$.

  $i = i + 1$.

  $\epsilon = \Delta(\tilde{\boldsymbol{\mu}}^i)$.

end while

5 Numerical Examples

In this section, we provide numerical results to show the efficiency of the proposed IPSUE algorithm. The first example is from circuit simulation used in [14]. It is characterized by its large parameter range. The second is a benchmark example of a microthruster device, from the `MORwiki` collection [31]. This model has four parameters. The final example is a finite element model of a waveguide filter, from [27]. All numerical tests were performed in MATLAB®2015a, on a laptop with Intel®Core™i5-7200U @ 2.5 GHZ, with 8 GB of RAM. In the numerical results, N_μ refers to the cardinality of the fixed training set Ξ, used in Algorithm 1, N_c, N_f are, respectively, the cardinality of the coarse and fine training sets used in Algorithm 2 and finally N_t denotes the cardinality of the parameter test set Ξ_t used for validating the accuracy of the final ROMs constructed by Algorithms 1 and 2. Also, we use the same test sets for comparing the performances of Algorithms 1 and 2. As mentioned earlier in Sect. 4.2, for the numerical examples, we have made use of inverse multiquadric and thin-plate spline kernels [32]. From our experience, the former was better able to interpolate the estimated error $\Delta(\tilde{\boldsymbol{\mu}})$ for cases where $\Delta(\tilde{\boldsymbol{\mu}})$ depends less smoothly on the parameter. The latter gave a better performance when the estimated error had a smoother variation as a function of the parameter. While there is no additional hyperparameter in the case of the thin-plate spline, the tuning parameter present in the inverse-multiquadric kernel can be used as an additional degree of freedom for capturing the local behavior of the function being interpolated.

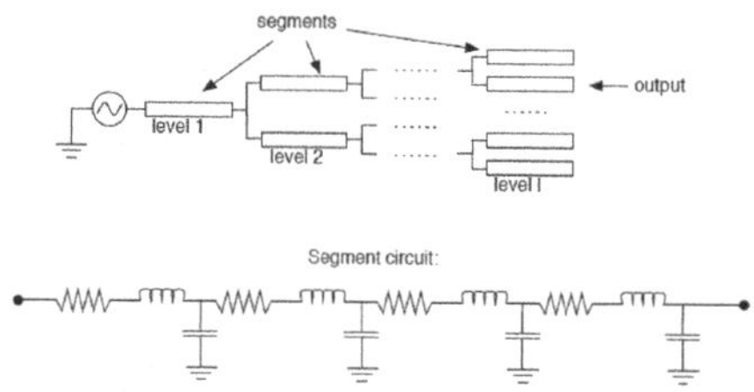

Fig. 1 RLC interconnect circuit

Table 1 Simulation settings for the RLC interconnect circuit

Setting	Value
n	6134
ϵ_{tol}	10^{-3}
N_μ	90 parameters
N_c	$\{21, 27\}$ parameters
N_f	200 parameters
N_t	900 parameters
η	3

5.1 RLC Interconnect Circuit

This example models the large-scale interconnects in integrated circuit (IC) design. It is represented in Fig. 1. The discretized model has dimension $n = 6134$. It is a non-parametric system in time domain, but in the frequency domain, the frequency f is considered as the parameter and the interpolation points are selected from a wide frequency range: $f \in [0, 3]$ GHz. Table 1 gives the simulation settings used for implementing Algorithms 1 and 2 to generate the reduced-order models for this example.

Test 1: Algorithm 1 *applied to RLC model*

To enable comparison, we use the same training set Ξ used in [14]. It consists of 90 samples covering the range of interest. The sampled frequencies are given by $f_i = 3 \times 10^{i/10}$, $s_i = 2\pi \jmath f_i$ with $i = 1, 2, \ldots, 90$. Algorithm 1 converges to the set tolerance in just three iterations. The obtained ROM is of dimension $r = 20$. On average, it takes 3.3 s for the algorithm to converge. For the sake of robustness, we test the ROM on a different set of test parameters Ξ_t with $N_t = 900$ parameters. Figure 2a shows the error of $\hat{\mathbf{H}}(s)$ for the parameters in Ξ_t.

Test 2: Algorithm 2 *applied to RLC model*

Next, we test Algorithm 2 on the RLC interconnect model. For this, we consider two different coarse training sets Ξ_c of cardinality 21, 27 sampled as $\Xi_c^j = 3 \times 10^{j/10}$, $j = 1, 2, \ldots, 21$ and $j = 1, 2, \ldots, 27$. We consider different samplings

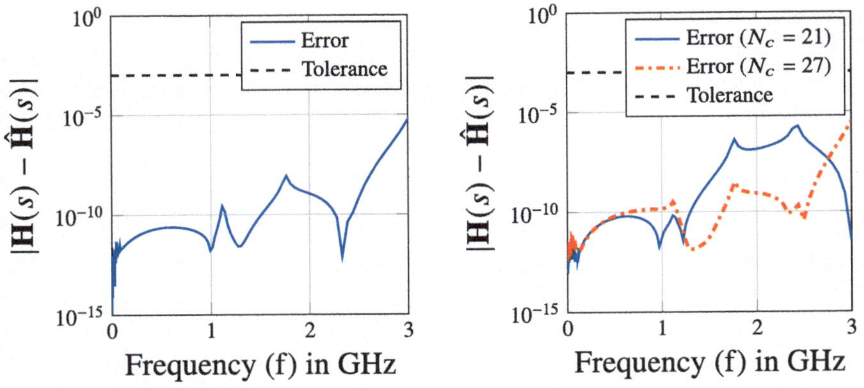

(a) Algorithm 1: Error $|\mathbf{H}(s) - \hat{\mathbf{H}}(s)|$ over the test set.

(b) Algorithm 2: Error $|\mathbf{H}(s) - \hat{\mathbf{H}}(s)|$ over the test set.

Fig. 2 Results for the RLC model

for the fine training set in order to numerically illustrate that the proposed algorithm is independent of the kind of sampling used. The fine training set Ξ_f consists of 200 logarithmically distributed parameters in the first case and in the second case contains 200 parameters distributed as $\Xi_f^j = 3 \times 10^{j/10}$, $j = 1, 2, \ldots, 200$. For the RBF interpolation, we use thin-plate splines as the kernel function. It is given by $\Phi(\tilde{\mu}_1, \tilde{\mu}_2) := (\|\tilde{\mu}_1 - \tilde{\mu}_2\|_2)^2 \log_e(\|\tilde{\mu}_1 - \tilde{\mu}_2\|_2)$. Algorithm 2 converges to the specified tolerance in just three iterations for both choices of Ξ_c, with $n_a^{(1)} = 1$. In the first case, the obtained ROM is of dimension $r = 21$ and takes 1.6 s to converge in average. The second case results in a ROM of dimension 21 and takes 1.7 s on average to converge to the defined tolerance. Figure 2b shows the error of $\hat{\mathbf{H}}(s)$ at parameters in the test set Ξ_t produced by the ROM obtained using Algorithm 2, with two different coarse training sets. Clearly, Algorithm 2 takes less time than Algorithm 1, while still producing a ROM that is uniformly below the tolerance, on an independent test set.

5.2 *Thermal Model*

The second example is the model of the heat transfer inside a microthruster unit [31]. It is obtained after spatial discretization using the finite element method and has dimension $n = 4257$. The governing equation is given as

$$\mathbf{E}\dot{\mathbf{x}}(t) = (\mathbf{A}_0 - \sum_{i=1}^{3} h_i \mathbf{A}_i)\mathbf{x}(t) + \mathbf{B}\mathbf{u}(t),$$

$$\mathbf{y}(t) = \mathbf{C}\mathbf{x}.$$

Table 2 Simulation settings for the thermal model

Setting	Value
n	4257
ϵ_{tol}	10^{-4}
N_{μ}	625 parameters, log-sampled
N_c	256 parameters, log-sampled
N_f	2401 parameters, log-sampled
N_t	1000 parameters, randomly sampled
η	1

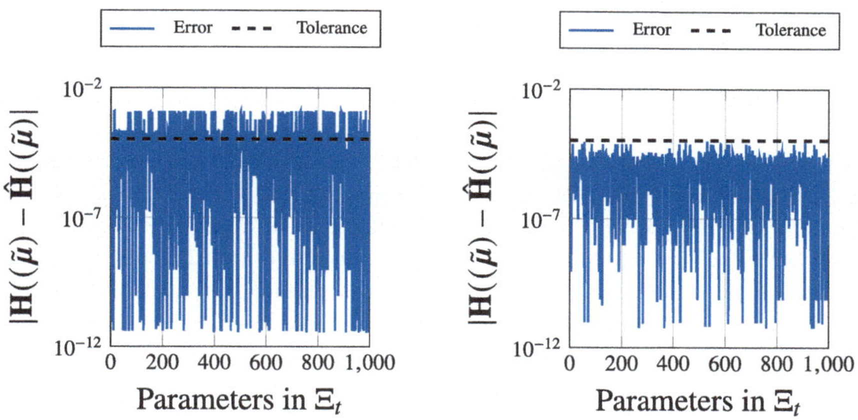

(a) Algorithm 1: Error $|\mathbf{H}(\tilde{\mu}) - \hat{\mathbf{H}}((\tilde{\mu})|$ over the test set.

(b) Algorithm 2: Error $|\mathbf{H}((\tilde{\mu}) - \hat{\mathbf{H}}((\tilde{\mu})|$ over the test set.

Fig. 3 Results for the thermal model

Here, $\mathbf{E}$, $\mathbf{A}_0$ are symmetric sparse matrices representing the heat capacity and heat conductivity, respectively. $\mathbf{A}_i$, $i \in \{1, 2, 3\}$ are diagonal matrices governing the boundary condition. The parameters $h_1, h_2, h_3 \in [1, 10^4]$ represent, respectively, the film coefficients of the top, bottom, and side of the microthruster unit. We transform the above system to the frequency domain and apply Algorithms 1 and 2. In the frequency domain, the system has four parameters $\tilde{\boldsymbol{\mu}} := (s, h_1, h_2, h_3)$ with $s = \jmath 2\pi f$. The frequency range of interest is $f \in [10^{-2}, 10^2]$ Hz. The tolerance for the ROM is set as 10^{-4} (Table 2).

Test 3: Algorithm 1 *applied to the thermal model*

Owing to the wide range of parameters, we consider a large fixed training set Ξ. To construct it, we consider five logarithmically spaced samples for each of the four parameters and form a grid consisting of 5^4 samples. For the test set Ξ_t, we form a grid of 8^4 logarithmically spaced parameters and randomly select 1000 parameters

Fig. 4 Dual-mode waveguide filter model from [27]

from it. The greedy algorithm takes 10 iterations to converge and results in a ROM of size $r = 86$. On an average over five runs, the greedy algorithm takes 254 s to converge. In Fig. 3a, we see the performance of the resulting ROM over Ξ_t. For several parameters, the ROM fails to meet the desired tolerance. This indicates that the training set was not fine enough to capture all the variations in the solutions over the parameter domain.

Test 4: Algorithm 2 *applied to the thermal model*

We now consider Algorithm 2 applied to the thermal model. The coarse training set Ξ_c has 4^4 parameters, with logarithmic sampling. The fine training set Ξ_f has 7^4 parameters. For the RBF interpolation, we make use of thin-plate splines as the kernel function. Further, we set $n_a^{(1)} = 1$ in Step 9 of Algorithm 2 so that the coarse training set is updated with one new parameter per iteration. The same test set as in Test 3 is used. The resulting ROM has order $r = 85$ and its error over the test set is below the tolerance, as shown in Fig. 3b. The algorithm took 162 s to converge. The runtime was measured as an average over five independent runs of the algorithm. Compared with Algorithm 1, Algorithm 2 is able to meet the required tolerance with a much smaller training set and also in shorter time.

5.3 Dual-Mode Circular Waveguide Filter

The next example is a MIMO system based on the model of a dual-mode circular waveguide filter from [27], see Fig. 4. It is a type of narrow bandpass filter widely used in satellite communication due to its power handling capabilities. Its operation is governed by the time-harmonic Maxwell's equations. After discretization in space, the governing equations of the filter can be represented in the form of Eq. (5). The system consists of just the frequency parameter $s := \jmath 2\pi f$, where $f \in [11.5\,,\,12]$ GHz is the operating frequency band of the filter. The affine form of the system matrix is $\mathscr{A}(s) := \mathcal{S} + s^2\mathcal{T}$ and $\mathbf{B}(s) := sQ$. We have $\mathcal{S}, \mathcal{T} \in \mathbb{R}^{n\times n}$ and $Q \in \mathbb{R}^{n\times 2}$, with $n = 36426$. The system has two inputs and two outputs. Table 3 summarizes the simulation settings. The quantity of interest is the scattering parameters, obtained via post-processing [28] from the system output $\mathbf{y}(s) := Q^{\mathsf{T}}\mathbf{X}(s)$. It is easy to see that $\mathbf{y}(s)$ has the same expression as $\mathbf{H}(\tilde{\mu})$ in Eq. (3) for $\tilde{\mu} = s$. The error estimator $\Delta(\tilde{\mu})$ in Sect. 3 can be directly applied to estimate the error of $\hat{\mathbf{y}}(s)$ computed by the ROM. See [14] for detailed analysis. Since the system is MIMO, the scattering

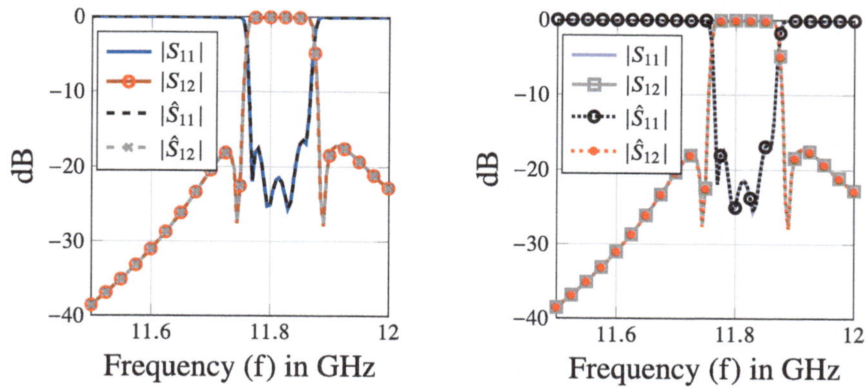

(a) Algorithm 1: Scattering parameters for the test set.

(b) Algorithm 2: Scattering parameters for the test set.

Fig. 5 Scattering parameters for the dual-mode filter

Table 3 Simulation settings for the dual-mode filter

Setting	Value
n	36426
ϵ_{tol}	10^{-5}
N_μ	51 parameters, uniformly spaced
N_c	17 parameters, uniformly spaced
N_f	500 parameters, sampled randomly
N_t	101 parameters, uniformly spaced
η	1

parameters at a given s are in the form of a complex-valued matrix given by

$$\mathbf{S} = \begin{bmatrix} S_{11} & S_{12} \\ S_{21} & S_{22} \end{bmatrix}.$$

Scattering parameters are important in characterizing the performance of filters [26].

Test 5: Algorithm 1 *applied to the dual-mode filter*

Applying Algorithm 1 with the fixed training set Ξ to the model results in a ROM of size $r = 10$ with the greedy algorithm taking five iterations to converge. Since this example is a MIMO system, we make use of Eq. (15). The average runtime over five independent runs of Algorithm 1 was found to be 46 s. Figure 5a plots the scattering parameters computed from FOM simulations and those obtained from the ROM at the parameters in the test set. We plot the absolute values of full-order scattering parameters S_{11}, S_{12} and the corresponding reduced ones $\hat{S}_{11}$, $\hat{S}_{12}$ on a Decibel scale.

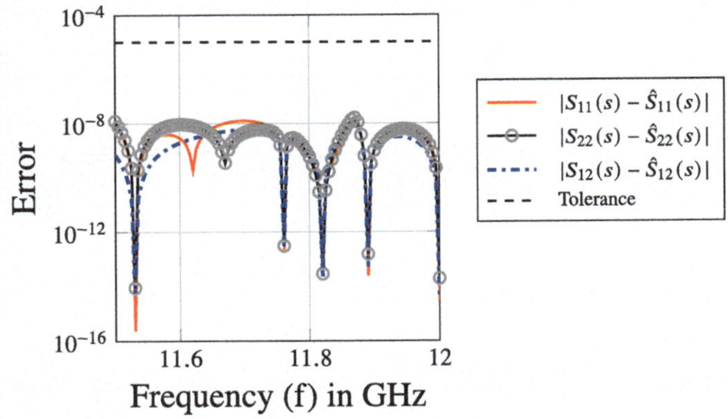

(a) Algorithm 1: Error of the scattering parameters computed from the ROM over the test set.

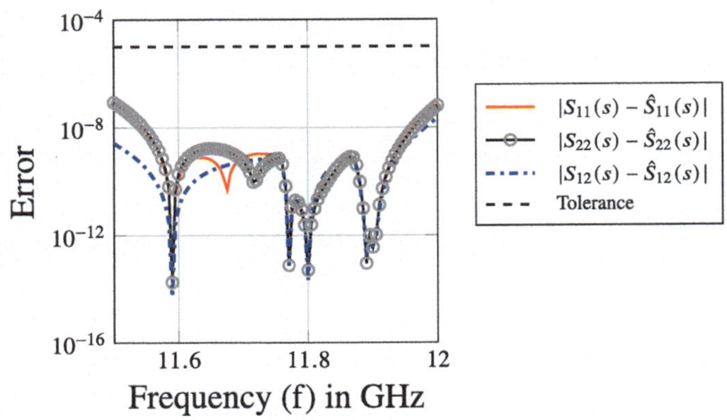

(b) Algorithm 2: Error of the scattering parameters computed from the ROM over the test set.

Fig. 6 Results for the dual-mode filter

In Fig. 6a, we plot the error of the scattering parameters $\hat{S}_{11}, \hat{S}_{12}, \hat{S}_{22}$ computed from the ROM, over the test set Ξ_t. Note that since $S_{12} = S_{21}$, we only show the error $|S_{12} - \hat{S}_{12}|$.

Test 6: Algorithm 2 *applied to the Dual-mode filter*

In Step 8 of Algorithm 2, we construct an RBF surrogate for each of $\Delta_{ij},\ i, j \in \{1, 2\}$ in Eq. (15) for this MIMO system. $n_a^{(1)}$ is set to be 1 and inverse multiquadric is used as the kernel function. It is given by $\Psi(\tilde{\mu}_1, \tilde{\mu}_2) := 1/(1 + (\gamma \|\tilde{\mu}_1 - \tilde{\mu}_2\|_2))^2$. γ is a user-defined parameter and we set $\gamma = 16$ in our experiments. We pick the maximum among the four surrogates and add the corresponding parameter to the coarse training set, i.e., in Step 9 of Algorithm 2, we replace $g(\tilde{\boldsymbol{\mu}})$ with $\max_{i,j\in\{1,2\}} g_{i,j}(\tilde{\boldsymbol{\mu}})$. Algorithm 2 results in a ROM that is of the same size as the ROM from Test 5 ($r = 10$). However, on average, Algorithm 2 only needs 24 s to converge, almost half that of the time required in Test 5. In Fig. 6b, we plot the errors of the scattering parameters computed from the ROM over the test set Ξ_t. Figure 5b plots the scattering parameters from the FOM simulations and those computed by the ROM. Both algorithms result in ROMs meeting the specified tolerance, but Algorithm 2 requires much shorter time to generate the ROM.

6 Conclusion

In this work, we have proposed IPSUE an adaptive algorithm for updating the training set and choosing the interpolation points for frequency-domain MOR methods. Our target applications are cases where the problem parameters vary over a wide range of values or the parameter space dimension is larger than two. In either of these cases, many interpolatory MOR algorithms may take a long time to generate the ROM. Moreover, a naive, heuristic sampling of the parameter training set may result in a ROM that is not robust. IPSUE offers a viable means to generate reliable ROMs that satisfy the user-defined tolerance and at the same time without being offline expensive. The illustrated numerical examples show that it is a promising approach. As future work, we plan to apply the algorithm to more complex models.

Acknowledgements The first author is affiliated to the International Max Planck Research School for Advanced Methods in Process and Systems Engineering (IMPRS-ProEng).

References

1. Antoulas, A.C.: Approximation of Large-Scale Dynamical Systems. Advances in Design and Control, vol. 6. SIAM Publications, Philadelphia, PA (2005). https://doi.org/10.1137/1.9780898718713
2. Antoulas, A.C., Sorensen, D.C., Gugercin, S.: A survey of model reduction methods for large-scale systems. Contemp. Math. **280**, 193–219 (2001)
3. Baur, U., Beattie, C.A., Benner, P., Gugercin, S.: Interpolatory projection methods for parameterized model reduction. SIAM J. Sci. Comput. **33**(5), 2489–2518 (2011). https://doi.org/10.1137/090776925
4. Baur, U., Benner, P., Feng, L.: Model order reduction for linear and nonlinear systems: a system-theoretic perspective. Arch. Comput. Methods Eng. **21**(4), 331–358 (2014). https://doi.org/10.1007/s11831-014-9111-2
5. Bechtold, T., Rudnyi, E., Korvink, J.: Error indicators for fully automatic extraction of heat-transfer macromodels for MEMS. Micromech. Microeng. **15**(3), 430–440 (2004). https://doi.org/10.1088/0960-1317/15/3/002
6. Benner, P., Goyal, P., Pontes Duff, I.: Identification of dominant subspaces for linear structured parametric systems and model reduction. e-prints 1910.13945, arXiv (2019). https://arxiv.org/abs/1910.13945. Math.NA

7. Benner, P., Gugercin, S., Willcox, K.: A survey of model reduction methods for parametric systems. SIAM Rev. **57**(4), 483–531 (2015). https://doi.org/10.1137/130932715
8. Benner, P., Mehrmann, V., Sorensen, D.C.: Dimension Reduction of Large-Scale Systems, Lecture Notes Computer Science and Engineering, vol. 45. Springer, Berlin/Heidelberg, Germany (2005)
9. Chellappa, S., Feng, L., Benner, P.: An adaptive sampling approach for the reduced basis method. e-prints 1910.00298, arXiv (2019). https://arxiv.org/abs/1910.00298. Math.NA
10. Drohmann, M., Carlberg, K.: The ROMES method for statistical modeling of reduced-order-model error. SIAM/ASA J. Uncertain. Quantif. **3**(1), 116–145 (2015). https://doi.org/10.1137/140969841
11. Fasshauer, G., McCourt, M.: Kernel-based Approximation Methods using MATLAB, Interdisciplinary Mathematical Sciences, vol. 19. World Scientific (2015)
12. Feng, L., Antoulas, A.C., Benner, P.: Some *a posteriori* error bounds for reduced-order modelling of (non-)parametrized linear systems. ESAIM: Math. Model. Numer. Anal. **51**(6), 2127–2158 (2017). https://doi.org/10.1051/m2an/2017014
13. Feng, L., Benner, P.: Reduced Order Methods for modeling and computational reduction, MS&A Series, vol. 9, chap. 6: A robust algorithm for parametric model order reduction based on implicit moment matching, pp. 159–186. Springer, Berlin, Heidelberg, New York (2014)
14. Feng, L., Benner, P.: A new error estimator for reduced-order modeling of linear parametric systems. IEEE Trans. Microw. Theory Techn. **67**(12), 4848–4859 (2019)
15. Feng, L., Korvink, J.G., Benner, P.: A fully adaptive scheme for model order reduction based on moment-matching. IEEE Trans. Compon. Packag. Manuf. Technol. **5**(12), 1872–1884 (2015). https://doi.org/10.1109/TCPMT.2015.2491341
16. Grimme, E.J.: Krylov projection methods for model reduction. Ph.D. thesis, University of Illinois at Urbana-Champaign, USA (1997)
17. Gugercin, S., Antoulas, A.C., Beattie, C.: $\mathcal{H}_2$ model reduction for large-scale linear dynamical systems. SIAM J. Matrix Anal. Appl. **30**(2), 609–638 (2008). https://doi.org/10.1137/060666123
18. Gugercin, S., Stykel, T., Wyatt, S.: Model reduction of descriptor systems by interpolatory projection methods. SIAM J. Sci. Comput. **35**(5), B1010–B1033 (2013). https://doi.org/10.1137/130906635
19. Haasdonk, B., Dihlmann, M., Ohlberger, M.: A training set and multiple bases generation approach for parameterized model reduction based on adaptive grids in parameter space. Math. Comput. Model. Dyn. Syst. **17**(4), 423–442 (2011)
20. Hesthaven, J.S., Stamm, B., Zhang, S.: Efficient greedy algorithms for high-dimensional parameter spaces with applications to empirical interpolation and reduced basis methods. ESAIM: Math. Model. Numer. Anal. **48**(1), 259–283 (2014)
21. Hetmaniuk, U., Tezaur, R., Farhat, C.: An adaptive scheme for a class of interpolatory model reduction methods for frequency response problems. Internat. J. Numer. Methods Engrg. **93**(10), 1109–1124 (2013). https://doi.org/10.1002/nme.4436
22. Hund, M., Mlinarić, P., Saak, J.: An $\mathcal{H}_2 \otimes \mathcal{L}_2$-optimal model order reduction approach for parametric linear time-invariant systems. Proc. Appl. Math. Mech. **18**(1), e201800084 (2018). https://doi.org/10.1002/pamm.201800084
23. Lee, H.J., Chu, C.C., Feng, W.S.: An adaptive-order rational Arnoldi method for model-order reductions of linear time-invariant systems. Linear Algebra Appl. **415**(2), 235–261 (2006). https://doi.org/10.1016/j.laa.2004.10.011. Special Issue on Order Reduction of Large-Scale Systems
24. Maday, Y., Stamm, B.: Locally adaptive greedy approximations for anisotropic parameter reduced basis spaces. SIAM J. Sci. Comput. **35**(6), A2417–A2441 (2013)
25. Paul-Dubois-Taine, A., Amsallem, D.: An adaptive and efficient greedy procedure for the optimal training of parametric reduced-order models. Internat. J. Numer. Methods Engrg. **102**(5), 1262–1292 (2015)
26. Pozar, D.M.: Microwave Engineering. Wiley, New York, USA (1998)

27. de la Rubia, V., Mrozowski, M.: A compact basis for reliable fast frequency sweep via the reduced-basis method. IEEE Trans. Microw. Theory Techn. **66**(10), 4367–4382 (2018)
28. de la Rubia, V., Razafison, U., Maday, Y.: Reliable fast frequency sweep for microwave devices via the reduced-basis method. IEEE Trans. Microw. Theory Techn. **57**(12), 2923–2937 (2009)
29. Santin, G., Haasdonk, B.: Kernel methods for surrogate modeling. e-prints 1907.10556, arXiv (2019). https://arxiv.org/abs/1907.10556. Math.NA
30. Schilders, W.H.A., van der Vorst, H.A., Rommes, J.: Model Order Reduction: Theory, Research Aspects and Applications. Springer, Berlin, Heidelberg (2008)
31. The MORwiki Community: MORwiki - Model Order Reduction Wiki. http://modelreduction.org
32. Wendland, H.: Scattered Data Approximation, Cambridge Monographs on Applied and Computational Mathematics, vol. 17. Cambridge University Press, Cambridge (2005)
33. Wolf, T., Panzer, H., Lohmann, B.: Gramian-based error bound in model reduction by Krylov subspace methods. IFAC Proc. Vol. **44**(1), 3587–3592 (2011). https://doi.org/10.3182/20110828-6-IT-1002.02809

A Link Between Gramian-Based Model Order Reduction and Moment Matching

C. Bertram and H. Faßbender

Abstract We analyze a family of Runge-Kutta-based quadrature algorithms for the approximation of the Gramians of linear time-invariant dynamical systems. The approximated Gramians are used to obtain an approximate balancing transformation similar to the approach used in balanced POD. It is shown that hereby rational interpolation is performed, as the approximants span certain Krylov subspaces. The expansion points are mainly determined by the time step sizes and the eigenvalues of the matrices given by the Butcher tableaus.

1 Introduction

Consider an asymptotically stable, minimal, linear time-invariant single-input single-output continuous-time dynamical system

$$\begin{aligned} \dot{x}(t) &= Ax(t) + Bu(t), \quad x(0) = x_0, \\ y(t) &= Cx(t) \end{aligned} \tag{1}$$

with $A \in \mathbb{R}^{n\times n}$, $B \in \mathbb{R}^{n\times 1}$, $C \in \mathbb{R}^{1\times n}$ and therefore $x(t) \in \mathbb{R}^n$, $u(t) \in \mathbb{R}$ and $y(t) \in \mathbb{R}$. Asymptotic stability of the system implies $\sigma(A) \subseteq \mathbb{C}_-$, i.e., all eigenvalues of A have negative real part. Throughout this work we have a large and sparse system matrix A in mind, so model order reduction makes sense. Yet all results are also true for any other square matrix.

The problem addressed here is to approximate the system (1) by another system

C. Bertram (✉) · H. Faßbender
Institute for Numerical Analysis, Technische Universität Braunschweig, Universitätsplatz 2, 38106 Braunschweig, Germany
e-mail: ch.bertram@tu-braunschweig.de

H. Faßbender
e-mail: h.fassbender@tu-braunschweig.de

P. Benner et al. (eds.), *Model Reduction of Complex Dynamical Systems*, International Series of Numerical Mathematics 171,
https://doi.org/10.1007/978-3-030-72983-7_6

$$\begin{aligned}\dot{\hat{x}}(t) &= \hat{A}\hat{x}(t) + \hat{B}u(t),\\ \hat{y}(t) &= \hat{C}\hat{x}(t)\end{aligned} \tag{2}$$

with possibly complex reduced system matrices $\hat{A} \in \mathbb{C}^{r\times r}$, $\hat{B} \in \mathbb{C}^{r\times 1}$, $\hat{C} \in \mathbb{C}^{1\times r}$ and $r \ll n$. Among the many approaches for model order reduction (see, e.g., [3] and the references therein) we will pursue a projection-based approach: two $n \times r$ matrices $V, W \in \mathbb{C}^{n\times r}$ with $W^{\mathsf{H}}V = I_r$ are computed which define the projector $\Pi = VW^{\mathsf{H}}$. The projection of the states of the original system generates the reduced-order model with system matrices

$$\hat{A} = W^{\mathsf{H}}AV, \qquad \hat{B} = W^{\mathsf{H}}B, \qquad \hat{C} = CV.$$

In practice, to obtain a real reduced system, the projection matrices can be kept real, but for ease of notation and explanation we allow for a complex-valued projection.

In particular we focus on a balancing related approach which is derived from numerical integration with Runge-Kutta methods. We demonstrate how the reduced system generated by the method presented in this work can also be obtained by rational interpolation. Thus the transfer functions of the original and the reduced-order system coincide at certain interpolation points. We give an explicit formulation of those interpolation points in terms of the time step sizes used in the Runge-Kutta method and the eigenvalues of the matrix which determines the Butcher tableau representing the Runge-Kutta method.

1.1 Balancing of LTI Systems

Balancing is closely related to the controllability Gramian $\mathcal{P}$ and the observability Gramian $\mathcal{Q}$ of the system (1) [1, 18, 21]. The Gramians are defined as

$$\mathcal{P} = \int_0^\infty e^{At}BB^{\mathsf{T}}e^{A^{\mathsf{T}}t}\,\mathrm{d}t \in \mathbb{R}^{n\times n}, \tag{3}$$

$$\mathcal{Q} = \int_0^\infty e^{A^{\mathsf{T}}t}C^{\mathsf{T}}Ce^{At}\,\mathrm{d}t \in \mathbb{R}^{n\times n}. \tag{4}$$

The Gramians $\mathcal{P}$ and $\mathcal{Q}$ are positive definite matrices as all eigenvalues of A have negative real part. Thus, their Cholesky decompositions $\mathcal{P} = SS^{\mathsf{T}}$ and $\mathcal{Q} = RR^{\mathsf{T}}$ can be determined. Let $U\Sigma T^{\mathsf{T}}$ be a singular value decomposition of $R^{\mathsf{T}}S$. Then $F = \Sigma^{-\frac{1}{2}}U^{\mathsf{T}}R^{\mathsf{T}}$ and $F^{-1} = ST^{-\mathsf{T}}\Sigma^{-\frac{1}{2}}$ define a balancing transformation. That is, the Gramians $\hat{\mathcal{P}} = F\mathcal{P}F^{\mathsf{T}}$ and $\hat{\mathcal{Q}} = F^{-\mathsf{T}}\mathcal{Q}F^{-1}$ of the transformed system

$$\begin{aligned}\dot{x}(t) &= FAF^{-1}x(t) + FBu(t), \quad x(0) = x_0,\\ y(t) &= CF^{-1}x(t)\end{aligned}$$

are equal and diagonal. Thus, in the balanced system it holds $\hat{\mathcal{P}} = \hat{\mathcal{Q}} = \Sigma$ with the Hankel singular values on the diagonal, which are an indicator for the importance of the corresponding state. They are invariant under state-space transformations, i.e., the same for $\mathcal{P}\mathcal{Q}$ and $\hat{\mathcal{P}}\hat{\mathcal{Q}} = F\mathcal{P}\mathcal{Q}F^{-1}$ due to similarity.

In the model order reduction method balanced truncation a projection is performed onto the r most important states, i.e., the states with large Hankel singular values. The projection $\Pi = VW^{\mathsf{T}}$ for the reduction process is derived from the partitioned singular value decomposition

$$R^{\mathsf{T}}S = \begin{bmatrix} U_r & U_0 \end{bmatrix} \begin{bmatrix} \Sigma_r & \\ & \Sigma_0 \end{bmatrix} \begin{bmatrix} T_r^{\mathsf{T}} \\ T_0^{\mathsf{T}} \end{bmatrix}$$

with $\Sigma_r \in \mathbb{R}^{r\times r}$, $U_r \in \mathbb{R}^{n\times r}$ and $T_r \in \mathbb{R}^{n\times r}$. The matrices V and W are obtained as

$$W = RU_r\Sigma_r^{-\frac{1}{2}}, \quad V = ST_r\Sigma_r^{-\frac{1}{2}} \tag{5}$$

and indeed $W^{\mathsf{T}}V = I_r$ holds. For more details on the Gramians and the energy associated with reaching/observing a state see, e.g., [1, ch. 4.3].

A bottleneck in this approach is the calculation of the Gramians $\mathcal{P}$, $\mathcal{Q}$ and their Cholesky factors. Different methods have been proposed for this situation, see, e.g., [4, 25] and the references therein. A key idea to make calculations for large systems computationally feasible is to approximate the Gramians with low-rank factors, i.e., $Z_{\mathrm{c}}Z_{\mathrm{c}}^{\mathsf{T}} \approx \mathcal{P}$ and $Z_{\mathrm{o}}Z_{\mathrm{o}}^{\mathsf{T}} \approx \mathcal{Q}$ with rectangular matrices $Z_{\mathrm{c}} \in \mathbb{R}^{n\times r_{\mathrm{c}}}$, $Z_{\mathrm{o}} \in \mathbb{R}^{n\times r_{\mathrm{o}}}$, which need not be triangular, and $r_{\mathrm{c}}, r_{\mathrm{o}} \ll n$. These approximate Cholesky factors Z_{c} and Z_{o} are then used instead of the actual Cholesky factors S and R to compute an approximate balancing transformation. This also includes a reduction of the system dimension as the number of columns in the approximate Cholesky factors is smaller than n. In balanced truncation with the actual Cholesky factors S and R of the Gramians as described above, stability of the system is preserved and there exists an error bound in terms of the truncated Hankel singular values [1, Thm. 7.9]. When the approximate Cholesky factors Z_{c} and Z_{o} are used, these properties are lost. However, in practice often the reduced models are stable even when approximate Cholesky factors are used, see, e.g., [12, Sec. 4.3].

In the method balanced proper orthogonal decomposition of snapshots (BPOD) as discussed in [24], the Gramians are approximated with finite sums (see [26] for a related approach). In particular the controllability Gramian is approximated via

$$\begin{aligned} \mathcal{P} = \int_0^\infty h(t)h(t)^{\mathsf{T}}\,\mathrm{d}t &\approx \int_0^T h(t)h(t)^{\mathsf{T}}\,\mathrm{d}t \\ &\approx \sum_{i=1}^{N} \delta_i h_i h_i^{\mathsf{H}} \end{aligned} \tag{6}$$

with $h_i \approx h(t_i)$, $h(t) = e^{At}B$, an end time $T \in \mathbb{R}_+$, times $t_1 < \cdots < t_N \in [0, T]$ and quadrature weights δ_i. The approximation of $h(t_i)$ is done by solving the ODE

$$\frac{\mathrm{d}}{\mathrm{d}t}h(t) = Ah(t), \qquad h(0) = B. \tag{7}$$

From (6) we find that the approximate Cholesky factor is given by

$$Z_\mathrm{c} = [h_1, \ldots, h_N]\,\mathrm{diag}(\sqrt{\delta_1}, \ldots, \sqrt{\delta_N}).$$

In [24, Prop. 2] it was shown that if approximate Cholesky factors with $\mathrm{rank}(Z_\mathrm{o}^\mathsf{T} Z_\mathrm{c}) = r$ are used in balanced truncation, then the matrix V from (5) contains the first columns of an approximate balancing transformation. It was shown in [22] that for certain quadrature methods for solving (7) the reduced system obtained by balanced POD matches some moments. We will proceed as in balanced POD to obtain an approximate balancing transformation. To obtain the approximate Cholesky factors of the Gramians we solve a system of ODEs which consists of the ODE (7) for approximating $h(t)$ and a second ODE $\frac{\mathrm{d}}{\mathrm{d}t}P(t) = h(t)h(t)^\mathsf{T}$ for approximating the time-dependent Gramian with Runge-Kutta methods. This allows us to show a connection between the Butcher tableau which characterizes the Runge-Kutta method and the expansion points at which the moments are matched.

1.2 Rational Interpolation

In rational interpolation [1, 9, 11] the projection matrices V and W are chosen so that the transfer function $G(s) = C(sI_n - A)^{-1}B$ of the original system (1) and the transfer function $\hat{G}(s) = \hat{C}(sI_r - \hat{A})^{-1}\hat{B}$ of the reduced system (2) (and some of their derivatives) coincide at certain interpolation points $s \in \mathbb{C} \cup \{\infty\}$. Rational interpolation is a powerful method: Almost every reduced LTI system (2) can be obtained via rational interpolation from (1), see [10].

A power series expansion around $s_0 \in \mathbb{C} \setminus \sigma(A)$ with $\|(s - s_0)(A - s_0 I_n)^{-1}\| < 1$ yields

$$G(s) = \sum_{j=0}^{\infty} m_j(s_0)(s - s_0)^j$$

with the so-called moments

$$m_j(s_0) = -C(A - s_0 I_n)^{-(j+1)}B = \frac{(-1)^j}{j!}\frac{\mathrm{d}^j}{\mathrm{d}s^j}G(s)\bigg|_{s=s_0}.$$

If either

$$\{(A - s_0 I_n)^{-1} B, \dots, (A - s_0 I_n)^{-k} B\} \subseteq \operatorname{span} V \tag{8}$$

$$\text{or} \quad \{(A^\mathsf{T} - \overline{s_0} I_n)^{-1} C^\mathsf{T}, \dots, (A^\mathsf{T} - \overline{s_0} I_n)^{-k} C^\mathsf{T}\} \subseteq \operatorname{span} W \tag{9}$$

then the first k moments around s_0 are matched, i.e., $m_j(s_0) = \hat{m}_j(s_0)$ for $j = 0, \dots, k-1$. If both conditions (8) and (9) are fulfilled, then even the first $2k$ moments around s_0 are matched.

For the expansion point $s_0 = \infty$ and $\|s^{-1}A\| < 1$ we use the power series expansion

$$G(s) = \sum_{j=1}^{\infty} m_j(\infty) s^{-j}$$

with the Markov parameters $m_j(\infty) = CA^{j-1}B$. If

$$\{B, AB, \dots, A^{k-1}B\} \subseteq \operatorname{span} V \tag{10}$$

$$\text{or} \quad \{C^\mathsf{T}, A^\mathsf{T} C^\mathsf{T}, \dots, (A^\mathsf{T})^{k-1} C^\mathsf{T}\} \subseteq \operatorname{span} W \tag{11}$$

then the first k Markov parameters are matched, i.e., $m_j(\infty) = \hat{m}_j(\infty)$ for $j = 1, \dots, k$. If both conditions (10) and (11) are fulfilled, then even the first $2k$ Markov parameters are matched.

The projection matrices can be kept real when the interpolation points occur in conjugated pairs as

$$\begin{aligned} &\operatorname{span}\{(A - s_0 I_n)^{-1} v, (A - \overline{s_0} I_n)^{-1} v\} \\ &= \operatorname{span}\{\operatorname{Re}((A - s_0 I_n)^{-1} v), \operatorname{Im}((A - s_0 I_n)^{-1} v)\} \end{aligned}$$

holds for real vectors v.

Of course combinations of the cases mentioned above and different expansion points are possible. To obtain a well approximating reduced system the choice of the expansion points is essential and many strategies exist to obtain them, see, e.g., [2, Sec. 2.2.2].

1.3 Organization of Paper

In the following we focus on the approximation of the controllability Gramian (3) by approximately solving the Lyapunov equation

$$A\mathcal{P} + \mathcal{P}A^\mathsf{T} + BB^\mathsf{T} = 0.$$

The observability Gramian (4) satisfies the Lyapunov equation

$$A^{\mathsf{T}}Q + QA + C^{\mathsf{T}}C = 0$$

and can be treated with the same methods as the controllability Gramian by exchanging A and B with A^{T} and C^{T}, so large parts of our discussion focus on the controllability Gramian only.

This paper is organized as follows. In Sect. 2 numerical integration with Runge-Kutta methods is introduced and applied to an ODE derived from the time-dependent Gramian. It is illustrated how the resulting system is solved efficiently and which space is spanned by the iterates. The numerical solution of the ODE is used for approximate balancing in Sect. 3. Using the results from the previous section it is proven that hereby moment matching is performed. In Sect. 4 we illustrate connections to balanced POD and the ADI iteration. Finally, in Sect. 5 some examples illustrate our findings.

2 Gramian Quadrature Algorithm

We now present a quadrature algorithm to obtain approximate Cholesky factors of the Gramians. It was first introduced in [5] and is recapitulated here in concise form. Consider the system of ordinary differential equations

$$\frac{\mathrm{d}}{\mathrm{d}t}P(t) = h(t)h(t)^{\mathsf{T}}, \qquad P(0) = 0 \in \mathbb{R}^{n\times n}, \tag{12}$$

$$\frac{\mathrm{d}}{\mathrm{d}t}h(t) = Ah(t), \qquad h(0) = B \in \mathbb{R}^{n\times 1}. \tag{13}$$

Equation (13) is a linear, homogeneous differential equation which has the solution $h(t) = e^{At}B$. Due to the fundamental theorem of calculus equation (12) is solved by the time-dependent Gramian

$$P(t) = \int_0^t e^{A\tau}BB^{\mathsf{T}}e^{A^{\mathsf{T}}\tau}\,\mathrm{d}\tau = \int_0^t h(\tau)h(\tau)^{\mathsf{T}}\,\mathrm{d}\tau.$$

We intend to solve the above system of ODEs numerically to obtain an approximation to the Gramian $\mathcal{P} = \lim_{t\to\infty} P(t)$.

2.1 Approximating the Gramian via Runge-Kutta Methods

There are numerous methods for the numerical solution of ordinary differential equations of the type $\frac{\mathrm{d}}{\mathrm{d}t}y(t) = f(t, y(t))$. Single-step methods make use of the fact that

$$y(t_j) = y(t_{j-1}) + \int_{t_{j-1}}^{t_j} f(t, y(t))\,\mathrm{d}t$$

holds in order to compute approximate solutions $y_j \approx y(t_j)$ iteratively. Here we consider s-stage Runge-Kutta methods (see, e.g., [6, 13–15]), a particular family of single-step methods. They are defined via

$$y_j = y_{j-1} + \omega_j \sum_{i=1}^{s} \beta_i k_i^{(j)}, \qquad j = 1, \dots, N, \tag{14}$$

$$k_i^{(j)} = f\Big(t_{j-1} + \gamma_i \omega_j,\; y_{j-1} + \omega_j \sum_{\ell=1}^{s} \lambda_{i\ell} k_\ell^{(j)}\Big), \qquad i = 1, \dots, s, \tag{15}$$

for certain $\beta_i \in \mathbb{C}$, $\gamma_i \in \mathbb{R}$, $i = 1, \dots, s$ and $\lambda_{i\ell} \in \mathbb{C}$, $i, \ell = 1, \dots, s$. Please note that we allow for complex-valued λ_{ij} and β_i unlike the usual definition of Runge-Kutta methods. Moreover, $\omega_j := t_j - t_{j-1} > 0$, $j = 1, \dots, N$, denotes the time step size. Often Runge-Kutta methods are given in short hand by the so-called Butcher tableau

$$\begin{array}{c|c} \gamma & \Lambda \\ \hline & \beta^\mathsf{T} \end{array} = \begin{array}{c|cccc} \gamma_1 & \lambda_{11} & \lambda_{12} & \dots & \lambda_{1s} \\ \gamma_2 & \lambda_{21} & \lambda_{22} & \dots & \lambda_{2s} \\ \vdots & \vdots & \vdots & \ddots & \vdots \\ \gamma_s & \lambda_{s1} & \lambda_{s2} & \dots & \lambda_{ss} \\ \hline & \beta_1 & \beta_2 & \dots & \beta_s \end{array}$$

with $\Lambda \in \mathbb{C}^{s\times s}$, $\beta \in \mathbb{C}^s$ and $\gamma \in \mathbb{R}^s$.

The most involved part in the iteration is the calculation of $k_i^{(j)}$ in (15). If in the Butcher tableau Λ is a strict lower triangular matrix, then the $k_i^{(j)}$ can be calculated explicitly one after another and the resulting method is called an explicit Runge-Kutta method. Otherwise they are only defined implicitly and a system of (in general nonlinear) equations with sn unknowns has to be solved to obtain them. One strategy to simplify the computation is by using lower triangular matrices Λ, resulting in so-called diagonally implicit Runge-Kutta (DIRK) methods. Another kind of methods, derived from DIRK methods, are the Rosenbrock-Wanner methods [15, IV.7]. There, the nonlinear function f is approximated by a linear function. If the function f to be integrated is linear, then the Rosenbrock-Wanner methods coincide with Runge-Kutta methods.

The ODEs (12) and (13) are solved with two possibly different s-stage Runge-Kutta methods as suggested in [5, Remark 1]. The ODE (12) is solved with a method based on a Butcher tableau with $\tilde{\Lambda} \in \mathbb{C}^{s\times s}$ and $\tilde{\beta} \in \mathbb{R}^s_{\geq 0}$. We only allow non-negative real entries in $\tilde{\beta}$ to ensure that the approximation to the Gramian is positive semidefinite, cf. (25). The ODE (13) is solved using Butcher tableaus with $\Lambda \in \mathbb{C}^{s\times s}$ and $\beta \in \mathbb{C}^s$.

Applying the iteration (14) to (13) we obtain

$$h_j = h_{j-1} + \omega_j \sum_{i=1}^{s} \beta_i k_i^{(j)}, \qquad j = 1, \ldots, N \tag{16}$$

with initial value $h_0 = B \in \mathbb{R}^{n\times 1}$. The slopes $k_i^{(j)}$ are given via (15) by

$$k_i^{(j)} = A\Big(h_{j-1} + \omega_j \sum_{\ell=1}^{s} \lambda_{i\ell} k_\ell^{(j)}\Big) \tag{17}$$

for $i = 1, \ldots, s$. Application of (14) to (12) yields

$$P_j = P_{j-1} + \omega_j \sum_{i=1}^{s} \tilde{\beta}_i \tilde{k}_i^{(j)}, \qquad j = 1, \ldots, N \tag{18}$$

with initial value $P_0 = 0 \in \mathbb{R}^{n\times n}$. To obtain the slopes $\tilde{k}_i^{(j)}$ we apply (15) to (12). This yields $\tilde{k}_i^{(j)} = \mathfrak{h}_i^{(j)}(\mathfrak{h}_i^{(j)})^{\mathsf{H}}$ with

$$\mathfrak{h}_i^{(j)} = h_{j-1} + \omega_j \sum_{\ell=1}^{s} \lambda_{i\ell} k_\ell^{(j)} \tag{19}$$

for $i = 1, \ldots, s$ and with $k_\ell^{(j)}$ from (17) as the ODEs (12) and (13) are coupled (i.e., h from (13) appears in (12)).

We now aggregate the vectors $\mathfrak{h}_i^{(j)}$ and $k_i^{(j)}$ for $i = 1, \ldots, s$ in matrices: Let $\mathcal{H}_j = [\mathfrak{h}_1^{(j)}, \ldots, \mathfrak{h}_s^{(j)}] \in \mathbb{C}^{n\times s}$ and $K_j = [k_1^{(j)}, \ldots, k_s^{(j)}] \in \mathbb{C}^{n\times s}$. Note that $k_i^{(j)} = A\mathfrak{h}_i^{(j)}$ holds. Thus

$$K_j = A\mathcal{H}_j. \tag{20}$$

To obtain $\mathcal{H}_j$ we rewrite (19) with the matrices K_j and $\mathcal{H}_j$

$$\begin{aligned} \mathcal{H}_j &= [h_{j-1}, \ldots, h_{j-1}] + \omega_j K_j \Lambda^{\mathsf{T}} \\ &= h_{j-1} \otimes \mathbb{1}_s^{\mathsf{T}} + \omega_j A\mathcal{H}_j \Lambda^{\mathsf{T}}, \end{aligned} \tag{21}$$

where $\mathbb{1}_s = [1, \ldots, 1]^{\mathsf{T}}$ is the s-dimensional vector containing only ones. In the latter equation $\mathcal{H}_j$ is the only unknown. In case we can compute $\mathcal{H}_j$ from (21), K_j can be determined via (20).

Finally, using $\mathcal{H}_j$ and K_j, we express the summations in (16) and (18) through matrix multiplications. Herewith the iteration reads

$$\begin{aligned} P_j &= P_{j-1} + \mathcal{H}_j \operatorname{diag}(\omega_j \tilde{\beta}) \mathcal{H}_j^{\mathsf{H}}, \\ h_j &= h_{j-1} + \omega_j K_j \beta. \end{aligned} \tag{22}$$

First $\mathcal{H}_j$ and K_j are determined using (21) and (20), then h_j and P_j are updated.

In order to see when $\mathcal{H}_j$ is uniquely determined, (21) is reformulated via vectorization as a linear system of equations with a system matrix of size $ns \times ns$

$$\left(I_{ns} - \omega_j(\Lambda \otimes A)\right) \operatorname{vec}(\mathcal{H}_j) = h_{j-1} \otimes \mathbb{1}_s \in \mathbb{C}^{ns\times 1}. \tag{23}$$

Let $\mu_1, \dots, \mu_s$ and $\lambda_1, \dots, \lambda_n$ be the eigenvalues of Λ and A respectively. Then the eigenvalues of $I_{ns} - \omega_j(\Lambda \otimes A)$ are given by $1 - \omega_j\mu_p\lambda_q$, $p = 1, \dots, s$, $q = 1, \dots, n$. Thus the solution of (23) is unique if and only if

$$\mu_p \neq \frac{1}{\omega_j \lambda_q} \tag{24}$$

for all $p = 1, \dots, s$ and $q = 1, \dots, n$.

As $\tilde{\beta} \in \mathbb{R}^s_{\geq 0}$ the approximant P_j is by construction a positive semidefinite matrix and can be expressed as $P_j = Z_j Z_j^{\mathsf{H}}$ for some complex-valued matrix Z_j. Thus we have

$$\begin{aligned} Z_j Z_j^{\mathsf{H}} &= Z_{j-1} Z_{j-1}^{\mathsf{H}} + \mathcal{H}_j \operatorname{diag}(\omega_j \tilde{\beta}) \mathcal{H}_j^{\mathsf{H}} \\ &= \left[Z_{j-1}, \mathcal{H}_j \operatorname{diag}(\omega_j \tilde{\beta})^{\frac{1}{2}}\right] \left[Z_{j-1}, \mathcal{H}_j \operatorname{diag}(\omega_j \tilde{\beta})^{\frac{1}{2}}\right]^{\mathsf{H}}. \end{aligned}$$

Instead of iterating on P_j as in (22), the above observation allows us to iterate on the low-rank factor

$$Z_j = [Z_{j-1}, \mathcal{H}_j \operatorname{diag}(\omega_j \tilde{\beta})^{\frac{1}{2}}] \in \mathbb{C}^{n\times js} \tag{25}$$

which gains s additional columns in every iteration step.

The procedure to obtain the Gramian approximation described in this section is summarized in Algorithm 1. We require that the eigenvalues of Λ satisfy (24) in order to ensure that all linear system solves have a unique solution and $\tilde{\beta} \in \mathbb{R}^s_{\geq 0}$ to ensure P_j is positive semidefinite.

2.2 Computation of $\mathcal{H}_j$ in Algorithm 1

The main part of Algorithm 1 is solving for $\mathcal{H}_j$ in line 3. Of course (23) can be used to determine $\mathcal{H}_j$. However, this means the solution of the ns-dimensional system (23). We present a more efficient way to obtain $\mathcal{H}_j$ with the solution of s linear systems of dimension n. A related approach has been used in [8].

Let $(\Lambda')^{\mathsf{T}} = S\Lambda^{\mathsf{T}}S^{-1} \in \mathbb{C}^{s\times s}$ be a Schur decomposition of Λ^{T}, so the diagonal entries of the upper triangular matrix $(\Lambda')^{\mathsf{T}}$ are the eigenvalues $\mu_1, \dots, \mu_s$ of Λ. Consider (21) and define $\mathcal{H}'_j = [\mathfrak{h}'^{(j)}_1, \dots, \mathfrak{h}'^{(j)}_s]$ via $\mathcal{H}_j = \mathcal{H}'_j S$. Then (21) can be reformulated as

Algorithm 1 Approximate Cholesky factor computation via an s-stage Runge-Kutta method

Input: $A \in \mathbb{R}^{n\times n}$ asymptotically stable, $B \in \mathbb{R}^{n\times 1}$, positive time step sizes $\{\omega_1, \dots, \omega_N\}$, Butcher tableau with $\tilde{\beta} \in \mathbb{R}^s_{\geq 0}$ and Butcher tableau with $\Lambda \in \mathbb{C}^{s\times s}$, $\beta \in \mathbb{C}^s$ which satisfies (24)
Output: $Z \in \mathbb{C}^{n\times sN}$ with $ZZ^{\mathsf{H}} \approx \mathcal{P}$
1: initialize $h_0 = B$, $Z_0 = [\;]$
2: **for** $j = 1, \dots, N$ **do**
3: solve $\mathcal{H}_j = [h_{j-1}, \dots, h_{j-1}] + \omega_j A\mathcal{H}_j\Lambda^{\mathsf{T}}$ for $\mathcal{H}_j \in \mathbb{C}^{n\times s}$
4: update $Z_j = [Z_{j-1}, \mathcal{H}_j \operatorname{diag}(\omega_j\tilde{\beta})^{\frac{1}{2}}]$
5: $h_j = h_{j-1} + \omega_j A\mathcal{H}_j\beta$
6: **end for**
7: $Z = Z_N$

$$\mathcal{H}'_j = (h_{j-1} \otimes \mathbb{1}_s^{\mathsf{T}})S^{-1} + \omega_j A\mathcal{H}'_j(\Lambda')^{\mathsf{T}}. \tag{26}$$

Let $[\alpha_1, \dots, \alpha_s] = \mathbb{1}_s^{\mathsf{T}} S^{-1}$ be the row vector containing the column sums of S^{-1}. Then we can rewrite (26) as

$$\mathcal{H}'_j = [\alpha_1 h_{j-1}, \dots, \alpha_s h_{j-1}] + \omega_j A\mathcal{H}'_j(\Lambda')^{\mathsf{T}}.$$

To obtain $\mathcal{H}'_j$, the following systems of linear equations have to be solved

$$(I_n - \omega_j\mu_i A)\mathfrak{h}_i'^{(j)} = \alpha_i h_{j-1} + \omega_j \sum_{l=1}^{i-1} \lambda'_{il} A\mathfrak{h}_l'^{(j)} \tag{27}$$

for $i = 1, \dots, s$. Finally, $\mathcal{H}_j$ is assembled via $\mathcal{H}_j = \mathcal{H}'_j S$.

Assume that linear systems with a system matrix of dimension $\tau \times \tau$ are solved with a method needing $O(\tau^3)$ flops. Then solving the ns-dimensional system (23) would need $O(s^3n^3)$ flops. In the procedure presented here a Schur decomposition of the $s \times s$ matrix Λ is necessary to obtain (27), at the costs of $O(s^3)$ flops. To solve the s systems of dimension $n \times n$ in (27) further $O(sn^3)$ flops are necessary. All in all the costs are reduced from $O(s^3n^3)$ flops to only $O(sn^3) + O(s^3)$ flops.

2.3 The Space Spanned by the Approximate Cholesky Factor Z

The main result of this section is that the columns of the approximate Cholesky factor $Z = Z_N$ obtained from Algorithm 1 span a (rational) Krylov subspace which is essentially determined by the eigenvalues of $\omega_i\Lambda$. To show this we first demonstrate how the iterate Z can be obtained in only one step of Algorithm 1 with certain Butcher tableaus assembled from Λ, β, $\tilde{\beta}$ and the time step sizes ω_j.

After N steps of Algorithm 1 we find the approximate Cholesky factor Z which is recursively defined via line 4. Expanding the for loop

$$Z = [\mathcal{H}_1, \ldots, \mathcal{H}_N] \begin{bmatrix} \operatorname{diag}(\omega_1 \tilde{\beta})^{\frac{1}{2}} & & \\ & \ddots & \\ & & \operatorname{diag}(\omega_N \tilde{\beta})^{\frac{1}{2}} \end{bmatrix} \tag{28}$$

is obtained. For $\mathcal{H}_1$ we have from line 3 in Algorithm 1

$$\begin{aligned} \mathcal{H}_1 &= \mathbb{1}_s^\mathsf{T} \otimes h_0 + \omega_1 A \mathcal{H}_1 \Lambda^\mathsf{T} \\ &= \mathbb{1}_s^\mathsf{T} \otimes h_0 + A \mathcal{H}_1 (\omega_1 \Lambda^\mathsf{T}). \end{aligned} \tag{29}$$

For $\mathcal{H}_2$ we find from line 3 and line 5 of Algorithm 1

$$\begin{aligned} \mathcal{H}_2 &= \mathbb{1}_s^\mathsf{T} \otimes h_1 + \omega_2 A \mathcal{H}_2 \Lambda^\mathsf{T} \\ &= \mathbb{1}_s^\mathsf{T} \otimes (h_0 + \omega_1 A \mathcal{H}_1 \beta) + \omega_2 A \mathcal{H}_2 \Lambda^\mathsf{T} \\ &= \mathbb{1}_s^\mathsf{T} \otimes h_0 + A \mathcal{H}_1 (\omega_1 [\beta, \ldots, \beta]) + A \mathcal{H}_2 (\omega_2 \Lambda^\mathsf{T}) \\ &= \mathbb{1}_s^\mathsf{T} \otimes h_0 + A [\mathcal{H}_1, \mathcal{H}_2] \begin{bmatrix} \omega_1 [\beta, \ldots, \beta] \\ \omega_2 \Lambda^\mathsf{T} \end{bmatrix}. \end{aligned} \tag{30}$$

Putting $\mathcal{H}_1$ from (29) and $\mathcal{H}_2$ from (30) together, one yields

$$[\mathcal{H}_1, \mathcal{H}_2] = \mathbb{1}_{2s}^\mathsf{T} \otimes h_0 + A [\mathcal{H}_1, \mathcal{H}_2] \begin{bmatrix} \omega_1 \Lambda^\mathsf{T} & \omega_1 [\beta, \ldots, \beta] \\ 0 & \omega_2 \Lambda^\mathsf{T} \end{bmatrix}.$$

Proceeding in this way up to iteration step N and setting $\hat{\mathcal{H}} = [\mathcal{H}_1, \ldots, \mathcal{H}_N]$ this leads to the equation

$$\hat{\mathcal{H}} = \mathbb{1}_{Ns}^\mathsf{T} \otimes h_0 + A \hat{\mathcal{H}} \hat{\Lambda}^\mathsf{T} \tag{31}$$

with

$$\hat{\Lambda}^\mathsf{T} := \begin{bmatrix} \omega_1 \Lambda^\mathsf{T} & \omega_1 [\beta, \ldots, \beta] & \cdots & \omega_1 [\beta, \ldots, \beta] \\ 0 & \omega_2 \Lambda^\mathsf{T} & \omega_2 [\beta, \ldots, \beta] & \omega_2 [\beta, \ldots, \beta] \\ \vdots & 0 & \ddots & \vdots \\ 0 & \cdots & 0 & \omega_N \Lambda^\mathsf{T} \end{bmatrix} \in \mathbb{C}^{Ns \times Ns}. \tag{32}$$

Thus, the result Z from (28) can also be interpreted as one step of Algorithm 1 with time step size $\boldsymbol{\omega}_1 = 1$, $\boldsymbol{\beta} = [\omega_1 \beta^\mathsf{T}, \ldots, \omega_N \beta^\mathsf{T}]^\mathsf{T}$, $\tilde{\boldsymbol{\beta}} = [\omega_1 \tilde{\beta}^\mathsf{T}, \ldots, \omega_N \tilde{\beta}^\mathsf{T}]^\mathsf{T}$ and $\boldsymbol{\Lambda} = \hat{\Lambda}$ from (32). It is therefore sufficient to analyze one step of Algorithm 1. The situation with more than one step is contained as a special case as described above.

Let all entries of $\tilde{\beta}$ be positive, i.e., $\tilde{\beta} \in \mathbb{R}^s_+$, then the diagonal matrix in (28) is regular and so the space spanned by the columns of Z equals the one spanned by the columns of $\hat{\mathcal{H}}$. We proceed with similarity transformations of $\hat{\Lambda}^\mathsf{T}$ as in Sect. 2.2 to uncouple the columns of $\hat{\mathcal{H}}$. Define $\hat{\mathcal{H}} = \hat{\mathcal{H}}'S$ with a similarity transformation $S \in \mathbb{C}^{Ns \times Ns}$ which transforms $\hat{\Lambda}^\mathsf{T}$ to its Jordan canonical form

$$(\hat{\Lambda}')^\mathsf{T} = S\hat{\Lambda}^\mathsf{T}S^{-1} = \begin{bmatrix} J_1 & & \\ & \ddots & \\ & & J_q \end{bmatrix} \tag{33}$$

with q Jordan blocks $J_l \in \mathbb{C}^{s_l \times s_l}$ of dimension s_l for $l = 1, \dots, q$. We further partition $\hat{\mathcal{H}}' = [\hat{\mathcal{H}}'_1, \dots, \hat{\mathcal{H}}'_q]$ and

$$\mathbb{1}^\mathsf{T}_{Ns} S^{-1} = [\alpha^{(1)}, \dots, \alpha^{(q)}] \tag{34}$$

according to the sizes of the Jordan blocks, i.e., $\hat{\mathcal{H}}'_l \in \mathbb{C}^{n \times s_l}$ and $(\alpha^{(l)})^\mathsf{T} \in \mathbb{C}^{s_l}$. Multiplication of (31) with S^{-1} from the right yields

$$[\hat{\mathcal{H}}'_1, \dots, \hat{\mathcal{H}}'_q] = [\alpha^{(1)}, \dots, \alpha^{(q)}] \otimes h_0 + A[\hat{\mathcal{H}}'_1, \dots, \hat{\mathcal{H}}'_q] \begin{bmatrix} J_1 & & \\ & \ddots & \\ & & J_q \end{bmatrix}.$$

Due to the partitioning this equation is equivalent to

$$\hat{\mathcal{H}}'_l = \alpha^{(l)} \otimes h_0 + A\hat{\mathcal{H}}'_l J_l \quad \text{for } l = 1, \dots, q.$$

The matrices $\hat{\mathcal{H}}'_l = [\hat{\mathfrak{h}}'^{(l)}_1, \dots, \hat{\mathfrak{h}}'^{(l)}_{s_l}]$ are determined by

$$\begin{aligned} (I_n - \hat{\mu}_l A)\hat{\mathfrak{h}}'^{(l)}_1 &= \alpha^{(l)}_1 h_0, \\ (I_n - \hat{\mu}_l A)\hat{\mathfrak{h}}'^{(l)}_i &= \alpha^{(l)}_i h_0 + A\hat{\mathfrak{h}}'^{(l)}_{i-1} \quad \text{for } i = 2, \dots, s_l \end{aligned} \tag{35}$$

with the eigenvalue $\hat{\mu}_l$ of $\hat{\Lambda}$ as the diagonal element of the Jordan block J_l.

Before we proceed with the main result of this section we state a technical lemma.

Lemma 1 *Let $(\hat{\Lambda}^\mathsf{T}, \mathbb{1}^\mathsf{T}_{Ns})$ be observable. Then there exists a transformation matrix S to Jordan canonical form in* (33) *such that $\alpha^{(l)} = [1, 0, \dots, 0]$ holds for $l = 1, \dots, q$ in* (34).

Proof For $l = 1, \dots, q$ define $e_l = [1, 0 \dots, 0] \in \mathbb{R}^{1 \times s_l}$. Assume that there exist polynomials p_l with

$$\alpha^{(l)} = e_l p_l(J_l). \tag{36}$$

Now replace the matrix S in (33) and (34) with $\tilde{S} = \operatorname{diag}(p_1(J_1), \dots, p_q(J_q))S$. As J_l commutes with rational functions in J_l the matrix $\tilde{S}$ is a similarity transformation to Jordan canonical form, too, and it holds

$$\begin{aligned}\mathbb{1}_{Ns}^{\mathsf{T}}\tilde{S}^{-1} &= \mathbb{1}_{Ns}^{\mathsf{T}}S^{-1}\operatorname{diag}(p_1(J_1), \dots, p_q(J_q))^{-1}\\ &= [\alpha^{(1)}, \dots, \alpha^{(q)}]\operatorname{diag}(p_1(J_1)^{-1}, \dots, p_q(J_q)^{-1})\\ &= [e_1, \dots, e_q].\end{aligned}$$

It remains to show that a polynomial p_l fulfilling (36) exists and $p_l(J_l)$ is invertible for $l = 1, \dots, q$. Define the upper shift matrix $r_l(J_l) = -\hat{\mu}_l I + J_l$ with ones above the diagonal and zeros everywhere else. It holds $e_l r_l(J_l)^{i-1} = [0, \dots, 0, 1, 0, \dots 0]$, a vector with a one at position i for $i = 1, \dots, s_l$. For the ith row of $p_l(J_l)$ we find with (36)

$$\begin{aligned}[0, \dots, 0, 1, 0, \dots 0]p_l(J_l) &= e_l r_l(J_l)^{i-1}p_l(J_l)\\ &= e_l p_l(J_l) r_l(J_l)^{i-1}\\ &= \alpha^{(l)} r_l(J_l)^{i-1}\\ &= [0, \dots, 0, \alpha_1^{(l)}, \dots, \alpha_{s_l-(i-1)}^{(l)}].\end{aligned}$$

This implies that $p_l(J_l)$ is an upper triangular matrix with entries $\alpha_1^{(l)}$ on the diagonal. As $(\hat{\Lambda}^{\mathsf{T}}, \mathbb{1}_{Ns}^{\mathsf{T}})$ is observable, so is $(J_l, \alpha^{(l)})$ and thus $\alpha_1^{(l)} \neq 0$. So $p_l(J_l)$ is invertible, which concludes the proof.

These preparations allow us to state the following lemma.

Lemma 2 *Let $Ns < n$ and $(\hat{\Lambda}^{\mathsf{T}}, \mathbb{1}_{Ns}^{\mathsf{T}})$ be observable. If $\hat{\mu}_l \neq 0$ then*

$$\operatorname{span}\hat{\mathcal{H}}_l' = \operatorname{span}\left\{(I_n - \hat{\mu}_l A)^{-i}h_0 \mid i = 1, \dots, s_l\right\}.$$

If $\hat{\mu}_l = 0$ then

$$\operatorname{span}\hat{\mathcal{H}}_l' = \operatorname{span}\left\{A^i h_0 \mid i = 0, \dots, s_l - 1\right\}.$$

Proof In this proof set $\hat{\mathfrak{h}}_i' := \hat{\mathfrak{h}}_i'^{(l)}$ for better readability. Due to the observability of $(\hat{\Lambda}^{\mathsf{T}}, \mathbb{1}_{Ns}^{\mathsf{T}})$ we find from (33) and (34) that $(J_l, \alpha^{(l)})$ is observable. Due to Lemma 1 we can assume $\alpha^{(l)} = [1, 0, \dots, 0]$.

Let $\hat{\mu}_l \neq 0$. Because of (35)

$$\operatorname{span}\hat{\mathfrak{h}}_1' = \operatorname{span}\{(I_n - \hat{\mu}_l A)^{-1}h_0\}$$

holds. From (35) we find for $1 < i \leq s_l$ as $\alpha_i^{(l)} = 0$

$$\begin{aligned}\hat{\mathfrak{h}}'_i &= (I_n - \hat{\mu}_l A)^{-1} A \hat{\mathfrak{h}}'_{i-1} \\ &= (I_n - \hat{\mu}_l A)^{-1}(-\hat{\mu}_l^{-1}(I_n - \hat{\mu}_l A) + \hat{\mu}_l^{-1} I_n)\hat{\mathfrak{h}}'_{i-1} \\ &= -\hat{\mu}_l^{-1}\hat{\mathfrak{h}}'_{i-1} + \hat{\mu}_l^{-1}(I_n - \hat{\mu}_l A)^{-1}\hat{\mathfrak{h}}'_{i-1}.\end{aligned}$$

Via induction this concludes the first part of the proof.

Now let $\hat{\mu}_l = 0$. From (35)

$$\operatorname{span} \hat{\mathfrak{h}}'_1 = \operatorname{span} h_0$$

is immediate. For $1 < i \leq s_l$ we have

$$\hat{\mathfrak{h}}'_i = A\hat{\mathfrak{h}}'_{i-1},$$

and the claim again results from induction.

We conclude that the space spanned by $\hat{\mathcal{H}}'$ (and thus also by $\hat{\mathcal{H}}$) mainly depends on the eigenvalues $\hat{\mu}_l$ of $\hat{\Lambda}$ and the dimensions s_l of their eigenspaces.

3 Approximate Balancing Transformation

We now present an algorithm which generates an approximate balancing transformation. The reduced system is obtained via projection using approximated Gramians. It can be seen as a variant of balanced POD where the Cholesky factors of the Gramians are approximated using the quadrature described in Sect. 2.1. This procedure is summarized in Algorithm 2.

Note that due to the use of Butcher tableaus with complex entries in general complex reduced system matrices are obtained. This is the reason for using conjugate transposition H instead of transposition T.

Algorithm 2 Approximate balancing transformation

Input: system matrices $A \in \mathbb{R}^{n\times n}$ asymptotically stable, $B \in \mathbb{R}^{n\times 1}$, $C \in \mathbb{R}^{1\times n}$, positive time step sizes $\{\omega_1, \dots, \omega_N\}$ and $\{\tau_1, \dots, \tau_N\}$, Butcher tableaus with $\tilde{\beta}_c, \tilde{\beta}_o \in \mathbb{R}^s_{\geq 0}$ and Butcher tableaus with $\Lambda_c, \Lambda_o \in \mathbb{C}^{s\times s}$, $\beta_c, \beta_o \in \mathbb{C}^s$ which satisfy (24)
Output: reduced system matrices $\hat{A} \in \mathbb{C}^{r\times r}$, $\hat{B} \in \mathbb{C}^{r\times 1}$, $\hat{C} \in \mathbb{C}^{1\times r}$ with $r = \operatorname{rank}(Z_o^{\mathsf{H}} Z_c)$
1: obtain Z_c with $Z_c Z_c^{\mathsf{H}} \approx \mathcal{P}$ from Algorithm 1 with A, B, Λ_c, β_c, $\tilde{\beta}_c$ and $\{\omega_1, \dots, \omega_N\}$
2: obtain Z_o with $Z_o Z_o^{\mathsf{H}} \approx \mathcal{Q}$ from Algorithm 1 with A^{T}, C^{T}, Λ_o, β_o, $\tilde{\beta}_o$ and $\{\tau_1, \dots, \tau_N\}$
3: calculate compact SVD $Z_o^{\mathsf{H}} Z_c = U\Sigma T^{\mathsf{H}}$
4: assemble projection matrices $V = Z_c T \Sigma^{-\frac{1}{2}}$, $W = Z_o U \Sigma^{-\frac{1}{2}}$
5: return $\hat{A} = W^{\mathsf{H}} A V$, $\hat{B} = W^{\mathsf{H}} B$, $\hat{C} = CV$

As will be shown next, the transfer function of the reduced system generated by Algorithm 2 interpolates the transfer function of the original system at expansion points which depend on the eigenvalues of the Butcher tableaus and the time step sizes. In particular the expansion points are the inverse eigenvalues of $\omega_i \Lambda_c$ for $i = 1, \ldots, N_c$ and the conjugated inverse eigenvalues of $\tau_i \Lambda_o$ for $i = 1, \ldots, N_o$.

Theorem 1 *Let the inputs of Algorithm 2 with $\tilde{\beta}_c, \tilde{\beta}_o \in \mathbb{R}^s_+$ be given. Define $\hat{\Lambda}_c^\mathsf{T}$ as in* (32) *with Λ_c, β_c and $\{\omega_1, \ldots, \omega_N\}$. Define $\hat{\Lambda}_o^\mathsf{T}$ as in* (32) *with Λ_o, β_o and $\{\tau_1, \ldots, \tau_N\}$. Let $\{\hat{\mu}_1, \ldots, \hat{\mu}_{q_c}\} = \cup_{i=1}^N \sigma(\omega_i \Lambda_c)$ and $\{\hat{\nu}_1, \ldots, \hat{\nu}_{q_o}\} = \cup_{i=1}^N \sigma(\tau_i \Lambda_o)$ be the eigenvalues of $\hat{\Lambda}_c$ and $\hat{\Lambda}_o$ with multiplicities $s_1, \ldots, s_{q_c}$ and $t_1, \ldots, t_{q_o}$.*

If $(\hat{\Lambda}_c^\mathsf{T}, \mathbb{1}_{Ns}^\mathsf{T})$ and $(\hat{\Lambda}_o^\mathsf{T}, \mathbb{1}_{Ns}^\mathsf{T})$ are observable and rank $Z_o^\mathsf{H} Z_c = Ns$ *holds, then the transfer function of the reduced system with system matrices $\hat{A}, \hat{B}, \hat{C}$ produced by Algorithm 2 satisfies*

$$\begin{aligned} \hat{G}^{(i)}(\hat{\mu}_{l_c}^{-1}) &= G^{(i)}(\hat{\mu}_{l_c}^{-1}) \quad \textit{for } i = 0, \ldots, s_{l_c} - 1, \\ \hat{G}^{(i)}(\overline{\hat{\nu}}_{l_o}^{-1}) &= G^{(i)}(\overline{\hat{\nu}}_{l_o}^{-1}) \quad \textit{for } i = 0, \ldots, t_{l_o} - 1 \end{aligned} \tag{37}$$

for $l_c = 1, \ldots, q_c$ and $l_o = 1, \ldots, q_o$. For any zero eigenvalues the corresponding interpolation in (37) *has to be read as interpolation at ∞. If some of the values $\hat{\mu}_i$ and $\overline{\hat{\nu}}_j$ coincide, even higher derivatives are interpolated.*

Proof The reduced system is generated via projection with the matrices V and W. Due to line 3 and line 4 of Algorithm 2 and as $Z_o^\mathsf{H} Z_c$ is regular $\operatorname{span}(V) = \operatorname{span}(Z_c)$ and $\operatorname{span}(W) = \operatorname{span}(Z_o)$ hold. With Lemma 2 we find for $\hat{\mu}_{l_c}, \hat{\nu}_{l_o} \neq 0$

$$\begin{aligned} \operatorname{span}\{(I_n - \hat{\mu}_{l_c} A)^{-i} B \mid i = 1, \ldots, s_{l_c}\} &\subseteq \operatorname{span}(V), \\ \operatorname{span}\{(I_n - \hat{\nu}_{l_o} A^\mathsf{T})^{-i} C^\mathsf{T} \mid i = 1, \ldots, t_{l_o}\} &\subseteq \operatorname{span}(W). \end{aligned}$$

Due to $(I_n - \hat{\mu}_{l_c} A)^{-1} = -\hat{\mu}_{l_c}^{-1}(A - \hat{\mu}_{l_c}^{-1} I_n)^{-1}$ and $(I_n - \hat{\nu}_{l_o} A^\mathsf{T})^{-1} = -\hat{\nu}_{l_o}^{-1}(A^\mathsf{T} - \hat{\nu}_{l_o}^{-1} I_n)^{-1}$ this means

$$\begin{aligned} \operatorname{span}\{(A - \hat{\mu}_{l_c}^{-1} I_n)^{-i} B \mid i = 1, \ldots, s_{l_c}\} &\subseteq \operatorname{span}(V), \\ \operatorname{span}\{(A^\mathsf{T} - \hat{\nu}_{l_o}^{-1} I_n)^{-i} C^\mathsf{T} \mid i = 1, \ldots, t_{l_o}\} &\subseteq \operatorname{span}(W). \end{aligned}$$

Further, if $\hat{\mu}_{l_c}, \hat{\nu}_{l_o} = 0$, then

$$\begin{aligned} \operatorname{span}\{A^i B \mid i = 0, \ldots, s_{l_c} - 1\} &\subseteq \operatorname{span}(V), \\ \operatorname{span}\{(A^\mathsf{T})^i C^\mathsf{T} \mid i = 0, \ldots, t_{l_o} - 1\} &\subseteq \operatorname{span}(W). \end{aligned}$$

Due to Sect. 1.2 this concludes the proof. □

It is interesting to see that using a Runge-Kutta method it is not possible to match moments around the expansion point zero, as this would require an infinite eigenvalue of Λ from the Butcher tableau or an infinite time step size, which is impossible.

In [22] complex time step sizes ω_j (τ_j respectively) are used in Runge-Kutta methods to achieve moment matching around complex expansion points. This is unfeasible in the method presented here as then the iterates P_j are in general not positive semidefinite and the approximate Cholesky factors Z_j would not exist. Instead, in the framework presented here, complex tableaus may be used.

4 Connection to Other Methods

We now show the connection of the method presented here to other methods involving Gramian approximations with low-rank Cholesky factors. We only consider the controllability Gramian $\mathcal{P}$. The approximation of the observability Gramian $\mathcal{Q}$ is done analogously, cf. Sect. 1.3. All methods have in common that the approximate Cholesky factors are computed directly, that is, no Cholesky decomposition of a large $n \times n$ matrix is necessary.

4.1 Balanced POD

We first consider balanced POD as introduced in [24] and summarized at the end of Sect. 1.1. A central task in BPOD is the numerical solution of the ODE (7). Unfortunately in [24] it is not stated which numerical method should be used for solving the ODE. In the following we assume a Runge-Kutta method with Λ_h and β_h is used to solve the ODE in the same way as (13) was solved in Sect. 2.1. In particular, for $h_0 = B$ and time step sizes $\omega_j = t_j - t_{j-1}$ this means

$$\begin{aligned} \mathcal{H}_j &= [h_{j-1}, \ldots, h_{j-1}] + \omega_j A \mathcal{H}_j \Lambda_h^{\mathsf{T}} \\ h_j &= h_{j-1} + \omega_j A \mathcal{H}_j \beta_h^{\mathsf{T}} \end{aligned} \tag{38}$$

just as in Algorithm 1, but in the BPOD method the approximate Cholesky factor is updated via

$$Z_j = [Z_{j-1}, h_j \delta_j^{\frac{1}{2}}]$$

instead of $Z_j = [Z_{j-1}, \mathcal{H}_j \operatorname{diag}(\omega_j \tilde{\beta})^{\frac{1}{2}}]$ as in Algorithm 1. We illustrate how the balanced POD iterates can be obtained using Algorithm 1 in case $h_j \delta_j h_j^{\mathsf{H}}$ and $\mathcal{H}_j \operatorname{diag}(\omega_j \tilde{\beta}) \mathcal{H}_j^{\mathsf{H}}$ coincide. Due to the dimension of h_j and $\mathcal{H}_j$ this is only possible for Butcher tableaus of size $s = 1$ or for $\tilde{\beta}$ having only one nonzero entry.

We first consider the case $s = 1$ and thus have $\mathcal{H}_j \in \mathbb{C}^{n \times 1}$. So (38) becomes

$$\begin{aligned} \mathcal{H}_j &= h_{j-1} + \omega_j A \mathcal{H}_j \Lambda_h^{\mathsf{T}} \\ h_j &= h_{j-1} + \omega_j A \mathcal{H}_j \beta_h^{\mathsf{T}}, \end{aligned}$$

i.e., $\mathcal{H}_j = h_j$ if $\Lambda_h = \beta_h$. This is, e.g., fulfilled in the backward Euler method with $\Lambda_h = \beta_h = 1$. If additionally $\tilde{\beta} = {}^{\delta_j}/_{\omega_j}$, balanced POD and Algorithm 1 produce the same iterates.

In case of arbitrary Butcher tableaus with s-dimensional Λ_h and β_h the way BPOD fits into the framework presented here is rather crude. Consider a Butcher tableau with the $s+1$-dimensional matrices

$$\Lambda = \begin{bmatrix} \Lambda_h & 0 \\ \beta_h^{\mathsf{T}} & 0 \end{bmatrix}, \quad \beta = \begin{bmatrix} \beta_h \\ 0 \end{bmatrix}, \quad \tilde{\beta} = \begin{bmatrix} 0 \\ {}^{\delta_j}/_{\omega_j} \end{bmatrix}.$$

Algorithm 1 generates the iterate

$$\underbrace{[\mathfrak{h}_1^{(j)}, \dots, \mathfrak{h}_s^{(j)}, \mathfrak{h}_{s+1}^{(j)}]}_{=\mathcal{H}_j} = [h_{j-1}, \dots, h_{j-1}] + \omega_j A[\mathfrak{h}_1^{(j)}, \dots, \mathfrak{h}_s^{(j)}, \mathfrak{h}_{s+1}^{(j)}] \begin{bmatrix} \Lambda_h^{\mathsf{T}} & \beta_h \\ 0 & 0 \end{bmatrix}.$$

Separating the first s columns from the last one yields

$$\begin{aligned} [\mathfrak{h}_1^{(j)}, \dots, \mathfrak{h}_s^{(j)}] &= [h_{j-1}, \dots, h_{j-1}] + \omega_j A[\mathfrak{h}_1^{(j)}, \dots, \mathfrak{h}_s^{(j)}]\Lambda_h^{\mathsf{T}} \\ \mathfrak{h}_{s+1}^{(j)} &= h_{j-1} + \omega_j A[\mathfrak{h}_1^{(j)}, \dots, \mathfrak{h}_s^{(j)}]\beta_h \end{aligned}$$

and so $h_j = \mathfrak{h}_{s+1}$. Due to the zero entries in $\tilde{\beta}$ we further find

$$\begin{aligned} \mathcal{H}_j \operatorname{diag}(\omega_j \tilde{\beta}) \mathcal{H}_j^{\mathsf{H}} &= \mathfrak{h}_j \omega_j \frac{\delta_j}{\omega_j} \mathfrak{h}_j^{\mathsf{H}} \\ &= h_j \delta_j h_j^{\mathsf{H}}, \end{aligned}$$

i.e., Algorithm 1 and BPOD produce the same iterates for this special choice of tableaus.

4.2 The ADI Iteration

It was shown in [5] that for certain Butcher tableaus Algorithm 1 is equivalent to the ADI iteration [17, 19, 20, 23, 27]. In particular, the Gramian approximation produced by Algorithm 1 for Butcher tableaus with $\beta = \tilde{\beta}$ and Λ satisfying

$$\operatorname{diag}(\beta)\overline{\Lambda} + \Lambda^{\mathsf{T}} \operatorname{diag}(\beta) - \beta\beta^{\mathsf{T}} = 0 \tag{39}$$

is equivalent to ADI approximants with parameters which are the negative inverses of the eigenvalues of $\omega_i \Lambda$. Runge-Kutta methods which fulfill (39) are given by the family of Gauß-Legendre methods (see [5], [16, Lem. 5.3]), i.e., the implicit midpoint rule with

$$\Lambda = \frac{1}{2}, \ \beta = 1$$

or the Gauß-Legendre method with $s = 2$ as in (41). A more generic way to construct Butcher tableaus which satisfy (39) is given by the lower triangular matrices

$$\Lambda = \begin{bmatrix} \mu_1 & 0 & \cdots & 0 \\ 2\operatorname{Re}(\mu_1) & \mu_2 & \ddots & 0 \\ \vdots & \vdots & \ddots & \vdots \\ 2\operatorname{Re}(\mu_1) & 2\operatorname{Re}(\mu_2) & \cdots & \mu_s \end{bmatrix}, \ \beta = \begin{bmatrix} 2\operatorname{Re}(\mu_1) \\ 2\operatorname{Re}(\mu_2) \\ \vdots \\ 2\operatorname{Re}(\mu_s) \end{bmatrix} \tag{40}$$

with parameters $\mu_1, \ldots, \mu_s \in \mathbb{C}_+$. With this tableau the connection to the ADI parameters is immediate as the eigenvalues can be read off the diagonal. An ADI iteration with parameters $\alpha_i \in \mathbb{C}_-$ is thus equivalent to one step of Algorithm 1 with step size $\omega_1 = 1$ using a Butcher tableau given by (40) with $\mu_i = -\alpha_i^{-1}$, see [5, Thm. 4]. From Lemma 2 and Theorem 1 it follows that the ADI iterates span a rational Krylov space and, if used in Algorithm 2, the moments at $-\alpha_i = \mu_i^{-1}$ are matched. See also [2, Sect. 2.4] for a different proof.

5 Examples

In this section we illustrate the findings from Theorem 1. We state the expansion points at which moments are matched for certain Runge-Kutta methods and visualize them in the complex plane.

Explicit Runge-Kutta methods are parameterized by Butcher tableaus with strictly lower triangular Λ. As such matrices have just zero eigenvalues only moments around ∞ are matched for explicit methods. An example is Euler's method given by the Butcher tableau with $\Lambda = 0, \ \beta = 1$.

For the backward Euler method we have $\Lambda = 1, \ \beta = 1$, so the moments are matched around the inverse time step sizes ω_j^{-1} and τ_j^{-1}.

Consider the Butcher tableaus from the Gauß-Legendre and Radau IA method of size $s = 2$. The Gauß-Legendre method is given by

$$\Lambda_{\mathrm{GL}} = \begin{bmatrix} \frac{1}{4} & \frac{1}{4} - \frac{1}{6}\sqrt{3} \\ \frac{1}{4} + \frac{1}{6}\sqrt{3} & \frac{1}{4} \end{bmatrix}, \ \beta_{\mathrm{GL}} = \begin{bmatrix} \frac{1}{2} \\ \frac{1}{2} \end{bmatrix}. \tag{41}$$

This method is equivalent to the Hammer-Hollingsworth method [7] which was used in [22]. The matrix Λ_{GL} has eigenvalues $\mu_{1/2} = \frac{1}{4} \pm \frac{\sqrt{3}}{12} i$. The Radau IA method is given by

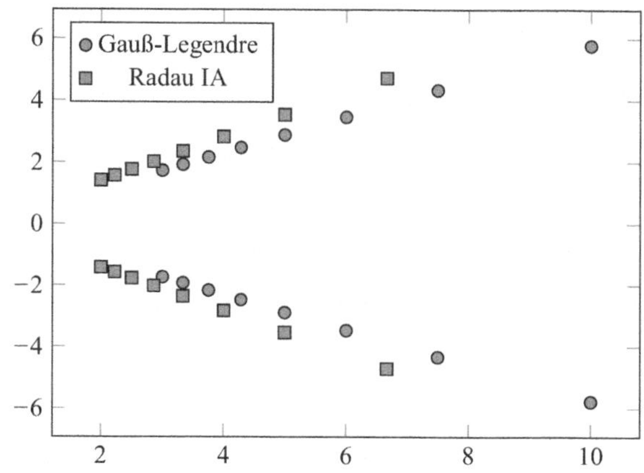

Fig. 1 Expansion points in the complex plane for Gauß-Legendre and Radau IA method

$$\Lambda_{\mathrm{R}} = \begin{bmatrix} \frac{1}{4} & -\frac{1}{4} \\ \frac{1}{4} & \frac{5}{12} \end{bmatrix}, \quad \beta_{\mathrm{R}} = \begin{bmatrix} \frac{1}{4} \\ \frac{3}{4} \end{bmatrix}.$$

It has eigenvalues $\lambda_{1/2} = \frac{1}{3} \pm \frac{\sqrt{2}}{6} i$.

When Algorithm 2 is executed with the Gauß-Legendre method for Z_{c} and the Radau IA method for Z_{o}, then the moments are matched around the expansion points

$$(\omega_j \mu_{1/2})^{-1} = \omega_j^{-1}(3 \mp \sqrt{3} i) \text{ and } (\tau_j \lambda_{1/2})^{-1} = \tau_j^{-1}(2 \mp \sqrt{2} i)$$

for $j = 1, \ldots, N$. These expansion points are visualized in the complex plane in Fig. 1 for $\omega_j = \tau_j = 0.3, 0.4, \ldots, 1$.

6 Conclusion

We have presented a method which generates approximate balancing transformations using approximate Cholesky factors of the Gramians obtained via numerical quadrature with Runge-Kutta methods. The moments of the reduced system coincide with the moments of the original systems at the inverses of the (conjugated) eigenvalues of the Butcher tableaus multiplied with the time step sizes, while explicit quadrature methods correspond to interpolation at infinity.

It remains an open question how the expansion points can be characterized if the SVD in Algorithm 2 is truncated, i.e., if balanced truncation is performed instead of an approximate balancing transformation. Then the reduced system is obtained via projection onto a subspace of a rational Krylov space and the direct connection between the poles of the rational Krylov space and the expansion points around which the moments are matched is lost.

References

1. Antoulas, A.C.: Approximation of large-scale dynamical systems. Soc. Ind. Appl. Math. (2005). ISBN: 978-0-898-71529-3
2. Baur, U., Benner, P., Feng, L.: Model order reduction for linear and nonlinear systems: a system-theoretic perspective. Arch. Comput. Methods Eng. **21**, 331–358 (2014). https://doi.org/10.1007/s11831-014-9111-2
3. Benner, P., Cohen, A., Ohlberger, M., Willcox, K. (eds.): Model Reduction and Approximation: Theory and Algorithms. SIAM (2017). ISBN: 978-1-611974-81-2
4. Benner, P., Hinze, M., ter Maten, E. (eds.): Model Reduction for Circuit Simulation. Lecture Notes in Electrical Engineering, vol. 74. Springer, Dordrecht, The Netherlands (2011). ISBN: 978-94-007-0089-5
5. Bertram, C., Faßbender, H.: Lyapunov and Sylvester equations: a quadrature framework. arXiv e-prints arXiv:1903.05383 (2019)
6. Butcher, J.: Numerical Methods for Ordinary Differential Equations. Wiley (2016). ISBN: 978-1-119-12150-3
7. Butcher, J.C.: Implicit Runge-Kutta processes. Math. Comput. **18**(85), 50–64 (1964). https://doi.org/10.2307/2003405
8. Butcher, J.C.: On the implementation of implicit Runge-Kutta methods. BIT **16**(3), 237–240 (1976). https://doi.org/10.1007/bf01932265
9. Freund, R.W.: Krylov-subspace methods for reduced-order modeling in circuit simulation. J. Comput. Appl. Math. **123**(1), 395–421 (2000). https://doi.org/10.1016/S0377-0427(00)00396-4. Numerical Analysis 2000. Vol. III: Linear Algebra
10. Gallivan, K., Vandendorpe, A., Dooren, P.V.: Model reduction via truncation: an interpolation point of view. Linear Algebra Appl. **375**, 115–134 (2003). https://doi.org/10.1016/S0024-3795(03)00648-7
11. Grimme, E.: Krylov projection methods for model reduction. Ph.D. thesis, University of Illinois at Urbana-Champaign (1997)
12. Gugercin, S., Sorensen, D., Antoulas, A.: A modified low-rank Smith method for large-scale Lyapunov equations. Numer. Algorithms **32**, 27–55 (2003). https://doi.org/10.1023/A:1022205420182
13. Hairer, E., Lubich, C., Wanner, G.: Geometric Numerical Integration: Structure-Preserving Algorithms for Ordinary Differential Equations, 2nd edn. Springer (2006). ISBN: 978-3-540-30666-5
14. Hairer, E., Norsett, S., Wanner, G.: Solving Ordinary Differential Equations I, 2nd edn. Springer (1993). ISBN: 978-3-540-78862-1
15. Hairer, E., Wanner, G.: Solving Ordinary Differential Equations II, 2nd edn. Springer (1996). ISBN: 978-3-642-05221-7
16. Iserles, A.: A First Course in the Numerical Analysis of Differential Equations, 2nd edn. Cambridge Texts in Applied Mathematics. Cambridge University Press (2008). https://doi.org/10.1017/CBO9780511995569
17. Kürschner, P.: Efficient low-rank solution of large-scale matrix equations. Ph.D. thesis, OvGU Magdeburg (2016)
18. Lall, S., Marsden, J.E., Glavaski, S.: A subspace approach to balanced truncation for model reduction of nonlinear control systems. Int. J. Robust Nonlinear Control **12**(6), 519–535 (2002). https://doi.org/10.1002/rnc.657
19. Li, J., White, J.: Low rank solution of Lyapunov equations. SIAM J. Matrix Anal. Appl. **24**(1), 260–280 (2002). https://doi.org/10.1137/S0895479801384937
20. Lu, A., Wachspress, E.: Solution of Lyapunov equations by alternating direction implicit iteration. Comput. Math. Appl. **21**(9), 43–58 (1991). https://doi.org/10.1016/0898-1221(91)90124-M
21. Moore, B.: Principal component analysis in linear systems: controllability, observability, and model reduction. IEEE Trans. Automat. Control **AC-26**, 17–32 (1981). https://doi.org/10.1109/TAC.1981.1102568

22. Opmeer, M.: Model order reduction by balanced proper orthogonal decomposition and by rational interpolation. IEEE Trans. Autom. Control **AC-57**, 472–477 (2012). https://doi.org/10.1109/TAC.2011.2164018
23. Peaceman, D., Rachford, H., Jr.: The numerical solution of parabolic and elliptic differential equations. J. Soc. Ind. Appl. Math. **3**(1), 28–41 (1955). https://doi.org/10.1137/0103003
24. Rowley, C.W.: Model reduction for fluids, using balanced proper orthogonal decomposition. I. J. Bifurcat. Chaos **15**(3), 997–1013 (2005). https://doi.org/10.1142/S0218127405012429
25. Simoncini, V.: Computational methods for linear matrix equations. SIAM Rev. **58**(3), 377–441 (2016). https://doi.org/10.1137/130912839
26. Willcox, K., Peraire, J.: Balanced model reduction via the proper orthogonal decomposition. AIAA J. **40**(11), 2323–2330 (2002). https://doi.org/10.2514/2.1570
27. Wolf, T., Panzer, H.: The ADI iteration for Lyapunov equations implicitly performs H2 pseudo-optimal model order reduction. Int. J. Control **89**(3), 481–493 (2016). https://doi.org/10.1080/00207179.2015.1081985

Comparing (Empirical-Gramian-Based) Model Order Reduction Algorithms

Christian Himpe

Abstract In this work, the empirical-Gramian-based model reduction methods: Empirical poor man's truncated balanced realization, empirical approximate balancing, empirical dominant subspaces, empirical balanced truncation, and empirical balanced gains are compared in a non-parametric and in two parametric variants, via ten error measures: Approximate Lebesgue L_0, L_1, L_2, L_∞, Hardy H_2, H_∞, Hankel, Hilbert-Schmidt-Hankel, modified induced primal, and modified induced dual norms, for variants of the thermal block model reduction benchmark. This comparison is conducted via a new meta-measure for model reducibility called MORscore.

1 Introduction

Model reduction research has made great strides in the past decades, spawning ever new methods and variants for specific requirements. Yet, this plethora of algorithms is not (or only very sparsely) evaluated against each other on common benchmarks. Such comparisons would enable a faster transfer of mathematically research to engineering and industrial applications.

In the following, prototypically, a comparison of empirical-Gramian-based methods is demonstrated for a standard benchmark system in a manner, which can be automated, for example, to test various variants of a method to determine the best suited for a certain problem. In the scope of this work, model reduction for affine-parametric, generalized, linear time-invariant systems is considered:

$$\begin{aligned} E\dot{x}(t) &= A(\theta)x(t) + Bu(t), \\ y(t) &= Cx(t), \end{aligned} \tag{1}$$

C. Himpe (✉)
Computational Methods in Systems and Control Theory, Max Planck Institute for Dynamics of Complex Technical Systems, Sandtorstr. 1, 39106 Magdeburg, Germany
e-mail: himpe@mpi-magdeburg.mpg.de

P. Benner et al. (eds.), *Model Reduction of Complex Dynamical Systems*, International Series of Numerical Mathematics 171,
https://doi.org/10.1007/978-3-030-72983-7_7

which consist of an ordinary differential equation in x, with a non-singular mass matrix $E \in \mathbb{R}^{N \times N}$, an affinely decomposable parametric system matrix $A(\theta) = A_0 + \sum_{p=1}^{P} \theta_p A_p \in \mathbb{R}^{N \times N}$, so that $E^{-1}A(\theta)$ is asymptotically stable for all parameters $\theta \in \Theta \subset \mathbb{R}^P$, and an input matrix $B \in \mathbb{R}^{N \times M}$, as well as a linear output function defined by the output matrix $C \in \mathbb{R}^{Q \times N}$.

In the following some fundamentals of projection-based model reduction are assumed; for a background on this topic the reader is referred to the seminal textbook [2].

2 Empirical Gramians for Linear Systems

System Gramians are system-theoretic operators encoding the input-output system properties of controllability and observability [34]. Empirical Gramians [36] are generalizations of these system Gramians, which are based on quadrature, and were introduced to apply linear, Gramian-based methods from linear system theory to nonlinear systems, while incorporating nonlinear information and avoiding (explicit) linearization. Since linear systems are a special case of nonlinear systems, with, admittedly, a very simple "nonlinearity", empirical Gramians can also be computed for linear systems. Note that for linear systems, the empirical Gramians correspond to the classic system Gramians up to numerical error; this is shown in [29, 36]. The quality of the empirical Gramians depends on simulated state and output trajectories for which the system is excited by perturbed input or initial state. These perturbations are defined by scales (c_m and d_q) and should reflect the operating region of the system. Following, we summarize the three fundamental empirical system Gramians in the special case of linear systems.

2.1 *Empirical Controllability Gramian*

The controllability Gramian quantifies the ability to drive a linear system to a steady state in finite time via the input [35]. For linear systems, the controllability Gramian matrix is defined as $W_C := \int_0^\infty e^{E^{-1}At} E^{-1} B B^\intercal E^{-\intercal} e^{A^\intercal E^{-\intercal} t} \, dt$, and classically computed as the (low-rank) solution to the Lyapunov equation $A W_C E^\intercal + E W_C A^\intercal = -B B^\intercal$. Based on the definition of W_C, the empirical controllability Gramian is given by

$$\widehat{W}_C := \sum_{m=1}^{M} \int_0^\infty x^m(t) x^m(t)^\intercal dt$$

with $x^m(t)$ being the solution of $E\dot{x}^m(t) = Ax^m(t) + B(c_m e_m \delta(t))$, suitable scales $c_m \in \mathbb{R}$, and the m-th canonical standard base vector $e_m \in \mathbb{R}^M$.

2.2 Empirical Observability Gramian

The observability Gramian matrix describes the ability to determine the state of a linear system via its output in finite time [35]. For linear systems, the observability Gramian matrix is defined as $W_O := \int_0^\infty \mathrm{e}^{A^\intercal E^{-\intercal} t} C^\intercal C \, \mathrm{e}^{E^{-1} A t} \, dt$, and is classically computed as the (low-rank) solution to the Lyapunov equation $A^\intercal W_O E + E^\intercal W_O A = -C^\intercal C$. Based on the definition of W_O, the (linear) empirical observability Gramian (via the dual system's controllability Gramian [59]) is given by

$$\widehat{W}_O := \sum_{q=1}^{Q} \int_0^\infty z^q(t) z^q(t)^\intercal dt,$$

with $z^q(t)$ being the solution of $E^\intercal \dot{z}^q(t) = A^\intercal z^q(t) + C^\intercal (d_q e_q \delta(t))$, suitable scales $d_q \in \mathbb{R}$, and the q-th canonical standard base vector $e_q \in \mathbb{R}^Q$.

2.3 Empirical Cross Gramian

The cross Gramian matrix combines controllability and observability information and hence delineates the minimality of a linear system [18]. For *square* linear systems (featuring the same number of inputs and outputs $M = Q$), the cross Gramian matrix W_X is defined as $W_X := \int_0^\infty \mathrm{e}^{E^{-1} A t} E^{-1} B C \, \mathrm{e}^{E^{-1} A t} \, dt$, and classically computed as the (low-rank) solution to the Sylvester equation $A W_X E + E W_X A = -BC$. Based on the definition of W_X, the (linear) empirical cross Gramian [6] is given by

$$\widehat{W}_X := \sum_{m=1}^{M} \int_0^\infty x^m(t) z^m(t)^\intercal dt$$

with $x^m(t)$ being the solution of $E \dot{x}^m(t) = A x^m(t) + B(c_m e_m \delta(t))$, $z^m(t)$ being the solution of $E^\intercal \dot{z}^m(t) = A^\intercal z^m(t) + C^\intercal (d_m e_m \delta(t))$, suitable scales $c_m, d_m \in \mathbb{R}$, and the m-th canonical standard base vector $e_m \in \mathbb{R}^M$.

For non-square systems $M \neq Q$, the non-symmetric cross Gramian W_Z, the cross Gramian of the average system $(A, \bar{B} = \sum_{m=1}^{M} B_{*,m}, \bar{C} = \sum_{q=1}^{Q} C_{q,*}, E)$, is proposed in [31]. The linear empirical non-symmetric cross Gramian is given by

$$\widehat{W}_Z := \sum_{q=1}^{Q} \sum_{m=1}^{M} \int_0^\infty x^m(t) z^q(t)^\intercal dt$$

with $x^m(t)$ being the solution of $E\dot{x}^m(t) = Ax^m(t) + \bar{B}(c_m e_m \delta(t))$, $z^q(t)$ being the solution of $E^\intercal \dot{z}^q(t) = A^\intercal z^q(t) + \bar{C}^\intercal (d_q \epsilon_q \delta(t))$, suitable scales $c_m, d_q \in \mathbb{R}$, and the m-th, q-th canonical standard base vectors $e_m \in \mathbb{R}^M$, $\epsilon_q \in \mathbb{R}^Q$.

2.4 Parametric Empirical Gramians

Empirical Gramians may also be applied to parametric systems. Here, the approach from [30] is utilized, which follows the general principle behind empirical Gramians: averaging over an operating region. Hence, given a preselected sampling Θ_h from the parameter space Θ, an average (controllability, observability, cross, or non-symmetric cross) Gramian is computable [6]:

$$\overline{W}_*(\Theta_h) := \sum_{\theta \in \Theta_h} W_*(\theta).$$

For low-dimensional parameter spaces, this could be some uniform grid in a region of interest; for higher dimensional parameter spaces, sparse grids can be utilized [5].

Even though this averaging process can lead to annihilation, it can be justified by the related accumulation process, typically used, i.e., in (balanced) proper orthogonal decomposition (POD) model reduction [59], which (compresses and) concatenates trajectories before assembling a Gramian matrix. So, given two discrete trajectory matrices X_1 and X_2, which are first concatenated and then a Gramian matrix is formed, as for the abstract computation of a POD,

$$\begin{bmatrix} X_1 & X_2 \end{bmatrix} \begin{bmatrix} X_1 & X_2 \end{bmatrix}^\intercal = X_1 X_1^\intercal + X_2 X_2^\intercal,$$

this is mathematically (but not numerically due to annihilation) equivalent to the sum of the individual trajectory Gramians.

3 Empirical-Gramian-Based Model Reduction

Following, five empirical-Gramian-based model reduction methods are summarized, of which either can be computed via the empirical controllability and observability Gramians $\{W_C, W_O\}$, or via the empirical cross Gramian W_X (empirical non-symmetric cross Gramian W_Z for non-square systems).

The considered empirical-Gramian-based model reduction methods are exclusively projection-based approaches, meaning from the empirical system Gramian matrices "projection" matrices are obtained—a reducing projection V and a reconstructing projection U, both of rank $n < N$:

$$U \in \mathbb{R}^{N\times n}, \quad V \in \mathbb{R}^{n\times N},$$

which appropriately applied to the system (1) yield a reduced-order system:

$$\begin{aligned}(VEU)\dot{\tilde{x}}(t) &= \big((VA_0U) + \sum_{p=1}^{P} \theta_p(VA_pU)\big)\tilde{x}(t) + (VB)u(t),\\ \tilde{y}(t) &= (CU)\tilde{x}(t),\end{aligned}$$

or in a more compact form, as the reduced system matrices can be precomputed:

$$\begin{aligned}\widetilde{E}\dot{\tilde{x}}(t) &= \widetilde{A}(\theta)\tilde{x}(t) + \widetilde{B}u(t),\\ \tilde{y}(t) &= \widetilde{C}\tilde{x}(t).\end{aligned}$$

An orthogonal projection $U = V^\intercal$, $VU = I$ is called (Bubnov-)Galerkin projection, a bi-orthogonal projection $U \neq V^\intercal$, $VU = I$ is called Petrov-Galerkin projection, and a projection $U \neq V$, $VU \neq I$ is just called oblique projection.

In the following, only the features of the considered model reduction techniques are briefly summarized, for a description and algorithm of these methods consult the referenced works in the respective subsections. Note, that even though error bounds and error indicators are mentioned below for each method, the purpose of this work is the heuristic comparison of methods against each other.

3.1 Empirical Poor Man

The Poor Man's Truncated Balanced Realization (PM) from [45] just utilizes either the (empirical) controllability Gramian, or the (empirical) observability Gramian, and uses Gramian's dominant singular vectors as Galerkin projection. Using the (time domain) controllability Gramian in this fashion is equivalent to the proper orthogonal decomposition (POD); using the observability Gramian is equivalent to the adjoint proper orthogonal decomposition [11] (aPOD).

Being a Galerkin projection, this method is stability preserving in the reduced-order model if the system is dissipative. As an error indicator, typically the normalized sum of kept singular values is used as well as the projection error of the data [42], which quantifies the reduced model's preserved energy in relation to the full model.

3.2 Empirical Approximate Balancing

Approximate balancing (AB) is a technique suggested in [46, M3], which uses the left and right singular vectors from a truncated SVD of the cross Gramian as oblique projection, yet, without the bi-orthogonality of the Petrov-Galerkin projections, but orthogonality of the reducing and reconstructing projections with respect to themselves. This method is based on the approximate balancing method from [54], but omits the eigenvector approximation. The counterpart variant based on controllability and observability Gramians is known as *modified proper orthogonal decomposition* [42], which uses singular vectors from truncated SVDs of W_C and W_O similarly as oblique projection.

Even though, this method is claimed to be "effective for non-normal systems" ([42, Sect. III.D]), for either method no error bounds or stability guarantees are available, but as indicated in [42, Fig. 8], an error indicator can be derived based upon the projection error. Due to the missing bi-orthogonality between the reducing and reconstructing projections, it is paramount to apply the projections to the mass matrix, even if $E = I$. Using empirical controllability, observability or cross Gramians yields the empirical approximate balancing method.

3.3 Empirical Dominant Subspaces

The dominant subspaces (DS) method constructs a Galerkin projection by directly combining the dominant controllability and observability subspaces [43], obtained from the respective (empirical) Gramians; while the variant based on the (empirical) cross Gramian is introduced in [8].

The column-rank of the projection is then determined by orthogonalization of the conjoined singular vectors of the system Gramians, weighted by their associated singular values. As an orthogonal projection, DS is stability preserving for dissipative systems. Furthermore, a Hardy-2 error bound exists for the controllability and observability Gramian-based DS [53] (in two variants), while a Lebesgue-2 error indicator is introduced in [8] for the cross-Gramian-based DS. To obtain and conjoin the system Gramians' singular vectors, various algorithms are available, here, a combination of truncated SVD and rank-revealing SVD is used for this task.

3.4 Empirical Balanced Truncation

Balanced truncation (BT) first transforms the system into a coordinate system in which controllability and observability are aligned, via a Petrov-Galerkin projection, so the respective controllability and observability Gramians are diagonal and equal. The diagonal entries, the Hankel singular values (HSVs), measure controllability and observability simultaneously, hence the subsystem associated to the small HSVs is truncated. This method from [40] is the gold standard of system-theoretic model reduction methods, due to, first, preserving stability in the reduced-order model [44], and second, error bounds in the Hardy-∞ norm [15, 19], Hardy-2 norm [2, 54], and Lebesgue-1 norm [37, 41].

To balance the Gramians $\{W_C, W_O\}$, the balanced POD ansatz [59] is employed, which corresponds to the square root method [56], but using SVD-based square-roots of the Gramians. Note that this does not lead to an exactly balanced system [58, MR3]. For the W_X (W_Z) balanced truncation variant, the method from [33] is used, which in turn is based on [49, 50].

3.5 Empirical Balanced Gains

Balanced gains (BG) is a variant of balanced truncation, of which the simplified variant from [14] is used here. In balanced gains, the system is balanced as for balanced truncation, but instead of the Hankel singular values, or the sum thereof, an alternate measure is utilized, based on an observation on the L_2-norm of the impulse response (of symmetric systems):

$$\begin{aligned}\|y\|_2^2 &= \operatorname{tr}(CW_CC^\intercal) = \operatorname{tr}(B^\intercal W_O B) = \operatorname{tr}(CW_X B)\\ &= \sum_{k=1}^{N} \hat{c}_k^\intercal \hat{c}_k \sigma_k = \sum_{k=1}^{N} \hat{b}_k \hat{b}_k^\intercal \sigma_k = \sum_{k=1}^{N} |\hat{b}_k \hat{c}_k| \sigma_k,\end{aligned}$$

for the k-th row $\hat{b}_k$ of the balanced input matrix $\widehat{B}$, and the k-th column $\hat{c}_k$ of the balanced output matrix $\widehat{C}$. Hence, the sequence of base vectors is given by the magnitude of the quantity d_k, instead of the HSVs σ_k:

$$d_k := \hat{c}_k^\intercal \hat{c}_k \sigma_k = \hat{b}_k \hat{b}_k^\intercal \sigma_k = |\hat{b}_k \hat{c}_k| \sigma_k.$$

This means compared to balanced truncation, the same modes are used, but in a different order. As the order of modes is not a requirement for stability preservation in the reduced-order model, it also holds for balanced gains, cf. [44, Corollary 2]. Empirical balanced gains is then given by the (simplified) balanced gains approach using empirical Gramians.

4 Approximate Norms

To comprehensively compare the reduced to the full order models, four signal norms, four system norms, and two induced norms are applied. For an elaborate discussion of these norms see [10, Ch. 5,6], [2, Ch. 5], [57, Ch. 2]. Due to numerical, efficiency, or practical reasons, only approximate norms of the error system are considered. Note, that the signal norms are computed from time-domain trajectories, and the system (and modified induced) norms are approximated by transformations of empirical Gramians, instead of frequency-domain sampling.

4.1 Signal Norms

The signal norms are based on time-domain evaluations of the system output y and the reduced system's output $\tilde{y}$, and are given as the Lebesgue norms of the output error $\|y - \tilde{y}\|$. Practically, vector norms of vectorized discrete output trajectories y_h, $\tilde{y}_h$ (Q outputs $\times$ K time steps data matrices) are computed.

4.1.1 Approximate L_0-"Norm"

The L_0 signal "norm" describes the sparsity of a *discrete-time* signal [52], and is approximated, based on [32], for an error signal by

$$\|y_h - \tilde{y}_h\|_{L_0} = \sum_{k=0}^{K} \sum_{q=1}^{Q} \left| \operatorname{sgn}\left(y_{h,q}(k) - \tilde{y}_{h,q}(k)\right)\right| \approx \sqrt[n]{\prod_{\ell=1}^{QK} |\operatorname{vec}(y_h - \tilde{y}_h)_\ell|}.$$

Technically, this is not a norm, due to the lack of absolute scalability, but for the intended purpose this function can be treated as a norm.

4.1.2 Approximate Lebesgue L_1-Norm

The Lebesgue L_1-norm of a signal quantifies the action or consumption of a process and its definition and approximation for an output error signal are given by

$$\|y - \tilde{y}\|_{L_1} = \int_0^\infty \|y(t) - \tilde{y}(t)\|_1 dt \approx \Delta t \, \|\operatorname{vec}(y_h - \tilde{y}_h)\|_1;$$

in terms of the model reduction error it can also be seen as the area under the error signal.

4.1.3 Approximate Lebesgue L_2-Norm

The Lebesgue L_2-norm of a signal measures its energy. Its definition and approximation for an output error signal are given by

$$\|y - \tilde{y}\|_{L_2} = \sqrt{\int_0^\infty \|y(t) - \tilde{y}(t)\|_2^2 dt} \approx \sqrt{\Delta t} \, \|\operatorname{vec}(y_h - \tilde{y}_h)\|_2,$$

which can be interpreted as the energy loss in the reduced-order model. As all methods tested in this work are energy-based, this norm is the canonical error measure.

4.1.4 Approximate Lebesgue L_∞-Norm

The Lebesgue L_∞-norm of a signal determines its peak, with definition and approximation for an output error signal given by

$$\|y - \tilde{y}\|_{L_\infty} = \sup_t \|y(t) - \tilde{y}(t)\|_\infty \approx \|\operatorname{vec}(y_h - \tilde{y}_h)\|_\infty,$$

which yields the maximum error between the full and reduced-order system's outputs.

4.2 System Norms

The system norms characterize frequency-domain errors of the reduced system's transfer function $G_r(\omega) := C_r(E_r\omega - A_r)^{-1}B_r$ compared to the full order transfer function $G(\omega) := C(E\omega - A)^{-1}B$, for frequencies $\omega \in \mathbb{C}$, $\operatorname{Re}(\omega) < 0$, and are either Hardy-norms and/or Schatten-norms of the Hankel operator H. The following four norms were selected based on [51, Sect. 2.2.7].

4.2.1 Approximate Hardy H_2-Norm

The Hardy H_2-norm can be interpreted as the root-mean-square of the frequency response to white noise, the L_2-norm of the impulse response (thus also known as impulse response norm), the maximum output amplitude for finite input, or average gain. To approximate the H_2-norm, the truncated balanced part of the output operator and controllability Gramian are utilized [23],[54, Remark 3.3]:

$$\|G - G_r\|_{H_2} = \sqrt{\int \operatorname{tr}\Big(\big(G(\imath\omega) - G_r(\imath\omega)\big)\big(G(\imath\omega) - G_r(\imath\omega)\big)^*\Big)d\omega} \approx \sqrt{\hat{\tilde{C}}_2 W_{C,22} \hat{\tilde{C}}_2^\intercal}.$$

4.2.2 Approximate Hardy H_∞-Norm

The Hardy H_∞-norm describes the worst-case frequency-domain error, which relates, via Parseval's equation, to the maximum L_2-gain, and thus to the time-domain L_2 error. Based on [19, Corollary 9.3], the H_∞ error can be approximated by the balanced truncation error bound, which in turn is approximated by the principal discarded Hankel singular value [25, Ch. 2.4]:

$$\|G - G_r\|_{H_\infty} = \sup\Big(\sigma_1\big(G(\imath\omega) - G_r\big(\imath\omega\big)\big)\Big) \approx 2\sum_{k=n+1}^{N} \sigma_k(H) \approx 2(N-n)\sigma_{n+1}(H)$$

and is related to the nuclear norm (Schatten-1 norm) of the Hankel operator H. Alternatively, the H_∞-norm could be approximated by the trace of the non-symmetric cross Gramian $\|G - G_r\|_{H_\infty} \approx -\frac{1}{2}\operatorname{tr}(W_{Z,22}) = -\bar{C}_2 A_{22}^{-1} \bar{B}_2$ [38].

4.2.3 Approximate Hilbert-Schmidt-Hankel Norm

The Hilbert-Schmidt-Hankel norm corresponds to the operator norm (Schatten-2 norm) of the Hankel operator, and as for the H_∞-norm, is approximated using only the principal discarded Hankel singular value:

$$\|G - G_r\|_{HSH} = \sqrt{\sum_{k=n+1}^{N} \sigma_k^2(H)} \approx \sqrt{(N-n)\sigma_{n+1}^2(H)}.$$

Scaled by a factor of π, the square root of this norm yields the enclosed area of the Nyquist plot [24].

4.2.4 Approximate Hankel Norm

The Hankel norm is given by the principal discarded singular value of the Hankel operator, which corresponds to the Schatten-∞ norm:

$$\|G - G_r\|_{Ha} = \sigma_{n+1}(H).$$

This norm is the lower bound for the model reduction error as by the Adamjan-Arov-Krein theorem [20, 21].

4.3 *Modified Induced Norms*

If the Hankel operator is used in its classic form, it maps from and to a function space of squarely integrable functions, and the (previous) Hankel norm is its induced norm. If one relaxes the Hankel operator to admit a function space of continuous functions as domain or range, the induced norms change as follows [60]. Note, that for single-input-single-output systems, these modified induced norms coincide with the Hardy-2 norm.

4.3.1 Induced Primal Norm

Expanding the Hankel operator's *domain* to continuous functions, the induced norm becomes the square root of the input-observability Gramian's spectral radius:

$$\|H - H_r\|_{\mathcal{H}_C} = \sqrt{\lambda_{max}(B_{22}^{\intercal} W_{O,22} B_{22})}.$$

4.3.2 Induced Dual Norm

Expanding the Hankel operator's *range* to continuous functions is equivalent to expanding the dual system's Hankel operator domain, thus the induced norm becomes the square root of the output-controllability Gramian's spectral radius:

$$\|H - H_r\|_{\mathcal{H}_O} = \sqrt{\lambda_{max}(C_{22} W_{C,22} C_{22}^{\intercal})}.$$

4.4 Parametric Norms

To obtain an error quantification for parametric systems, the previous norms are extended with respect to the considered system's parameter space. Given a (state-space) error norm $\|\cdot\|_X$, the associated *parametric state-space error norm* is given by the composition with a parameter space norm $\|\cdot\|_Y$. In [4] (see also [7]), this composite state-parameter norms are defined via a norm as a mapping $\|\cdot\|_{Y\otimes X} : M \times \Theta \to \mathbb{R}_+$, with the Cartesian product of output, response or operator domain M and parameter domain Θ, respectively. To approximate these parametric norms, a sampling of the parameter space $\Theta_s \subset \Theta$ is drawn, and given this finite, discrete parameter sample Θ_s, an approximate norm is computed. We follow [22], in evaluating the parametric $L_1 \otimes X$, $L_2 \otimes X$, and $L_\infty \otimes X$ norms:

$$\begin{aligned}
\|y(\theta) - \tilde{y}(\theta)\|_{L_1\otimes X} &= \int_\Theta \|y(\theta) - \tilde{y}(\theta)\|_X d\theta &&\approx \sum_{\theta\in\Theta_s} \|y(\theta) - \tilde{y}(\theta)\|_X, \\
\|y(\theta) - \tilde{y}(\theta)\|_{L_2\otimes X} &= \sqrt{\int_\Theta \|y(\theta) - \tilde{y}(\theta)\|_X^2 d\theta} &&\approx \sqrt{\sum_{\theta\in\Theta_s} \|y(\theta) - \tilde{y}(\theta)\|_X^2}, \\
\|y(\theta) - \tilde{y}(\theta)\|_{L_\infty\otimes X} &= \max_{\theta\in\Theta} \|y(\theta) - \tilde{y}(\theta)\|_X &&\approx \max_{\theta\in\Theta_s} \|y(\theta) - \tilde{y}(\theta)\|_X,
\end{aligned}$$

for X being any of the signal, system or induced norms. To estimate the quality of a parametric reduced-order model fairly, it is a basic requirement to have disjoint training and test parameter sets $\Theta_h \cap \Theta_s = \emptyset$. Typically, this is implicitly ensured by a (sparse) grid parameter sampling for the training and randomly drawn test parameters from a suitable distribution.

5 MORscore

The comparison of (relative) model reduction errors for varying reduced orders, see for example, Fig. 1, is a useful vehicle to evaluate the performance of model reduction techniques for a specific system in a certain norm. Yet, there are multiple relevant features in these error graphs characterizing the associated model order reduction algorithm, such as lowest attained error or fastest error decay. Now, a one-by-one comparison for multiple methods, in various norms is too tedious for potentially many systems. A similar problem arises in comparing optimization codes, which is managed by so-called *relative minimization profiles* (RMP) [13, Sect. 5]. These RMPs standardize such comparisons in various measures, such as best computed objective, and inspired the following scoring. To make many-way model reduction comparisons feasible, a scalar score is introduced next, summarizing a method's features in a specific norm based on the error graph.

Definition *(MORscore)* Given an error graph $(n, \varepsilon(n)) \in \mathbb{N}_{>0} \times (0, 1]$, relating a reduced-order n to a relative output error of a model reduction method M for a system Σ in norm $\| \cdot \|$, the normalized error graph $(\varphi_n, \varphi_{\varepsilon(n)})$ is determined by the maximum reduced-order $n_{\max} \in \mathbb{N}_{>0}$, and machine precision $\epsilon_{\text{mach}} \in (0, 1] \subset \mathbb{R}$ via mappings:

$$\varphi_n : \mathbb{N}_{>0} \to [0, 1], \quad n \mapsto \frac{n}{n_{\max}},$$
$$\varphi_\varepsilon : (0, 1] \to [0, 1], \quad \varepsilon \mapsto \frac{\log_{10}(\varepsilon)}{\lfloor \log_{10}(\epsilon_{\text{mach}}) \rfloor},$$

and the **MORscore** μ is defined as the area under this normalized error graph,

$$\mu_{(n_{\max}, \epsilon_{\text{mach}})}(M, \Sigma, \| \cdot \|) := \text{area}(\varphi_n, \varphi_\varepsilon).$$

By φ_n the discrete reduced orders $1, 2 \dots n_{\max}$ are mapped to the real interval $[0, 1]$ by normalization. And by φ_ε the *relative* model reduction error ε is mapped to the real interval $[0, 1]$ by normalizing the 10-base logarithm of the error by the 10-base logarithm of the maximum accuracy ϵ_{mach} of the utilized number system; i.e., double precision floating point numbers have an accuracy of approximately $\epsilon_{\text{mach}}(\text{dp}) \approx 10^{-16}$, so $\lfloor \log_{10}(\epsilon_{\text{mach}}(\text{dp})) \rfloor = -16$. Practically, the area is computed via the trapezoid rule.[1] Note, that the maximum tested reduced-order $n_{\max}$ should be (far) below the original model order, since the error decay flattens at some reduced order. Hence, given a system of large order, and two model reduction methods, both yielding their minimal error reduced models at low orders, a MORscore up to the full order would show only little difference. Selecting the largest reduced order which attains the minimal error as $n_{\max}$, the MORscore is a lot more meaningful.

---

[1] Specifically via: https://www.mathworks.com/help/matlab/ref/trapz.html.

Altogether, the MORscore is specified by normalization and describes the model reduction performance of a method for a system in a norm by single number, as typical for (desktop) computer performance benchmarks. A larger MORscore $\mu \in (0, 1)$ means better model reduction performance, since the more area covered, the faster and lower the error decay. Contrary to β-RMPs [13, Def. 5.2], no computational budget is prescribed here, nonetheless, the MORscore could be extended in this manner by limited computational time or even a prescribed $n_{\max}$.

6 Benchmark Comparison

For a thorough comparison, the presented empirical-Gramian-based model reduction methods are tested in ten (approximate) norms for different configurations of a benchmark system. In coordination with the model reduction software projects: pyMOR [39], MORLAB [9], M.E.S.S [48], a thermal block benchmark is tested. A summary of the components for this comparison is given below.

Methods

Each of the five methods summarized in Sect. 3 can be computed via the empirical controllability and observability Gramians $\{W_C, W_O\}$, or the empirical (non-symmetric) linear cross Gramian W_Z. Hence overall, ten empirical-Gramian-based model reduction techniques are compared:

- Empirical Poor Man (PM), via W_C or W_O,
- Empirical Approximate Balancing (AB), via $\{W_C, W_O\}$ or W_Z,
- Empirical Dominant Subspaces (DS), via $\{W_C, W_O\}$ or W_Z,
- Empirical Balanced Truncation (BT), via $\{W_C, W_O\}$ or W_Z,
- Empirical Balanced Gains (BG), via $\{W_C, W_O\}$ or W_Z.

Parameterization

In Sect. 6.2, a parametric benchmark with a four-dimensional parameter space is tested. The benchmark is compared in three configurations:

- Non-Parametric (parameters treated as constants),
- Single Parameter (parameters treated as single parameter),
- Multiple Parameters (parameters treated separately).

Measures

The model reduction methods are compared via their MORscore for varying reduced orders in the following normalized norms $\frac{\|y-y_r\|}{\|y\|}$ from Sect. 4:

- Approximate Lebesgue L_0-"norm",
- Approximate Lebesgue L_1-norm,
- Approximate Lebesgue L_2-norm,
- Approximate Lebesgue L_∞-norm,
- Approximate Hardy H_2-norm,
- Approximate Hardy H_∞-norm,
- Approximate Hilbert-Schmidt-Hankel-norm,
- Approximate Hankel-norm,
- Approximate modified induced primal norm,
- Approximate modified induced dual norm,

as well as the number of unstable ROMs up to the maximum order (denoted by the symbol $\mathcal{L}$). Lyapunov stability is assessed via the real part of the largest real eigenvalue of the pencil $(\widetilde{E}, \widetilde{A}(\theta))$. In the parametric case, these counts are averaged, similar to the considered norms, in an L_1, L_2 and L_∞ sense over the sampled parameters.

6.1 `emgr` – EMpirical GRamian Framework

All tested methods are based on empirical system Gramian matrices. To compute these empirical Gramians for the subsequent numerical experiments, the empirical Gramian framework `emgr` [26] is employed, which has a unified interface [28] for the empirical controllability, observability, and (linear) cross Gramians. The convergence of the empirical Gramians to the classic algebraic Gramians for linear systems is shown in [25]. Practically, the current version `emgr` 5.7 [27] is used.

6.2 Thermal Block Benchmark

For the comparison of the empirical-Gramian-based model order reduction methods, a recurring benchmark example (due to the well reducible diffusion process), modeling the heat equation on the unit-square [55, Thermal Block] is utilized.

This thermal block benchmark system models dynamic heating of a two-dimensional, square domain $\Omega = (0, 1) \times (0, 1)$ with four enclosed circular regions $\omega_{i=1\dots4}$ of equal radius, one per quadrant, and each of individual parametric heat conductivity (diffusivity) $\kappa(x)$. The left boundary of the domain $\partial\Omega_1 := \{0\} \times (0, 1)$ is the inflow, realized by a Neumann boundary condition, the top and bottom boundaries $\partial\Omega_2 := (0, 1) \times \{0\}$, $\partial\Omega_4 := (0, 1) \times \{1\}$ are insulated, via zero Neumann conditions, while the right boundary $\partial\Omega_3 := \{1\} \times (0, 1)$ prescribes Dirichlet-zero boundary conditions. Lastly, the four quantities of interests $\mathcal{Y}_i$ are the average temperature of each circle ω_i, respectively. The overall partial differential equation (PDE) system is thus given by

$$
\begin{aligned}
\partial_t u(x,t) &= -\kappa(x)\Delta_x u(x,t), && x \in \Omega, \\
\partial_x u(x,t) &= F(x,t), && x \in \partial\Omega_1, \\
\partial_x u(x,t) &= 0, && x \in \partial\Omega_2 \cup \partial\Omega_4, \\
u(x,t) &= 0, && x \in \partial\Omega_3, \\
\mathcal{Y}_i(t) &= \int_{\omega_i} u(x,t)dx, \\
\kappa(x) &= \begin{cases} \theta_i & x \in \omega_i,\ i = 1\ldots 4, \\ \theta_0 & \text{otherwise.} \end{cases}
\end{aligned}
$$

This PDE is discretized in space using the finite element method (FEM), via the FEniCs software package [1], yielding an ordinary differential equation system of the form (1). The resulting linear input-output system has one input and four outputs, while the state-space has dimension 7488, and the parameter space is four dimensional, with $\theta_{i=1\ldots4} \in [1, 10] \subset \mathbb{R}$ as in [3], while the background diffusivity constant is set to $\theta_0 = 1$. For more a detailed description of this benchmark, and the software stack used for its creation, see also [47].

6.3 *Numerical Results*

In the following, three variants of the thermal block benchmark are tested:

1. No parameter: $\frac{1}{5}\theta_1 = \frac{2}{5}\theta_2 = \frac{3}{5}\theta_3 = \frac{4}{5}\theta_4 \equiv \sqrt{10}$,
2. One parameter: $\frac{1}{5}\theta_1 = \frac{2}{5}\theta_2 = \frac{3}{5}\theta_3 = \frac{4}{5}\theta_4 \in [1, 10]$,
3. Four parameters: $\theta \in [1, 10]^4$.

For the parametric variants, the $(3 \cdot \dim(\theta))$ training samples of the parameter space are taken from a logarithmically uniform grid, whereas (ten) test samples are drawn randomly from a logarithmically uniform distribution over the parameter range. The empirical Gramians are build from trajectories excited by impulses, while the ROMs are tested by random input. The decompositions for the empirical-Gramian-based model reduction methods are approximated up to order one-hundred.

Practically, the following numerical results are conducted using MATLAB 2020a on an Intel(R) Core(TM) i3-7130U CPU @ 2.70GHz with 8GB RAM.

6.3.1 Fixed Parameter

In the first set of numerical experiments, the thermal block benchmark is tested with a single fixed parameter. Exemplary in Fig. 1, the model reduction error in the approximate L_2-norm for the ten considered methods are compared for reduced models of orders one to fifty. This figure illustrates how complex a visualization already in a single norm is. The proposed MORscores are listed in Table 1, which is similarly not directly decipherable by a human observer, yet, algorithmically it can be processed.

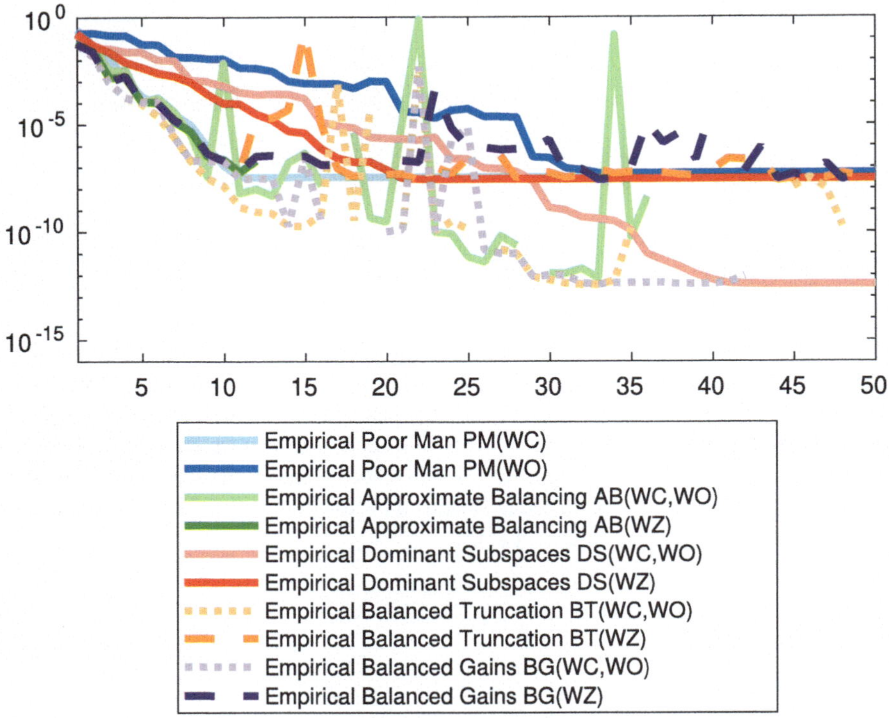

Fig. 1 Relative error of reduced-order models in the L_2-norm compared to the full order model for varying reduced orders

Table 1 MORscore $\mu(50, \epsilon_{\text{mach}}(\text{dp}))$ for the non-parametric benchmark

	L_0	L_1	L_2	L_∞	H_2	H_∞	HSH	Ha	$\mathcal{H}_C$	$\mathcal{H}_O$	$\mathcal{L}$
PM(W_C)	0.42	0.42	0.41	0.39	0.63	0.49	0.51	0.52	0.54	0.06	0
PM(W_O)	0.29	0.29	0.29	0.28	0.10	0.38	0.38	0.38	0.10	0.45	0
AB(W_C,W_O)	0.33	0.33	0.32	0.30	0.46	0.03	0.04	0.04	0.44	0.39	37
AB(W_Z)	0.08	0.08	0.08	0.08	0.35	0.02	0.02	0.02	0.35	0.04	38
DS(W_C,W_O)	0.45	0.45	0.44	0.43	0.34	0.51	0.52	0.52	0.30	0.25	0
DS(W_Z)	0.39	0.38	0.38	0.36	0.34	0.39	0.39	0.39	0.34	0.08	0
BT(W_C,W_O)	0.38	0.38	0.37	0.35	0.45	0.36	0.36	0.36	0.43	0.18	25
BT(W_Z)	0.40	0.38	0.37	0.35	0.28	0.30	0.30	0.30	0.28	0.08	37
BG(W_C,W_O)	0.43	0.43	0.42	0.41	0.43	0.35	0.35	0.35	0.42	0.17	25
BG(W_Z)	0.35	0.34	0.33	0.31	0.28	0.30	0.30	0.30	0.28	0.08	37

In the approximate signal norms the maximum MORscores are achieved by the DS(W_C,W_O), closely followed by BG(W_C,W_O). Notably the BT variants used are not in lead, which in this case is related to many unstable reduced-order models, due to the low-rank approximation of the Gramians and using an SVD-based square root method for balancing, and thus nullifying the stability preservation of the original balanced truncation method. While the Galerkin methods do not produce unstable ROMs, all Petrov-Galerkin methods produce at least 25 unstable ROMs.[2] The H_2-norm is lead by the PM(W_C) method, whereas the H_∞, Ha and HSH norms are headed by DS(W_C,W_O), closely followed by PM(W_C). Finally, in modified induced norms $\mathcal{H}_C$ and $\mathcal{H}_O$, PM(W_C) and PM(W_O) perform best, respectively.

Overall for this benchmark, the methods using W_C and/or W_O outperformed methods using W_Z, likely due to the non-square system, which requires additional averaging in the non-symmetric cross Gramian.

6.3.2 Single Parameter

The MORscores for the single parameter benchmark are given in Table 2 ($L_1 \otimes X$), Table 3 ($L_2 \otimes X$) and Table 4 ($L_\infty \otimes X$). Generally, all methods perform worse compared to the non-parametric benchmark, since the averaging of empirical Gramians over parameter samples decreases specific accuracy while increasing general applicability.

The signal norms are lead by BT(W_C,W_O) and directly followed by BG(W_C,W_O), PM(W_C), DS(W_C,W_O), and DS(W_Z). In the H_2 and $\mathcal{H}_C$ norms, the methods BT(W_C,W_O), PM(W_C), and AB(W_C,W_O) are in the lead, while in the system norms H_∞, HSH, Ha, PM(W_C) heads the MORscores. The $\mathcal{H}_O$ norm is topped by PM(W_O) and AB(W_C,W_O) methods. Balanced gains (BG) seem to work well for this benchmark, while approximate balancing (AB) performs worst overall.

As for the non-parametric benchmark, the Galerkin methods consistently produce stable ROMs, and the Petrov-Galerkin methods tend to assemble unstable ROMs.

6.3.3 Multiple Parameters

The MORscores for the multiple parameter benchmark are given in Table 5 ($L_1 \otimes X$), Table 6 ($L_2 \otimes X$) and Table 7 ($L_\infty \otimes X$), and correspond overall to the single parameter setting, yet, with again slightly lower scores. Curiously, balanced gains performance drops more than balanced truncation.

[2] Unstable ROMs are treated as relative error of one, $\varepsilon = 1$, in the method's MORscores.

6.3.4 MORscore Discussion

Summarizing, the presented MORscore tables can improve heuristic comparisons of model reduction methods. An automated evaluation could include filtering extreme values per norm, as demonstrated in the previous evaluations, or (generalized) means [12] per methods across norms.

Specifically for the comparison of the empirical-Gramian-based model reduction methods on the thermal block benchmark, the arithmetic means of MORscores across norms yields the PM(W_C) and DS(W_C,W_O) methods as top scoring for the non-parametric benchmark, and the PM(W_C) = POD for the parametric benchmark variants, as in [6].

Beyond this sample comparison, the proposed MORscore could find application in model reduction software development signaling regressions, or defining highscore boards of competing methods for benchmark problems.

7 Conclusion

This work should be considered an exemplary quantitative comparison using MORscores, and by no means exhaustive. Specifically, other relevant (empirical) Gramian-based methods not tested here are (empirical) singular perturbation approximation [17], and (empirical) Hankel norm approximation [16], yet both methods are not purely projection based but require a numerically potentially expensive post-processing of a balanced realization. Also, the empirical Gramians have various variants [26] that could be tested, as well as different balancing algorithms [58]. Nevertheless, this work can serve as a template for benchmarking model reduction methods by their **MORscore**.

Code Availability Section

The source code of the presented numerical examples can be obtained from:

http://runmycode.org/companion/view/3760

and is authored by: Christian Himpe.

Acknowledgements Supported by the German Federal Ministry for Economic Affairs and Energy (BMWi), in the joint project: "MathEnergy—Mathematical Key Technologies for Evolving Energy Grids", sub-project: Model Order Reduction (Grant number: 0324019**B**).

Appendix

Single Parameter Benchmark MORscores

Table 2 MORscore $\mu(50, \epsilon_{\text{mach}}(\text{dp}))$ for the single parameter benchmark $(L_1 \otimes X)$

	L_0	L_1	L_2	L_∞	H_2	H_∞	HSH	Ha	$\mathcal{H}_C$	$\mathcal{H}_O$	$\mathcal{L}$
PM(W_C)	0.26	0.25	0.25	0.23	0.37	0.42	0.44	0.44	0.37	0.07	0
PM(W_O)	0.18	0.18	0.18	0.17	0.10	0.23	0.24	0.24	0.10	0.18	0
AB(W_C,W_O)	0.15	0.15	0.14	0.14	0.35	0.03	0.04	0.04	0.36	0.18	37.5
AB(W_Z)	0.06	0.06	0.06	0.06	0.24	0.02	0.02	0.02	0.23	0.05	38.1
DS(W_C,W_O)	0.24	0.23	0.23	0.22	0.19	0.30	0.31	0.32	0.19	0.15	0
DS(W_Z)	0.24	0.23	0.23	0.22	0.24	0.29	0.19	0.30	0.24	0.07	0
BT(W_C,W_O)	0.25	0.25	0.24	0.24	0.36	0.28	0.28	0.28	0.36	0.14	14.8
BT(W_Z)	0.18	0.18	0.18	0.17	0.20	0.19	0.19	0.19	0.20	0.10	33.2
BG(W_C,W_O)	0.26	0.26	0.26	0.25	0.33	0.23	0.23	0.23	0.33	0.12	18.5
BG(W_Z)	0.12	0.12	0.12	0.11	0.19	0.18	0.18	0.18	0.19	0.09	33.2

Table 3 MORscore $\mu(50, \epsilon_{\text{mach}}(\text{dp}))$ for the single parameter benchmark $(L_2 \otimes X)$

	L_0	L_1	L_2	L_∞	H_2	H_∞	HSH	Ha	$\mathcal{H}_C$	$\mathcal{H}_O$	$\mathcal{L}$
PM(W_C)	0.22	0.22	0.22	0.20	0.34	0.39	0.40	0.41	0.34	0.04	0
PM(W_O)	0.15	0.15	0.15	0.14	0.07	0.20	0.21	0.21	0.07	0.15	0
AB(W_C,W_O)	0.11	0.11	0.10	0.10	0.32	0.00	0.01	0.01	0.33	0.15	118.66
AB(W_Z)	0.03	0.03	0.03	0.02	0.21	0.00	0.00	0.00	0.20	0.02	120.56
DS(W_C,W_O)	0.20	0.20	0.20	0.19	0.16	0.27	0.28	0.29	0.16	0.12	0
DS(W_Z)	0.20	0.20	0.20	0.19	0.21	0.26	0.26	0.27	0.21	0.04	0
BT(W_C,W_O)	0.21	0.21	0.21	0.20	0.33	0.25	0.25	0.25	0.33	0.10	47.03
BT(W_Z)	0.15	0.14	0.14	0.13	0.17	0.16	0.16	0.16	0.17	0.07	105.00
BG(W_C,W_O)	0.23	0.22	0.22	0.21	0.30	0.20	0.20	0.20	0.30	0.09	58.52
BG(W_Z)	0.08	0.08	0.08	0.07	0.16	0.15	0.15	0.15	0.16	0.05	105.00

Table 4 MORscore $\mu(50, \epsilon_{\text{mach}}(\text{dp}))$ for the single parameter benchmark $(L_\infty \otimes X)$

	L_0	L_1	L_2	L_∞	H_2	H_∞	HSH	Ha	$\mathcal{H}_C$	$\mathcal{H}_O$	$\mathcal{L}$
PM(W_C)	0.24	0.23	0.23	0.21	0.37	0.42	0.44	0.44	0.37	0.07	0
PM(W_O)	0.17	0.17	0.17	0.16	0.10	0.23	0.24	0.24	0.10	0.18	0
AB(W_C,W_O)	0.12	0.12	0.12	0.11	0.35	0.03	0.04	0.04	0.36	0.18	40
AB(W_Z)	0.05	0.05	0.05	0.05	0.24	0.02	0.02	0.02	0.23	0.05	41
DS(W_C,W_O)	0.22	0.22	0.21	0.20	0.19	0.30	0.31	0.32	0.19	0.15	0
DS(W_Z)	0.22	0.22	0.22	0.21	0.24	0.29	0.29	0.30	0.24	0.07	0
BT(W_C,W_O)	0.23	0.23	0.22	0.21	0.36	0.28	0.28	0.28	0.36	0.14	17
BT(W_Z)	0.16	0.16	0.16	0.15	0.20	0.19	0.19	0.19	0.20	0.10	34
BG(W_C,W_O)	0.24	0.24	0.24	0.23	0.33	0.23	0.23	0.23	0.33	0.12	19
BG(W_Z)	0.10	0.09	0.09	0.09	0.19	0.18	0.18	0.18	0.19	0.09	34

Multi Parameter Benchmark MORscores

Table 5 MORscore $\mu(50, \epsilon_{\text{mach}}(\text{dp}))$ for the multi parameter benchmark $(L_1 \otimes X)$

	L_0	L_1	L_2	L_∞	H_2	H_∞	HSH	Ha	$\mathcal{H}_C$	$\mathcal{H}_O$	$\mathcal{L}$
PM(W_C)	0.23	0.23	0.23	0.22	0.30	0.33	0.34	0.35	0.29	0.08	0
PM(W_O)	0.18	0.17	0.17	0.16	0.10	0.24	0.24	0.24	0.10	0.18	0
AB(W_C,W_O)	0.07	0.07	0.07	0.06	0.29	0.03	0.04	0.04	0.28	0.18	44.8
AB(W_Z)	0.07	0.07	0.07	0.07	0.18	0.02	0.02	0.02	0.18	0.07	33.3
DS(W_C,W_O)	0.21	0.21	0.20	0.19	0.20	0.30	0.32	0.33	0.20	0.16	0
DS(W_Z)	0.19	0.19	0.19	0.18	0.20	0.24	0.25	0.25	0.21	0.09	0
BT(W_C,W_O)	0.24	0.23	0.23	0.22	0.29	0.22	0.22	0.22	0.29	0.20	5.0
BT(W_Z)	0.08	0.08	0.07	0.07	0.15	0.14	0.14	0.14	0.15	0.11	29.8
BG(W_C,W_O)	0.18	0.18	0.17	0.17	0.27	0.19	0.19	0.19	0.27	0.18	9.0
BG(W_Z)	0.05	0.05	0.05	0.05	0.13	0.12	0.12	0.12	0.13	0.11	36.3

Table 6 MORscore $\mu(50, \epsilon_{\text{mach}}(\text{dp}))$ for the multi parameter benchmark $(L_2 \otimes X)$

	L_0	L_1	L_2	L_∞	H_2	H_∞	HSH	Ha	$\mathcal{H}_C$	$\mathcal{H}_O$	$\mathcal{L}$
PM(W_C)	0.20	0.19	0.19	0.18	0.27	0.30	0.31	0.32	0.26	0.05	0
PM(W_O)	0.14	0.14	0.14	0.13	0.07	0.21	0.21	0.21	0.07	0.15	0
AB(W_C,W_O)	0.03	0.02	0.02	0.02	0.26	0.00	0.01	0.01	0.25	0.15	141.75
AB(W_Z)	0.03	0.03	0.03	0.03	0.15	0.00	0.00	0.00	0.15	0.04	105.49
DS(W_C,W_O)	0.18	0.17	0.17	0.16	0.17	0.27	0.29	0.30	0.17	0.13	0
DS(W_Z)	0.16	0.16	0.16	0.15	0.17	0.21	0.22	0.22	0.18	0.05	0
BT(W_C,W_O)	0.20	0.19	0.19	0.18	0.26	0.19	0.19	0.19	0.26	0.17	16.12
BT(W_Z)	0.03	0.03	0.03	0.03	0.12	0.11	0.11	0.11	0.12	0.08	94.29
BG(W_C,W_O)	0.14	0.13	0.13	0.13	0.24	0.16	0.16	0.16	0.24	0.15	29.53
BG(W_Z)	0.01	0.00	0.00	0.00	0.10	0.08	0.09	0.09	0.10	0.08	114.90

Table 7 MORscore $\mu(50, \epsilon_{\text{mach}}(\text{dp}))$ for the multi parameter benchmark $(L_\infty \otimes X)$

	L_0	L_1	L_2	L_∞	H_2	H_∞	HSH	Ha	$\mathcal{H}_C$	$\mathcal{H}_O$	$\mathcal{L}$
PM(W_C)	0.21	0.21	0.20	0.20	0.30	0.33	0.34	0.35	0.29	0.08	0
PM(W_O)	0.16	0.16	0.16	0.15	0.10	0.24	0.24	0.24	0.10	0.18	0
AB(W_C,W_O)	0.04	0.04	0.04	0.04	0.29	0.03	0.04	0.04	0.28	0.18	47
AB(W_Z)	0.04	0.04	0.04	0.04	0.18	0.02	0.02	0.02	0.18	0.07	38
DS(W_C,W_O)	0.19	0.19	0.18	0.18	0.20	0.30	0.32	0.33	0.20	0.16	0
DS(W_Z)	0.18	0.17	0.17	0.17	0.20	0.24	0.25	0.25	0.21	0.09	0
BT(W_C,W_O)	0.21	0.20	0.20	0.19	0.29	0.22	0.22	0.22	0.29	0.20	7
BT(W_Z)	0.05	0.05	0.04	0.04	0.15	0.14	0.14	0.14	0.15	0.11	32
BG(W_C,W_O)	0.15	0.15	0.14	0.14	0.27	0.19	0.19	0.19	0.27	0.18	15
BG(W_Z)	0.02	0.02	0.02	0.02	0.13	0.12	0.12	0.12	0.13	0.11	40

References

1. Alnæs, M.S., Blechta, J., Hake, J., Johansson, A., Kehlet, B., Logg, A., Richardson, C., Ring, J., Rognes, M.E., Wells, G.N.: The FEniCS project version 1.5. Arch. Numer. Softw. **3**(100), 9–23 (2015). https://doi.org/10.11588/ans.2015.100.20553
2. Antoulas, A.C.: Approximation of Large-Scale Dynamical Systems, vol. 6, Adv. Des. Control. SIAM Publications, Philadelphia, PA (2005). https://doi.org/10.1137/1.9780898718713
3. Ballani, J., Kressner, D.: Reduced basis methods: from low-rank matrices to low-rank tensors. SIAM J. Sci. Comput. **38**(4), A2045–A2067 (2016). https://doi.org/10.1137/15M1042784
4. Baur, U., Beattie, C.A., Benner, P., Gugercin, S.: Interpolatory projection methods for parameterized model reduction. SIAM J. Sci. Comput. **33**(5), 2489–2518 (2011). https://doi.org/10.1137/090776925
5. Baur, U., Benner, P.: Parametrische Modellreduktion mit dünnen Gittern. In: Lohmann, B., Kugi, A. (eds.) Tagungsband GMA-FA 1.30, 'Modellierung, Identifikation und Simulation in der Automatisierungstechnik', Workshop in Anif, 24.-26.9.2008, pp. 262–271 (2008). ISBN: 978-3-9502451-1-0, available from http://csc.mpi-magdeburg.mpg.de/mpcsc/benner/pub/BaurBenner-GMA-Proceedings2008.pdf
6. Baur, U., Benner, P., Haasdonk, B., Himpe, C., Martini, I., Ohlberger, M.: Comparison of methods for parametric model order reduction of time-dependent problems. In: Benner, P., Cohen, A., Ohlberger, M., Willcox, K. (eds.) Model Reduction and Approximation: Theory and Algorithms, pp. 377–407. SIAM (2017). https://doi.org/10.1137/1.9781611974829.ch9
7. Benner, P., Gugercin, S., Willcox, K.: A survey of model reduction methods for parametric systems. SIAM Rev. **57**(4), 483–531 (2015). https://doi.org/10.1137/130932715
8. Benner, P., Himpe, C.: Cross-Gramian-based dominant subspaces. Adv. Comput. Math. **45**(5), 2533–2553 (2019). https://doi.org/10.1007/s10444-019-09724-7
9. Benner, P., Werner, S.W.R.: MORLAB – Model Order Reduction LABoratory (version 5.0), 2019. see also: http://www.mpi-magdeburg.mpg.de/projects/morlab. https://doi.org/10.5281/zenodo.3332716
10. Boyd, S., Barratt, C.: Linear Controller Design: Limits and Performance. Prentice-Hall (1991)
11. Bui-Thanh, T., Willcox, K.: Model reduction for large-scale CFD applications using balanced proper orthogonal decomposition. In: 17th AIAA Computational Fluid Dynamics Conference, pp. 1–15 (2005). https://doi.org/10.2514/6.2005-4617
12. Bullen, P.S.: Handbook of Means and Their Inequalities, vol. 560, Mathematics and Its Applications. Springer (2003). https://doi.org/10.1007/978-94-017-0399-4
13. Curtis, F.E., Mitchell, T., Overton, M.L.: A BFGS-SQP method for nonsmooth, nonconvex, constrained optimization and its evaluation using relative minimization profiles. Optim. Methods Softw. **32**(1), 148–181 (2017). https://doi.org/10.1080/10556788.2016.1208749
14. Davidson, A.: Balanced systems and model reduction. Electron. Lett. **22**(10), 531–532 (1986). https://doi.org/10.1049/el:19860362
15. Enns, D.F.: Model reduction with balanced realizations: an error bound and a frequency weighted generalization. In: Proceedings of the 23rd IEEE Conference on Decision and Control, vol. 23, pp. 127–132 (1984). https://doi.org/10.1109/CDC.1984.272286
16. Fernandez, T., Djouadi, S.M., Foster, J.: Empirical Hankel norm model reduction with application to a prototype nonlinear convective flow. In: Proceedings of the American Control Conference, pp. 3771–3776 (2010). https://doi.org/10.1109/ACC.2010.5531560
17. Fernando, K.V., Nicholson, H.: Singular perturbational model reduction of balanced systems. IEEE Trans. Autom. Control **27**(2), 466–468 (1982). https://doi.org/10.1109/TAC.1982.1102932
18. Fernando, K.V., Nicholson, H.: On the structure of balanced and other principal representations of SISO systems. IEEE Trans. Autom. Control **28**(2), 228–231 (1983). https://doi.org/10.1109/TAC.1983.1103195
19. Glover, K.: All optimal Hankel-norm approximations of linear multivariable systems and their L^{∞}-error norms. Internat. J. Control **39**(6), 1115–1193 (1984). https://doi.org/10.1080/00207178408933239

20. Glover, K.: Model reduction: a tutorial on Hankel-norm methods and lower bounds on L^2 errors. In: IFAC Proceedings Volume (10th Triennial IFAC Congress on Automatic Control), vol. 20(5), pp. 293–298 (1987). https://doi.org/10.1016/S1474-6670(17)55515-9
21. Glover, K., Partington, J.R.: Bounds on the achievable accuracy in model reduction. In: Curtain, R.F. (ed.) Modelling, Robustness and Sensitivity Reduction in Control Systems, vo. 30, NATO ASI Series (Series F: Computer and Systems Sciences), pp. 95–118. Springer (1987). https://doi.org/10.1007/978-3-642-87516-8_7
22. Grundel, S., Hornung, N., Klaassen, B., Benner, P., Clees, T.: Computing surrogates for gas network simulation using model order reduction. In: Koziel, S., Leifsson, L. (eds.) Surrogate-Based Modeling and Optimization, pp. 189–212. Springer, New York (2013). https://doi.org/10.1007/978-1-4614-7551-4_9
23. Halvarsson, B.: Comparison of some Gramian based interaction measures. In: 2008 IEEE Int Symposium on Computer-Aided Control System Design, pp. 128–143 (2008). https://doi.org/10.1109/CACSD.2008.4627362
24. Hanzon, B.: The area enclosed by the (oriented) Nyquist diagram and the Hilbert-Schmidt-Hankel norm of a linear system. IEEE Trans. Autom. Control **37**(6), 835–839 (1992). https://doi.org/10.1109/9.256345
25. Himpe, C.: Combined State and Parameter Reduction for Nonlinear Systems with an Application in Neuroscience. Ph.D. thesis, Westfälische Wilhelms-Universität Münster, 2017. Sierke Verlag Göttingen, ISBN 9783868448818. https://doi.org/10.14626/9783868448818
26. Himpe, C.: emgr - the empirical Gramian framework. Algorithms **11**(7), 91 (2018). https://doi.org/10.3390/a11070091
27. Himpe, C.: emgr – EMpirical GRamian framework (version 5.7). https://gramian.de (2019). https://doi.org/10.5281/zenodo.2577980
28. Himpe, C., Ohlberger, M.: A unified software framework for empirical Gramians. J. Math. **1–6**, 2013 (2013). https://doi.org/10.1155/2013/365909
29. Himpe, C., Ohlberger, M.: Cross-Gramian based combined state and parameter reduction for large-scale control systems. Math. Prob. Eng. **2014**, 843869 (2014). https://doi.org/10.1155/2014/843869
30. Himpe, C., Ohlberger, M.: The empirical cross Gramian for parametrized nonlinear systems. In: IFAC-PapersOnLine (Proceedings of the 8th Vienna International Conference on Mathematical Modelling), vol. 48(1), pp. 727–728 (2015). https://doi.org/10.1016/j.ifacol.2015.05.163
31. Himpe, C., Ohlberger, M.: A note on the cross Gramian for non-symmetric systems. Syst. Sci. Control Eng. **4**(1), 199–208 (2016). https://doi.org/10.1080/21642583.2016.1215273
32. Arash (https://math.stackexchange.com/users/92185/arash). Geometric mean limit of ℓ_p norm of sums. Mathematics Stack Exchange, 2013. (version: 2013-09-13). https://math.stackexchange.com/q/492953
33. Jiang, Y.-L., Qi, Z.-Z., Yang, P.: Model order reduction of linear systems via the cross Gramian and SVD. IEEE Trans. Circuits Syst. II: Express Briefs **66**(3), 422–426 (2019). https://doi.org/10.1109/TCSII.2018.2864115
34. Kalman, R.E.: Contributions to the theory of optimal control. Boletin Sociedad Matematica Mexicana **5**, 102–119 (1960). http://liberzon.csl.illinois.edu/teaching/kalman_paper.pdf
35. Kalman, R.E.: Mathematical description of linear dynamical systems. SIAM J. Control Optim. **1**, 182–192 (1963). https://doi.org/10.1137/0301010
36. Lall, S., Marsden, J.E., Glavaški, S.: Empirical model reduction of controlled nonlinear systems. In: IFAC Proceedings Volumes (14th IFAC World Congress), vol. 32(2), pp. 2598–2603 (1999). https://doi.org/10.1016/S1474-6670(17)56442-3
37. Lam, J., Anderson, B.D.O.: L_1 impulse response error bound for balanced truncation. Syst. Control Lett. **18**(2), 129–137 (1992). https://doi.org/10.1016/0167-6911(92)90017-M
38. Liu, W.Q., Sreeram, V., Teo, K.L.: Model reduction and H_∞ norm computation for state-space symmetric systems. In: Proceedings of the 37th IEEE Conference on Decision and Control, pp. 2195–2200 (1998). https://doi.org/10.1109/CDC.1998.758666
39. Milk, R., Rave, S., Schindler, F.: pyMOR - generic algorithms and interfaces for model order reduction. SIAM J. Sci. Comput. **38**(5), S194–S216 (2016). https://doi.org/10.1137/15M1026614

40. Moore, B.C.: Principal component analysis in linear systems: controllability, observability, and model reduction. IEEE Trans. Autom. Control **AC–26**(1), 17–32 (1981). https://doi.org/10.1109/TAC.1981.1102568
41. Obinata, G., Anderson, B.D.O.: Model Reduction for Control System Design. Communications and Control Engineering. Springer, London, UK (2001). https://doi.org/10.1007/978-1-4471-0283-0
42. Or, A.C., Speyer, J.L., Kim, J.: Reduced balancing transformations for large nonnormal state-space systems. J. Guid. Control Dyn. **35**(1), 129–137 (2012). https://doi.org/10.2514/1.53777
43. Penzl, T.: Algorithms for model reduction of large dynamical systems. Linear Algebra Appl. **415**(2–3), 322–343 (2006). (Reprint of Technical Report SFB393/99-40, TU Chemnitz, 1999.). https://doi.org/10.1016/j.laa.2006.01.007
44. Pernebo, L., Silverman, L.M.: Model reduction via balanced state space representations. IEEE Trans. Autom. Control **27**(2), 382–387 (1982). https://doi.org/10.1109/TAC.1982.1102945
45. Phillips, J.R., Silveira, L.M.: Poor man's TBR: a simple model reduction scheme. IEEE Trans. Comput.-Aided Design Integr. Circuits Syst. **24**(1), 43–55 (2005). https://doi.org/10.1109/TCAD.2004.839472
46. Rahrovani, S., Vakilzadeh, M.K., Abrahamsson, T.: On Gramian-based techniques for minimal realization of large-scale mechanical systems. Top. Modal Anal. **7**, 797–805 (2014). https://doi.org/10.1007/978-1-4614-6585-0_75
47. Rave, S., Saak, J.: A non-stationary thermal-block benchmark model for parametric model order reduction. In: Benner, P., et al. (eds.) Model Reduction of Complex Dynamical Systems, International Series of Numerical Mathematics, p. 171. https://doi.org/10.1007/978-3-030-72983-7_16
48. Saak, J., Köhler, M., Benner, P.: M-M.E.S.S.-2.0 – the matrix equations sparse solvers library, August 2019. see also: https://www.mpi-magdeburg.mpg.de/projects/mess. https://doi.org/10.5281/zenodo.3368844
49. Safonov, M.G., Chiang, R.Y.: A Schur method for balanced model reduction. In: Proceedings of the American Control Conference, pp. 1036–1040 (1988). https://doi.org/10.23919/ACC.1988.4789873
50. Safonov, M.G., Chiang, R.Y.: A Schur method for balanced-truncation model reduction. IEEE Trans. Autom. Control **34**(7), 729–733 (1989). https://doi.org/10.1109/9.29399
51. Schelfhout, G.: Model Reduction for Control Design. Ph.D. Thesis, Dept. Electrical Engineering, KU Leuven, 3001 Leuven–Heverlee, Belgium (1996)
52. Schuler, S., Ebenhauer, C., Allgöwer, F.: ℓ_0-system gain and ℓ_1-optimal control. In: IFAC Proceedings Volumes (18th IFAC World Congress), vol. 44(1), pp. 9230–9235 (2011). https://doi.org/10.3182/20110828-6-IT-1002.00755
53. Shi, G., Shi, C.-R.J.: Model-order reduction by dominant subspace projection: error bound, subspace computation, and circuit applications. IEEE Trans. Circuits Syst. I: Regul. Pap. **52**(5), 975–993 (2005). https://doi.org/10.1109/TCSI.2005.846217
54. Sorensen, D.C., Antoulas, A.C.: The Sylvester equation and approximate balanced reduction. Numer. Lin. Alg. Appl. **351–352**, 671–700 (2002). https://doi.org/10.1016/S0024-3795(02)00283-5
55. The MORwiki Community. MORwiki - Model Order Reduction Wiki. http://modelreduction.org
56. Tombs, M.S., Postlethwaite, I.: Truncated balanced realization of a stable non-minimal state-space system. Internat. J. Control **46**(4), 1319–1330 (1987). https://doi.org/10.1080/00207178708933971
57. Toscano, R.: Structured Controllers for Uncertain Systems. Advances in Industrial Control. Springer London (2013). https://doi.org/10.1007/978-1-4471-5188-3
58. Varga, A.: Minimal realization procedures based on balancing and related techniques. In: Pichler, F., Diaz, R.M. (eds.) Computer Aided Systems Theory – EUROCAST'91, vol. 585, Lecture Notes in Computer Science, pp. 733–761. Springer (1991). https://doi.org/10.1007/BFb0021056

59. Willcox, K., Peraire, J.: Balanced model reduction via the proper orthogonal decomposition. AIAA J. **40**(11), 2323–2330 (2002). https://doi.org/10.2514/2.1570
60. Wilson, D.A.: The Hankel operator and its induced norms. Internat. J. Control **42**(1), 65–70 (1985). https://doi.org/10.1080/00207178508933346

Optimization-Based Parametric Model Order Reduction for the Application to the Frequency-Domain Analysis of Complex Systems

Rupert Ullmann, Stefan Sicklinger, and Gerhard Müller

Abstract A parametric model order reduction approach for the frequency-domain analysis of complex industry models is presented. Linear time-invariant subsystem models are reduced for the use in domain integration approaches in the context of structural dynamics. These subsystems have a moderate number of resonances in the considered frequency band but a high-dimensional input parameter space and a large number of states. A global basis approach is chosen for model order reduction, in combination with an optimization-based greedy search strategy for the model training. Krylov subspace methods are employed for local basis generation, and a goal-oriented error estimate based on residual expressions is developed as the optimization objective. As the optimization provides solely local maxima of the non-convex error in parameter space, an in-situ and a-posteriori error evaluation strategy is combined. On the latter, a statistical error evaluation is performed based on Bayesian inference. The method finally enables parametric model order reduction for industry finite element models with complex modeling techniques and many degrees of freedom. After discussing the method on a beam example, this is demonstrated on an automotive example.

R. Ullmann (✉)
BMW Research, New Technologies, Innovations, Parkring 19, 85748 Garching, Germany
e-mail: rupert.ullmann@bmwgroup.com

R. Ullmann · G. Müller
Chair of Structural Mechanics, TU Munich, Arcisstraße 21, 80333 München, Germany

S. Sicklinger
TU Munich, Arcisstraße 21, 80333 München, Germany

P. Benner et al. (eds.), *Model Reduction of Complex Dynamical Systems*,
International Series of Numerical Mathematics 171,
https://doi.org/10.1007/978-3-030-72983-7_8

1 Introduction

Numerical methods as uncertainty quantification (UQ) or globalized optimization provide new opportunities for a robust analysis and synthesis of the vibroacoustic quality of vehicles. Typically, such multi-query approaches involve system evaluations for an extensive number of different parameter values. In the automotive industry, these methods thus are infeasible, as numerical models often have a fine finite element (FE) discretization, which needs to be valid for different analysis types. Corresponding models, therefore, have many degrees of freedom ($n \approx 10^6$). The latter is also the case for the band-limited frequency-domain analysis of linear time-invariant models, which is discussed in the following. Although the system may have a moderate number of resonant modes in that frequency band, which theoretically allows for coarser meshes, remeshing the model for any particular application is time expensive and thus impossible in practice. Parametric model order reduction (pMOR) is one remedy to enable multi-query methods for such a large-scale full order model (FOM).

The transmission of structure-borne sound must be analyzed for whole vehicle assemblies composed of different subsystems. This motivates the combination of domain decomposition, respectively, integration, with pMOR by projection on subsystem level. Coupling the input-to-output behavior of several reduced-order models (ROM) in the frequency domain, efficient numerical algorithms can be obtained. As the coupling step is not in the focus of this study, refer to [23] for an overview in the context of dual-domain integration or [42] in the context of co-simulation. Complete vehicle models contain hundreds of design or uncertain model parameters. Domain decomposition allows for a localized treatment of these parameters on a subsystem level. A practical pMOR method then needs to preserve multiple subsystem parameters for variation in the ROM, not in the order of hundreds but still up to a high-dimensional order of $d = 15$. At the same time, considering large-scale subsystem FOMs, the pMOR approach must require a minimum number of FOM system evaluations for the projection matrix construction. For later coupling, pMOR also must be efficient with respect to the input-to-output behavior, defined by the transfer function matrix $\mathbf{H}$, for multiple-input-multiple-output (MIMO) subsystems.

The authors of [9] provide an extensive overview of pMOR. Following the scheme of that publication and an offline-online separation principle, pMOR methods can be classified according to the approaches chosen for the three necessary steps: parameter sampling for identifying samples which should be included in the basis, thus the training, basis construction itself in the offline phase and ROM generation at requested parameters in the online phase. The method discussed here uses approaches for these steps, as follows. For the first step of training, which is challenging for the combination of large-scale FOMs and high-dimensional parameter spaces, a grid-free greedy optimization-procedure is derived in Sect. 3. For the second step of basis generation, Krylov subspace methods are chosen, see Sect. 2.1. These allow controlling the error directly on the subsystems' $\mathbf{H}$ in a bounded frequency range. The third step of ROM generation contains the basic concept of handling ROM parameter variations in the online phase, in which a UQ is performed, for example. Two basic schemes are available: local and global approaches. Local approaches generate reduced models at certain parameter samples in the offline phase. In the online phase, these local ROMs are interpolated over the parameter space for any parameter sample at which the system should be evaluated employing generically chosen interpolating basis functions. There are different concepts for interpolation: interpolating local projection matrices [2, 10, 44], reduced system matrices [3, 18, 36], or reduced transfer functions [6, 7]. In the context of the latter, an interpolation in the pole-residue form recently received attention [47, 48]. Local pMOR methods are attractive as no information about the parametric dependency is necessary, thus they can be applied to black-box subsystems. However, due to the same reason, they suffer from the "curse of dimensionality" for parameter variations in the ROM, which can be reduced by the application of sparse grids [7, 22], for example. As another remedy to that for high-dimensional parameter spaces, there is a second class of methods, which is discussed in the following section.

2 Basics of the Global Basis and Krylov Subspace Method

A projection matrix $\mathbf{V}_{\mathrm{g}}$ is found in global approaches, which is valid over the whole parameter space at any parameter sample $\mathbf{p}_j \in [\mathbf{p}_{\mathrm{l}}, \mathbf{p}_{\mathrm{u}}]$. $\mathbf{V}_{\mathrm{g}}$ can be obtained by assembling a sufficiently high number of local bases $\mathbf{V}_z$ at different parameter points

$$\mathbf{V}_{\mathrm{g}} = [\mathbf{V}_1, \mathbf{V}_2, ..., \mathbf{V}_r] . \tag{1}$$

Global basis methods are established in a class of techniques under the name of reduced basis methods, in particular. In these methods, parameter sampling is performed using a greedy search strategy and a global basis is constructed, mainly using Proper Orthogonal Decomposition (POD), refer to [43], for example.

2.1 *Krylov Subspaces*

In contrast to classic reduced basis methods, Krylov subspace methods are chosen for the local basis generation of $\mathbf{V}_z$ in the following. By means of Krylov subspace methods, moments of the transfer function matrix $\mathbf{H}(s, \mathbf{p})$ can be matched in the FOM and ROM. The moments $\mathbf{m}_{\mathbf{p},j}$ are defined at an expansion point s_0 in the Laplace domain and for a fixed $\mathbf{p}_z$ as

$$\mathbf{H}(s, \mathbf{p}_z) = \sum_{j=1}^{\infty} \frac{1}{j!} \frac{\partial \mathbf{H}(s, \mathbf{p}_z)}{\partial s^j}\bigg|_{s_0} (s - s_0)^{j-1} = \sum_{j=1}^{\infty} -\mathbf{m}_{\mathbf{p}_z,j} (s - s_0)^{j-1} . \tag{2}$$

Purely imaginary expansion points $s_0 = i\omega_0$ are chosen in the following, for which $\mathbf{H}(s, \mathbf{p})$ of the underlying vibroacoustic second-order system is obtained by

$$\mathbf{H}(s = i\omega, \mathbf{p}) = \mathbf{C}\left(\mathbf{K}(\mathbf{p}) + i\text{sign}(\omega)\,\mathbf{S}(\mathbf{p}) + i\omega\mathbf{D}(\mathbf{p}) - \omega^2\mathbf{M}(\mathbf{p})\right)^{-1}\mathbf{B}. \quad (3)$$

$\mathbf{C}$ and $\mathbf{B}$ are the output and input matrices. $\mathbf{K}$, $\mathbf{M}$, $\mathbf{D}$, $\mathbf{S}$ are the stiffness, mass, viscous damping, and structural damping matrix, which are assumed to be symmetric. The dynamic stiffness matrix $\mathbf{K}_\text{d} = -\omega_0^2\mathbf{M} + i\omega_0\mathbf{D} + \mathbf{K} + i\text{sign}(\omega_0)\,\mathbf{S}$ is singular at $\omega_0 = 0$ for subsystems without Dirichlet boundary conditions. As consequence, no moments around $\omega_0 = 0$ can be matched. Numerically stable moment matching is obtained from Bubnov-Galerkin projection with the projection matrix $\mathbf{V}_z$,

$$\mathcal{G}_q(\mathbf{P}_1, \mathbf{P}_2; \mathbf{Q}) = \text{span}\{\mathbf{R}_0, \mathbf{R}_1, \ldots \mathbf{R}_{r-1}\} = \text{span}(\mathbf{V}_z). \quad (4)$$

$\mathcal{G}_q$ is a second-order Krylov subspace [40] of order q, defining the vector sequence

$$\begin{aligned} \mathbf{R}_0 &= \mathbf{Q} \\ \mathbf{R}_1 &= \mathbf{P}_1\mathbf{R}_0 \\ \mathbf{R}_j &= \mathbf{P}_1\mathbf{R}_{j-1} + \mathbf{P}_2\mathbf{R}_{l-2} \text{ for } j \geq 2. \end{aligned}$$

By choosing $\mathbf{P}_1 = -\mathbf{K}_\text{d}^{-1} \cdot \mathbf{D}_\text{d}$, $\mathbf{P}_2 = -\mathbf{K}_\text{d}^{-1} \cdot \mathbf{M}$, and $\mathbf{Q} = -\mathbf{K}_\text{d}^{-1} \cdot \mathbf{B}$, while $\mathbf{D}_\text{d} = 2i\omega_0\mathbf{M} + \mathbf{D}$, at least q moments of the FOM and ROM can be matched at s_0. For many cases of damping, like $\mathbf{S}$ as well as $\mathbf{D}$ are zero or defined as Rayleigh-damping $\mathbf{S}, \mathbf{D} = \alpha\mathbf{K} + \beta\mathbf{M}$, first-order Krylov subspaces can be used, see, [24], for example. In any case, modified Arnoldi (like) algorithms are used for constructing the local projection matrices in the following, involving an orthogonalization by the modified Gram-Schmidt algorithm and a subsequent QR decomposition. In the latter, inexact deflation is considered, and moment matching is not exactly fulfilled anymore. Columns in $\mathbf{V}_z$ with an euclidean norm smaller than the deflation length l_defl are removed. This reduces the column size of the projection matrix for many subsystem inputs, which are present in subsystem coupling applications.

Equation (2) defines moments of $\mathbf{H}(s, \mathbf{p})$ with respect to s. In view of pMOR, there are methods extending the concept of moments to additional parameter dimensions, see [8, 17] for one approach. Although conceptually promising, established algorithms need to be at least modified.

2.2 *Affine Matrix Decomposition*

Global methods showed up to be suitable for ROM generation in higher dimensional parameter spaces [9]. With an efficient training strategy, potentially less FOM system solutions are needed. Simply speaking, one can see the reason for that in the shifted interpolation problem compared to local approaches. For an efficient global basis method, the interpolation scheme of **K**, **M**, **D** and **S** must be found for their dependency on **p** a-priori in the form of an affine matrix decomposition

$$\mathbf{A}(\mathbf{p}) = \sum_{j=1}^{h} f_j(\mathbf{p})\,\mathbf{A}_j. \tag{5}$$

$f_j(\mathbf{p})$ are scalar functions of the parameter vector $\mathbf{p}$ and $\mathbf{A}_j \in \mathbb{C}^{n\times n}$ are parameter-independent matrices. In vectorized form with $\mathbf{a}(\mathbf{p}) = \mathrm{vec}\,(\mathbf{A}(\mathbf{p}))$ one can reformulate Eq. (5) as

$$\mathbf{a}(\mathbf{p}) = \boldsymbol{\Omega}\hat{\mathbf{f}}(\mathbf{p}). \tag{6}$$

$\hat{\mathbf{f}}(\mathbf{p}) \in \mathbb{C}^{h\times 1}$ contains the h basis functions for interpolation and $\boldsymbol{\Omega} \in \mathbb{C}^{n_a\times h}$ the corresponding interpolation coefficients, respectively $\mathbf{A}_j$ per column. Knowing the affine decomposition of a matrix allows to compute its projection by

$$\mathbf{A}_{\mathrm{R}}(\mathbf{p}) = \mathbf{V}_{\mathrm{g}}^{H}\mathbf{A}(\mathbf{p})\mathbf{V}_{\mathrm{g}} = \sum_{j=1}^{h} f_j(\mathbf{p})\,\mathbf{V}_{\mathrm{g}}^{H}\mathbf{A}_j\mathbf{V}_{\mathrm{g}} = \sum_{j=1}^{h} f_j(\mathbf{p})\,\mathbf{A}_{\mathrm{R},j}. \tag{7}$$

Consequently, it is possible to precompute the projections $\mathbf{A}_{\mathrm{R},j}$ once the projection basis is found. Afterward, $\mathbf{A}_{\mathrm{R}}(\mathbf{p})$ can be evaluated in the online phase for any parameter value using the same FOM interpolation rule along with $\mathbf{A}_{\mathrm{R},j}$.

In summary, not an interpolation scheme between different ROMs has to be approximated after projection as in local pMOR, but the interpolation rules for the system matrices of the FOM. In contrast to the unknown analytic relationship between the reduced systems in local pMOR, there is a low-rank parametric dependency of the FOM system matrices for many physical quantities, which can be derived analytically. In this case, additional knowledge is available to avoid the curse of dimensionality in the ROM evaluations.

Commercial FE codes usually do not provide white-box access to their code for intrusive changes. Nevertheless, an exact interpolation problem Eq. (6) can be reconstructed from parameter samples of the FOM system matrices, when the parametric dependency is known from theory. This is the case for many parameters, which are relevant for vibroacoustic FE models. For many FE formulations, material parameters like modulus of elasticity E, or mass density ρ have a linear influence on the stiffness, respectively, mass matrix. The same holds for the damping matrix in the case of the structural damping coefficient η, or the Rayleigh-damping coefficients α and β. A linear dependence also exists for the parameters of discretized components in the model, like linear springs or viscous dampers. The basis functions are known for many cases of geometric parametrization, in addition. For shells, which are modeled by triangular plate elements, the system matrices depend on the thickness through a cubic polynomial. For Euler-Bernoulli and Timoshenko beam elements, the influence of element length and cross-sectional dimensions is known. The derivation of Eq. (6) is also possible for geometric parameters in the case of more general element formulations. Fröhlich et al. [21] derived affine matrix decompositions for shape variations of solid elements. An application is found in Sect. 4.1.

For the case that neither the underlying FE code is accessible nor additional knowledge about the parametric dependency, and, therefore, the interpolating basis functions are available, inexact interpolation must be applied. This can be achieved by polynomial basis functions, for example, or regression methods. Another, conceptually different, approach to approximate Eq. (6) is given by the Discrete Empirical Interpolation Method [4, 9, 14].

3 OGPA: Optimization-based Greedy Parameter Sampling

For the construction of the local projection matrices, proper sample positions in the high-dimensional parameter space and the expansion points in the Laplace domain have to be determined in the training phase of global basis construction. Greedy algorithms are a practical approach to determine such sampling points in $\mathbf{p}$ and s. In a greedy search, the samples, respectively local ROMs are found one after the other. The best location for the next local model is determined per step, based on maximizing error measures on a discrete training set of parameter samples. Such approach does not provide a point selection, which is strictly optimal with respect to some norm, as the methods of [5, 25, 29]. However, no integration of error measures over the whole parameter space is necessary or the repetitive factorization of $\mathbf{K}_\mathrm{d}$ at all sampling points; thus greedy approaches are attractive for the application to large-scale industry FOMs.

To ensure the generation of efficient ROMs by a greedy approach, the training set must represent the typically non-convex error measure in the parameter space accurately enough. This is challenging for the high-dimensional parameter spaces of industry FOMs. Regular sampling grids show a complexity of $O(n_{\mathrm{sam}}^d)$ for n_{sam} samples per parameter dimension, leading again to the curse of dimensionality, now for the training phase of global basis generation. Sampling a 15-dimensional parameter space coarsely by three samples per parameter dimension already requires $14 \cdot 10^6$ grid points, for example. Remedies are available for that. Training sets can be generated using non-regular sampling strategies, Latin Hypercube Sampling [34], for example. Instead of using a static training set, several adaptive greedy approaches were developed as an alternative, which follow different refinement strategies, see [15, 26, 27], for example. In that context, [41] introduced a multi-stage procedure, [38] used surrogate models to identify regions for sampling refinement. Adaptive hierarchical greedy approaches are also available via sparse grids, see [13, 49], which can be applied in global reduced basis methods, see [16, 39].

3.1 Grid-Free Sampling

In order to meet that curse of dimensionality for the training phase of global basis construction, a grid-free sampling approach is proposed in the following. Opposed to the above discussed greedy approaches, the maximization of the error (estimate) does not rely on the evaluation on a fixed or adaptive grid. Instead, the i-th expansion point position $\mathbf{p}_i$ in the parameter space is found in a grid-free way. To determine the parameter sample, which is added to the global basis next, the constrained nonlinear optimization problem

$$\underset{{}^{i}\mathbf{p}_{\text{opt}} \in [\mathbf{p}_{\text{l}}, \mathbf{p}_{\text{u}}]}{\operatorname{argmax}} \quad \|\varepsilon(\mathbf{p})\| \tag{8}$$

is solved for an initial (random) guess of the parameter position ${}^{i}\mathbf{p}_0$. The objective function $\varepsilon(\mathbf{p})$ is an error function of the ROM. To start the greedy procedure, an initial ROM with $\mathbf{V}_{\text{g}} = \mathbf{V}_0$ is required. Afterward, a gradient-based optimization is employed for the solution of Eq. (8) in each greedy iteration. The idea of a greedy search via an optimization-based determination of $\mathbf{p}$ was introduced first by [11, 12]. Later, the concept was followed by [30, 46] in the context of reduced basis methods, for example.

Following the same basic idea of an optimization-based greedy parameter sampling (OGPA), the approach presented here is derived for band-limited frequency-domain analyses and Krylov subspace methods. The error evaluation, respectively, estimation, is in the frequency domain and is developed goal-oriented for the MIMO subsystem transfer function matrix; thus for the use in subsystem coupling. Gradients are derived efficiently based on an adjoint formulation for the use with many parameters.

3.1.1 Local Error Indicators for Sampling

Accounting for the fact that large-scale industry FOMs are considered, the use of the true ROM transfer function error as optimization objective in Eq. (8) is expensive as each evaluation of the true error needs a FOM solution. At the same time, it is sufficient to assess the correct trend of the error, not the absolute amplitude, to solve the optimization problem of Eq. (8). In general, the use of an error estimate results in a trade-off between the accuracy of the reduced ROM for a fixed ROM size, hence the efficiency, and the required computational efforts, which are necessary for error calculations in the basis generation. Bui-Thanh et al. [12] showed this effect for their approach in the time domain by comparing the use of the true error function and a cheaper estimation by the residual introduced in the FOM by the ROM solution. Utilizing the latter, more expansion points were necessary to obtain a prescribed error. This indicates that a point placement based on error estimates is less optimal. However, in any case, multiple local optimizations are required for basis generation, each needing multiple iterations up to a few hundred. Consequently, many error evaluations are necessary, and a suitable error measure must be primarily cheap for the application to large-scale vehicle FOMs. Evaluating the true error is not possible.

Residual expressions are an attractive choice for the error approximation as they make use of the FOM matrices but without the necessity of factorizations of the latter. For the Galerkin projection of symmetric subsystems with $\mathbf{C}^H = \mathbf{B}$ and symmetric $\mathbf{K}_\mathrm{d}$, the error in the transfer function matrix is related to the residual as follows

$$\varepsilon_\mathrm{H} = \mathbf{H} - \mathbf{H}_\mathrm{R} = \mathbf{r}_\mathrm{B}^H \mathbf{K}_\mathrm{d}^{-1} \mathbf{r}_\mathrm{B}. \tag{9}$$

n_i is the number of subsystem inputs, $\mathbf{r}_\mathrm{B} \in \mathbb{C}^{n \times n_i}$ is obtained by

$$\mathbf{r}_\mathrm{B} = \mathbf{B} - \mathbf{K}_\mathrm{d} \mathbf{V}_\mathrm{g} \mathbf{x}_\mathrm{R,B} \tag{10}$$

while $\mathbf{x}_\mathrm{R,B} = \mathbf{K}_\mathrm{d,R}^{-1} \mathbf{V}_\mathrm{g}^H \mathbf{B}$ is the ROM solution and $\mathbf{K}_\mathrm{d,R} = \mathbf{V}_\mathrm{g}^H \mathbf{K}_\mathrm{d} \mathbf{V}_\mathrm{g}$ the projected dynamic stiffness matrix. Taking a submultiplicative matrix norm of Eq. (9), one results in

$$\|\varepsilon_\mathrm{H}\| \leq \left\|\mathbf{r}_\mathrm{B}^H\right\| \left\|\mathbf{K}_\mathrm{d}^{-1}\right\| \|\mathbf{r}_\mathrm{B}\|. \tag{11}$$

This error bound can be used to calculate various approximations to the error $\|\varepsilon_\mathrm{H}\|$, see [19, 20] for some recent work. A more basic approach is to omit the first part $\left\|\mathbf{r}_\mathrm{B}^H\right\| \left\|\mathbf{K}_\mathrm{d}^{-1}\right\|$ and just to assume a proportional behavior between $\|\varepsilon_\mathrm{H}\| \propto \|\mathbf{r}_\mathrm{B}\|$. On the one hand, such approach can be only a rough approximation to the trend of $\|\varepsilon_\mathrm{H}\|$ and it is not possible to determine the quantitative error from that expression. The amplification by $\left\|\mathbf{K}_\mathrm{d}^{-1}\right\|$ is not considered; consequently this error estimate may fail at system resonances with a low modal damping coefficient. On the other hand, it is a computationally cheap approach, as one does not need to evaluate or approximate $\left\|\mathbf{K}_\mathrm{d}^{-1}\right\|$. In the context of large-scale FOMs, therefore, this approach is considered in the following by evaluating the Frobenius norm of the force residual $\mathbf{r}_\mathrm{B}$

$$r = -\log_{10}(\|\mathbf{r}_\mathrm{B}\|_\mathrm{F}). \tag{12}$$

The Frobenius norm was chosen as it is submultiplicative and gradients can be calculated for it. The logarithm of the norm is finally calculated. This accounts for the fact that the norm of the residual usually varies by orders of magnitude in the parameter space. As a result, convergence criteria would be hard to determine for the optimization algorithm without taking the logarithm.

Gradients of r are provided to the optimization algorithm. To derive the gradients for the general case of a MIMO system, consider the case of a SISO system with $\mathbf{H} = \mathbf{c}\mathbf{K}_\mathrm{d}^{-1}\mathbf{b}$ in a first step. As $\mathbf{c}^T$ and $\mathbf{b}$ are column vectors, the residual $\mathbf{r}_\mathrm{b}$ is also a column vector and gradients are provided by

$$\frac{\mathrm{d}r}{\mathrm{d}p_j} = \frac{\partial r}{\partial p_j} + \frac{\partial r}{\partial \mathbf{x}_\mathrm{R,b}} \frac{\mathrm{d}\mathbf{x}_\mathrm{R,b}}{\mathrm{d}p_j}. \tag{13}$$

The evaluation of Eq. (13) requires the calculation of the total derivative ${}^{\mathrm{d}\mathbf{x}_{\mathrm{R,b}}}/_{\mathrm{d}p_j}$. An additional equation is available for that, which is given by the residuum of the ROM's governing equations

$$\mathbf{r}_{\mathrm{R}} = \mathbf{b}_{\mathrm{R}} - \mathbf{K}_{\mathrm{d,R}}\mathbf{x}_{\mathrm{R,b}} = 0. \tag{14}$$

Calculating the total derivative of Eq. (14) with respect to the design parameter p_j, rearranging the expression for ${}^{\mathrm{d}\mathbf{x}_{\mathrm{R,b}}}/_{\mathrm{d}p_j}$ and inserting it in Eq. (13), one arrives at

$$\frac{\mathrm{d}r}{\mathrm{d}p_j} = \frac{\partial r}{\partial p_j} - \frac{\partial r}{\partial \mathbf{x}_{\mathrm{R,b}}}\left[\frac{\partial \mathbf{r}_{\mathrm{R}}}{\partial \mathbf{x}_{\mathrm{R,b}}}\right]^{-1}\frac{\partial \mathbf{r}_{\mathrm{R}}}{\partial p_j} = \frac{\partial r}{\partial p_j} + \boldsymbol{\Psi}^T\frac{\partial \mathbf{r}_{\mathrm{R}}}{\partial p_j}. \tag{15}$$

Introducing the substitution by $\boldsymbol{\Psi}^T$, the adjoint approach for gradient calculation is followed. This is an effective choice, as there are usually multiple input parameters, but exactly one objective, r, see [33]. With $\frac{\mathrm{d}\|\mathbf{r}_{\mathrm{b}}\|_{\mathrm{F}}^2}{\mathrm{d}p_j} = 2\Re\left(\mathbf{r}_{\mathrm{b}}^H\frac{\mathrm{d}\mathbf{r}_{\mathrm{b}}}{\mathrm{d}p_j}\right)$, the adjoint approach finally provides the gradient as

$$\frac{\mathrm{d}r}{\mathrm{d}p_j} = k\left(\frac{\partial\|\mathbf{r}_{\mathrm{b}}\|_{\mathrm{F}}^2}{\partial p_j} + \boldsymbol{\Psi}^T\frac{\partial \mathbf{r}_{\mathrm{R}}}{\partial p_j}\right) = k\Re\left(2\mathbf{r}_{\mathrm{b}}^H\frac{\partial \mathbf{K}_{\mathrm{d}}}{\partial p_j}\mathbf{V}_{\mathrm{g}}\mathbf{x}_{\mathrm{R,b}} + \boldsymbol{\Psi}^T\frac{\partial \mathbf{K}_{\mathrm{d,R}}}{\partial p_j}\mathbf{x}_{\mathrm{R,b}}\right). \tag{16}$$

$k = \frac{1}{2\|\mathbf{r}_{\mathrm{b}}\|_{\mathrm{F}}^2\log(10)}$ is obtained from applying the chain rule to the square root and the logarithm of $\|\mathbf{r}_{\mathrm{b}}\|_{\mathrm{F}}^2$. The adjoint $\boldsymbol{\Psi}$ is obtained from the solution of

$$\left[\frac{\partial \mathbf{r}_{\mathrm{R}}}{\partial \mathbf{x}_{\mathrm{R,b}}}\right]^T\boldsymbol{\Psi} = -\left[\frac{\partial\|\mathbf{r}_{\mathrm{b}}\|_{\mathrm{F}}^2}{\partial \mathbf{x}_{\mathrm{R,b}}}\right]^T, \tag{17}$$

which results in

$$\boldsymbol{\Psi} = \left(\mathbf{K}_{\mathrm{d,R}}^T\right)^{-1}\left(-2\mathbf{r}_{\mathrm{b}}^H\mathbf{K}_{\mathrm{d}}\mathbf{V}_{\mathrm{g}}\right)^T. \tag{18}$$

Starting from the gradient formulation for SISO systems, the gradients can be derived for the general MIMO case. For the latter, the residual $\mathbf{r}_{\mathrm{B}}$ of Eq. (10) is a matrix with as many columns as the number of inputs n_i. In order to extend the above framework for gradient calculation, the squared Frobenius norm is considered as the sum of the squared column vector lengths of the residual $\left\|\mathbf{r}_{\mathrm{B},i}\right\|_{\mathrm{F}}^2$. Following that, the squared Frobenius norm is obtained from the trace of a matrix-matrix product $\|\mathbf{r}_{\mathrm{B}}\|_{\mathrm{F}}^2 = \sum_{i=1}^{n_i}\left\|\mathbf{r}_{\mathrm{B},i}\right\|_{\mathrm{F}}^2 = \mathrm{tr}\left(\mathbf{r}_{\mathrm{B}}^H\mathbf{r}_{\mathrm{B}}\right)$. With that considerations, the adjoint $\boldsymbol{\Psi}$ is calculated for the n_i objectives of the squared column lengths $\left\|\mathbf{r}_{\mathrm{B},i}\right\|_{\mathrm{F}}^2$. Consequently $\boldsymbol{\Psi}$ becomes a matrix, which is obtained from n_i right hand sides

$$\boldsymbol{\Psi} = \left(\mathbf{K}_{\mathrm{d,R}}^T\right)^{-1}\left(-\mathbf{r}_{\mathrm{B}}^H\mathbf{K}_{\mathrm{d}}\mathbf{V}_{\mathrm{g}}\right)^T. \tag{19}$$

With that, Eq. (16) is finally extended for the general MIMO case to

$$\frac{\mathrm{d}r}{\mathrm{d}p_j} = k\,\Re\left(\mathrm{tr}\left(\mathbf{r}_{\mathrm{B}}^{H}\frac{\partial \mathbf{K}_{\mathrm{d}}}{\partial \nu}\mathbf{V}_{\mathrm{g}}\mathbf{x}_{\mathrm{R,B}} + \boldsymbol{\Psi}^{T}\frac{\partial \mathbf{K}_{\mathrm{d,R}}}{\partial p_j}\mathbf{x}_{\mathrm{R,B}}\right)\right) \tag{20}$$

with $k = 1/\|\mathbf{r}_{\mathrm{B}}\|_{\mathrm{F}}^{2}\log(10)$.

The solution of the adjoint system Eq. (19) is needed, along with the derivatives of the system matrices $\frac{\partial \mathbf{K}_{\mathrm{d}}}{\partial p_j}$ and $\frac{\partial \mathbf{K}_{\mathrm{d,R}}}{\partial p_j}$ to calculate the gradients of the objective function r. As, the affine matrix decomposition is known for the dynamic stiffness, derivatives of $\mathbf{K}_{\mathrm{d}}$ and $\mathbf{K}_{\mathrm{d,R}}$ can be obtained analytically by differentiating $f_j(\mathbf{p})$ in Eq. (5). The affine matrices $\mathbf{A}_j$ remain unchanged. The calculation of the adjoint solutions is computationally cheap, as they are obtained from the reduced model.

For many systems, the sum of matrix-matrix and matrix-vector multiplications of FOM size is computationally more demanding than the solution of the reduced system for the adjoints. This is especially the case, when the Hessian matrix should be calculated and provided to the optimization algorithms. Adjoint formulations can be also found for the Hessian matrix, see [37] for example. But, hundreds of matrix-matrix multiplications of FOM size are required to calculate the Hessian matrix at one objective evaluation already for a small number of parameters. For large-scale industry FOMs, the evaluation of the Hessian thus is prohibitively expensive.

3.1.2 Optimization Strategy

Employing Krylov subspaces for local bases, not only the optimal position in parameter space has to be found. To determine the expansion points for each local basis $\mathbf{V}_z$, suitable points in the Laplace, respectively, frequency domain as $s_0 = i\omega_0$, are also required. This raises the question of how the frequency dependency of the system should be included in the optimization process. Motivated by the approach of [12] for unsteady problems, one option is to follow a separation principle: Firstly, to integrate the error over the relevant frequency range $\varepsilon(\mathbf{p}) = \int_{\omega} \varepsilon(\mathbf{p}, \omega)\, d\omega$ and to find the parameter sample by optimization of Eq. (8), at which the local ROM is constructed. Secondly, to determine the frequency point(s) for moment matching by an appropriate method. Even if cheaper error estimates are used instead of the true error, however, this strategy prohibits the use of large-scale FOMs. It is computationally too expensive to consider the error integral over the relevant frequency range as optimization objective. In fact, Eq. (3) belongs to a steady problem with an additional parametric dependence on the frequency ω. Consequently, ω defines an additional parameter dimension and is included in $\tilde{\mathbf{p}}$

$$\tilde{\mathbf{p}} = \left[\mathbf{p}, \omega\right]^{T}. \tag{21}$$

Using this definition of the parameter vector in combination with a one-expansion-point-per-local-basis strategy, Eq. (8) provides both: the position of the local basis in parameter space and the frequency position of the expansion point are obtained without integration over the frequency domain.

In combination with the error estimate of Sect. 3.1.1, an efficient gradient-based optimization of Eq. (8) is possible. Sequential Quadratic Programming (SQP) is used as optimization algorithm, due to the efficiency reasons discussed above. The algorithm does not employ the true Hessian, but an estimation of the Hessian via the Broyden-Fletcher-Goldfarb-Shanno approach, see [35] for example. However, the optimization objective r, Eq. (12), is a potentially highly non-convex function. A single gradient-based optimization in Eq. (8) thus results in a local maximum of the error function, not the global one. Globalized optimization methods as a remedy, [28], for example, lead to an increased computational effort in each greedy iteration step, as they typically require a significantly high number of objective evaluations. In the case of the application to large-scale FOMs, such globalized approaches become impractical even for the residual error estimation as objective. Therefore, the global maximum of the error in the parameter range remains unknown, which implies the following consequences.

Firstly, this impacts the evaluation of the ROM accuracy and no definite evidence over the whole relevant parameter space is possible. This holds for both, the instantaneous ROM accuracy during the basis construction as well as the final one after basis generation. As a consequence of the grid-free approach, no training set is available from basis generation, which samples the parameter space densely and which allows estimating the error bound directly from the sampling or some interpolation in between. Adaptive approaches are one remedy to that unknown error bound of the ROM in parameter space. In such a method, there is an error control in the online phase, and $\mathbf{V}_g$ is enriched if needed. See [1], for an example. Another strategy is followed here, which provides a strict separation between offline and online phase, in contrast. As no a-priori error bound is available, an additional validation set is introduced. On this validation set, the true error in the transfer functions is evaluated after basis generation, but before the online phase. Only statistical measures are employed as the maximum error remains unknown, which is discussed in Sect. 3.2 more in detail.

Another consequence of the unknown global maximum of the non-convex error function in each greedy iteration is a semi-optimal expansion point placement in the parameter space. While in a greedy search, one aims to place the expansion points at the global maximum of the error per iteration, only local maxima are obtained as discussed. As an alternative to distinct globalized optimization, heuristics are available in literature. Urban et al. [46] discussed several possibilities for a preselection of candidates for the initial point ${}^i\tilde{\mathbf{p}}_0$ at the ith iteration of the greedy search. Motivated by this, a greedy search for ${}^i\tilde{\mathbf{p}}_0$ on a random parameter sample set precedes the local optimization. Based on this sample set, which is small ($n_{\mathrm{pre}} \leq 150$) and is changing in each iteration, the parameter sample with the smallest r is selected as ${}^i\tilde{\mathbf{p}}_0$.

The semi-optimal expansion point placement does not solely result from the non-global optimization, but also the use of the error estimate instead of the true error. The latter leads to a placement of expansion points even not necessarily at local maxima of the true error. To address that, a two-step validation procedure for ${}^i\tilde{\mathbf{p}}_{\mathrm{opt}}$ is introduced in the iterations of the greedy search. In a second step after each optimization, the

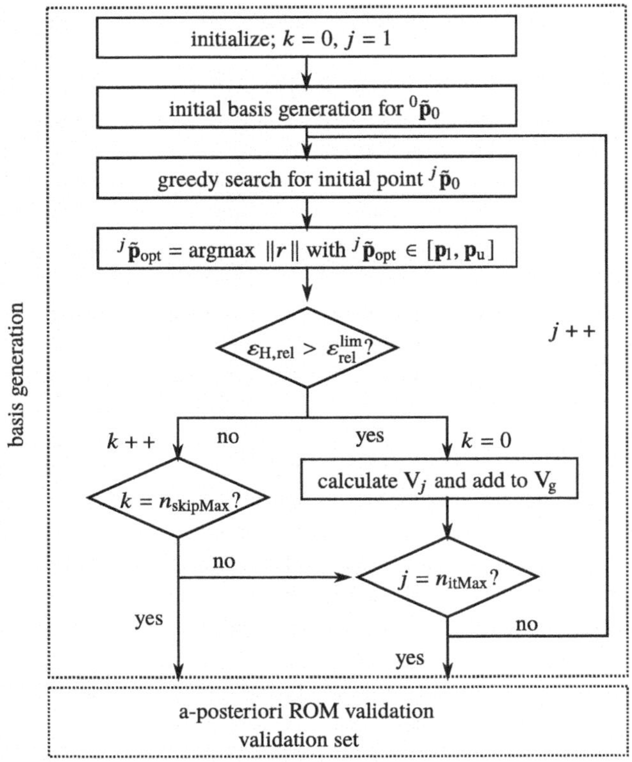

Fig. 1 Flow diagram for the pMOR basis generation and validation

maximum norm of the true relative error at this candidate parameter sample point is evaluated in-situ. Only in the case the true error is above $\varepsilon_{\text{rel}}^{\text{lim}}$, a local basis is calculated at ${}^i\tilde{\mathbf{p}}_{\text{opt}}$ and added to $\mathbf{V}_g$.

Introducing the two-step procedure for expansion point selection, it is also possible to define a stopping criterion for the greedy-based procedure. If n_{skipMax} successive iterations result in local maxima with $\varepsilon_{\text{H,rel}} < \varepsilon_{\text{rel}}^{\text{lim}}$ this can be considered as an indicator for an accurate ROM with the prescribed tolerance over the parameter range. ROM training is ended in this case, which is considered as a lucky breakdown. Afterward, the ROM quality can be evaluated further in the a-posteriori ROM evaluation, see Sect. 3.2. Otherwise, if n_{itMax} iterations are performed, but still a parameter sample is obtained from optimization with $\varepsilon_{\text{H,rel}} > \varepsilon_{\text{rel}}^{\text{lim}}$, there is a bad breakdown. Obviously, it was not possible to find a ROM in $n_{\text{itMax}} - 1$ iterations, which covers the parameter space with the prescribed error bound. In this case, the basis generation can be restarted with different settings, or the ROM can be evaluated for its qualification for the desired application in the a-posteriori ROM evaluation. The procedure for basis generation is summarized in Fig. 1.

3.2 A-Posteriori Model Quality Evaluation

To assess the quality of the ROM over the whole admissible parameter space robustly, the true relative error is evaluated on another additional sample set, which is decoupled from basis generation. In line with the concept of offline and online phase, a validation set is used, on which the true error is analyzed after basis generation, but before the online phase. To account for the high-dimensional parameter space, which requires many samples on the one hand, and the expensive FOM evaluations on the other hand, only a small validation set can be considered. The basic idea here is not trying to assess the global maximum of the error, as this would require knowledge of the underlying deterministic error expression or prohibitively large validation sets to sample the whole parameter space sufficiently dense as discussed above. Instead, the insufficient knowledge is met by assuming the error function as a black-box with stochastic behavior.

The validation set is generated by random sampling of $\mathbf{p}$, and statistic measures can be applied to the errors $\varepsilon_{\mathrm{H,rel}}$ obtained on the sample set. Histograms are provided for the following examples in order to show the potential of the proposed pMOR method descriptively. Providing such a measure, like mean and variance, however, is challenging in daily industrial applications. For each parameter sample, the FOM needs to be solved to obtain the true error $\varepsilon_{\mathrm{H,rel}}$. At the same time, the convergence rate of the statistic measures may be small in high-dimensional parameter spaces, and a large validation set may be required. Confidence levels of the obtained statistic measures can only be estimated by extensive sampling approaches, in addition.

As a remedy, one can employ probability theory to obtain a robust and cost-effective measure for the true error $\boldsymbol{\varepsilon}_{\mathrm{H,rel}}$ of pROMs in high-dimensional parameter spaces. Utilizing Bayesian inference, one can generate an estimate for the statistical behavior of $\boldsymbol{\varepsilon}_{\mathrm{H,rel}}$ for relatively small sample sets and can provide confidence levels at the same time. Bayesian inference allows determining a subjective probability of a good outcome, including the confidence level, respectively, the probability for that assumed subjective probability of success. Success, respectively, a good outcome is defined as $\varepsilon_{\mathrm{H,rel}} \leq \varepsilon_{\mathrm{rel}}^{\mathrm{lim}}$, $\varepsilon_{\mathrm{H,rel}} > \varepsilon_{\mathrm{rel}}^{\mathrm{lim}}$ is classified as bad outcome for a parameter sample, respectively, as an overshoot. Consequently, a binomial setting is followed here. Information is dropped, as one does not consider the actual magnitude of the error except the question if it is below the error threshold or not. This leads to a relaxed requirement for ROM model quality, which only needs to be ensured from a statistical point of view. This will be demonstrated in Sect. 4.1. Even high errors of the ROM model may be accepted, as long as there are only few overshoots, which can be a sufficient requirement for many multi-query methods.

The application of such binomial setting as quality indicator for simulation outputs in engineering problems is not new. But no applications to the assessment of (p)ROMs are available to the authors' knowledge. Lehar and Zimmermann [31] introduced the principle in order to estimate failure probabilities for vehicle crash simulations. The remaining paragraphs of this section follow closely that publication. Zimmermann and von Hoessle [50] use Bayesian inference to determine solution spaces in complex engineering models. In the most general form, Bayes theorem is

$$P(A|B) = \frac{P(B|A)\,P(A)}{P(B)}. \tag{22}$$

In the setup of (p)MOR, B is an observed result from an experiment, which is defined here as m good outcomes with $\varepsilon_{\mathrm{H,rel}} \leq \varepsilon_{\mathrm{rel}}^{\mathrm{lim}}$ within N samples calculated during the experiment in total. A is the probability of a good outcome and $P(A|B)$ the probability of that probability for the observed experimental result. $P(A)$ is the prior distribution of probability A. As there is no knowledge about this distribution before the experiment, it is assumed to be uniform between 0 and 1 initially, leading to probabilities which are all equally likely. Performing the experiment then allows to update that initial guess by means of Bayes theorem. For a binomial setup with initial uniform distribution, Eq. (22) can be rewritten as

$$p(a|m,N) = \frac{p(m,N|a)\,p(a)}{\int_0^1 p(m,N|g)\,p(g)dg} = \frac{\binom{N}{m} a^m (1-a)^{N-m}\,p(a)}{\int_0^1 \binom{N}{m} g^m (1-g)^{N-m}\,p(g)dg}. \tag{23}$$

As a is not known upfront, it is not practical to determine the confidence level of a single probability value for a good sample, a. Typically this would lead to small confidence levels. To get robust results, the confidence level has to be determined for confidence intervals of a instead, $a_{\mathrm{l}} < a < a_{\mathrm{u}}$. Equation (23) can be reformulated and simplified for confidence intervals of a as [31]

$$P(a_{\mathrm{l}} < a < a_{\mathrm{u}}|m,N) = \frac{\int_{a_{\mathrm{l}}}^{a_{\mathrm{u}}} t^m (1-t)^{N-m}\mathrm{d}t}{\int_0^1 g^m (1-g)^{N-m}\mathrm{d}g}. \tag{24}$$

Note, for the prior uniform distribution the probability of the probability a is constant $p(a) = p(s)$ and was eliminated from Eq. (24). Equation (24) finally allows to determine the confidence level that the probability a for $\varepsilon_{\mathrm{H,rel}} \leq \varepsilon_{\mathrm{rel}}^{\mathrm{lim}}$ at any requested parameter combination p is in the confidence interval $a_{\mathrm{l}} < a < a_{\mathrm{u}}$. The user predetermines the latter, and the true error, calculated at N uniformly distributed parameter values is required as experimental measurement. No additional inputs or assumptions are necessary on the distribution of the error $\varepsilon_{\mathrm{H,rel}}$. The relation between the parameters $\mathbf{p}$ and the error $\varepsilon_{\mathrm{H,rel}}$ is handled as black-box. The most significant advantage, however, is the fast convergence of the width of confidence intervals for prescribed values of confidence level, as pointed out by [31]. In the case of an actual a close to 1 (or 0), only small sets of samples are necessary to obtain high confidence levels for rather narrow confidence intervals.

4 Numerical Examples

The pMOR approach is discussed in two numerical examples in the following. Firstly, more general results are discussed on a simple beam example. Secondly, it is demonstrated for an automotive rear axle carrier, how the numerically efficient methods for basis generation and validation enable the application to large-scale industry models.

4.1 Cantilever Solid Beam

A beam example with high-dimensional parameter space is considered, as illustrated in Fig. 2. The fully parameterized example is available for reproduction along with its affine decomposition in [45]. A MIMO setting is chosen with two interface nodes, each having six degrees of freedom (DOFs), which results in 12 inputs and outputs. The local coordinate system of the first interface node is rotated intrinsically by $\varphi_z = 30°$ and $\varphi_x = 20°$, the one of the second node by $\varphi_z = 15°$ and $\varphi_x = 25°$. Although the geometry of the model is conceptually simple, it has relevance for industrial applications: modeling techniques for standard industry models are used, which are provided by the commercial FE software ABAQUS. Linear hexagon 3D elements with eight nodes and incompatible modes (C3D8I) are chosen for discretization of the beam. A rather coarse mesh of three elements in each cross-sectional dimension and 40 elements per section length results in 13164 degrees of freedom (DOFs). The nodes of the inputs are tied to the corresponding surfaces of the beam via kinematic couplings. As there is no ability to access the program code of the ABAQUS FE program but the affine functions $f_j(\mathbf{p})$ are known from theory, the affine decomposition is reconstructed from parameter samples of the FOM system matrices. Nine variable parameters are chosen in the following: the cross-sectional width b, and per section the material density ρ_i, the Elastic modulus E_i, the structural damping coefficient η_i as well as the section length L_i. The affine matrix decompositions of $\mathbf{K}$ and $\mathbf{S}$ consist of 20 terms for the selected parameters, the one of $\mathbf{M}$ of two terms, see [45] for the detailed formula.

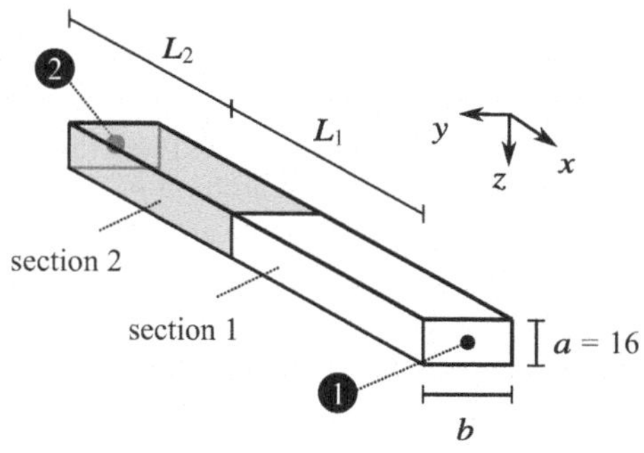

Fig. 2 Schematic drawing of the cantilever beam example

Table 1 Lower and upper bounds of the parameter values for the beam example. All units are omitted as they are in Newton, millimeter and ton

	E_1	ρ_1	η_1	E_2	ρ_2	η_2	b	L_1	L_2
$\mathbf{p}_l$	$2.5 \cdot 10^4$	$7 \cdot 10^{-10}$	0.005	$0.85 \cdot 10^4$	$1 \cdot 10^{-9}$	0.005	20	200	200
$\mathbf{p}_u$	$7.5 \cdot 10^4$	$10 \cdot 10^{-10}$	0.05	$3.5 \cdot 10^4$	$5 \cdot 10^{-9}$	0.05	24	300	400

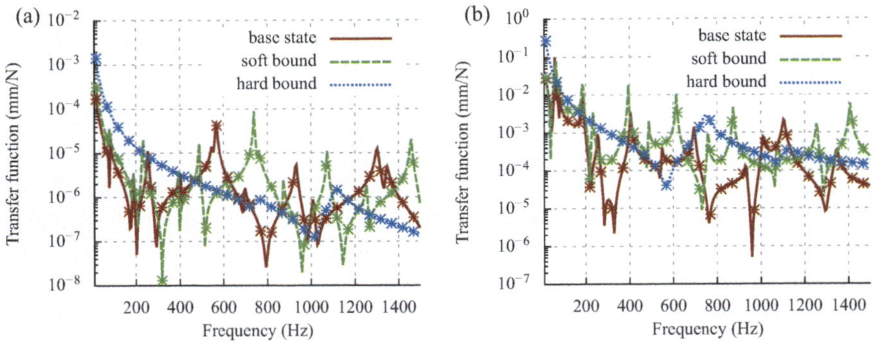

Fig. 3 Magnitude of the transfer function at **a** the diagonal element for node 1 DOF 1; **b** the off diagonal element, relating node 1 DOF 1 to node 2 DOF 2. The solid line belongs to the transfer function of the FOM, the starred line to the one of the ROM. A third setting with arbitrary parameter values is visualized in addition to the soft and hard bound setting

The parameter ranges are specified by the bounds in Table 1 and lead to a significant variance in the system response over parameter range, which needs to be covered by the global basis. This is visualized for the transfer function by two parameter settings in Fig. 3: a set which is called *soft bound*, and another called *hard bound*. In the soft setting, the upper bound values are chosen for ρ_i and L_i, the lower bound values for b, E_i and η_i. In the hard setting, the values are defined in the opposite way. A 10D parameter space has to be covered, with the frequency ω as one parameter dimension, for which a range of $\omega \in \left[0\,\mathrm{rad\,s}^{-1},\ 3000\pi\ \mathrm{rad\,s}^{-1}\right]$ is chosen.

PMOR with a training by OGPA is applied for these parameter settings. A maximum relative error of the ROM of $\varepsilon_{\mathrm{rel}}^{\mathrm{lim}} = 5 \cdot 10^{-3}$ is prescribed, a deflation length of $l_{\mathrm{defl}} = 1 \cdot 10^{-9}$ and a Krylov order of $o = 3$ is specified. As initial expansion point ${}^{0}\tilde{\mathbf{p}}_0$, the center point in parameter space is chosen. In addition, the maximum number of residual evaluations per local optimization is limited to 200; thus non-converged local optimizations may be accepted. Prior to each optimization the starting point is determined in a greedy search on $n_{\mathrm{pre}} = 50$ randomly distributed values. For these settings, 19 expansion points are added in 20 local optimizations, respectively, iterations of the automatic greedy search. Afterward, the greedy search is ended in a lucky breakdown after $n_{\mathrm{skipMax}} = 12$ consecutive iterations with discarded expansion point. In total, 32 local optimizations are performed, leading to a ROM size of $m = 235$.

Table 2 Overview over the optimized parameter samples for the greedy iterations in the beam example. All units omitted. Green color-coded values belong to the lower bound, blue color-coded values to the upper bound of parameter range

it	E_1	ρ_1 $\times 10^{-10}$	η_1	E_2	ρ_2 $\times 10^{-9}$	η_2	b	L_1	L_2	f_0
0	50000	8.5	0.0275	21750	3.0	0.0275	22	250	300	755.0
1	75000	7.0	0.005	8500	1.2	0.005	24	200	367.7	60.8
2	75000	7.0	0.005	8500	1.0	0.005	22.8	261.5	313.2	42.2
3	75000	7.0	0.005	32900.0	1.0	0.005	21.2	206.6	347.2	20
4	75000	7.0	0.005	35000	1.0	0.005	21.1	215.4	321.3	20
5	75000	7.0	0.05	35000	1.0	0.05	24	200	200	20
6	25000	7.0	0.005	35000	2.3	0.005	24	300	400	437.0
7	75000	7.0	0.005	8500	5.0	0.005	20	200	400	863.3
8	49567.7	7.0	0.005	21779.4	2.8	0.005	20	300	400	127.6
9	75000	7.0	0.005	8500	2.5	0.005	24	200	400	305.5
10	25000	7.0	0.005	35000	2.2	0.005	20	200	202.5	475.7
11	75000	7.0	0.005	8500	5.0	0.005	24	300	251.6	130.3
12	75000	9.1	0.005	8500	1.0	0.005	20	294.2	200	316.9
13	25000	10	0.005	8500	5.0	0.005	20	300	400	1500
14	75000	7.6	0.005	8500	5.0	0.005	20	300	306.7	1500
15	25000	7.0	0.005	35000	3.3	0.005	20	300	282.8	201.9
16	75000	7.0	0.005	8500	2.1	0.005	24	200	377.9	1335.6
17	25000	7.0	0.005	35000	5.0	0.005	20	261.0	400	140.3
18	71360.2	7.0	0.005	8500	5.0	0.005	23.6	300	230.7	1500
19	25000	10	0.005	35000	1.1	0.005	20	300	255.2	1500

The positions of the in total 20 expansion points in parameter space are found in Table 2. Analyzing the expansion point locations, which are determined by the local optimizations, one can identify a general tendency of OGPA for problems in structural dynamics. The algorithm places expansion points at the boundaries of certain parameter dimensions, which is an observation that is common for greedy algorithms [32]. When the frequency parameter is not considered, only 34 out of 171 parameter samples are no bound values. For the damping parameters η_1 and η_2, the lower bound values are obtained except for one iteration, as used for the *soft* setting. Based on heuristic experience, this is a general tendency of OGPA and mainly lower bound damping values are chosen for any structural model. Thus, the lower bound damping parameters can be directly used for basis construction, when the dimension of the parameter space should be reduced for training.

As the model setting is MIMO, the ROM size reduction by inexact deflation is significant. During the basis generation, 485 out of 720 candidates for Krylov modes are removed from the final basis.

While for the *hard* and *soft* parameter setting, the ROM quality visually is good (see Fig. 3), the latter is assessed statistically on the validation set a-posteriori after basis generation ended. As the FOM has a moderate size of 13164 DOFs for this first example, the true relative error can be evaluated on a large, randomly distributed validation set of size $n_{\mathrm{sam}} = 2 \cdot 10^4$. Solely one error overshoot with a value of

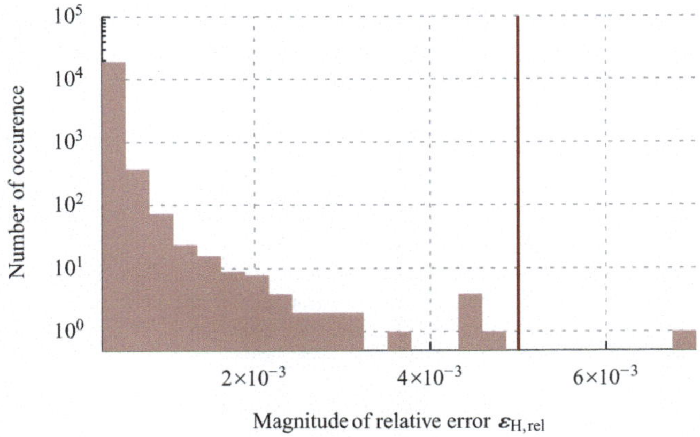

Fig. 4 Histogram of the maximum relative error $\varepsilon_{H,rel}$ per parameter sample. ε_{rel}^{lim}, which was used as error threshold in the training phase is indicated by the red vertical line

$\varepsilon_{rel}^{max} = 6.5 \cdot 10^{-3}$ is observed, which is slightly larger than the error tolerance of $\varepsilon_{rel}^{lim} = 5 \cdot 10^{-3}$ for training. The distribution of the relative error is visualized in Fig. 4. As such large validation sets are not feasible for large-scale industry FOMs, the error evaluation based on Bayesian inference was introduced in Sect. 3.2. For the large-scale validation set, one obtains

$$P\,(99.9\% < a < 100\% | 19999, 20000) = 100\%.$$

As the ROM is accurate, the Bayesian framework indicates this good ROM quality already with a small sample set. A validation set of 200 parameter points would have given $P\,(98.5\% < a < 100\%|199, 200) = 80.5\%$, a set of 500 samples $P\,(98.5\% < a < 100\%|499, 500) = 99.6\%$. In all cases, it was assumed conservatively that the error overshoot was within the first samples.

The availability of a rather large validation set also allows to discuss the performance of OGPA in the context of other approaches for training. Therefore, the setting of OGPA is changed for further efficiency enhancements and the maximum number of residual evaluations per local optimization is limited to 50 (M-O50:50). The results for this setting of OGPA are discussed in the context of two base-line scenarios: a worst-case scenario, in which the expansion points are placed arbitrarily in the parameter space (M-WC) and a quasi best case one, in which an extensive greedy search is performed on the validation set by using the true relative error (M-BC). In addition, they are compared to more classic greedy search strategies with comparable costs than M-O50:50: once on a set of 400 random parameter samples (M-RF400), which is fixed for all iterations; once on a set of 100 random parameter samples (M-RC100), which is smaller but changing completely in each iteration for a better coverage of the parameter space. The same residual error indicator as for OGPA (Eq. (12)) is used for all these greedy approaches.

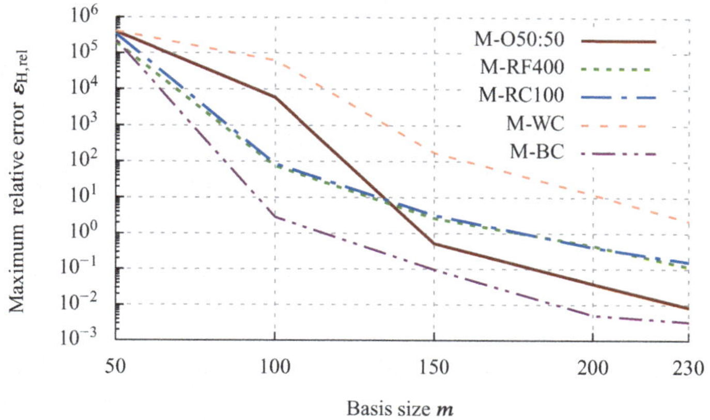

Fig. 5 Comparison of the decrease of the maximum relative error $\varepsilon_{\mathrm{H,rel}}$ on the validation set for an increasing basis size in the context of different methods for training

In a first evaluation, the maximum relative error is analyzed in relation to the basis size. The decay of the maximum error, which is analyzed in many publications on pMOR approaches, is visualized in Fig. 5 for the discussed methods. The latter provide slightly different results for each execution, as they involve a random parameter sample selection (except M-BC). As consequence, the corresponding algorithms were executed five times and averaged results are provided in Fig. 5. The maximum error values of the methods M-O50:50, M-RC100, and M-RF400, which involve a point placement based on the error estimate of Sect. 3.1.1, are in between the results for the worst and *best* case point placement in general. This shows that the error indicator provides valuable information for training, although only the rough trend of the true error ε_{H} is provided as discussed in Sect. 3.1.1. OGPA clearly outperforms a more classic greedy search with the same error indicator for sufficiently large basis sizes in Fig. 5. At the same time, only one quarter of residual evaluations is needed per greedy iteration for M-O50:50 compared to M-RF400, however, additional calculation efforts for gradient calculation are necessary. Thus, OGPA can be valuable especially for applications, in which the maximum relative error is important.

When the a-posteriori error analysis is performed by evaluating error overshoots with Bayesian inference according to Sect. 3.2, OGPA is again providing the best results around the converged basis size of $m = 230$. On average, there are five error overshoots within 20000 samples. Anyhow, an error analysis by Bayesian inference leads to another perspective on ROM quality and thus often relaxed requirements. This is reflected in a less distinct superiority of OGPA: a small number of 22 overshoots is also obtained with a classic greedy search (M-RF400); even a purely random point placement results in only 203 overshoots on average. Evaluating the latter with Bayesian inference, still confidence intervals of $P\,(98.5\% < a < 100\% | 19797, 20000) = 94.7\%$ can be obtained. In other words, simpler training strategies than OGPA may be chosen for applications, in which

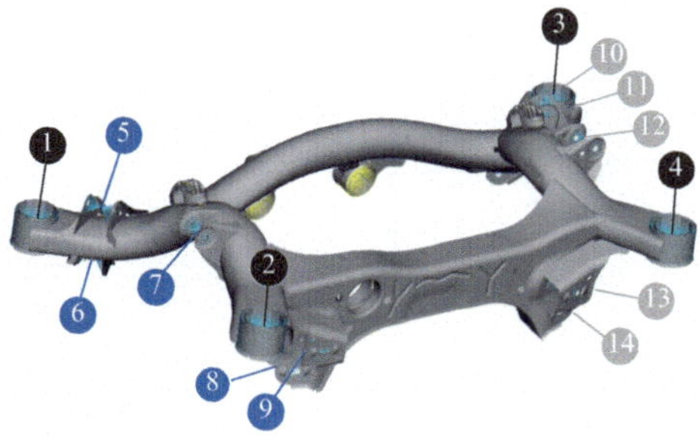

Fig. 6 The rear axle carrier and the labeled interface nodes. Each node provides six inputs and outputs

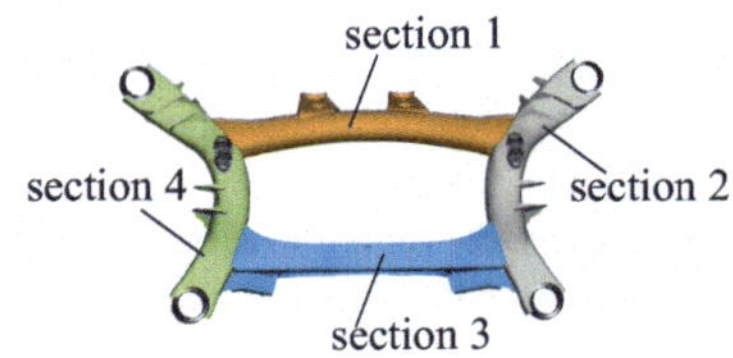

Fig. 7 Partitioning of the rear axle carrier into different sections

the maximum error is not crucial, as long as there are not prohibitively many error overshoots.

4.2 Rear Axle Carrier

As industrial example, an automotive rear axle carrier (RAC) is analyzed. The subsystem has an interface of 14 connection points to other subsystems, resulting in both, 84 inputs and outputs (see Fig. 6). A FE discretization using triangular plate-like shell elements is used with 258708 DOFs in total. As visualized in Fig. 7, the RAC is partitioned into four segments. The shell thickness and the material density is varied independently in each segment from $t_i \in [1.3\,\text{mm}, 2.3\,\text{mm}]$ and $\rho_i \in \left[5.5 \cdot 10^{-9}\,\text{tmm}^{-3}, 13 \cdot 10^{-9}\,\text{tmm}^{-3}\right]$. The frequency band is defined as $f \in [20\,\text{Hz}, 500\,\text{Hz}]$, leading to a 9D parameter space. The system matrices are assembled again in ABAQUS and the affine decompositions, Eq. (6), are reconstructed from the system matrices at corresponding parameter samples. For the cubic dependency of $\mathbf{K}$ and $\mathbf{M}$ on the single t_i, nine affine terms are needed for both of the two matrices. The linear influence of ρ_i on $\mathbf{M}$ can be considered with the same decomposition. In addition, the model contains a nonzero structural damping matrix $\mathbf{S}$, which is constructed from the affine decomposition of $\mathbf{K}$ with fixed $\eta_i = 0.005$. As in the cantilever beam example, a pre-evaluation showed that an optimization of the structural damping coefficients results in the lower bound value for each greedy iteration.

The amplitudes of two transfer functions of $\mathbf{H} \in \mathbb{C}^{84\times 84}$ are visualized exemplary in Fig. 8. Again, a soft setting with $\rho_i = 13 \cdot 10^{-9}\,\text{tmm}^{-3}$, $t_i = 1.3\,\text{mm}$ and a hard setting with $\rho_i = 5.5 \cdot 10^{-9}\,\text{tmm}^{-3}$, $t_i = 2.3\,\text{mm}$ is defined for visualization.

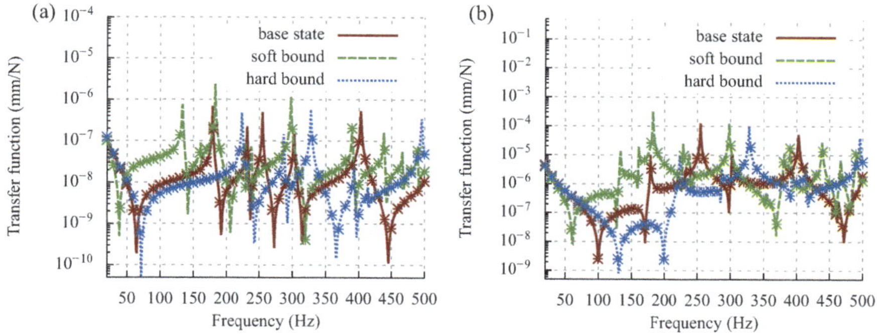

Fig. 8 Absolute value of the transfer function at **a** the diagonal entry for node 3 DOF 5; **b** at the off diagonal entry relating node 3 DOF 5 and node 7 DOF 1. The solid line belongs to the transfer function of the FOM, the starred line to the one of the ROM

For basis generation, a maximum relative ROM error $\varepsilon_{\text{rel}}^{\text{lim}} = 5 \cdot 10^{-3}$ is specified, for the deflation tolerance $l_{\text{defl}} = 10^{-7}$ and for the Krylov order $q = 2$ is defined. As another result of Sect. 4.1, the parameter values of the soft setting along with $f = 20$ Hz are chosen as ${}^{0}\tilde{\mathbf{p}}_0$ for the initial local basis. Again, the maximum number of residual evaluations per local optimization is limited to 200 and a greedy search on $n_{\text{pre}} = 150$ randomly distributed values is performed for each ${}^{i}\tilde{\mathbf{p}}_0$. pMOR with OGPA results in a basis size of $m = 836$. In addition to the one of ${}^{0}\tilde{\mathbf{p}}_0$, 20 additional local bases in the parameter space are identified. The resulting ${}^{i}\tilde{\mathbf{p}}_{\text{opt}}$ is omitted four times in the in-situ error evaluation until the last expansion point is found. The in-situ control was performed considering $\varepsilon_{\text{rel}}^{\text{lim}}$ for all entries of $\mathbf{H}$ with $|H_{ij}| > 5 \cdot 10^{-9}$. For entries with a smaller amplitude, a relaxed error threshold $\varepsilon_{\text{rel}}^{\text{rlx}} = 2 \cdot 10^{-1}$ was used for control.

After 24 greedy iterations, no more local basis is added in six subsequent optimizations, leading to a lucky breakdown after 30 greedy iterations in total.

Except for the frequency dimension, again ${}^{i}\tilde{\mathbf{p}}_{\text{opt}}$ contain mostly values from the bounds of the parameter space, only 29 out of 160 sample values are no bound values. Inexact deflation plays a significant role for basis reduction in that example. Without any deflation, 21 expansion points would result in $m = 3528$, thus almost 76% of the possible columns in $\mathbf{V}_{\text{g}}$ are removed during basis generation.

For the a-posteriori ROM error evaluation, a validation set of size $n_{\text{sam}} = 1000$ is chosen, which is smaller than the one in Sect. 4.1 due to the higher computational costs for the FOM evaluations. For 996 samples, the error was below $\varepsilon_{\text{H,rel}} = \varepsilon_{\text{rel}}^{\text{lim}}$, see Fig. 9 for a visualization. With four error overshoots, by means of Bayesian inference one finally obtains a statistical ROM quality of

$$P\,(99\% < a < 100\% | 996, 1000) = 97.1\%$$

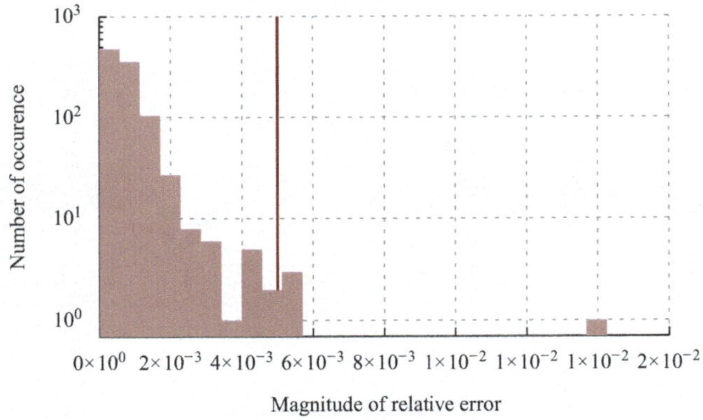

Fig. 9 Histogram of the relative transfer function error for the ROM of the rear axle carrier in logarithmic representation. $\varepsilon_{\rm rel}^{\rm lim}$ is indicated by the red vertical line

5 Summary

A global basis approach for parametric model order reduction was developed, which is suited to large-scale industry FOMs with a moderate number of resonances in the analyzed frequency band but a high-dimensional input parameter space. Although a global basis method is employed, the combination of a large-scale FOM and high-dimensional input parameter space is challenging. Only a few FOM system solves are possible, but the model needs to be trained for the analytically unknown, non-convex error function in the high-dimensional parameter space. Numerically efficient procedures were introduced for basis generation and validation, to enable pMOR for such models. An optimization-based greedy search strategy was employed (OGPA) to meet the "curse of dimensionality" in the training phase of global basis generation. In the context of a combination with domain integration and frequency-domain analysis, OGPA was combined with Krylov subspace methods for local basis generation. A goal-oriented error estimate was developed as the optimization objective, based on a residual expression as an indicator for the error in the MIMO system transfer function matrix. During the optimization-based greedy search, local maxima of the error estimate are obtained. The global maximum of the error, however, remains unknown during the training and local bases are placed solely at local maxima. As remedies, firstly the model training in the offline phase was modified for an improved candidate expansion point placement. A pre-selection of the starting points for optimization was performed based on an additional greedy search on a small set of random parameter samples. A two-step validation procedure for expansion points was introduced. Secondly, an alternative a-posteriori measure for ROM quality was discussed: a certain number of error overshoots in the later online phase may be accepted, as long as the ROM quality is statistically sufficient. For the evaluation of the latter, an additional validation sampling set was introduced, on which the true error in the transfer function is evaluated after basis generation, but before online phase. In combination with Bayesian inference, small validation sets are possible, and confidence intervals are provided. The efficiency of the pMOR approach was finally shown for both, a beam and an industry example of an automotive rear axle carrier.

References

1. Aliyev, N., Benner, P., Mengi, E., Schwerdtner, P., Voigt, M.: Large-scale computation of L_{∞}-norms by a greedy subspace method. SIAM J. Matrix Anal. Appl. **38**(4), 496–1516 (2017). https://doi.org/10.1137/16M1086200
2. Amsallem, D., Farhat, C.: Interpolation method for adapting reduced-order models and application to aeroelasticity. AIAA J. **46**(7), 1803–1813 (2008). https://doi.org/10.2514/1.35374
3. Amsallem, D., Farhat, C.: An online method for interpolating linear parametric reduced-order models. SIAM J. Sci. Comput. **33**(5), 2169–2198 (2011). https://doi.org/10.1137/100813051
4. Antil, H., Heinkenschloss, M., Sorensen, D.C.: Application of the Discrete Empirical Interpolation Method to Reduced Order Modeling of Nonlinear and Parametric Systems. In: Quarteroni, A., Rozza, G. (eds.) Reduced Order Methods for Modeling and Computational Reduction, pp. 101–136. Springer International Publishing, Cham (2014). https://doi.org/10.1007/978-3-319-02090-7_4
5. Baur, U., Beattie, C., Benner, P., Gugercin, S.: Interpolatory projection methods for parameterized model reduction. SIAM J. Sci. Comput. **33**(5), 2489–2518 (2011). https://doi.org/10.1137/090776925
6. Baur, U., Benner, P.: Modellreduktion für parametrisierte Systeme durch balanciertes Abschneiden und Interpolation - Model Reduction for Parametric Systems Using Balanced Truncation and Interpolation. Autom. **57**(8) (2009). https://doi.org/10.1524/auto.2009.0787
7. Baur, U., Benner, P., Greiner, A., Korvink, J., Lienemann, J., Moosmann, C.: Parameter preserving model order reduction for MEMS applications. Math. Comput. Model. Dyn. Syst. **17**(4), 297–317 (2011). https://doi.org/10.1080/13873954.2011.547658
8. Benner, P., Feng, L.: A robust algorithm for parametric model order reduction based on implicit moment matching. In: Quarteroni, A., Rozza, G. (eds.) Reduced Order Methods for Modeling and Computational Reduction, pp. 159–185. Springer International Publishing, Cham (2014). https://doi.org/10.1007/978-3-319-02090-7_6
9. Benner, P., Gugercin, S., Willcox, K.: A survey of projection-based model reduction methods for parametric dynamical systems. SIAM Rev. **57**(4), 483–531 (2015). https://doi.org/10.1137/130932715
10. Borggaard, J., Pond, K.R., Zietsman, L.: Parametric reduced order models using adaptive sampling and interpolation. IFAC Proc. Vol. **47**(3), 7773–7778 (2014). https://doi.org/10.3182/20140824-6-ZA-1003.02664
11. Bui-Thanh, T.: Model-Constrained Optimization Methods for Reduction of Parameterized Large-Scale Systems. Ph.D. Thesis, Massachusetts Institute of Technology (2007)
12. Bui-Thanh, T., Willcox, K., Ghattas, O.: Model reduction for large-scale systems with high-dimensional parametric input space. SIAM J. Sci. Comput. **30**(6), 3270–3288 (2008). https://doi.org/10.1137/070694855
13. Bungartz, H.J., Griebel, M.: Sparse grids. Acta Numerica **13**, 147–269 (2004). https://doi.org/10.1017/S0962492904000182
14. Chaturantabut, S., Sorensen, D.C.: Nonlinear model reduction via discrete empirical interpolation. SIAM J. Sci. Comput. **32**(5), 2737–2764 (2010). https://doi.org/10.1137/090766498
15. Chellappa, S., Feng, L., Benner, P.: An Adaptive Sampling Approach for the Reduced Basis Method. ArXiv191000298 Cs Math (2019)

16. Chen, P., Quarteroni, A.: A new algorithm for high-dimensional uncertainty quantification based on dimension-adaptive sparse grid approximation and reduced basis methods. J. Comp. Phys. **298**, 176–193 (2015). https://doi.org/10.1016/j.jcp.2015.06.006
17. Daniel, L., Siong, O., Chay, L., Lee, K., White, J.: A multiparameter moment-matching model-reduction approach for generating geometrically parameterized interconnect performance models. IEEE Trans. Comput.-Aided Des. Integr. Circuits Syst. **23**(5), 678–693 (2004). https://doi.org/10.1109/TCAD.2004.826583
18. Degroote, J., Vierendeels, J., Willcox, K.: Interpolation among reduced-order matrices to obtain parameterized models for design, optimization and probabilistic analysis. Int. J. Numer. Meth. Fluids **63**(2), 207–23 (2009). https://doi.org/10.1002/fld.2089
19. Feng, L., Antoulas, A.C., Benner, P.: Some a posteriori error bounds for reduced-order modelling of (non-)parametrized linear systems. ESAIM Math. Model. Numer. Anal. **51**(6), 2127–2158 (2017). https://doi.org/10.1051/m2an/2017014
20. Feng, L., Benner, P.: A new error estimator for reduced-order modeling of linear parametric systems. IEEE Trans. Microwave Theory Techn. **67**(12), 4848–4859 (2019). https://doi.org/10.1109/TMTT.2019.2948858
21. Fröhlich, B., Gade, J., Geiger, F., Bischoff, M., Eberhard, P.: Geometric element parameterization and parametric model order reduction in finite element based shape optimization. Comput. Mech. **63**(5), 853–868 (2019). https://doi.org/10.1007/s00466-018-1626-1
22. Geuss, M., Butnaru, D., Peherstorfer, B., Bungartz, H.J., Lohmann, B.: Parametric model order reduction by sparse-grid-based interpolation on matrix manifolds for multidimensional parameter spaces. In: 2014 European Control Conference (ECC), pp. 2727–2732. IEEE, Strasbourg, France (2014). https://doi.org/10.1109/ECC.2014.6862414
23. Gosselet, P., Rey, C.: Non-overlapping domain decomposition methods in structural mechanics. Arch. Comput. Methods Eng. **13**(4), 515–572 (2006). https://doi.org/10.1007/BF02905857
24. Gugercin, S.: Projection methods for model reduction of large-scale dynamical systems. Ph.D. Thesis, Rice University (2003)
25. Gugercin, S., Antoulas, A.C., Beattie, C.: Rational Krylov Methods for Optimal $\mathcal{H}_2$ Model Reduction (2006)
26. Haasdonk, B., Dihlmann, M., Ohlberger, M.: A training set and multiple bases generation approach for parameterized model reduction based on adaptive grids in parameter space. Math. Comp. Model. Dyn. Sys. **17**(4), 423–442 (2011). https://doi.org/10.1080/13873954.2011.547674
27. Hesthaven, J.S., Stamm, B., Zhang, S.: Efficient greedy algorithms for high-dimensional parameter spaces with applications to empirical interpolation and reduced basis methods. ESAIM Math. Model. Numer. Anal. **48**(1), 259–283 (2014). https://doi.org/10.1051/m2an/2013100
28. Horst, R., Pardalos, P.M., Thoai, N.V.: Introduction to Global Optimization. No. 3 in Nonconvex Optimization and Its Applications. Kluwer Academic Publishers, Dordrecht (1995)
29. Hund, M., Mlinarić, P., Saak, J.: An $\mathscr{H}_2 \otimes \mathscr{L}_2$ -Optimal Model Order Reduction Approach for Parametric Linear Time-Invariant Systems. Proc. Appl. Math. Mech. **18**(1) (2018). https://doi.org/10.1002/pamm.201800084
30. Iapichino, L., Volkwein, S.: Optimization strategy for parameter sampling in the reduced basis method. IFAC-PapersOnLine **48**(1), 707–712 (2015). https://doi.org/10.1016/j.ifacol.2015.05.020
31. Lehar, M., Zimmermann, M.: An inexpensive estimate of failure probability for high-dimensional systems with uncertainty. Struct. Saf. **36–37**, 32–38 (2012). https://doi.org/10.1016/j.strusafe.2011.10.001
32. Maday, Y., Stamm, B.: Locally adaptive greedy approximations for anisotropic parameter reduced basis spaces. ArXiv12043846 Math (2012)
33. Martins, J.R.R.A., Hwang, J.T.: Review and unification of methods for computing derivatives of multidisciplinary computational models. AIAA J. **51**(11), 2582–2599 (2013). https://doi.org/10.2514/1.J052184
34. McKay, M.D., Beckman, R.J., Conover, W.J.: A comparison of three methods for selecting values of input variables in the analysis of output from a computer code. Technometrics **21**(2), 239 (1979). https://doi.org/10.2307/1268522

35. Nocedal, J., Wright, S.J.: Numerical Optimization, 2nd edn. Springer Series in Operations Research. Springer, New York (2006)
36. Panzer, H., Mohring, J., Eid, R., Lohmann, B.: Parametric Model Order Reduction by Matrix Interpolation. - Autom. **58**(8) (2010). https://doi.org/10.1524/auto.2010.0863
37. Papadimitriou, D.I., Giannakoglou, K.C.: Direct, adjoint and mixed approaches for the computation of Hessian in airfoil design problems. Int. J. Numer. Meth. Fluids **56**(10), 1929–1943 (2008). https://doi.org/10.1002/fld.1584
38. Paul-Dubois-Taine, A., Amsallem, D.: An adaptive and efficient greedy procedure for the optimal training of parametric reduced-order models. Int. J. Numer. Methods Eng. **102**(5), 1262–1292 (2015). https://doi.org/10.1002/nme.4759
39. Peherstorfer, B., Zimmer, S., Bungartz, H.J.: Model reduction with the reduced basis method and sparse grids. In: Garcke, J., Griebel, M. (eds.) Sparse Grids and Applications, vol. 88, pp. 223–242. Springer, Berlin, Heidelberg (2012). https://doi.org/10.1007/978-3-642-31703-3_11
40. Salimbahrami, B., Lohmann, B., Bechtold, T., Korvink, J.: A two-sided Arnoldi algorithm with stopping criterion and MIMO selection procedure. Math. Comput. Model. Dyn. Syst. **11**(1), 79–93 (2005). https://doi.org/10.1080/13873950500052595
41. Sen, S.: Reduced-basis approximation and a posteriori error estimation for many-parameter heat conduction problems. Numer. Heat Transf. Part B: Fundam. **54**(5), 369–389 (2008). https://doi.org/10.1080/10407790802424204
42. Sicklinger, S., Belsky, V., Engelmann, B., Elmqvist, H., Olsson, H., Wüchner, R., Bletzinger, K.U.: Interface Jacobian-based co-simulation. Int. J. Numer. Methods Eng. **98**(6), 418–444 (2014). https://doi.org/10.1002/nme.4637
43. Sirovich, L.: Turbolence and the dynamics of coherent structures part I: coherent structures. Q. Appl. Math. **45**(3), 561–571 (1987)
44. Son, N.T.: A real time procedure for affinely dependent parametric model order reduction using interpolation on Grassmann manifolds. Int. J. Numer. Meth. Eng. 818–833 (2012). https://doi.org/10.1002/nme.4408
45. Ullmann, R.: A 3D solid beam benchmark for model order reduction. Mendeley Data V1 (2020). https://doi.org/10.17632/cprx2kx2ws.1
46. Urban, K., Volkwein, S., Zeeb, O.: Greedy sampling using nonlinear optimization. In: Quarteroni, A., Rozza, G. (eds.) Reduced Order Methods for Modeling and Computational Reduction, pp. 137–157. Springer International Publishing, Cham (2014). https://doi.org/10.1007/978-3-319-02090-7_5
47. Yue, Y., Feng, L., Benner, P.: An Adaptive Pole-Matching Method for Interpolating Reduced-Order Models. ArXiv190800820 Cs Math (2019)
48. Yue, Y., Feng, L., Benner, P.: Reduced-order modelling of parametric systems via interpolation of heterogeneous surrogates. Adv. Model. Simul. Eng. Sci. **6**(1), 10 (2019). https://doi.org/10.1186/s40323-019-0134-y
49. Zenger, C.: Sparse grids. In: Parallel Algorithms for Partial Differential Equations, pp. 241–251. Vieweg (1991)
50. Zimmermann, M., von Hoessle, J.E.: Computing solution spaces for robust design. Int. J. Numer. Methods Eng. **94**(3), 290–307 (2013). https://doi.org/10.1002/nme.4450

On Extended Model Order Reduction for Linear Time Delay Systems

Sajad Naderi Lordejani, Bart Besselink, Antoine Chaillet, and Nathan van de Wouw

Abstract This chapter presents a so-called extended model-reduction technique for linear delay differential equations. The presented technique preserves the infinite-dimensional nature of the system and facilitates the preservation of properties such as system parameterizations (uncertainties). It is proved in this chapter that the extended model-reduction technique also preserves stability properties and provides a guaranteed a-priori bound on the reduction error. The reduction technique relies on the solution of matrix inequalities that characterize controllability and observability properties for time delay systems. This work presents conditions on the feasibility of these inequalities, and studies the applicability of the extended model reduction to a spatio-temporal model of neuronal activity, known as delay neural fields. Lastly, it discusses the relevance of this technique in the scope of model reduction of uncertain time delay systems, which is supported by a numerical example.

S. Naderi Lordejani (✉) · N. van de Wouw
Mechanical Engineering Department, Eindhoven University of Technology, Eindhoven, The Netherlands
e-mail: s.naderilordejani@tue.nl

N. van de Wouw
e-mail: n.v.d.wouw@tue.nl

B. Besselink
Bernoulli Institute for Mathematics, Computer Science and Artificial Intelligence, University of Groningen, Groningen, The Netherlands
e-mail: b.besselink@rug.nl

A. Chaillet
L2S, CentraleSupélec, Université Paris Saclay, IUF, Gif sur Yvette, France
e-mail: antoine.chaillet@centralesupelec.fr

N. van de Wouw
Department of Civil, Environmental & Geo-Engineering, University of Minnesota, Minneapolis, USA

P. Benner et al. (eds.), *Model Reduction of Complex Dynamical Systems*, International Series of Numerical Mathematics 171,
https://doi.org/10.1007/978-3-030-72983-7_9

1 Introduction

Models in terms of delay differential equations have extensively been used to describe engineering systems such as, e.g., mechanical and electric/electronic systems [1, 25]. Systems of delay differential equations have also been used to model phenomena in, for instance, economics and biology [18]. Such models can, however, be complex in the sense that they consist of a large number of delay equations. This complexity can handicap simulation, analysis, or controller synthesis and implementation. This work presents a model order reduction technique to address the issue of model complexity of time delay systems.

In the course of the past four decades, a myriad of model order reduction techniques have been proposed for linear delay-free systems. Balanced truncation [22] is probably the most popular of these (see [15] for an overview). Parallel to these efforts, the model-reduction problem of time delay systems has also been studied, though to a much lesser extent. A common approach in the model complexity reduction of time delay systems is approximating the time delay system by a finite-dimensional model of, potentially, low order [2, 19, 20]. This approach has been motivated by the fact that currently analysis and design based on finite-dimensional models is in general more appealing, as it allows for the use of well-developed classical systems and control theory. Nonetheless, delay-structure preserving methods, i.e., methods that preserve the infinite-dimensional nature of the time delay system during model reduction, have also gained considerable attention [3–5, 16, 23, 28, 32, 33]. This attention is because reliable analysis and controller synthesis techniques are available today also for time delay systems [11, 21]. In addition, for a particular order of the reduced model, a reduced model in terms of delay differential equations has the potential to be more accurate than a finite-dimensional approximation of the same order [26]. In addition to the delay structure, in many cases, it is beneficial to preserve other desirable properties of the original model in the reduced-order model. Important examples are stability properties, structures of physical interconnections (e.g., the interconnection of a system and a controller) and the presence of uncertainties and model parameters. This chapter presents such a robust/parameterized model-reduction techniques for linear time delay systems.

This chapter is an extension of the work in [24], which introduced a so-called *extended* balanced truncation procedure for time delay systems. This procedure was motivated by the technique of extended-balanced truncation for finite-dimensional systems in [27, 29]. Following [4, 23], the work [24] defined bounds on the controllability and observability energy functionals of time delay systems, and constructed a model-reduction procedure based upon those. These bounds were characterized by matrices which are solutions to a set of matrix inequalities. Compared to the results in [23], extended balanced truncation comes with additional degrees of freedom in the computation of (bounds on) these functionals through the use of slack variables. It has been shown that the proposed technique is useful for the structured model reduction of closed-loop time delay systems and also for delay systems with polytopic parametrizations/uncertainties. It preserves both asymptotic stability and the infinite-dimensional nature of the time delay system, while also providing an a-priori computable, guaranteed, delay-dependent error bound.

The contributions of this chapter are fourfold. First, the feasibility of the matrix inequalities in [24] is studied in detail by presenting necessary and sufficient conditions on the feasibility of those matrix inequalities. Crucial results in [24] on the error bound and preservation of stability were lacking mathematical proofs. As a second contribution, this work presents the missing proofs for those results. Third, it studies and numerically illustrates the effectiveness of the extended balancing approach for parameterized/robust model reduction of time delay systems. Lastly, this work studies the applicability of the extended model-reduction technique to models in neuroscience. Namely, a method for dropping the spatial dependency in a particular model of neural fields is presented. This leads to a high-order time delay system, and the extended model-reduction technique is then applied to reduce the order of the resulting neural model without spatial dependency. This contribution is presented as a numerical example.

Outline. After introducing notation, a problem statement is given in Sect. 2. Section 3 introduces and gives a characterization of the observability and controllability energy functionals of a time delay system. Section 4 recapitulates the proposed model order reduction procedure in [24] and provides novel detailed proofs, and easy-to-check feasibility conditions for it are discussed in Sect. 5. The application of this technique to delay neural fields and robust/parameterized model reduction of delay systems is elaborated on in Sects. 6, and 7, respectively, and conclusions are presented in Sect. 8.

Notation. The set of real (non-negative) numbers is indicated by $\mathbb{R}$ ($\mathbb{R}_{\geq 0}$), and the Euclidean norm of a vector $x \in \mathbb{R}^n$ is denoted by $|x|$, which is defined as $|x| := \sqrt{x^T x}$. The notation $\mathcal{L}_2([a, b], \mathbb{R}^n)$ is the space of functions $x : [a, b] \to \mathbb{R}^n$ which have a bounded norm $\|x\|_2 = (\int_a^b |x(t)|^2\, dt)^{1/2}$, whereas $\mathcal{L}_\infty([a, b], \mathbb{R}^n)$ is the space of bounded, piecewise continuous functions mapping $[a, b]$ onto $\mathbb{R}^n$. The Banach space of absolutely continuous functions which map the interval $[-\tau, 0]$ onto $\mathbb{R}^n$ is indicated by $\mathcal{C}_n = \mathcal{C}([-\tau, 0], \mathbb{R}^n)$. Furthermore, $\mathcal{W}_n = \mathcal{W}([-\tau, 0], \mathbb{R}^n)$ refers to the space of bounded functions $\varphi \in \mathcal{C}_n$ with square-integrable derivative in a weak sense, i.e., $\dot{\varphi} \in \mathcal{L}_2([-\tau, 0], \mathbb{R}^n)$ for $\varphi \in \mathcal{W}_n$. [12, 18]. A block-diagonal matrix with A_1, ..., A_m on the diagonal is represented as $\text{blkdiag}\{A_1, \cdots, A_m\}$, and I_m is the $m \times m$ identity matrix. The notation $P > 0$, for $P \in \mathbb{R}^{n \times n}$, means that P is a symmetric, positive definite matrix. Matrix transposition and conjugate transposition are shown by the superscripts T and H, respectively. A star $*$ in a symmetric matrix represents a symmetric term.

2 Problem Statement

In this chapter, we consider a time delay system Ω of the form

$$\Omega : \begin{cases} \dot{x}(t) = Ax(t) + A_d x(t-\tau) + Bu(t), \\ y(t) = Cx(t) + C_d x(t-\tau) + Du(t), \\ x_0 = \varphi. \end{cases} \quad (1)$$

Here, $x(t) \in \mathbb{R}^n$ is the state vector, $u(t) \in \mathbb{R}^m$ and $y(t) \in \mathbb{R}^p$ are the external input and the output, respectively, while τ is a constant time delay. We assume that for all $\tau \in [0, \bar{\tau}]$, with a constant $\bar{\tau} > 0$, the system is asymptotically stable for zero input. For $t \in \mathbb{R}$, the function segment $x_t : [-\tau, 0] \to \mathbb{R}^n$ denotes the state of Ω at the time instance t, where $x_t(\theta) = x(t+\theta)$ for $\theta \in [-\tau, 0]$. The initial condition of the system is given by $\varphi \in \mathcal{C}_n$, such that $x(t) = \varphi(t)$, $t \in [-\tau, 0]$.

The objective is to approximate Ω by an asymptotically stable model $\hat{\Omega}$ of order $k < n$ which has the same delay structure as Ω. Moreover, the input-output behavior of $\hat{\Omega}$ should be close enough, in some measurable sense, to that of Ω. In addition, the model-reduction procedure itself should be applicable to time delay systems with polytopic uncertainties/parameterizations and it should facilitate structured model order reduction (that is, a model order reduction procedure which preserves physical interconnection structures in a system) for time delay systems.

It is noted that since the state of Ω belongs to $\mathcal{C}_n$, it has an infinite-dimensional nature in addition to the, potentially large, finite number of dynamical equations (i.e., state equations) describing it. In this chapter, model order reduction is pursued with respect to only the latter aspect.

3 Observability and Controllability Inequalities

Following [23, 24], we will discuss a model-reduction procedure for time delay systems based on so-called energy functionals.

First, the observability energy functional characterizes the output energy of (1) for a non-zero initial condition and zero input, and it can thus be regarded as a measure of observability. More precisely, we have the following definition taken from [4] (see [16] for a similar definition).

Definition 1 The observability functional of the system (1) is the functional $L_o : \mathcal{C}_n \to \mathbb{R}_{\geq 0}$ defined as

$$L_o(\varphi) = \int_0^\infty |y(t)|^2 dt, \quad (2)$$

where $y(\cdot)$ is the output of the system (1) for the initial condition $x_0 = \varphi$ and zero input $u = 0$.

In addition to the observability functional, the development of a balancing-based model-reduction procedure requires information on the controllability properties of the time delay system. In this regard, we consider the following definition of the controllability functional as a measure of controllability, see again [4] (and [16]).

Definition 2 The controllability functional of the system (1) is the functional $L_c : \mathcal{D}_n \to \mathbb{R}_{\geq 0}$ defined as

$$L_c(\varphi) = \inf\left\{\int_{-\infty}^{0} |u(t)|^2 dt \,\middle|\, u \in \mathcal{L}_2 \cap \mathcal{L}_\infty\big((-\infty, 0], \mathbb{R}^m\big),\ \lim_{T\to\infty} x_{-T} = 0,\ x_0 = \varphi\right\}, \tag{3}$$

where x_t is the solution of (1) for u that satisfies the above and $\mathcal{D}_n \subset \mathcal{C}_n$ is the domain of L_c, that is the space of function segments φ for which $L_c(\varphi)$ is well defined.

Generally, the a-priori computation of the observability and controllability functionals (2) and (3) is a challenging task [16]. The following lemmas from [24] present quadratic functionals characterized by computable matrices which can provide a tight upper and lower bound of $L_o(\varphi)$ and $L_c(\varphi)$, respectively.

Lemma 1 *Consider the asymptotically stable system Ω in* (1)*. Let there exist matrices $Q > 0$, $Q_d > 0$, $\bar{Q} > 0$ and $S > 0$, and a scalar α_o for which*

$$M_o = \begin{bmatrix} SA + A^T S + Q_d - \bar{Q} & \bar{Q} + SA_d & Q - S + \alpha_o A^T S & C^T \\ * & Q_d - \bar{Q} & \alpha_o A_d^T S & C_d^T \\ * & * & -2\alpha_o S + \tau^2 \bar{Q} & 0 \\ * & * & * & -I_p \end{bmatrix} < 0 \tag{4}$$

holds. Then the functional $E_o : \mathcal{W}_n \times \mathcal{L}_2([-\tau, 0], \mathbb{R}^n) \to \mathbb{R}_{\geq 0}$ given by

$$E_o(\varphi, \dot{\varphi}) = \varphi^T(0) Q \varphi(0) + \int_{-\tau}^{0} \varphi^T(s) Q_d \varphi(s)\, ds + \tau \int_{-\tau}^{0}\int_{\theta}^{0} \dot{\varphi}^T(s) \bar{Q} \dot{\varphi}(s)\, ds d\theta, \tag{5}$$

satisfies

$$E_o(\varphi, \dot{\varphi}) \geq L_o(\varphi), \tag{6}$$

for each $\varphi \in \mathcal{W}_n$ and with the functional L_o as in Definition 1.

Proof The proof of this lemma can be found in [24]. □

Lemma 2 *Consider the time delay system in* (1)*. Let there exist matrices $P > 0$, $P_d > 0$, $\bar{P} > 0$ and $R > 0$, and a positive scalar α_c which satisfy*

$$M_c = \begin{bmatrix} AR + RA^T + P_d - \bar{P} & \bar{P} + A_d R & P - R + \alpha_c RA^T & B \\ * & -P_d - \bar{P} & \alpha_c RA_d^T & 0 \\ * & * & -2\alpha_c R + \tau^2 \bar{P} & \alpha_c B \\ * & * & * & -I_m \end{bmatrix} < 0. \tag{7}$$

Then the functional $E_c : \mathcal{W}_n \times \mathcal{L}_2([-\tau, 0], \mathbb{R}^n) \to \mathbb{R}_{\geq 0}$ *given by*

$$E_c(\varphi, \dot{\varphi}) = \varphi^T(0) U \varphi(0) + \int_{-\tau}^{0} \varphi^T(s) U_d \varphi(s)\, ds + \tau \int_{-\tau}^{0} \int_{\theta}^{0} \dot{\varphi}^T(s) \bar{U} \dot{\varphi}(s)\, ds d\theta, \quad (8)$$

with $U = R^{-1} P R^{-1}$, $U_d = R^{-1} P_d R^{-1}$, $\bar{U} = R^{-1} \bar{P} R^{-1}$, *satisfies*

$$E_c(\varphi, \dot{\varphi}) \leq L_c(\varphi), \quad (9)$$

for all $\varphi \in \mathcal{D}_n \cap \mathcal{W}_n$ *and* L_c *as in Definition 2.*

Proof The proof has been omitted for the sake of brevity.

Remark 1 The variables S, α_o in (2), and R, α_c in (7) are referred to as the slack variables. By contrast, Q, Q_d, $\bar{Q}$ and U, U_d, $\bar{U}$ (also P, P_d, $\bar{P}$) which characterize the energy functionals (5) and (8), respectively, are referred to as the main decision variables.

The next section recalls the proposed model-reduction procedure in [24] and provides proofs for the technical results not provided in [24].

4 Model order reduction by truncation

Consider a partitioning of $x(t)$ and x_t (and φ) as

$$x(t) = \begin{bmatrix} x_1(t) \\ x_2(t) \end{bmatrix}, \; x_t = \begin{bmatrix} x_{1,t} \\ x_{2,t} \end{bmatrix}, \; \varphi = \begin{bmatrix} \varphi_1 \\ \varphi_2 \end{bmatrix}, \quad (10)$$

where $x_1(t) \in \mathbb{R}^k$ and $\varphi_1 \in \mathcal{W}_k$, with $k < n$ and together with the corresponding partitioning of the system matrices

$$\begin{aligned} A &= \begin{bmatrix} A_{11} & A_{12} \\ A_{21} & A_{22} \end{bmatrix}, \; A_d = \begin{bmatrix} A_{d,11} & A_{d,12} \\ A_{d,21} & A_{d,22} \end{bmatrix}, \; B = \begin{bmatrix} B_1 \\ B_2 \end{bmatrix}, \\ C &= \begin{bmatrix} C_1 & C_2 \end{bmatrix}, \; C_d = \begin{bmatrix} C_{d,1} & C_{d,2} \end{bmatrix}. \end{aligned} \quad (11)$$

A reduced-order approximation of (1), denoted by $\hat{\Omega}$, is obtained by truncation of the dynamics that correspond to x_2, leading to

$$\hat{\Omega} : \begin{cases} \dot{\zeta}(t) = A_{11}\zeta(t) + A_{d,11}\zeta(t - \tau) + B_1 u(t), \\ \hat{y}(t) = C_1 \zeta(t) + C_{d,1}\zeta(t - \tau) + D u(t), \\ \zeta_0 = \hat{\varphi}, \end{cases} \quad (12)$$

where $\zeta(t) \in \mathbb{R}^k$ and $\hat{y}(t) \in \mathbb{R}^p$ approximates $y(t)$, and $\hat{\varphi} \in \mathcal{W}_k$ is the initial condition of $\hat{\Omega}$.

The system $\hat{\Omega}$ approximates x_1 in the partitioned coordinates, and it clearly captures the delay structure of the original system Ω. In the sequel, it is shown that this type of model approximation can preserve other properties of the original model in the reduced-order model provided that the matrices S and R have a certain structure. First, we define an extended-balanced realization of Ω.

Definition 3 A realization as in (1) is said to be extended balanced if there exist matrices $S > 0$, $Q > 0$, $Q_d > 0$, $\bar{Q} > 0$, and a scalar α_o satisfying (4), matrices $R > 0$, $P > 0$, $P_d > 0$, $\bar{P} > 0$, and a scalar α_c satisfying (7), and, additionally, S and R are such that

$$S = R = \Sigma = \text{blkdiag}\{\sigma_1 I_{m_1}, \sigma_2 I_{m_2}, \cdots, \sigma_q I_{m_q}\}. \tag{13}$$

Here, the constants $\sigma_i > 0$, which satisfy $\sigma_i > \sigma_{i+1}, i \in \{1, ..., q-1\}$, are extended singular values of multiplicities m_i and $\Sigma_{i=1}^q m_i = n$.

Since S and R are symmetric, positive definite matrices, the system (1) can always be transformed into an extended-balanced form by exploiting the standard balancing transformation [10].

Lemma 3 *Let there exist symmetric matrices $S > 0$, $Q > 0$, $Q_d > 0$ and $\bar{Q} > 0$, and a scalar α_o satisfying (4), and symmetric matrices $R > 0$, $P > 0$, $P_d > 0$ and $\bar{P} > 0$, and a scalar α_c satisfying (7). Then, there exists a coordinate transformation $x(t) = Tz(t)$, with $T \in \mathbb{R}^{n\times n}$, such that the realization in the new coordinates is extended balanced.*

An interesting feature of the presented model order reduction is that it guarantees the preservation of stability properties, as stated in the following theorem.

Theorem 1 *Let the system* (1)*, which is asymptotically stable for zero input, be in an extended-balanced realization and consider the reduced-order system* (12) *obtained by truncation for $k \geq 1$. Then, the reduced-order system $\hat{\Omega}$ is asymptotically stable for zero input.*

Proof As the system (1) is an extended-balanced realization, there exist a diagonal matrix $S > 0$, and matrices $Q > 0$, $Q_d > 0$ and $\bar{Q} > 0$, and a scalar α_o such that (4) holds. Thus, for any full-column rank matrix Ψ of appropriate dimensions it holds that

$$\Psi^T M_o \Psi < 0 \tag{14}$$

with M_o as in (4). Since S is diagonal (recall Definition 3), we can write it in a block-diagonal form as $S = \text{blkdiag}\{S_1, S_2\}$, where $S_1 \in \mathbb{R}^{k\times k}$ corresponds to the reduced model $\hat{\Omega}$ and S_2 to the truncated dynamics. Now, we choose $\Psi = \text{blkdiag}\{\psi, \psi, \psi\}$, with $\psi = [I_k \;\; 0_{k\times(n-k)}]^T$. With this choice of Ψ and exploiting the block-diagonal structure of S, (14) implies that

$$\Psi^T M_o \Psi = \begin{bmatrix} A_{11}^T S_1 + S_1 A_{11} + Q_{d,11} - \bar{Q}_{11} & * & * \\ A_{d,11}^T S_1 + \bar{Q}_{11} & -Q_{d,11} - \bar{Q}_{11} & * \\ Q_{11} - S_1 + \alpha_o S_1 A_{11} & \alpha_o S_1 A_{d,11} & -2\alpha_o S_1 + \tau^2 \bar{Q}_{11} \end{bmatrix} < 0, \quad (15)$$

where $Q_{11} > 0$, $Q_{d,11} > 0$, $\bar{Q}_{11} > 0$ are the upper left $k \times k$ blocks of $Q > 0$, $Q_d > 0$ and $\bar{Q} > 0$, respectively. Now, using results from [11, Chapter 3], it is easily verified that (15) is a sufficient condition for the asymptotic stability of the reduced-order system for all time delays in the interval $[0, \tau]$. It should be mentioned that one may use the inequality (7) to prove this theorem in a similar way. □

The availability of an a-priori computable error bound is an appealing property of the presented model order reduction technique. The next theorem presents this property.

Theorem 2 *Let the asymptotically stable system* Ω *as in* (1) *be in an extended-balanced realization, as defined in Definition 3, and consider the reduced-order system* $\hat{\Omega}$*, as in* (12)*, obtained by truncation for* $k = \Sigma_{i=1}^r m_i$ *for some* $r > 0$*. Moreover, let* $\alpha_o = \alpha_c = \alpha$*. Then, for any common input function* $u \in \mathcal{L}_2 \cap \mathcal{L}_\infty([0, T], \mathbb{R}^m)$ *and initial conditions* $\varphi = 0$ *and* $\hat{\varphi} = 0$ *for* (1) *and* (12)*, respectively,*

$$\int_0^T \left| y(t) - \hat{y}(t) \right|^2 dt \leq \varepsilon^2 \int_0^T |u(t)|^2 \, dt,$$

for all $T \geq 0$ *and where the error bound* ε *is given as*

$$\varepsilon = 2 \sum_{i=r+1}^{q} \sigma_i, \quad (16)$$

with σ_i *as in* (13).

Before presenting a proof for this theorem, we give a technical lemma which can be proved based on results in [9].

Lemma 4 *Consider a system of the form* (1)*. If* $x_{t_0} \in \mathcal{W}_n$ *at* $t_0 \in \mathbb{R}_{\geq 0}$ *and* $u \in \mathcal{L}_\infty([t_0, t_1], \mathbb{R}^m)$ *for* $t_1 \geq t_0$*, then* $x_t \in \mathcal{W}_n$ *for all* $t \in [t_0, t_1]$.

Now, we prove Theorem 2.

Proof To prove this theorem, we take a one-step reduction approach. To this end, we first take a reduced-order system of the form (12) which is obtained by truncating the states corresponding to the final extended singular value σ_q, leading to a reduced-order model with $k = n - m_q$. Next, we define auxiliary states

$$z(t) := \begin{bmatrix} x_1(t) - \zeta(t) \\ x_2(t) \end{bmatrix}, \quad w(t) := \begin{bmatrix} x_1(t) + \zeta(t) \\ x_2(t) \end{bmatrix}. \quad (17)$$

Using (1) and (12) for zero initial conditions, the definitions in (17) lead to the dynamics

$$\begin{aligned}\dot{z}(t) &= Az(t) + A_d z(t-\tau) + \bar{B}\bar{u}(t),\\ \delta y(t) &= Cz(t) + C_d z(t-\tau),\end{aligned} \tag{18}$$

and

$$\dot{w}(t) = Aw(t) + A_d w(t-\tau) + 2Bu(t) - \bar{B}\bar{u}(t), \tag{19}$$

where $\bar{u}^T(t) = [\,\zeta^T(t)\ \zeta^T(t-\tau)\ u^T(t)\,]$, $\bar{B}^T = [\,0\ \bar{B}_2^T\,]$, with $\bar{B}_2 = [\,A_{21}\ A_{d,21}\ B_2\,]$, and $\delta y(t) = y(t) - \hat{y}(t)$ is the output of the error system. Now, based on the auxiliary dynamics and the observability and controllability functionals in (5) and (8), a functional is introduced as

$$V(z_t, w_t, \dot{z}_t, \dot{w}_t) = E_o(z_t, \dot{z}_t) + \sigma_q^2 E_c(w_t, \dot{w}_t), \tag{20}$$

which is well defined as $z, w \in \mathcal{W}_n$ (u is assumed to be piecewise continuous and bounded) due to Lemma 4. Similar to the proof of Lemma 1 in [24], it can be shown that the time-derivative of V along the trajectories of (18) and (19) is upper bounded by

$$\begin{aligned}\dot{V}(z_t, w_t, \dot{z}_t, \dot{w}_t) \le{}& \xi_z^T(t)\bar{M}_o\xi_z(t) + \sigma_q^2\xi_w^T(t)\bar{M}_c\xi_w(t) - |\delta y(t)|^2\\ &+ \left(2\sigma_q\right)^2|u(t)|^2 + 2\left(z^T(t) + \alpha_o\dot{z}^T(t)\right)S\bar{B}\bar{u}(t)\\ &- 2\sigma_q^2\left(w^T(t) + \alpha_c\dot{w}^T(t)\right)R^{-1}\bar{B}\bar{u}(t),\end{aligned} \tag{21}$$

where $\bar{M}_o$ is obtained by applying a Schur complement to M_o defined in (4) and

$$\bar{M}_c := \text{blkdiag}\{R, R, R, I_m\}^{-T} M_c \text{blkdiag}\{R, R, R, I_m\}^{-1},$$

with M_c and R as in (7), and

$$\begin{aligned}\xi_z^T(t) &:= [\,z^T(t)\ z^T(t-\tau)\ \dot{z}^T(t)\,],\\ \xi_w^T(t) &:= [\,w^T(t)\ w^T(t-\tau)\ \dot{w}^T(t)\ u^T(t)\,].\end{aligned}$$

Given that $\bar{M}_o < 0$ and $\bar{M}_c < 0$ due to (4) and (7), (21) further implies that

$$\begin{aligned}\dot{V}(z_t, w_t, \dot{z}_t, \dot{w}_t) \le{}& -|\delta y(t)|^2 + \left(2\sigma_q\right)^2|u(t)|^2 + 2\left(z^T(t) + \alpha_o\dot{z}^T(t)\right)S\bar{B}\bar{u}(t)\\ &- 2\sigma_q^2\left(w^T(t) + \alpha_c\dot{w}^T(t)\right)R^{-1}\bar{B}\bar{u}(t).\end{aligned} \tag{22}$$

Also, recalling that S and R have diagonal structures due to the extended-balanced form of the high-order system (see Definition 3), the time-derivative of V in (22) satisfies

$$\begin{aligned}&\dot{V}(z_t, w_t, \dot{z}_t, \dot{w}_t) \le -|\delta y(t)|^2 + \left(2\sigma_q\right)^2|u(t)|^2\\ &\quad + 2\left(z_2^T(t) + \alpha_o\dot{z}_2^T(t)\right)S_2\bar{B}_2\bar{u}(t) - 2\sigma_q^2\left(w_2^T(t) + \alpha_c\dot{w}_2^T(t)\right)R_2^{-1}\bar{B}_2\bar{u}(t),\end{aligned} \tag{23}$$

where S_2 and R_2 are the lower right $m_q \times m_q$ blocks of S and R, respectively.

Next, using the facts that $S_2 - \sigma_q^2 R_2^{-1} = 0$, $w_2 = z_2 = x_2$, for $\alpha_o = \alpha_c = \alpha$, we obtain

$$\dot{V}(z_t, w_t, \dot{z}_t, \dot{w}_t) \leq -|\delta y(t)|^2 + (2\sigma_q)^2 |u(t)|^2,$$

. Now, integrating the above over the interval $[0, T]$ gives

$$V(z_T, w_T, \dot{z}_T, \dot{w}_T) - V(z_0, w_0, \dot{z}_0, \dot{w}_0) \leq -\int_0^T |\delta y(t)|^2 \, dt + \left(2\sigma_q\right)^2 \int_0^T |u(t)|^2 \, dt.$$

The asymptotic stability of the original system implies that $0 \leq V(z_T, w_T, \dot{z}_T, \dot{w}_T) < \infty$. Moreover, $V(z_0, w_0, \dot{z}_0, \dot{w}_0) = 0$, because of the zero initial condition. Therefore the left-hand side of the above inequality exists and it is positive for all $T \geq 0$, thus

$$\int_0^T \left|y(t) - \hat{y}(t)\right|^2 dt \leq \left(2\sigma_q\right)^2 \int_0^T |u(t)|^2 \, dt.$$

As a result, the one-step reduction error bound is

$$\varepsilon = 2\sigma_q. \tag{24}$$

Next, following an analysis similar to the one presented in [13], which is based on the triangle inequality, it can be shown that extending the above to multiple one-step reductions leads to (16). □

The next section studies the feasibility of the matrix inequalities (4) and (7).

5 Feasibility of the Matrix Inequalities

In this section, we discuss feasibility conditions for the proposed model order reduction method. As this method relies on the matrix inequalities (4) and (7), we give easy-to-check conditions (both necessary and sufficient) for existence of solutions to these matrix inequalities for a common scalar $\alpha_c = \alpha_o = \alpha$, as required for the application of Theorem 2.

First, the following lemma shows that the feasibility of the inequalities is always guaranteed for sufficiently small delays provided $A + A_d$ is Hurwitz.

Lemma 5 *Let* (1) *be asymptotically stable for* $\tau = 0$. *Then, there exists a positive scalar* ϵ *for which the matrix inequalities in* (4) *and* (7) *are feasible for all* $\tau \in [0, \epsilon)$.

Proof The fact that the system (1) is asymptotically stable for $\tau = 0$ implies that $A_c := A + A_d$ is Hurwitz. Therefore, there exists a matrix $Q = Q^T > 0$ such that

$$A_c^T Q + Q A_c + C_c^T C_c < 0, \tag{25}$$

where $C_c = C + C_d$. The strict inequality in (25) guarantees the existence of a (large) $\bar{\alpha} > 0$ such that

$$QA_c + A_c^T Q + C_c^T C_c + \left(QA_d - Q_d + C_c^T C_d\right) \times \left(\bar{\alpha}Q + Q_d - C_d^T C_d\right)^{-1}\left(QA_d - Q_d + C_c^T C_d\right)^T < 0. \tag{26}$$

Following a Schur complement, this inequality implies that

$$\begin{bmatrix} QA_c + A_c^T Q + C_c^T C_c & QA_d - Q_d + C_c^T C_d \\ * & -\bar{\alpha}Q - Q_d + C_d^T C_d \end{bmatrix} + \tau^2\bar{\alpha}\begin{bmatrix} A_c^T \\ A_d^T \end{bmatrix} Q \begin{bmatrix} A_c & A_d \end{bmatrix} < 0 \tag{27}$$

for all $\tau \in [0, \epsilon_o)$ provided ϵ_o is sufficiently small. It can be shown that this inequality is equivalent to (4) for $S = Q, \alpha = \tau^2\bar{\alpha}$ and $\bar{Q} = \bar{\alpha}Q$. Thus, inequality (4) also holds for all $\tau \in [0, \epsilon_o)$. A similar argument can be performed about the feasibility of (7), i.e., we can show that there exists a sufficiently small ϵ_c such that (7) becomes feasible for all $\tau \in [0, \epsilon_c)$. The definition $\epsilon := \min\{\epsilon_o, \epsilon_c\}$ completes the proof of Lemma 5. □

Next, we present *necessary* conditions for the feasibility of (4) and (7) in terms of upper bounds on the delay τ.

Lemma 6 *Let $A_m := A - A_d$ be a non-Hurwitz matrix and $\bar{\lambda}_m$ be an eigenvalue of A_m which has the largest modulus in the right-half complex plane. Then, a necessary condition for* (4) *and* (7) *to hold is that*

$$\tau < \frac{2}{|\bar{\lambda}_m|}. \tag{28}$$

Lemma 7 *Let A in* (1) *be a non-Hurwitz matrix and $\bar{\lambda}$ be an eigenvalue of A which has the largest modulus in the closed right-half complex plane. Then, a necessary condition for* (4) *and* (7) *to hold is that*

$$\tau < \sqrt{\frac{2}{|\bar{\lambda}|^2 + \underline{\sigma}_d^2}}, \tag{29}$$

where $\underline{\sigma}_d$ is the smallest singular value of A_d.

Proof We present proofs for Lemmas 6 and 7 jointly, and based only on (4). First, we eliminate the slack variables from (4) by multiplying it from the left and right by

$$\begin{bmatrix} I_n & 0 & A^T & 0 \\ 0 & I_n & A_d^T & 0 \end{bmatrix}, \text{ and } \begin{bmatrix} I_n & 0 & A^T & 0 \\ 0 & I_n & A_d^T & 0 \end{bmatrix}^T,$$

respectively. This procedure results in

$$\begin{bmatrix} QA + A^TQ - \bar{Q} + Q_d + \tau^2 A^T \bar{Q} A & QA_d + \bar{Q} + \tau^2 A^T \bar{Q} A_d \\ * & -\bar{Q} - Q_d + \tau^2 A_d^T \bar{Q} A_d \end{bmatrix} < 0. \quad (30)$$

This inequality further implies that

$$\begin{bmatrix} A_m^T Q + QA_m - 4\bar{Q} + \tau^2 A_m^T \bar{Q} A_m & QA_d + 2\bar{Q} + Q_d + \tau^2 A_m^T \bar{Q} A_d \\ * & -\bar{Q} - Q_d + \tau^2 A_d^T \bar{Q} A_d \end{bmatrix} < 0. \quad (31)$$

Namely, this can be shown by the left and right multiplication of (30) by

$$\begin{bmatrix} I_n & -I_n \\ 0 & I_n \end{bmatrix}, \text{and} \begin{bmatrix} I_n & -I_n \\ 0 & I_n \end{bmatrix}^T,$$

respectively. Considering its upper left block, the inequality in (31) now implies that

$$A_m^T Q + QA_m - 4\bar{Q} + \tau^2 A_m^T \bar{Q} A_m < 0.$$

Let v be an eigenvector of A_m for the eigenvalue $\lambda_m = \mu_m + j\omega_m$. Then, left and right multiplication of this inequality by v^H and v implies

$$2\mu_m v^H Q v + (\tau^2 |\lambda_m|^2 - 4) v^H \bar{Q} v < 0. \quad (32)$$

Now, we consider only eigenvalues in the right-half complex plane. Namely, if $\mu_m \geq 0$, the satisfaction of (32) requires that $\tau < 2/|\lambda_m|$ and $\bar{Q} > 0$. This result establishes (28).

Next, we prove Lemma 7. The feasibility of (30) implies that

$$A^T Q + QA - \bar{Q} + Q_d + \tau^2 A^T \bar{Q} A < 0, \quad (33)$$

$$-\bar{Q} - Q_d + \tau^2 A_d^T \bar{Q} A_d < 0, \quad (34)$$

respectively, as follows from considering the block-diagonal elements. From (34), we obtain that $-\bar{Q} + \tau^2 A_d^T \bar{Q} A_d < Q_d$. Using this result in (33), we conclude the necessity of the following inequality:

$$A^T Q + QA - 2\bar{Q} + \tau^2 A^T \bar{Q} A + \tau^2 A_d^T \bar{Q} A_d < 0. \quad (35)$$

Now, if we take v as an eigenvector of A corresponding to an eigenvalue λ which lies in the closed right-half complex plane, (35) implies that

$$2\text{Re}(\lambda) v^H Q v + \left(\tau^2 |\lambda|^2 + \tau^2 \underline{\sigma}_d^2 - 2\right) v^H \bar{Q} v < 0.$$

Since $\text{Re}(\lambda) \geq 0$, this relation cannot be feasible without the satisfaction of (29). □

Remark 2 We note that the conditions provided by Lemmas 6 and 7 are only necessary conditions and not sufficient, i.e., they imply the infeasibility of the matrix inequalities if those conditions do not hold.

Remark 3 The condition of Lemmas 6 and 7 are more beneficial and practical when the model order reduction problem of feedback control systems with delays in the feedback channel is concerned, especially for systems with an unstable plant, leading to a non-Hurwitz A. In these system, the matrix $A - A_d$ is often non-Hurwitz.

Next, we present a result that is helpful in solving the matrix inequalities. Namely, given the couplings among α and the slack matrices in (4) and (7) (assuming that $\alpha_o = \alpha_c = \alpha$, in view of Theorem 2), these inequalities are nonlinear. To still enable solving these inequalities by using existing techniques for linear matrix inequalities, we perform a line search over α. For the line search to become more efficient, bounds on the search space for α should be provided. The following lemma provides such lower bound.

Lemma 8 *Consider A, and define $A_m := A - A_d$ and let λ and λ_m be arbitrary eigenvalues of A and A_m, respectively. Then, a necessary condition for the matrix inequalities* (4) *and* (7) *to hold is that*

$$\alpha > \max\{\tau^2 \mathrm{Re}(\lambda), \frac{\tau^2}{4}\mathrm{Re}(\lambda_m)\}. \tag{36}$$

Proof Here, we use only (4) to derive this inequality. The term $(M_o)_{33}$ (the (3,3) component of M_o) implies that

$$\bar{Q} < \frac{2\alpha}{\tau^2}S, \tag{37}$$

which follows from the fact that $(M_o)_{33}$ is a diagonal element. Using this result along with the fact that $(M_o)_{11} < 0$, we can conclude that

$$SA + A^T S + Q_d - \frac{2\alpha}{\tau^2}S < 0. \tag{38}$$

Let $Av = \lambda v$, i.e., v is an eigenvector corresponding to the eigenvalue λ of A. Then, left and right multiplication of the above inequality with v and v^H, respectively, implies that

$$v^H SAv + v^H A^T Sv - \frac{2\alpha}{\tau^2}v^H Sv + v^H Q_d v < 0.$$

This, in turn, leads to

$$\left(2\mathrm{Re}(\lambda) - \frac{2\alpha}{\tau^2}\right) v^H Sv < 0.$$

Since $S > 0$, this inequality holds only for $\alpha > \mathrm{Re}(\lambda)\tau^2$. Following a similar procedure, it can be shown that the satisfaction of (4) also requires $\alpha > \mathrm{Re}(\lambda_m)\tau^2/4$, with λ_m an eigenvalue of A_m. The fact that these hold for all eigenvalues of A and A_m leads to (36). □

Remark 4 Clearly, the lower bound in (36) becomes zero when A and $A - A_d$ are both Hurwitz, given the fact that $\alpha < 0$ is not allowed because of the fact that $(M_o)_{33}$ must be negative definite.

6 Example: Delay Neural Fields

This section presents a numerical example. The involved matrix inequalities are solved using the software CVX [14].

In this example, we study the application of the extended model-reduction technique to a model which describes the spatio-temporal interactions between neural populations in the brain. For comparison, we have also applied the position balancing technique in [16] to this model. Contrary to the bounds on the energy functions used in this chapter, position balancing relies on matrices that characterize the exact observability and energy functionals for a restricted class of functionals. These matrices represent the solution to a set of differential equations which are solved approximately [17].

Consider the delayed-neural fields model (see [6] for a survey) in the form of integro-differential equations:

$$l_i \dot{x}_i(r,t) = -x_i(r,t) + s_i\left(\sum_{j=1}^{n}\int_{\mathcal{R}} w_{ij}(r,r')x_j(r',t-\tau_{ij}(r,r'))\,dr' + I_i(r,t)\right), \tag{39}$$

for $i = 1, ..., q$, where q is the number of considered neuronal populations. The compact set $\mathcal{R} \subset \mathbb{R}$ describes the spatial domain containing the neuronal populations; it is assumed here to be uni-dimensional for simplicity. Moreover, $r \in \mathcal{R}$ is the spatial variable and $x_i(r,t)$ represents the neuronal activity of population i at time $t \geq 0$ and position $r \in \mathcal{R}$; $w_{ij} : \mathcal{R} \times \mathcal{R} \to \mathbb{R}$ is a bounded function such that $w_{ij}(r,r')$ describes the synaptic strength between location r' in population j and location r in population i. The constant $l_i > 0$ is the time decay constant of population i; $I_i : \mathcal{R} \times \mathbb{R} \to \mathbb{R}$ denotes the exogenous input to population i; $\tau_{ij} : \mathcal{R} \times \mathcal{R} \to [0, \bar{\tau}]$, $\bar{\tau} \geq 0$, is the self (for $i = j$) or mutual (for $i \neq j$) time delay resulting from the non-instantaneous communication between neurons, due to the finite velocity of signals along the axons. The continuously differentiable function $s_i : \mathbb{R} \to \mathbb{R}$ describes the excitability of population i.

To be able to rewrite (39) in the form (1), we first assume that the self time delays are zero ($\tau_{ii} = 0$) and the mutual delays are all fixed and equal, i.e., $\tau_{ij}(r,r') = \tau$ for all $i \neq j$ and all $r, r' \in \mathcal{R}$. With this assumption, and after linearizing the system around an operating profile $x_i^*(r)$ for the input $I_i(r,t) = I_i^*(r)$ (see [7] for details), the approximate model has the form

$$L\dot{\tilde{x}}(r,t) = -\tilde{x}(r,t) + S\int_{\mathcal{R}} \big(W_1(r,r')\tilde{x}(r',t) + W_2(r,r')\tilde{x}(r',t-\tau)\big)\,dr' + S\tilde{I}(r,t), \tag{40}$$

where $\tilde{x}^T := [\tilde{x}_1, ..., \tilde{x}_q]$ with $\tilde{x}_i = x_i - x_i^*$, $\tilde{I}^T := [\tilde{I}_1, ..., \tilde{I}_q]$ with $\tilde{I}_i = I_i - I_i^*$, $L = \text{diag}\{l_1, ..., l_q\}$ and $W_1 = \text{diag}\{w_{ii}\}$, for $i = 1, ..., q$, and $W_2 = [w_{ij}] - W_1$, for all $i, j = 1, ..., q$. Finally, $S = \text{diag}\{\bar{s}_1, ..., \bar{s}_q\}$, where $\bar{s}_i$ results from the linearization of the function s_i.

In the absence of delays ($\tau_{ij}(r,r') = 0$), an approach was proposed in [30] to analytically reduce the dynamics of the infinite-dimensional dynamics (39) to a finite-dimensional differential equation by assuming that the kernels w_{ij} can be decomposed on a finite basis of spatial functions. Following this idea, we assume that $W_i(r,r')$, $i = 1, 2$, is a so-called Pincherle-Goursat Kernel, i.e., there exist $X_i(r) \in \mathbb{R}^{q\times N_i}$ and $Y_i(r) \in \mathbb{R}^{q\times N_i}$, $N_i \in \mathbb{N}$, such that

$$W_i(r,r') = X_i(r)Y_i^T(r'). \tag{41}$$

We note that $X_i(r)$ contains the basis vectors of W_i. We further assume that there exists $\tilde{i}(t) \in \mathbb{R}^{N_1+N_2}$, for which the decomposition $\tilde{I}(r,t) = X(r)\tilde{i}(t)$, with $X = [X_1, X_2]$, holds. Given the structure of W_i in (41) and of $\tilde{I}$, we approximate the solution $\tilde{x}$ as $\tilde{x}(r,t) = \tilde{X}(r)v(t)$, where

$$\tilde{X}(r) = \big[X_1(r)\ X_2(r)\ X_e(r)\big] \tag{42}$$

can be regarded as a reduction basis (albeit depending on the spatial variable). In (42), $X_e(r) \in \mathbb{R}^{q\times N_e}$ denotes a potential enrichment of this reduction basis over the elements X_1 and X_2, which result from the structure of W_i. Moreover, $v(t) \in \mathbb{R}^n$, $n = N_1 + N_2 + N_e$, is an unknown vector the driving dynamics which is yet to be obtained.

Remark 5 We note that the structure of this approximation separates the effect of the spatial and temporal variables.

Then, the substitution of (41) and the approximation (42) into (40) leads to

$$L\tilde{X}(r)\dot{v}(t) = -\tilde{X}(r)v(t) + SX_1(r)K_1v(t) + SX_2(r)K_2v(t-\tau) + S\tilde{X}(r)\tilde{i}(t), \tag{43}$$

where $K_i = \int_{\mathcal{R}} Y_i^T(r')\tilde{X}(r')\,dr'$, $i = 1, 2$. This equation holds for every r, so we can multiply both sides of (43) by $\tilde{X}^T(r)$ from the left. Then, integration of both sides of the resulting equation over $\mathcal{R}$ leads to

$$M_l\dot{v}(t) = (M_1K_1 - M)v(t) + M_2K_2v(t-\tau) + M_s\tilde{i}(t),$$

where

Table 1 Parameters of the neural field

Parameter	Value	Parameter	Value
L	diag{10, 20}	w_{11}	0
S	diag{20, 20}	w_{12}	$-30\exp\left(-\frac{\lvert r-r'-1.32\times10^{-2}\rvert^2}{0.06}\right)$
$\mathcal{R}$	$[0, 2.5]\cup[12.5, 15]\times10^{-3}$	w_{21}	$38\exp\left(-\frac{\lvert r-r'-1.25\times10^{-3}\rvert^2}{0.06}\right)$
τ	0.03 sec	w_{22}	$-2.55\exp\left(-\frac{\lvert r-r'\rvert^2}{0.03}\right)$

$$M_l = \int_{\mathcal{R}} \tilde{X}^T(r)L\tilde{X}(r)dr, \quad M_1 = \int_{\mathcal{R}} \tilde{X}^T(r)SX_1(r)dr, \quad M_2 = \int_{\mathcal{R}} \tilde{X}^T(r)SX_2(r)dr,$$
$$M_s = \int_{\mathcal{R}} \tilde{X}^T(r)SX(r)dr, \quad M = \int_{\mathcal{R}} \tilde{X}^T(r)\tilde{X}(r)dr. \tag{44}$$

Clearly, if M_l is invertible, this equation can be written in the form (1) by defining

$$A = M_l^{-1}(M_1K_1 - M), \quad A_d = M_l^{-1}M_2K_2, \quad B = M_l^{-1}M_sF,$$
$$C = \int_{\mathcal{R}} \bar{C}(r)\tilde{X}(r)\,dr, \quad C_d = 0, \quad D = 0.$$

Here, we have considered $\tilde{i}(t) = Fu(t)$ with $F \in \mathbb{R}^{(N_1+N_2)\times m}$ and $u(t) \in \mathbb{R}^m$ as the input. We note that F is defined such that the elements of u are independent. Moreover, $\bar{C}(r) \in \mathbb{R}^p$ is the distributed output matrix. Namely, we consider outputs of the form $y(t) = \int_{\mathcal{R}} \bar{C}(r)\tilde{x}(r,t)\,dr$. Given the complexity of w_{ij} and the enrichment basis X_e, the dimension of $\tilde{X}(r)$ and, subsequently, the order n of the time delay system describing the dynamics of $v(t)$ can be large.

In this example, we consider a neural field with the parameters reported in Table 1. The input is given by $\tilde{I}_1(r,t) = 0$ and $\tilde{I}_2(r,t) = (1+r)\exp(-r^2/0.03)u(t)$ and the output is characterized by $\bar{C}(r) = [1, 0.1]$. After computing $X_1(r)$ and $X_2(r)$, where a truncated Taylor series expansion has been exploited (for details, see Appendix A) and considering $X_e = 0$, we obtain a system of the form (1) of order $n = 9$, and $F^T = [1, 1, 0, ..., 0]$. The frequency response function of this system between the input u and the output y is represented by $G_v(j\omega)$.

The corresponding singular values resulting from the application of the extended model order reduction technique in comparison to those from the position balancing technique are plotted in Fig. 1. In the same figure, we have reported the reduction error ε, for the extended technique, as a function of the reduction order k. It is observed that the singular values from the position balancing technique are smaller than those from the extended method. However, we note that the position balancing technique does not provide an a-priori error bound, neither does it guarantee the stability of the reduced system. We observe a quick decay in the singular values from the extended technique after $k = 2$. Thus, we may approximate the dynamics of $v(t)$ by a model, with the frequency response function represented by $\hat{G}_v(j\omega)$, of order $k = 2$ and expect an accurate model approximation. In Fig. 2, the frequency

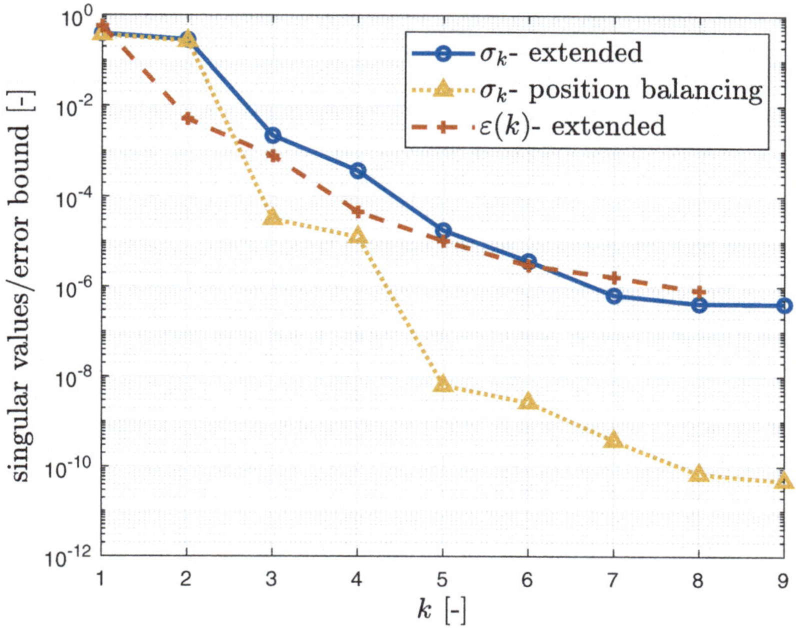

Fig. 1 The singular values σ_k and the error bound ε as a function of the reduction order k

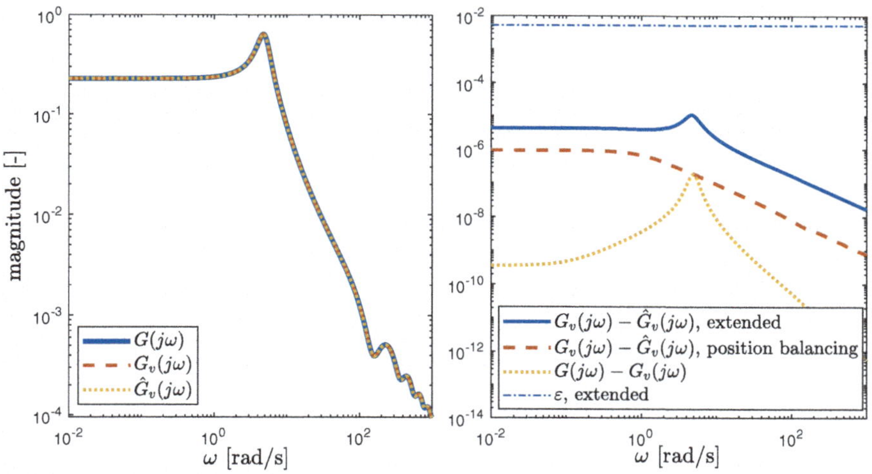

Fig. 2 Comparison between the transfer functions of the original, reduced and error systems in the neural field example

response function $G(j\omega)$ of the original (linearized) model in (40) is compared to $G_\upsilon(j\omega)$ and $\hat{G}_\upsilon(j\omega)$. In the same figure, we have presented transfer functions of the error systems $G(j\omega) - G_\upsilon(j\omega)$ and $G_\upsilon(j\omega) - \hat{G}_\upsilon(j\omega)$, of both techniques. From this figure, we can clearly observe the high accuracy of the approximation from the extended technique. The approximate model from the position balancing method is slightly more accurate. We note that $G(j\omega)$ is obtained by performing a spatial discretization over a grid of 200 cells, and the same grid has been used to numerically compute the matrices in (44). The error between $G(j\omega)$ and $G_\upsilon(j\omega)$ stems from the limited resolution of the discretization and also the Taylor series expansion.

Remark 6 In addition to slightly outperforming the presented model-reduction technique in terms of accuracy, position balancing relies on the computation of delay Lyapunov equations which only require asymptotic stability of the model (instead of solutions to matrix inequalities as in (4) and (7)). Nonetheless, we stress that position balancing does neither provide guarantees on stability preservation nor gives an a-priori bound on the reduction error.

We stress that the assumption made here requires a strong separation between spatial and temporal evolution of (40) as well as a spatially uniform delays. Further work is needed to relax these requirements.

7 Application to Parameterized Model Reduction

An extended model-reduction procedure as presented in the previous sections is particularly suited for system-theoretic applications such as structured and parameterized model reduction. In this chapter, we focus on the latter application and refer to [24] for a detailed discussion on the former.

Namely, a large class of parameterized time delay systems can be written in the form of time delay systems with a polytopic parameterization of the form

$$\Omega_\delta : \begin{cases} \dot{x}(t) = A_\delta x(t) + A_{d\delta} x(t-\tau) + B_\delta u(t), \\ y(t) = C_\delta x(t) + C_{d\delta} x(t-\tau) + D_\delta u(t), \\ x_0 = \varphi. \end{cases} \tag{45}$$

where the subscript δ denotes a polytopic parameterization such that a parameterized matrix M_δ is defined as $M_\delta := \sum_{i=1}^{d} \delta_i M_i$, where $M_i, i = 1, ..., d$, is a given matrix and $\delta \in \Delta$ with $\Delta = \{\delta \in \mathbb{R}^d | \delta_i \geq 0, \sum_{i=1}^{d} \delta_i = 1\}$. It is assumed that for all $\delta \in \Delta$, this system has the same stability properties as the system in (1).

Although the methods in [23] and [4] can be generalized to enable the reduction of this type of systems, those can result in low-quality model approximations and conservative error bounds, if not infeasible. On the other hand, the extended model reduction improves both the feasibility and the accuracy of model approximation for this type of systems. This is due to the fact that in an extended model-reduction method, we can assign a polytopic structure to the main decision variables to increase the degrees of freedom in the model-reduction procedure. In conventional methods, such as those in [23] and [4], the main decision variables Q and P are directly used in computing the balancing transformation, and assigning a polytopic structure to those complicates the reduction procedure (see [31], for parameterized model reduction of delay-free systems to get an idea about complexities that can arise when assigning parametric structures to P and Q).

In the extended technique, for the parameterized system in (45), the inequality (4) is adapted to the following form:

$$M_{o\delta} = \begin{bmatrix} SA_\delta + A_\delta^T S + Q_{d\delta} - \bar{Q}_\delta & \bar{Q}_\delta + SA_{d\delta} & Q_\delta - S + \alpha_o A_\delta^T S & C_\delta^T \\ * & -Q_{d\delta} - \bar{Q}_\delta & \alpha_o A_{d\delta}^T S & C_{d\delta}^T \\ * & * & -2\alpha_o S + \tau^2 \bar{Q}_\delta & 0 \\ * & * & * & -I_p \end{bmatrix} < 0. \tag{46}$$

By virtue of the properties of the polytopic uncertainty/parameterization, it can be shown that $M_{o\delta} = \sum_{i=1}^{d} \delta_i M_{oi}$ (note that $S = \sum_{i=1}^{d} \delta_i S$) with

$$M_{oi} = \begin{bmatrix} SA_i + A_i^T S + Q_{di} - \bar{Q}_i & \bar{Q}_i + SA_{di} & Q_i - S + \alpha_o A_i^T S & C_i^T \\ * & -Q_{di} - \bar{Q}_i & \alpha_o A_{di}^T S & C_{di}^T \\ * & * & -2\alpha_o S + \tau^2 \bar{Q}_i & 0 \\ * & * & * & -I_p \end{bmatrix}, i = 1, ..., d. \tag{47}$$

This implies that if there exist matrices $Q_i > 0$, $\bar{Q}_i > 0$, $Q_{di} > 0$, $i = 1, ..., d$, and $S > 0$, and a scalar α_o such that $M_{oi} < 0$ for $i = 1, ..., d$, then $M_{o\delta} < 0$. This result together with a similarly adapted inequality $M_{c\delta} < 0$ (an adaption to the inequality (7)) provides matrices S and R required for reducing (45) by pursuing the same procedure as in Sect. 4.

Remark 7 It is noted that in this parameterized model order reduction technique, S and α_o must satisfy d (the number of parameters) inequalities of the form (47) simultaneously.

Remark 8 The error bound obtained from the parameterized technique is robust in the sense that it holds for all $\delta \in \Delta$. Moreover, it can be shown that the reduced system is asymptotically stable and it has the same parameterization as the original one.

7.1 Example

Next, we present an example. In this example, we consider a wave equation which has a damping factor in the forward direction. The wave equation, together with the considered boundary conditions and the initial condition, is given by

$$\frac{\partial}{\partial t}q_1(t,\xi)+c\frac{\partial}{\partial x}q_1(t,\xi)=0.025fq_1(t,\xi), \tag{48}$$

$$\frac{\partial}{\partial t}q_2(t,\xi)-c\frac{\partial}{\partial x}q_2(t,\xi)=0, \tag{49}$$

$$q_1(t,0)=\beta_1 q_2(t,0)+u(t), \tag{50}$$

$$q_2(t,l)=\beta_2 q_1(t,l), \tag{51}$$

$$q_1(0,\xi)=0, \tag{52}$$

$$q_2(0,\xi)=0, \tag{53}$$

where $t\geq 0$ and $\xi\in[0,l]$ are the temporal and spatial variables, respectively. Here, $l=1000$ m is the length of the spatial domain. Moreover, $q_i(t,\xi)\in\mathbb{R}$, $i=1,2$, are the distributed variables, $c=1000$ m/s is the speed of the traveling wave components, and f is a damping factor. We take f to be uncertain, but we assume that the upper and lower bounds of it are known as $f\in[0.5,10.5]$. Moreover, $\beta_1=1$ and $\beta_2=0.7$, and $u(t)$ is the input. The output is given by

$$y(t)=q_1(t,l). \tag{54}$$

From the literature, it is know that this system can be modeled by delay-difference equations [8]. However, in this study, for the sake of illustration, we discretize the first PDE describing q_1 (48) to obtain an approximative model of it in terms of ODEs, whereas we write the other PDE (49) in terms of an equivalent delay equation, that is, we can show that $q_2(t,0)=q_2(t-\tau,l)$, with $\tau=l/c$.

To perform the discretization, the spatial domain of the first PDE is discretized into n cells of length $\Delta\xi$. In the discretization scheme, $Q_i(t)$, for $i=1,2,\ldots,n$, approximates the spatial average of $q_1(t,\xi)$ over the ith cell and satisfies

$$\dot{Q}_i(t)=\gamma_1 Q_{i-1}(t)-\gamma_2 Q_i(t),\quad i=1,2,\ldots,n \tag{55}$$

with $\gamma_1=c/\Delta\xi$ and $\gamma_2=c/\Delta\xi-0.025f$. In this formulation, we approximate $Q_0(t)\approx q_1(t,0)$. Following the fact that $q_2(t,0)=q_2(t-\tau,l)$, and by using the boundary conditions (50) and (51), we can further write $Q_0(t)\approx\beta_1\beta_2 Q_n(t-\tau)+u(t)$, where the approximation $q_1(t,l)\approx Q_n(t)$ has been used. Finally, using (55) together with these relations and the approximation $y(t)\approx Q_n(t)$, we obtain a model of the form (1) with $C=[0,0,\cdots,1]$, $C_d=0$ and

$$A=\begin{bmatrix}-\gamma_2 & 0 & & 0\\ \gamma_1 & \ddots & \ddots & \\ & \ddots & \ddots & 0\\ 0 & & \gamma_1 & -\gamma_2\end{bmatrix},\quad A_d=\begin{bmatrix}0 & \gamma_1\beta_1\beta_2\\ 0 & 0\end{bmatrix},\quad B=\begin{bmatrix}\gamma_1\\ 0\end{bmatrix}.$$

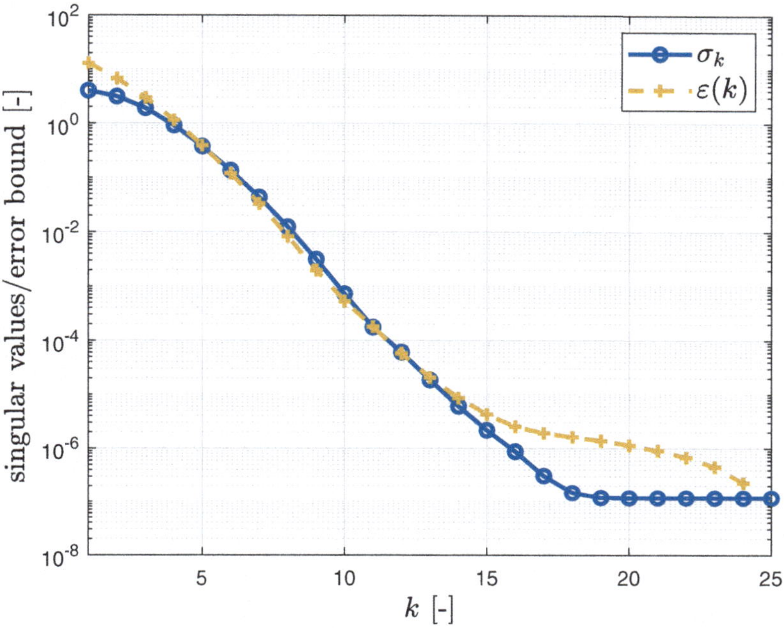

Fig. 3 The singular values σ_k and the error bound ε as a function of the reduction order k, for the robust model reduction

We can then write this model as a time delay system with polytopic uncertainties, due to uncertainties in f, in the form of (45) with $d = 2$. The order of this model, determined by the resolution of the discretization, is chosen to be $n = 25$. The frequency response function of the discretized model between input u and output y is denoted by $G(j\omega)$.

The presented robust/parameterized model order reduction method has been applied to this model. Figure 3 presents the resulting extended singular values σ_i in comparison to the error bound ε as a function of the order k of the reduced system. Based on this figure, we choose $k = 4$. Note that σ_i and ε are independent of the uncertain variable. Figure 4 reports the frequency response function of the original model $G(j\omega)$ of order $n = 25$ in comparison to the reduced-order model $\hat{G}(j\omega)$ of order $k = 4$ for the extremal values $f = 0.5$ and $f = 10.5$. We observe that for both extremal values of f, the model-reduction results are quite accurate. We also observe, in the subfigure on the right-hand side of Fig. 4, that in both cases, the $\mathcal{H}_\infty$-norm of the error system $G(j\omega) - \hat{G}(j\omega)$ is smaller than the a-priori obtained error bound, as expected.

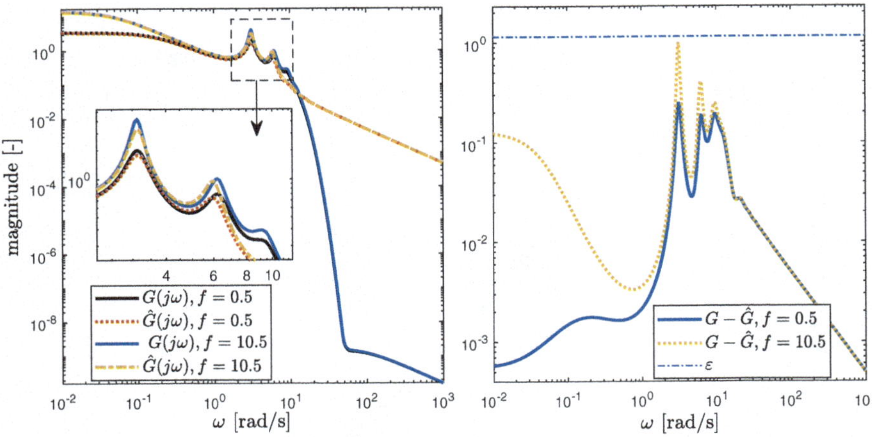

Fig. 4 (Left) comparison between the frequency response function of the original system G and the reduced-order one $\hat{G}$, and (right) error bound in comparison to the frequency response function of the error system $G - \hat{G}$ for the extremal values of the uncertain parameter f

8 Conclusions

In this chapter, by introducing slack variables in the computation of bounds on the energy functionals, we have obtained an extended model-reduction technique for linear time delay systems. This technique exhibits more flexibility compared to its existing counterparts, making it interesting for purposes such as parameterized and structured model reduction. Moreover, the proposed technique preserves stability properties and also provides a computable error bound. We have numerically evaluated the performance of the proposed method by applying it to a model of neural fields in the brain and to a model with polytopic uncertainties.

Appendix A. Derivation of $X(r)$

We consider $w_{ij}(r, r')$, for $i, j = 1, 2$. This function can be written in the following general form

$$\begin{aligned} W_{ij}(r, r') &= k_{ij} \exp\left(-\frac{|r - r' - \mu_{ij}|^2}{2\sigma_{ij}}\right) \\ &= k_{ij} \exp\left(-\frac{|r|^2}{2\sigma_{ij}}\right) \exp\left(-\frac{|r' + \mu_{ij}|^2}{2\sigma_{ij}}\right) \exp\left(\frac{r(r' + \mu_{ij})}{\sigma_{ij}}\right) \end{aligned}$$

for some constants k_{ij}, σ_{ij} and μ_{ij}. We wish to decompose $w_{ij}(r, r')$ into a multiplication of only-r and only-r' dependent functions. However, the term $\exp(r(r' + \mu_{ij})/\sigma_{ij})$ cannot be directly decomposed into such a desirable form. To cope with this issue, we use the Taylor series approximation of order ρ of this term to obtain

$$\exp\left(\frac{r(r' + \mu_{ij})}{\sigma_{ij}}\right) \approx \left[\, 1 \; r \; r^2 \; ... \; r^\rho \,\right]\left[\, 1 \; \frac{(r'+\mu_{ij})}{\sigma_{ij}} \; \frac{(r'+\mu_{ij})^2}{2\sigma_{ij}^2} \; \cdots \; \frac{(r'+\mu_{ij})^\rho}{\rho!\sigma_{ij}^\rho} \,\right]^T,$$

where ρ is the order of approximation. With this approximation, we can now write

$$w_{ij}(r, r') \approx f_{ij}(r) g_{ij}^T(r')$$

where

$$f_{ij}(r) = \left[\, f_{ij,0}(r) \cdots f_{ij,\rho}(r) \,\right],$$
$$g_{ij}(r) = \left[\, g_{ij,0}(r) \cdots g_{ij,\rho}(r) \,\right]$$

with

$$f_{ij,m}(r) = r^m \exp\left(-\frac{|r|^2}{2\sigma_{ij}}\right),$$
$$g_{ij,m}(r') = k_{ij} \frac{\left(r' + \mu_{ij}\right)^m}{m!\sigma_{ij}^m} \exp\left(-\frac{\left|r' + \mu_{ij}\right|^2}{2\sigma_{ij}}\right), \quad m = 0, 2, ..., \rho.$$

With this representation of $w(r, r')$, we may choose

$$X_1 = \begin{bmatrix} 0 \\ f_{22} \end{bmatrix}, \quad X_2 = \begin{bmatrix} f_{12} & 0 \\ 0 & f_{21} \end{bmatrix},$$
$$Y_1 = \begin{bmatrix} 0 \\ g_{22} \end{bmatrix}, \quad Y_2 = \begin{bmatrix} 0 & g_{21} \\ g_{12} & 0 \end{bmatrix}.$$

With this choice of X_1 and X_2, we obtain $N_1 = \rho$ and $N_2 = 2\rho$. We also note that this choice of X_1 and X_2 leads to $w_{11} = 0$.

References

1. Aarsnes, U.J.F., van de Wouw, N.: Dynamics of a distributed drill string system: characteristic parameters and stability maps. J. Sound Vib. **417**, 376–412 (2018)
2. Amghayrir, A., Tanguy, N., Brehonnet, P., Vilbe, P., Calvez, L.C.: Laguerre-Gram reduced-order modeling. IEEE Trans. Autom. Control **50**(9), 1432–1435 (2005)
3. Beattie, C., Gugercin, S.: Interpolatory projection methods for structure-preserving model reduction. Syst. Control Lett. **58**(3), 225–232 (2009)

4. Besselink, B., Chaillet, A., van de Wouw, N.: Model reduction for linear delay systems using a delay-independent balanced truncation approach. In: Proceeding of the 56th IEEE Conference on Decision and Control, pp. 3793–3798 (2017)
5. Breiten, T.: Structure-preserving model reduction for integro-differential equations. SIAM J. Control Optim. **54**(6), 2992–3015 (2016)
6. Bressloff, P.: Spatiotemporal dynamics of continuum neural fields. J. Phys. A: Math. Theor. **45**(3) (2012)
7. Chaillet, A., Detorakis, G.I., Palfi, S., Senova, S.: Robust stabilization of delayed neural fields with partial measurement and actuation. Automatica **83**, 262–274 (2017)
8. Cooke, K.L., Krumme, D.W.: Differential-difference equations and nonlinear initial-boundary value problems for linear hyperbolic partial differential equations. J. Math. Anal. Appl. **24**(2), 372–387 (1968)
9. Curtain, R.F., Zwart, H.: An Introduction to Infinite-Dimensional Linear Systems Theory. Springer, New York (1995)
10. Dullerud, G.E., Paganini, F.: A Course in Robust control Theory: A Convex Approach, 1st edn. 2000. corr. 2nd printing. softcover version of original hardcover edition 2000 edn. No. 36 in Texts in Applied Mathematics. Springer New York, New York, NY (2010)
11. Fridman, E.: Introduction to Time-delay Systems: Analysis and Control. Birkhauser Boston (2014)
12. Fridman, E., Dambrine, M., Yeganefar, N.: On input-to-state stability of systems with time-delay: a matrix inequalities approach. Automatica **44**(9), 2364–2369 (2008)
13. Glover, K.: All optimal Hankel-norm approximations of linear multivariable systems and their L^∞-error bounds. Int. J. Control **39**(6), 1115–1193 (1984)
14. Grant, M., Boyd, S.: CVX: Matlab software for disciplined convex programming, version 2.1. http://cvxr.com/cvx (2018)
15. Gugercin, S., Antoulas, A.C.: A survey of model reduction by balanced truncation and some new results. Int. J. Control **77**(8), 748–766 (2004)
16. Jarlebring, E., Damm, T., Michiels, W.: Model reduction of time-delay systems using position balancing and delay Lyapunov equations. Math. Control Signals Syst. **25**(2), 147–166 (2013)
17. Jarlebring, E., Poloni, F.: Iterative methods for the delay Lyapunov equation with T-Sylvester preconditioning. Appl. Numer. Math. **135**, 173–185 (2019)
18. Kolmanovskii, V., Myshkis, A.: Applied Theory of Functional Differential Equations. Springer, Dordrecht (1992)
19. Lam, J.: Model reduction of delay systems using Pade approximants. Int. J. Control **57**(2), 377–391 (1993)
20. Michiels, W., Jarlebring, E., Meerbergen, K.: Krylov-based model order reduction of time-delay systems. SIAM J. Matrix Anal. Appl. **32**(4), 1399–1421 (2011)
21. Michiels, W., Niculescu, S.: Stability, Control, and Computation for Time-Delay Systems. Society for Industrial and Applied Mathematics, Philadelphia, PA (2014)
22. Moore, B.: Principal component analysis in linear systems: controllability, observability, and model reduction. IEEE Trans. Autom. Control **26**(1), 17–32 (1981)
23. Naderi Lordejani, S., Besselink, B., Chaillet, A., van de Wouw1, N.: Model order reduction for linear time delay systems: a delay-dependent approach based on energy functionals. Automatica **112**, 108701 (2020)
24. Naderi Lordejani, S., Besselink, B., van de Wouw, N.: An extended model order reduction technique for linear delay systems. In: Proceeding of the 58th IEEE Conference on Decision and Control (2019)
25. Ruehli, A.E., Cangellaris, A.C.: Progress in the methodologies for the electrical modeling of interconnects and electronic packages. Proc. IEEE **89**(5), 740–771 (2001)
26. Saadvandi, M., Meerbergen, K., Jarlebring, E.: On dominant poles and model reduction of second order time-delay systems. Appl. Numer. Math. **62**(1), 21–34 (2012)
27. Sandberg, H.: An extension to balanced truncation with application to structured model reduction. IEEE Trans. Autom. Control **55**(4), 1038–1043 (2010)

28. Scarciotti, G., Astolfi, A.: Model reduction of neutral linear and nonlinear time-invariant time-delay systems with discrete and distributed delays. IEEE Trans. Autom. Control **61**(6), 1438–1451 (2016)
29. Scherpen, J., Fujimoto, K.: Extended balanced truncation for continuous time LTI systems. In: Proceedings of the European Control Conference (2018)
30. Veltz, R., Faugeras, O.: Local/global analysis of the stationary solutions of some neural field equations. SIAM J. Appl. Dyn. Syst. **9**(3), 954–998 (2010)
31. Wittmuess, P., Tarin, C., Keck, A., Arnold, E., Sawodny, O.: Parametric model order reduction via balanced truncation with Taylor series representation. IEEE Trans. Autom. Control **61**(11), 3438–3451 (2016)
32. van de Wouw, N., Michiels, W., Besselink, B.: Model reduction for delay differential equations with guaranteed stability and error bound. Automatica **55**, 132–139 (2015)
33. Xu, S., Lam, J., Huang, S., Yang, C.: $\mathcal{H}_\infty$ model reduction for linear time-delay systems: continuous-time case. Int. J. Control **74**(11), 1062–1074 (2001)

Applications of Model Order Reduction

A Practical Method for the Reduction of Linear Thermo-Mechanical Dynamic Equations

Artur Jungiewicz, Christoph Ludwig, Shuwen Sun, Utz Wever, and Roland Wüchner

Abstract Linear thermo-mechanical equations are widely used for the dynamic modeling of electric motors/generators or gas turbines. In order to use them in the context of a digital twin, real-time capable versions of the models must be achieved. In principle, model order reduction techniques for coupled thermo-elastic physics are known. However, commercial tools and even open-source tools allow only limited access to the necessary information in terms of coupling terms. This paper aims at providing an algorithm for the reduction of thermo-elastic equations in the framework of given software tools. After sampling and running some simple test cases in the offline stage, the reduced coupling term can be obtained and directly applied in the online stage to solve the reduced thermo-mechanical equations without intrusion into the commercial FEM software. Moreover, the numerical residual due to sampling and grouping techniques is also discussed in the paper. Basically, an algorithm similar to operator inference [15] is applied for extracting the coupling term. The complete workflow for extracting the coupling matrix is demonstrated on the open-source software Code_Aster.

Keywords Thermo-mechanics · Linear model order reduction · Krylov subspace · Conventional software tools

---

A. Jungiewicz
Siemens AG, Berlin, Germany
e-mail: Artur.Jungiewicz@siemens.com

C. Ludwig · U. Wever (✉)
Siemens AG, Munich, Germany
e-mail: Utz.Wever@siemens.com

C. Ludwig
e-mail: Christoph.Ludwig@siemens.com

S. Sun · R. Wüchner
Technical University of Munich, Munich, Germany
e-mail: Shuwen.Sun@tum.de

R. Wüchner
e-mail: Wuechner@tum.de

P. Benner et al. (eds.), *Model Reduction of Complex Dynamical Systems*,
International Series of Numerical Mathematics 171,
https://doi.org/10.1007/978-3-030-72983-7_10

1 Introduction

An increasing number of disruptive innovations with high economic and social impact shape our digitalizing world. Simulation technologies are key enablers of digitalization, since they facilitate digital twins that mirror physical products and systems into the digital world. However, digital twins require a paradigm shift in computational engineering: Instead of expert centric tools, such as common CAx software, engineering and operation require largely autonomous digital assist systems that continuously interact with the physical environment through background simulation, optimization, and control. This new type of digital engineering tools must efficiently integrate models and data from different product life cycle phases and master the resulting exploding computational complexities [6].

Within this paper we concentrate on technologies supporting the operation phase of a component. These technologies allow to monitor the entire state of a system at any point in time. A simulation model runs in parallel to the operation and is synchronized by sensor values at discrete time points. In some reports this construct is called the "digital twin" of a real system. Further benefits of the digital twin are, e.g., inspection and service planning, lifetime prediction [11], advanced fault detection and control and optimization during operation [10].

An overview of model order reduction techniques are given in [3, 5, 8]. The reduction of linear thermal equation [13] and structural mechanical equation [16] by Krylov methods are well known. Practically, the mass and the stiffness matrix may be extracted from commercial and open-source software and Arnoldi iterations can be applied. Within this paper, we want to discuss coupled linear thermo-elastic equations. While the basic algorithms for model order reduction have already been discussed [4, 14], reduction in the framework of commercial tools with limited access to necessary information is still an active field of research. Here, we basically use the ideas derived in [15]. They reconstruct system matrices in the projected space from data, which is called operator inference. In the present paper we use the method for extracting the necessary (reduced) coupling matrix. The projection onto the subspace is performed by the Krylov method and the data are provided by the underlying simulator.

The paper is organized as follows: In Sect. 2, the coupled equations for the thermo-elastic model are described. The partial differential equations are presented and their structure after spatial discretization is discussed. Section 3.1 discusses methods how to reduce the coupled equations. Krylov reduction is considered for the thermal and mechanical part of the equations. Especially, those reduction strategies are discussed in the framework of software tools, where often the desired coupling information cannot be extracted. Section 3.2 describes a new algorithm for generating the desired coupling matrix. This scheme has to be performed once and can be interpreted as a further preprocessing step (additionally to the reduction itself). It is a general algorithm which can be applied for any software, where only mass and stiffness matrix can be extracted for a single physics [7, 9]. The overall algorithm is summarized in Sect. 3.3. Section 4 describes the full workflow for the reduction of linear thermo-elastic equations in the case of using Code_Aster [7] as the underlying simulation software. The accuracy of the reduction method is demonstrated on a typical application.

2 The Thermo-Mechanical Model

In this section we derive the basic equations for heat transfer and structural mechanics. Furthermore, the coupling of the equations is discussed.

2.1 Structural Mechanics

Let us consider a solid body of the form $\Omega \subset \mathbb{R}^3$ composed of a material with Young's modulus E, $E \geq 0$ and Poisson ratio ν, $-1 \leq \nu \leq 0.5$, and with the boundary $\partial\Omega = \Gamma_D \cup \Gamma_N$. The body is subject to volume forces $\boldsymbol{f} \in \mathbb{R}^3$ in the body Ω and surface forces $\boldsymbol{g} \in \mathbb{R}^3$ on the boundary $\Gamma_N \subset \partial\Omega$. The boundary $\Gamma_D \subset \partial\Omega$ should be fixed, i.e., all displacements are zero. Displacements $\boldsymbol{u} \in \Omega \to \mathbb{R}^3 \in \mathcal{E}(\Omega, \mathbb{R}^3)$ with some appropriate function space $\mathcal{E}(\Omega, \mathbb{R}^3)$ are determined by the equations of linear elasticity (see, e.g., [12]):

$$\begin{aligned} &\rho \boldsymbol{u}_{tt} - \operatorname{div}(\boldsymbol{A}\boldsymbol{e}(\boldsymbol{u})) = \boldsymbol{F} && \text{in } \Omega \times [t_0, T], \\ &(\boldsymbol{A}\boldsymbol{e}(\boldsymbol{u})) \cdot \boldsymbol{n} = \boldsymbol{g} && \text{on } \Gamma_N \times [t_0, T], \\ &\boldsymbol{u} = \boldsymbol{0} && \text{on } \Gamma_D \times [t_0, T], \end{aligned} \tag{1}$$

where the strain $\boldsymbol{e}(\boldsymbol{u})$ is given the symmetrized gradient of displacements

$$\boldsymbol{e}(\boldsymbol{u}) = \frac{1}{2}(\nabla \boldsymbol{u} + \nabla \boldsymbol{u}^\top) \in \mathbb{R}^{3\times 3}, \tag{2}$$

and the stress $\boldsymbol{A}\boldsymbol{e}(\boldsymbol{u})$ is determined by

$$\begin{aligned} \boldsymbol{A}\boldsymbol{e}(\boldsymbol{u}) &= 2\mu \boldsymbol{e}(\boldsymbol{u}) + \lambda \operatorname{trace}(\boldsymbol{e}(\boldsymbol{u})) \cdot \boldsymbol{I} \\ &= 2\mu \boldsymbol{e}(\boldsymbol{u}) + \lambda \operatorname{div}(\boldsymbol{u}) \boldsymbol{I}. \end{aligned} \tag{3}$$

Here, $\lambda = \frac{\nu E}{(1+\nu)(1-2\nu)}$ and $\mu = \frac{E}{2(1+\nu)}$ are the so-called Lamé constants and $\boldsymbol{I}$ is the identity matrix.

Equation (1) gives the strong formulation for linear elasticity. Discretization of the weak formulation by finite elements leads to the linear system, cf. [17]:

$$\boldsymbol{M}_u \ddot{\boldsymbol{u}} + \boldsymbol{K}_u \boldsymbol{u} = \boldsymbol{F}, \quad \boldsymbol{K}_u, \boldsymbol{M}_u \in \mathbb{R}^{N_1 \times N_1}, \quad \boldsymbol{u}, \boldsymbol{F} \in \mathbb{R}^{N_1}, \tag{4}$$

where N_1 denotes the dimension of the ansatz space, $\boldsymbol{M}_u$ is mass matrix, $\boldsymbol{K}_u$ is the stiffness matrix, and, by abuse of notation, $\boldsymbol{F}$ the vector of acting forces and $\boldsymbol{u}$ the vector of displacements.

Note that, in this work, an undamped system is considered. In general, a damping matrix $\boldsymbol{D}_u \in \mathbb{R}^{N_1 \times N_1}$ may be introduced in Eq. (4) which might even be time dependent (the same holds for the stiffness matrix $\boldsymbol{K}_u$).

2.2 Heat Transfer

The heat equation is given by

$$\begin{aligned} &T_t + \nabla \cdot (\kappa(\boldsymbol{x})\nabla T) + s = 0 && \text{in } \Omega, \\ &\kappa(\boldsymbol{x})\nabla T \cdot \boldsymbol{n} = g && \text{on } \Gamma_N, \\ &T = f && \text{on } \Gamma_D. \end{aligned} \tag{5}$$

Spatial discretization of the weak formulation of (5) leads to the linear system:

$$\boldsymbol{M}_T \dot{\boldsymbol{T}} + \boldsymbol{K}_T \boldsymbol{T} = \boldsymbol{Q}, \quad \boldsymbol{M}_T, \boldsymbol{K}_T \in \mathbb{R}^{N_2 \times N_2}, \quad \boldsymbol{T}, \boldsymbol{Q} \in \mathbb{R}^{N_2}, \tag{6}$$

where $\boldsymbol{M}_T$ is the thermal mass matrix and $\boldsymbol{K}_T$ the thermal stiffness matrix.

2.3 Coupling of Equations

The coupling of temperature and the displacements takes place by extending the definition of the strain:

$$\boldsymbol{e}(\boldsymbol{u}) = \frac{1}{2}(\nabla \boldsymbol{u} + \nabla \boldsymbol{u}^\top) + \alpha \Delta T \boldsymbol{I}, \tag{7}$$

where ΔT is the temperature difference w.r.t. a given reference temperature and $\boldsymbol{I}$ is the identity matrix. Then, according to Eq. (3), also the mechanical stress depends on the temperature. After assembly of the finite elements, a coupling block appears in the overall stiffness matrix. The coupled dynamic equation is given by

$$\begin{pmatrix} \boldsymbol{0} & \boldsymbol{0} \\ \boldsymbol{0} & \boldsymbol{M}_u \end{pmatrix} \begin{pmatrix} \ddot{\boldsymbol{T}} \\ \ddot{\boldsymbol{u}} \end{pmatrix} + \begin{pmatrix} \boldsymbol{M}_T & \boldsymbol{0} \\ \boldsymbol{0} & \boldsymbol{0} \end{pmatrix} \begin{pmatrix} \dot{\boldsymbol{T}} \\ \dot{\boldsymbol{u}} \end{pmatrix} + \begin{pmatrix} \boldsymbol{K}_T & \boldsymbol{0} \\ \boldsymbol{K}_{Tu} & \boldsymbol{K}_u \end{pmatrix} \begin{pmatrix} \boldsymbol{T} \\ \boldsymbol{u} \end{pmatrix} = \begin{pmatrix} \boldsymbol{Q}(t) \\ \boldsymbol{F}(t) \end{pmatrix}, \tag{8}$$

where $\boldsymbol{K}_{Tu} \in \mathbb{R}^{N_1 \times N_2}$ is the coupling matrix. Due to the unsymmetric coupling in the common stiffness matrix, Eq. (8) may be decoupled:

$$M_T\dot{T} + K_T T = Q(t), \tag{9}$$
$$M_u\ddot{u} + K_u u = F(t) - K_{Tu}T. \tag{10}$$

Nearly all software tools are able to extract the thermal and structural mass and stiffness matrix [1, 7, 9]. However, the coupling matrix K_{Tu} is often not available. Instead, the thermal load is given as a complete vector:

$$F_{Tu} = F_{Tu}(T(t)) = K_{Tu}T. \tag{11}$$

3 Derivation of the Reduction Algorithm

In this section we describe the complete scheme for obtaining the reduced model. First we recall the known algorithm for reducing linear thermo-mechanic equations Sect. 3.1. The main contribution of this paper is presented in Sect. 3.2, where we describe the extraction of the coupling matrix. Finally, the overall algorithm is summarized in Sect. 3.3.

3.1 Model Order Reduction

In many real world applications, the number N_1 and N_2 of degrees of freedom of the discretized systems (9) and (10) is large and its numerical integration is not possible in real time. Model order reduction strategies [3] introduce a reduced state of significantly lower dimension.

For the structural mechanics we have the reduction $v \in \mathbb{R}^{n_1}$ with $n_1 \ll N_1$:

$$u \approx \Psi v, \quad \Psi \in \mathbb{R}^{N_1 \times n_1}. \tag{12}$$

One way to obtain the reduction matrix Ψ for the mechanical system (4) is to use a Krylov subspace method [2, 16], where it consists of an orthonormal basis of the Krylov subspace

$$\text{span}\left\{K_\omega^{-1}F, K_\omega^{-1}MK_\omega^{-1}F, \ldots, (K_\omega^{-1}M)^{r-1}K_\omega^{-1}F\right\}, \tag{13}$$

which may be computed by the Arnoldi algorithm.

Inserting the reduction (12), obtained by (13) into the differential equation (4) and multiplying from left by $\Psi^\top$, one obtains the *reduced equation*

$$\Psi^\top M_u \Psi \ddot{v} + \Psi^\top K_u \Psi v = \Psi^\top F(t). \tag{14}$$

Similar to the structural mechanics, the temperature is reduced. With

$$\boldsymbol{T} \approx \boldsymbol{\Phi}\boldsymbol{S}, \quad \boldsymbol{S} \in \mathbb{R}^{n_2}, \quad \boldsymbol{\Phi} \in \mathbb{R}^{N_2 \times n_2}, \quad n_2 \ll N_2, \tag{15}$$

it holds

$$\boldsymbol{\Phi}^\top \boldsymbol{M}_T \boldsymbol{\Phi}\dot{\boldsymbol{S}} + \boldsymbol{\Phi}^\top \boldsymbol{K}_T \boldsymbol{\Phi}\boldsymbol{S} = \boldsymbol{\Phi}^\top \boldsymbol{Q}(t). \tag{16}$$

The reduced coupled equation for (9) and (10) is

$$\boldsymbol{\Phi}^\top \boldsymbol{M}_T \boldsymbol{\Phi}\dot{\boldsymbol{S}} + \boldsymbol{\Phi}^\top \boldsymbol{K}_T \boldsymbol{\Phi}\boldsymbol{S} = \boldsymbol{\Phi}^\top \boldsymbol{Q}(t), \tag{17}$$

$$\boldsymbol{\Psi}^\top \boldsymbol{M}_u \boldsymbol{\Psi}\ddot{\boldsymbol{v}} + \boldsymbol{\Psi}^\top \boldsymbol{K}_u \boldsymbol{\Psi}\boldsymbol{v} = \boldsymbol{\Psi}^\top \boldsymbol{F}(t) - \boldsymbol{\Psi}\boldsymbol{K}_{Tu}\boldsymbol{\Phi}\boldsymbol{S}. \tag{18}$$

Introducing the abbreviations

$$\begin{aligned}
\hat{\boldsymbol{M}}_T &= \boldsymbol{\Phi}^\top \boldsymbol{M}_T \boldsymbol{\Phi} \in \mathbb{R}^{n_2 \times n_2}, \\
\hat{\boldsymbol{K}}_T &= \boldsymbol{\Phi}^\top \boldsymbol{K}_T \boldsymbol{\Phi} \in \mathbb{R}^{n_2 \times n_2}, \\
\hat{\boldsymbol{M}}_u &= \boldsymbol{\Psi}^\top \boldsymbol{M}_u \boldsymbol{\Psi} \in \mathbb{R}^{n_1 \times n_1}, \\
\hat{\boldsymbol{K}}_u &= \boldsymbol{\Psi}^\top \boldsymbol{K}_u \boldsymbol{\Psi} \in \mathbb{R}^{n_1 \times n_1}, \\
\hat{\boldsymbol{K}}_{Tu} &= \boldsymbol{\Psi}^\top \boldsymbol{K}_{Tu} \boldsymbol{\Phi} \in \mathbb{R}^{n_1 \times n_2},
\end{aligned}$$

we obtain

$$\begin{aligned}
\hat{\boldsymbol{M}}_T \dot{\boldsymbol{S}} + \hat{\boldsymbol{K}}_T \boldsymbol{S} &= \boldsymbol{\Phi}^\top \boldsymbol{Q}(t), \\
\hat{\boldsymbol{M}}_u \ddot{\boldsymbol{v}} + \hat{\boldsymbol{K}}_u \boldsymbol{v} &= \boldsymbol{\Psi}^\top \boldsymbol{F}(t) - \hat{\boldsymbol{K}}_{Tu}\boldsymbol{S}.
\end{aligned} \tag{19}$$

The presented algorithm for the reduction of the thermo-elastic equations makes use of the fact that the coupling matrix $\boldsymbol{K}_{Tu}$ is available. However, as it was pointed out in the previous section, this is often not the case. That is, only the combined thermal load $\boldsymbol{F}_{Tu}(\boldsymbol{T}(t)) \in \mathbb{R}^{N_1}$ (11) is available . Then Eq. (19) read

$$\begin{aligned}
\hat{\boldsymbol{M}}_T \dot{\boldsymbol{S}} + \hat{\boldsymbol{K}}_T \boldsymbol{S} &= \boldsymbol{\Phi}^\top \boldsymbol{Q}(t), \\
\hat{\boldsymbol{M}}_u \ddot{\boldsymbol{v}} + \hat{\boldsymbol{K}}_u \boldsymbol{v} &= \boldsymbol{\Psi}^\top \boldsymbol{F}(t) - \boldsymbol{\Psi}^\top \boldsymbol{F}_{Tu}(\boldsymbol{T}(t)).
\end{aligned} \tag{20}$$

The reduced Eq. (19) may be computed without any problems. However, Eq. (20) is not suited for real-time evaluation. The term $\boldsymbol{\Psi}^\top \boldsymbol{F}_{Tu}(\boldsymbol{T}(t)) \in \mathbb{R}^{n_1}$ depends on the temperature $\boldsymbol{T}(t)$ which, in turn, depends on the source vector $\boldsymbol{Q}(t)$ and is therefore not known in advance. The equation

$$\boldsymbol{\Psi}^\top \boldsymbol{F}_{Tu}(\boldsymbol{T}(t)) = \hat{\boldsymbol{K}}_{Tu}\boldsymbol{S} \tag{21}$$

also performs a decoupling and thus allows the evaluation of the coupling term at the actual (reduced) temperature. Therefore, we present an algorithm for extracting the reduced coupling matrix $\hat{\boldsymbol{K}}_{Tu}$ from the vectors $\boldsymbol{F}_{Tu}(\boldsymbol{T}(t))$ which can be performed offline.

3.2 Extraction of the Coupling Matrix

As already mentioned, the main difficulty lies in how the coupling matrix $\boldsymbol{K}_{Tu}$ can be obtained since most software tools do not provide an option for users to extract this matrix. In the following we present an algorithm similar to the ideas presented in [15] for estimating the coupling matrix $\boldsymbol{K}_{Tu}$ in a pre-processing step to solve the coupling problem. The basic idea is to use the equation

$$\boldsymbol{F}_{Tu} = \boldsymbol{K}_{Tu}\boldsymbol{T}, \tag{22}$$

setting up a sample of temperatures

$$\boldsymbol{T}^m = (\boldsymbol{T}_1, \ldots, \boldsymbol{T}_m) \in \mathbb{R}^{N_2 \times m}, \tag{23}$$

computing the corresponding thermal loads

$$\boldsymbol{F}^m_{Tu} = (\boldsymbol{F}_{Tu,1}, \ldots, \boldsymbol{F}_{Tu,m}) \in \mathbb{R}^{N_1 \times m}, \tag{24}$$

and solving the least-squares problem

$$\min_{\boldsymbol{K}_{Tu} \in \mathbb{R}^{N_1 \times N_2}} \|\boldsymbol{F}^m_{Tu} - \boldsymbol{K}_{Tu}\boldsymbol{T}^m\|^2_F, \tag{25}$$

where $\|.\|_F$ is the Frobenius norm. In order to obtain a unique solution of the least-squares problem (25), the number of samples m should fulfill the inequality $m > N_2$. Solving the least-squares problem (25) has two significant drawbacks:

- The dimension of the least-squares problem (25) is at least $N_1 \times N_2$. Decoupling of the problem by determining the $\boldsymbol{K}_{Tu}$ column wise is possible, but still very expensive in terms of computation time.
- The temperature samples must be linear independent in order to obtain a reasonable condition number for problem (25).

In the following we try to overcome these drawbacks by setting up a more practical identification for the coupling matrix. Following [15], the main idea within this paper is to directly estimate the reduced coupling matrix. The number of coefficients to be estimated reduces drastically:

$$\boldsymbol{K}_{Tu} \in \mathbb{R}^{N_1 \times N_2} \quad \Rightarrow \quad \hat{\boldsymbol{K}}_{Tu} = \boldsymbol{\Psi}^\top \boldsymbol{K}_{Tu}\boldsymbol{\Phi} \in \mathbb{R}^{n_1 \times n_2}. \tag{26}$$

The following Eq. (27) is used for the identification of the reduced coupling matrix:

$$\boldsymbol{\Psi}^\top \boldsymbol{F}_{Tu} = \boldsymbol{\Psi}^\top \boldsymbol{K}_{Tu}\boldsymbol{T} = \boldsymbol{\Psi}^\top \boldsymbol{K}_{Tu}\boldsymbol{\Phi}\boldsymbol{S} = \hat{\boldsymbol{K}}_{Tu}\boldsymbol{S}. \tag{27}$$

The data required for the least-squares identification are again generated by the underlying software tool. That is, for a sample of temperatures the corresponding thermal load is evaluated:

$$\boldsymbol{T}_i, \quad i = 1, \ldots, k \quad \Rightarrow \quad \boldsymbol{F}_{Tu,i}, \quad i = 1, \ldots, k. \tag{28}$$

From the given temperature samples the reduced temperature is derived:

$$\boldsymbol{S}_i = \boldsymbol{\Phi}^\top \boldsymbol{T}_i, \quad i = 1, \ldots, k. \tag{29}$$

Similar to (23) and (24), sample matrices of the reduced temperature and the reduced thermal loads are set up:

$$\boldsymbol{S}^k = (\boldsymbol{S}_1, \ldots, \boldsymbol{S}_k) \in \mathbb{R}^{n_2 \times k} \tag{30}$$

$$\hat{\boldsymbol{F}}^k_{Tu} = (\boldsymbol{\Psi}^\top \boldsymbol{F}_{Tu,1}, \ldots, \boldsymbol{\Psi}^\top \boldsymbol{F}_{Tu,k}) \in \mathbb{R}^{n_1 \times k}. \tag{31}$$

Now we are able to set up the least-squares identification problem for the reduced coupling matrix:

$$\min_{\hat{\boldsymbol{K}}_{Tu} \in \mathbb{R}^{n_1 \times n_2}} \|\hat{\boldsymbol{F}}^k_{Tu} - \hat{\boldsymbol{K}}_{Tu} \boldsymbol{S}^k\|^2_F, \tag{32}$$

where $\|.\|_F$ is again the Frobenius norm. In order to obtain a unique solution of the least-squares problem (32), the number of samples k should fulfill the inequality $k \geq n_2$, which is of course much smaller than in the case of (25).

The least-squares problem (32) holds for arbitrary temperature samples $\boldsymbol{T}_i$, $i = 1, \ldots, k$ and the corresponding thermal loads $\boldsymbol{F}_{Tu,i}$, $i = 1, \ldots, k$. Therefore, we should choose the matrix $\boldsymbol{S}^k = \boldsymbol{\Phi}^\top \boldsymbol{T}^k$ in such a way that the least-squares problem (32) has a reasonable condition number. Note, that $\boldsymbol{\Phi}$ as an orthonormal matrix does not downgrade the condition number and it is sufficient to choose a reasonable sample matrix $\boldsymbol{T}^k$.

A possible approach is setting up a transient thermal simulation and choosing the temperature field at different time steps as samples. However, as already discussed in [15], in practical problems the states at different time steps may be linearly dependent and a large condition number of the matrix $\boldsymbol{S}^k$ will be encountered in the computation. Thus, instead of pursuing this transient approach, it is better to choose a set of temperature fields directly and gather them in the temperature sample matrix $\boldsymbol{T}^k$. An obvious and elegant choice is $\boldsymbol{T}^k = \boldsymbol{T}^{n_2} = \boldsymbol{\Phi}$. This would result in $\boldsymbol{S}^{n_2}$ being the identity matrix and yield a trivial identification problem (32). However, if $\boldsymbol{\Phi}$ is calculated outside the commercial tool it may prove difficult to correctly import and prescribe these vectors as temperature fields. To do so, internal settings such as node ordering or dualization of boundary conditions would have to be taken into account. Therefore, we propose to choose simple temperature fields only prescribing nonzero temperature at a few nodes. Such an approach can usually be performed easily via a GUI or a script.

More precisely, we choose temperature samples in the following way: First, unit temperature is prescribed at a few nodes. In each subsequent case, a different set of nodes is selected and unit temperature is prescribed. The remaining nodes are assumed to have zero temperature. This procedure is repeated until all nodes are at least covered one time. An example of this method is given by

$$\boldsymbol{T}^k = \begin{pmatrix} 1\,1\,1\,1\,0\,0\,0\,0\,0\,0 \cdots 0\,0\,0\,0 \\ 0\,0\,0\,1\,1\,1\,1\,0\,0\,0 \cdots 0\,0\,0\,0 \\ 0\,0\,0\,0\,0\,0\,1\,1\,1\,1 \cdots 0\,0\,0\,0 \\ \vdots \\ 0\,0\,0\,0\,0\,0\,0\,0\,0\,0 \cdots 1\,1\,1\,1 \end{pmatrix}^{\top} . \tag{33}$$

Here, the ith column of $\boldsymbol{T}^k$ denotes the ith testing case. In total, k different test cases are generated. In principle, $\boldsymbol{T}^k$ has a good condition number and the least-squares problem (32) is well posed.

Another improvement of least-squares problem may be achieved by considering also the geometric order of nodes. Because of the initial ordering of the nodes, a naive grouping of nodes normally gathers the nodes only in one direction of the given structure and thus, generate similar thermal loads. It is highly recommended to consider also other directions of the structure in the generation of testing cases. Due to the fact, that we are solving least-squares problems, we are not limited in the number of test cases. The error of this simple method highly depends on which testing cases are chosen. The extreme situation is that the identity matrix is selected for the testing cases but better sampling techniques may be chosen to reduce computational effort. The better the sampling technique catches the properties of the mechanics and interactions between the nodes, the better the result of the approximated coupling term will be. More techniques can be applied here in future work for a better result. In Sect. 4, the influence of sampling and grouping of test cases on the approximation residual is illustrated using an example setup.

3.3 Algorithm

The whole process of reducing the thermo-mechanical equations (8) is summarized in Algorithm 1. The main parts are the separate reduction of the thermal and the mechanical equation, and the generation of the coupling matrix. The flow chart is displayed in Fig. 1.

After running Algorithm 1, the reduced system (19) may be solved in a natural way: Solving alternatively the reduced thermal and the mechanical equations. Note that it is possible to use different time steps for thermal and mechanical equations and synchronize them after a few time steps. For numerical time integration we use the Newmark-beta method with $\alpha = 0.25$ and $\beta = 0.5$.

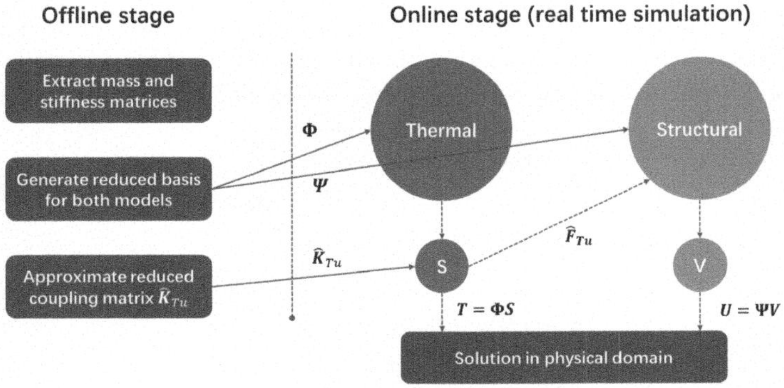

Fig. 1 The flow chart of the algorithm

Algorithm 1: Algorithm for the reduction of the coupled thermo-elastic equations (9) and (10) (Offline stage)

procedure REDUCTION($\boldsymbol{M}_u$,$\boldsymbol{K}_u$,$\boldsymbol{M}_T$,$\boldsymbol{K}_T$,$\boldsymbol{F}$,$\boldsymbol{Q}$)
    Input : $\boldsymbol{M}_u$,$\boldsymbol{K}_u$ Mechanical mass and stiffness matrix
    Input : $\boldsymbol{M}_T$,$\boldsymbol{K}_T$ Thermal mass and stiffness matrix
    Input : $\boldsymbol{F}, \boldsymbol{Q}$ Mechanical and thermal load ▷All inputs can be extracted directly from commercial software
    Output: $\boldsymbol{\Psi}$,$\boldsymbol{\Phi}$ Projection of mechanical and thermal equations
    Output: $\hat{\boldsymbol{K}}_{Tu}$ Reduced coupling matrix
    Reduce: $[\boldsymbol{\Psi}, \boldsymbol{H}_u] = \text{Arnoldi}(\boldsymbol{K}_u^{-1}\boldsymbol{M}_u, \boldsymbol{K}_T^{-1}\boldsymbol{F})$
    Reduce: $[\boldsymbol{\Phi}, \boldsymbol{H}_T] = \text{Arnoldi}(\boldsymbol{K}_T^{-1}\boldsymbol{M}_T, \boldsymbol{K}_T^{-1}\boldsymbol{Q})$
    Compute thermal load: $\boldsymbol{T}_i \Rightarrow \boldsymbol{F}_{Tu,i}, \quad i = 1, \ldots, k$ according to (33) and (28)
    Reduce temperature and thermal load according to (29), (30) and (31) $\Rightarrow \boldsymbol{S}^k, \hat{\boldsymbol{F}}^k_{Tu}$
    Compute reduced coupling matrix: $\min_{\hat{\boldsymbol{K}}_{Tu} \in \mathbb{R}^{n_1 \times n_2}} \| \hat{\boldsymbol{F}}^k_{Tu} - \hat{\boldsymbol{K}}_{Tu}\boldsymbol{S}^k \|_F^2$
end procedure

4 Implementation and Results

In this section, we demonstrate the performance of the method on a Finite Element model [7] and compare the full model with the reduced one. In a first section, the modeling process and the details of the model will be described. Afterward, the results for the Finite Element model will be presented in detail. Especially the influence of how to generate the coupling matrix is discussed.

4.1 Modeling

The validation of the described approach for the reduction of linear thermo-elastic equations was performed with the open-source FEM software package Code_Aster

[7]. The geometry from the included testcase **tpna01b** was used and thermo-mechanical coupling was introduced. The geometry model and its mesh are presented in Fig. 2. The mesh is composed of 525 nodes and 320 hexahedron elements. The material parameters are listed in Table 1.

For the thermal system, homogeneous Dirichlet boundary conditions are applied at the face $ABCD$. The heat flux through $ABEF$ and $DCHG$ is fixed to $20.0\,\mathrm{W\,m^{-2}}$ and to $50.0\,\mathrm{W\,m^{-2}}$ through $BCHE$ and $ADGF$.

In the mechanical model, the nodes on face $ABCD$ are fixed for all spatial directions. A harmonic force is acting in vertical direction with magnitude of 0.5 N and frequency of 2 Hz. It is located at the highest points of the arc. Additionally, it is assumed that displacement data are measured at a single node located at the highest point of the arc (see again Fig. 2).

Table 1 Material parameters of the model

Material parameters	Value
Young modulus E (GPa)	2100000
Poisson's ratio ν	0.3
Density ρ (kg/m^3)	7800
Linear dilation coefficient α (K^{-1})	0.001
Thermal conductivity λ (W/(m^3 K))	100.0
Heat capacity ρC_p (J/(m^3 K))	100.0

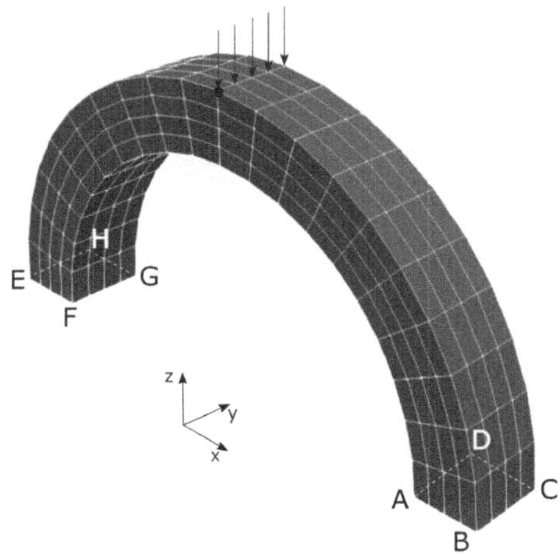

Fig. 2 Geometry and mesh of the model **tpna01b**. The acting outer load is marked as arrows in the middle of the arc. Furthermore, the position for observing the displacements are marked as a small black circle

4.2 Results

Finally, the presented algorithm is validated by means of comparison with a reference solution obtained from Code_Aster. For the current application, the degrees of freedom of the model are reduced from 500 to 6 for temperature and from 1500 to 21 for the displacements.

First, the temperature evolution from 0 s to 1 s at a randomly picked node is compared with the reference solution, see Fig. 3. It can be seen that, after some initial oscillations, the solutions are in good agreement. Figure 4 presents the temperature at all nodes after 1 s. The absolute difference between these results is approximately 4.9×10^{-5} corresponding to a relative error of about 0.01 %.

Additionally, the thermal expansion load in the reduced space, calculated by Algorithm 1, is compared with the reference solution obtained from a coupled simulation performed in Code_Aster.

Figure 5 presents a comparison between the values of the thermal load at each DOF in the reduced space. In addition, the thermal load history at the sixth DOF is visualized in Fig. 6.

As mentioned in Sect. 3.2, not only the condition number of the test temperature matrix (33) is important, but also the connectivity of the nodes (for obtaining differences in the response in terms of the thermal load). In the configuration of Fig. 7 we have chosen a more naive sample of temperatures shown in (33) which does not consider the connectivity of the nodes. It can be observed that the configuration in Fig. 6 is preferable. In this case, the node connectivity is considered by adding 20 extra

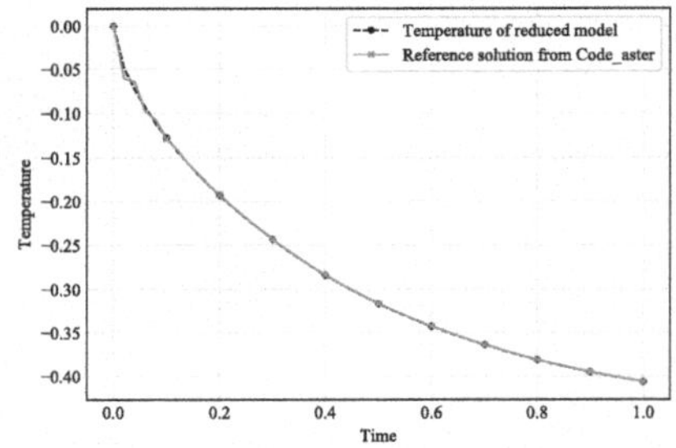

Fig. 3 The temperature history of node 51 from 0 to 1 s

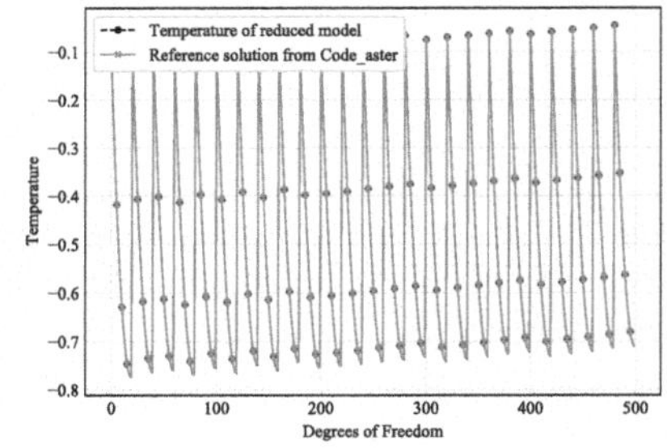

Fig. 4 The temperature distribution of all nodes at 1 s

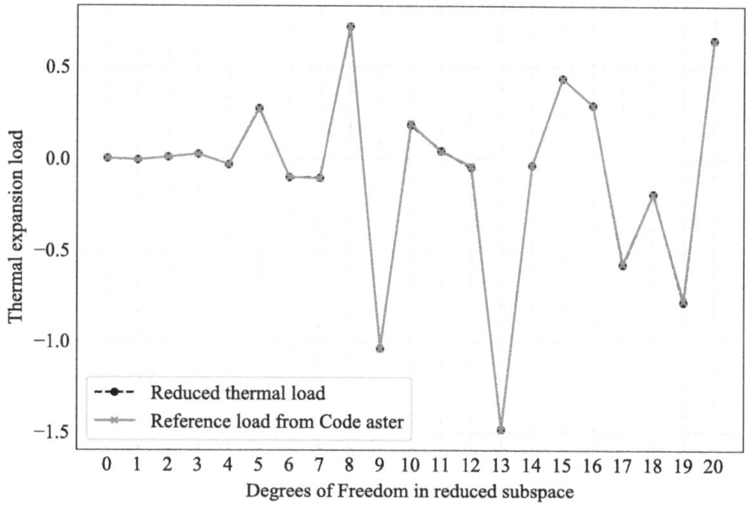

Fig. 5 The value of thermal expansion load applied on each reduced dof at 1 s

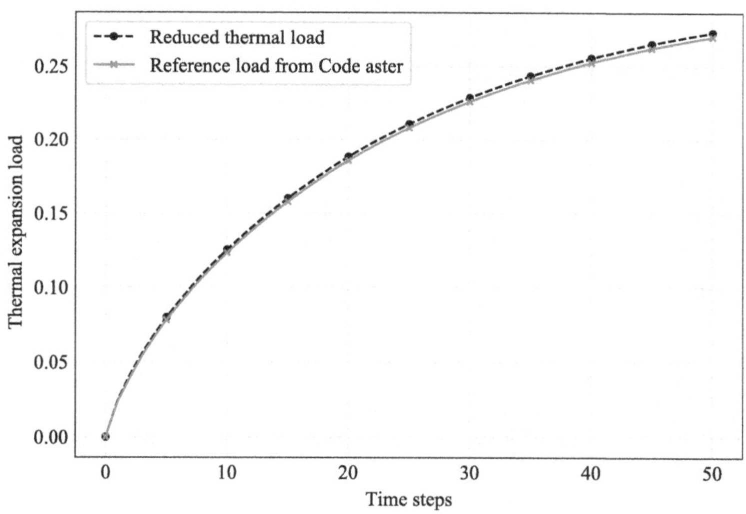

Fig. 6 The load history applied on the sixth degree of freedom in the reduced subspace

cases with unit temperatures applied on the nodes lying on the same longitudinal direction.

Finally, displacements at the observation point obtained from the reduced-order model are compared to the reference solution from Code_Aster. Two different loading conditions have been investigated: Thermal loading/no mechanical loading and thermal loading with an additional harmonic mechanical load acting at the top of the

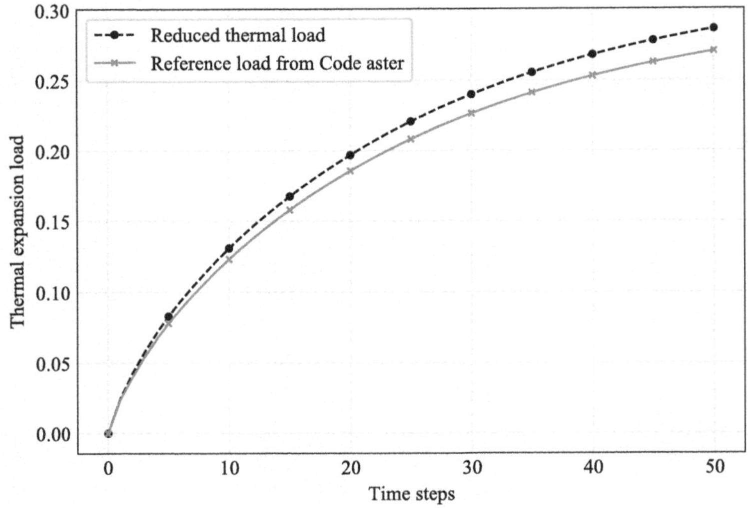

Fig. 7 The thermal expansion load when worse grouping technique is applied compared with Fig. 6

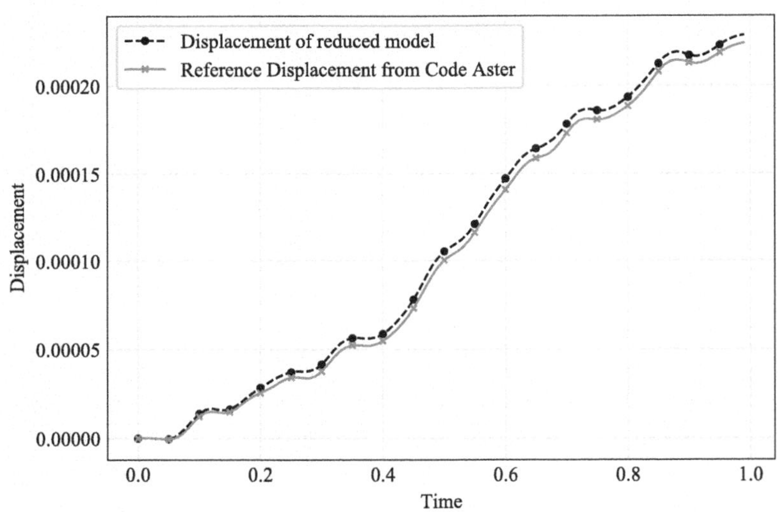

Fig. 8 The displacement of the observation point in x direction from 0 to 1s with purely thermal load

arc. The displacements at the observation node for the former case is presented in Fig. 8, whereas for the latter case in Fig. 9.

Figure 10 (only thermal load) and Fig. 11 (thermal and mechanical load) present the displacements at a second randomly chosen observation point.

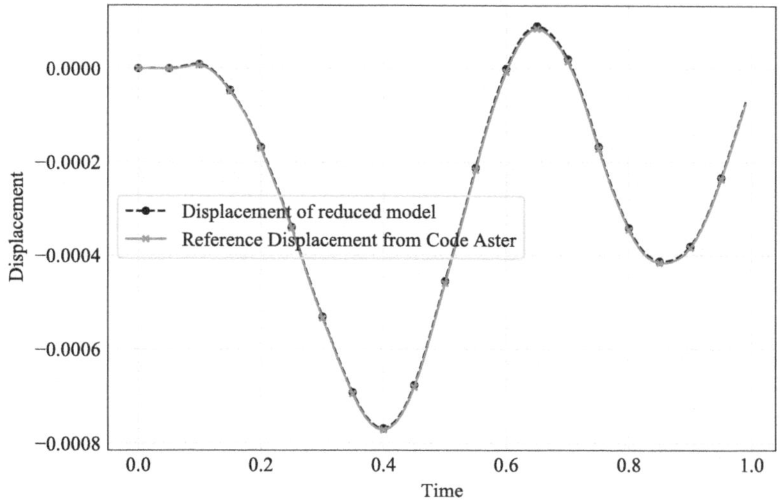

Fig. 9 The displacement of the observation point in x direction from 0 s to 1 s with additional harmonic mechanical load

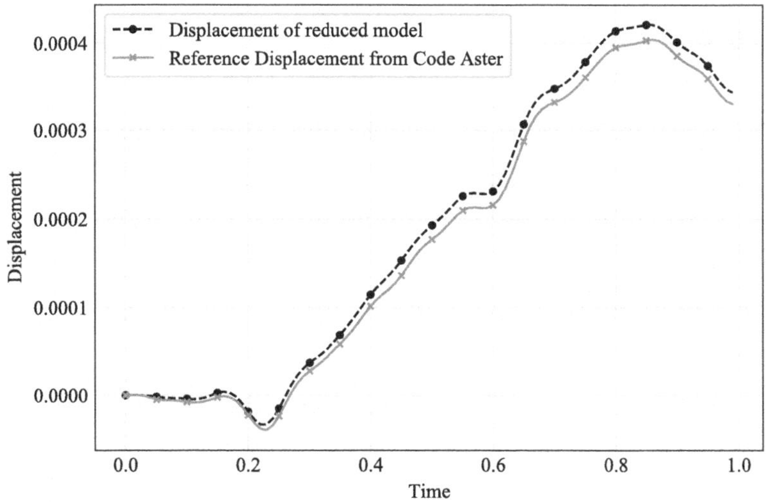

Fig. 10 The displacement of a second observation point in x direction from 0 to 1s with pure thermal load

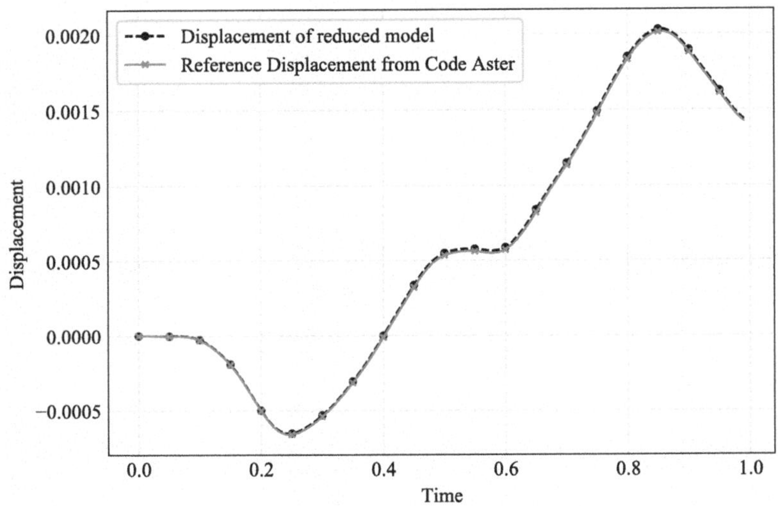

Fig. 11 The displacement of a second observation point in x direction from 0 s to 1 s with additional harmonic mechanical load

5 Conclusions

The reduction of thermo-elastic dynamic equations is a task of ongoing interest in industry. While the basic algorithms and techniques for reduction are known, at least in the linear case, it is not yet known how to use them in the framework of existing software tools. It turned out that many software tools are able to extract the mass matrix and stiffness matrix for a single physics. However, for coupled physics such as thermo-elasticity also coupling information is needed for reduction. Many software tools solve this by adding an additional right-hand side (thermal load) to the mechanical equation. In this paper we have presented an algorithm similar to that of [15] for constructing the coupling matrix from solver information such as the thermal load. The method works reliably, because we directly estimate the reduced coupling matrix and thus obtain a small and well-conditioned linear least square problem. This algorithm has to be performed only once and thus can be seen as an additional offline preprocessing step. The performance of the method is demonstrated on a half arc where we achieve good agreement of the original and the reduced equations.

Further work will be spent to apply the algorithm within other commercial software tools.

Acknowledgements The authors would like to thank the reviewers for their careful revision of the paper. Furthermore, they want to thank the reviewer for the hint of applying directly the thermal modes for the training of the coupling matrix.

References

1. Siemens AG. Nastran. https://www.plm.automation.siemens.com/de_de/products/simcenter/nastran/
2. Bai, Z.: Krylov subspace techniques for reduced-order modeling of large-scale dynamical systems. Appl. Numer. Math. **43**, 9–44 (2002)
3. Baur, U., Benner, P., Feng, L.: Model order reduction for linear and nonlinear systems: a system-theoretic perspective. Arch. Comput. Methods Eng. **21**, 331–358 (2014)
4. Benner, P., Feng, L.: Model order reduction for coupled problems. Appl. Comput. Math. **14**(1), 3–22 (2015)
5. Benner, P., Gugercin, S., Willcox, K.: Survey of projection-based model reduction methods for parametric dynamical systems. SIAM Rev. **57**, 483–531 (2015)
6. Eigner, M., Gilz, T., Zafirov, R.: Interdisziplinäre Produktentwicklung - Modellbasiertes Systems Engineering. PLM Portal (2012)
7. Electricité de France. Finite element *code_aster*, analysis of structures and thermomechanics for studies and research. Open source on www.code-aster.org, 1989–2017
8. Farhat, C., Chapman, T., Avery, P.: Structure-preserving, stability, and accuracy properties of the energy-conserving sampling and weighting method for the hyper reduction of nonlinear finite element dynamic models. Int. J. Numer. Methods Eng. **102**(5), 1077–1110 (2015)
9. Ansys GmbH. Ansys. http://www.ansys.com/Solutions/Solutions-by-Application/Structures
10. Hartmann, D., Herz, M., Wever, U.: Model order reduction a key technology for digital twins. In: Reduced-Order Modeling (ROM) for Simulation and Optimization. Springer (2017)
11. Heinrich, C., Khalil, M., Martynov, K., Wever, U.: Online remaining lifetime estimation for structures. Mech. Syst. Signal Process. **119**, 312–327 (2018)
12. Hughes, T.J.: Linear Static and Dynamic Finite Element Analysis. Dover Publication Inc. (2000)
13. Lohmann, B., Salimbahrami, B.: Ordnungsreduktion mittels Krylov-Unterraummethoden (Model order reduction using Krylov subspace methods). Automatisierungstechnik **52**, 30–38 (2004)
14. Naumann, A., Lang, N., Partzsch, M., Beitelschmidt, M., Benner, P., Voigt, A., Wensch, J.: Computation of thermo-elastic deformations on machine tools - a study of numerical methods. Prod. Eng. **10**(3), 253–263 (2016)
15. Peherstorfer, B., Willcox, K.: Data-driven operator inference for nonintrusive projection-based model reduction. Comp. Meth. Appl. Mech. Eng. **306**, 196–215 (2016)
16. Salimbahrami, B., Lohmann, B.: Order reduction of large scale second-order systems using Krylov subspace methods. Linear Algebra Appl. **415**, 385–405 (2005)
17. Yoon, S.Y., Lin, Z., Allaire, P.E.: Control of surge in centrifugal compressors by active magnetic bearings: theory and implementation. In: Control of Surge in Centrifugal Compressors by Active Magnetic Bearings: Theory and Implementation. Springer (2013)

Reduced-Order Methods in Medical Imaging

Saifon Chaturantabut, Thomas Freeze, Elias Salomão Helou, and Charles H. Lee

Abstract With technological advances and increasing demand for finer resolution images, tomographic medical imaging can be a huge computational problem, consequently, the processing time for image construction can be prohibitively large and impractical for real-time applications. The main bottlenecks are retrieving data and solving the inverse problem to attain medical images. Proper Orthogonal Decomposition (POD) is a model-reduction technique that can be used to compress a large set of images into an orthonormal basis whose elements can accurately generate any images in the original collection with the fewest possible modes. Applicability of POD on tomographic images is not possible without the linearity of the inverse problem. Due to its structure, the first few POD elements contain all the dominant features of the entire image collection. Thus, instead of performing the inverse Radon transform on all medical tomographic images, one needs to process only the primary POD modes once and reuse them to construct all tomographic images, rendering its computational savings. In this article, we improve the POD method further by implementing its *hybrid* version. Namely, the computation of the covariance matrix and associated eigenvalues and eigenvectors can be expensive for large-size images. This process can be sped by working with the down-sampled data and use the resulting coefficients to reconstruct the full-resolution images. Image reconstruction of fish eggs in a test tube will be presented. Errors and computational savings will also be discussed.

The original version of this chapter was revised: Author name "Nicole Hemming-Schroeder" has been removed. The correction to this chapter is available at https://doi.org/10.1007/978-3-030-72983-7_20.

S. Chaturantabut
Thammasat University, Bangkok, Thailand

T. Freeze · C. H. Lee (✉)
California State University - Fullerton, Fullerton, USA
e-mail: charleshlee@fullerton.edu

E. S. Helou
Universidade de São Paulo, São Paulo, Brazil

P. Benner et al. (eds.), *Model Reduction of Complex Dynamical Systems*,
International Series of Numerical Mathematics 171,
https://doi.org/10.1007/978-3-030-72983-7_11

1 Introduction

One way of creating a tomographic image is to emit X-rays through an object from a variety of different angles and spatial positions. Both the emission intensity of each ray and the intensity of the beam received by sensors on the other side of the object are then recorded. The attenuation of the ray as it transverses the matter depends on the attenuation coefficient along the X-ray beam path. These attenuation readings give us line integrals of the function which represents a cross-sectional image of the object. Medical tomography calls for methods to solve both the computational and storage-related problems involved in handling these large data sets.

Numerous studies have used the model-reduction technique called proper orthogonal decomposition (POD). The method was first introduced by Sirovich in 1987 [24] and extracts the dominant features of a data set. Subsequent applications of POD include reducing and controlling fluid flow in chemical vapor deposition reactor [16, 17], amplifying weak signal-to-noise ratios of antenna arrays [13], the compression of hyperspectral data from satellites [21], cancer detection and classification [1, 11, 22, 23], and maximizing stock return [12]. In medical imaging, POD has been used to speed up the reconstruction process of many applications. In electrical impedance tomography (EIT) [14, 15], POD was shown to speed up the computation of reconstruction without decreasing the quality of the reconstructed images significantly. In [4], the use of POD in Hyperspectral tomography (HT) was shown to significantly reduce the computational cost, enhance the fidelity of the tomographic reconstructions, and improve the stability of the reconstruction in the presence of measurement noise. In [5], POD was used for the preprocessing of datasets in vortex detection in 4D MRI Data. In [18], POD was used to generate a basis model for simulated blood patterns for a given vascular location with various anatomical configurations. This basis was then further used to improve the noisy data of blood flow images.

In this study, we use POD to reduce a large set of medical tomography data into a much smaller set of representative modes that can be used to reconstruct any image in the set with a high degree of accuracy. Our method uses an inexpensive technique to down sample and extract the weights of a smaller dataset to use in the computation of a large dataset in order to reduce the number of applications of the inverse mathematical transformation needed to create the medical tomographic images. Rigorous details will be furnished in Sect. 2.2.

2 Methods

In the present section we describe the most well known and widespread tomographic image reconstruction technique and the basic theory of proper orthogonal decomposition.

2.1 Medical Tomography

As an X-ray beam moves across matter, it is attenuated following the Beer-Lambert law:

$$\frac{I_e}{I_d} = e^{\int_L f(s)\mathrm{d}s},$$

where I_e is the X-ray beam's emitted intensity, I_d is the X-ray beam's detected intensity, L is the line segment joining emitter to detector and $f : \mathbb{R}^2 \to \mathbb{R}$ is the linear attenuation factor.

Therefore, on one hand, it is possible to experimentally measure $\int_L f(s)\mathrm{d}s$. On the other hand, because the attenuation factors provide information about the object's interior, it is interesting to know f. In what follows we will describe one way of recovering f from its integrals with respect to arc length along straight lines. Since it is possible to measure integral data, we parametrize this information, accordingly defining the so-called Radon Transform (RT) as follows:

$$p_\theta(t) := \mathcal{R}[f](\theta, t) := \int_{\mathbb{R}} f(t\boldsymbol{\xi}_\theta + s\boldsymbol{\xi}_\theta^\perp)\mathrm{d}s,$$

where $\boldsymbol{\xi}_\theta := \binom{\cos\theta}{\sin\theta}$ and $\boldsymbol{\xi}_\theta^\perp := \binom{-\sin\theta}{\cos\theta}$. For a fixed θ, the function p_θ is also known as a projection. A geometric representation of the RT is given at the left in Fig. 1. In this figure, the Shepp-Logan phantom, which is an image composed by the linear combination of the indicator function of 10 ellipses (a complete description can be found in [10]), is centralized in axes x and y. The t axis, whose slope is determined by the angle θ, is also shown. For the point $t = t'$, the perpendicular dashed line represents the integration path, and the graph of $p_\theta(t)$ is plotted. The representation of the RT in the plane $\theta \times t$ is called sinogram. The sinogram of the Shepp-Logan phantom is presented at the right in Fig. 1.

Because it is useful in the Radon inversion problem, we define here the Fourier Transform (FT) and its inverse. Let $f : \mathbb{R}^n \to \mathbb{C}$ and define, for $\omega \in \mathbb{R}^n$:

$$\hat{f}(\boldsymbol{\omega}) := \mathcal{F}[f](\boldsymbol{\omega}) := \int_{\mathbb{R}^n} f(\boldsymbol{x})e^{-i\boldsymbol{x}\cdot\boldsymbol{\omega}}\mathrm{d}\boldsymbol{x}.$$

Then, under reasonable conditions:

$$f(\boldsymbol{x}) = \mathcal{F}^{-1}[\hat{f}](\boldsymbol{x}) := \frac{1}{(2\pi)^n}\int_{\mathbb{R}^n} \hat{f}(\boldsymbol{\omega})e^{i\boldsymbol{\omega}\cdot\boldsymbol{x}}\mathrm{d}\boldsymbol{\omega}.$$

The dimension n is not explicitly stated in the notation, but should always be clear from the domain of the function being transformed.

One of the key mathematical results about the Radon Transform is the Fourier Slice Theorem (FST), which relates the Fourier Transform of the RT of an image to the Fourier Transform of the image itself. It can be stated as the following equality

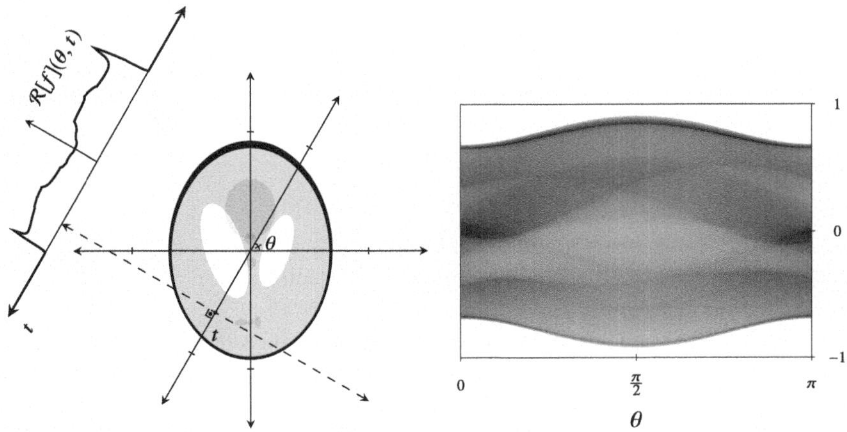

Fig. 1 Left: geometric representation of the line integral defining the Radon transform $\mathcal{R}[f](\theta, t)$. Right: sinogram, i.e., the Radon transform $\mathcal{R}[f]$ of the image f on the left depicted on the $\theta \times t$ plane

(the proof is not difficult and is left as an exercise or may be found in [9, 10, 20]):

$$\hat{p}_\theta(\omega) = \hat{f}\left(\omega \boldsymbol{\xi}_\theta\right). \tag{1}$$

Recall that $p_\theta(t) = \mathcal{R}[f](\theta, t)$. Therefore, what the FST states is that the one-dimensional FT of a projection p_θ corresponds to a "slice" of the two-dimensional FT of the original image.

Therefore, because the projections p_θ can, in principle, be measured using X-rays, it is possible to fill the Fourier space with data in order to use the Fourier inversion formula and to obtain the desired image. Methods that use this idea are called Fourier reconstruction methods and can be useful. However, because this process will require interpolation in the Fourier space and because the Fourier sampling thus obtained is sparse in the higher frequencies, which determine the finer details of the image, Fourier reconstruction methods may present undesirable image artifacts. An alternative is provided by the Filtered Backprojection algorithm, which is presented and discussed in what follows. The two-dimensional Fourier inversion formula can be written in polar coordinates as follows:

$$f(\boldsymbol{x}) = \frac{1}{(2\pi)^2} \int_{[0,\pi]} \int_{\mathbb{R}} |\omega| \hat{f}(\omega \boldsymbol{\xi}_\theta) e^{i\omega \boldsymbol{x}\cdot\boldsymbol{\xi}_\theta} \mathrm{d}\omega \mathrm{d}\theta.$$

Now, using the FST (1) we have

$$f(\boldsymbol{x}) = \frac{1}{2\pi} \int_{[0,\pi]} \frac{1}{2\pi} \int_{\mathbb{R}} |\omega| \hat{p}_\theta(\omega) e^{i\omega \boldsymbol{x}\cdot\boldsymbol{\xi}_\theta} \mathrm{d}\omega \mathrm{d}\theta,$$

which is an inversion formula for the RT since it only depends on data given by the Radon Transform in order to recover f. This formula is called the Filtered Back-projection (FBP) algorithm. In order to estimate the computational cost of the FBP algorithm, we will assume that the RT was sampled in n_θ angles, each of which was sampled in n_t positions. The most common sampling procedure, which will assume here, use parallel projections, that is, the RT is sampled at the pairs (θ_κ, t_ℓ) where

$$(\kappa, \ell) \in \{1, 2, \ldots, n_\theta\} \times \{1, 2, \ldots, n_t\},$$

and

$$\theta_\kappa = \pi \frac{(\kappa - 1)}{n_\theta - 1} \quad \text{and} \quad t_\ell = -1 + 2\frac{(\ell - 1)}{n_t - 1}.$$

The FBP algorithm can be split in two operations: a filtering step followed by a backprojection step. The filtering is described by the following integration:

$$g(\theta, t) := \frac{1}{2\pi} \int_{\mathbb{R}} |\omega| \hat{p}_\theta(\omega) e^{i\omega t} \mathrm{d}\omega,$$

which must be computed for every θ_κ where the Radon Transform was sampled. Each of these integrations can be computed using the Fast Fourier Transform (FFT) algorithm which uses $O(n_t \log n_t)$ flops. Because there are n_θ of such integrations, the total asymptotic flops count becomes $O(n_\theta n_t \log n_t)$. Notice that prior to the computation of the above integral, each of the projections p_{θ_κ} must be Fourier transformed, which is a step that will also take $O(n_\theta n_t \log n_t)$ flops. We will not be concerned with the constant implicit in the big O notation, we simply state that the filtering step consumes $O(n_\theta n_t \log n_t)$ flops. Now, the backprojection step is defined as

$$\mathcal{B}[g](\boldsymbol{x}) := \int_{[0,\pi]} g(\theta, \boldsymbol{x} \cdot \boldsymbol{\xi}_\theta) \mathrm{d}\theta.$$

We will assume that the image being reconstructed will be estimated in a cartesian grid of $n_p \times n_p$ samples. Therefore, the above integration will be computed n_p^2 times, one for each image sample. Since there are n_θ samples θ_κ, the total flop count in this case will be $O(n_\theta n_p^2)$.

Summing up, the total flops count is $O(n_\theta n_p^2) + O(n_\theta n_t \log n_t)$. It is usual to sample the RT such that the sampling rates are proportional among themselves, that is $n_p \sim n_t \sim n_\theta$ and we can, therefore, see that the computation cost is dominated by the backprojection step giving an overall $O(n_p^3)$ asymptotic flop count.

A three-dimensional reconstruction can be obtained by stacking n_r two-dimensional reconstructions. In this case, the dataset is going to have n_r sinograms, each of which contains $n_\theta \times n_t$ samples of the RT of the function f_i, $i \in \{1, 2, \ldots, n_r\}$, that represents slice i of the three-dimensional image.

We finish this brief introduction by mentioning some literature that approaches the problem of reducing the computational load of the backprojection step in many cases obtaining equivalent formulations of the operation that can be computed using around $O(n_p^2 \log n_p)$ flops. The list is not meant to be exhaustive. One approach uses a transformation to log-polar coordinates, which recast the backprojection as a convolution [2]. Also possible is to interpolate Radon samples to a linogram coordinate system [6, 7]. Another approach considers the use of hierarchical decompositions in either the Radon [8] or in the image [3] domains. It is also possible to use a *backprojection slice theorem* in order to reduce the flops count of the most computationally intensive part of the FBP algorithm [19].

2.2 Proper Orthogonal Decomposition

Proper Orthogonal Decomposition (POD) has been used in many applications to construct a low-dimensional subspace that captures the dominant behavior in various applications. One of the most important properties of POD is that it can construct an approximation that minimizes the error in 2-norm for a given fixed basis rank.

Consider a set of snapshots $\{\mathbf{X}(1), \mathbf{X}(2), \ldots, \mathbf{X}(N_s)\} \subset \mathbb{R}^N$. In general, each snapshot may depend on certain parameter value, time instance, or spatial location. Suppose we want to approximate a snapshot $\mathbf{X}(j)$ by using a set of orthonormal vectors $\{\Phi^{(1)}, \Phi^{(2)}, \ldots, \Phi^{(K)}\} \subset \mathbb{R}^N$, which has rank $K < N$. Then, the approximation can be written in the form of

$$\mathbf{X}(j) \approx \sum_{k=1}^{K} \alpha_k^{(j)} \Phi^{(k)}, \qquad j = 1, \ldots, N_s. \tag{2}$$

or, equivalently in a matrix form $\mathbf{X}(j) \approx \mathbf{\Phi}\alpha^{(j)}$, where $\alpha_k^{(j)}$ is the kth component of vector $\alpha^{(j)} = \mathbf{\Phi}^T\mathbf{X}(j) \in \mathbb{R}^K$, and $\mathbf{\Phi} = [\Phi^{(1)}, \Phi^{(2)}, \ldots, \Phi^{(K)}] \in \mathbb{R}^{N\times K}$. The above approximation can be considered as applying orthogonal projection $\mathbf{\Phi}\mathbf{\Phi}^T$, i.e.,

$$\mathbf{X}(j) \approx \mathbf{\Phi}\mathbf{\Phi}^T\mathbf{X}(j), \qquad j = 1, \ldots, N_s. \tag{3}$$

POD provides an orthonormal basis that minimizes this approximation error in 2-norm for a given basis rank $K \leq r$, where $r := \text{rank}(\{\mathbf{X}(1), \mathbf{X}(2), \ldots, \mathbf{X}(N_s)\})$. That is, POD basis is the optimal solution to the following minimization problem:

$$\mathbf{\Phi}_{POD} = \arg \min_{\mathbf{\Phi}\in\mathbb{R}^{N\times K}} \sum_{j=1}^{N_s} \|\mathbf{X}(j) - \mathbf{\Phi}\mathbf{\Phi}^T\mathbf{X}(j)\|_2^2. \tag{4}$$

It can be shown [25] that POD basis defined above can be obtained from the left singular vector of the snapshot matrix $\mathbf{X} = [\mathbf{X}(1), \mathbf{X}(2), \ldots, \mathbf{X}(N_s)]$. Let $\mathbf{X} = \mathbf{U}\Sigma\mathbf{V}^T$ be the singular value decomposition of $\mathbf{X}$, where matrices $\mathbf{U} = [U^{(1)}, \ldots, U^{(r)}] \in \mathbb{R}^{N\times r}$ and $\mathbf{V} = [V^{(1)}, \ldots, V^{(r)}] \in \mathbb{R}^{N_s\times r}$ are matrices with orthonormal columns and

$\Sigma = diag(\sigma_1, \ldots, \sigma_r) \in \mathbb{R}^{r \times r}$ is a diagonal matrix with $\sigma_1 \geq \sigma_2 \geq \cdots \geq \sigma_r > 0$. Then the POD basis matrix of dimension K is the first K columns of the left singular matrix $\mathbf{U}$, i.e., $\mathbf{\Phi}_{POD} = [U^{(1)}, \ldots, U^{(K)}] \in \mathbb{R}^{N \times K}$, for $K \leq r$. Moreover, it is well known [25] that this minimum error is given by

$$\sum_{j=1}^{N_s} \|\mathbf{X}(j) - \mathbf{\Phi}_{POD}\mathbf{\Phi}_{POD}^T\mathbf{X}(j)\|_2^2 = \sum_{\ell=K+1}^{r} \sigma_\ell^2, \tag{5}$$

which is the sum of the neglected singular values $\sigma_{K+1}^2, \ldots, \sigma_r^2$ from the SVD of $\mathbf{X}$.

When the dimension N of snapshots is not too large, we can directly obtain the POD basis from the SVD of the snapshot matrix directly. However, in practice, N can be extremely large and computing POD basis through SVD might not be efficient. In this case, it is common to use a technique called the **method of snapshots**, which is based on finding the eigen-decomposition of the covariance matrix of $\mathbf{X}$ defined by $\mathbf{\Omega} := \mathbf{X}^T\mathbf{X} \in \mathbb{R}^{N_s \times N_s}$. For $N_s < N$ and $N_s = r = \text{rank}(\mathbf{X})$, recall that the singular value decomposition of $\mathbf{X}$ is given by $\mathbf{X} = \mathbf{U}\Sigma\mathbf{V}^T$. Then $\mathbf{\Omega} = (\mathbf{U}\Sigma\mathbf{V}^T)^T(\mathbf{U}\Sigma\mathbf{V}^T) = \mathbf{V}\Sigma^2\mathbf{V}^T$. Note that, since $\mathbf{V} \in \mathbb{R}^{N_s \times N_s}$ is an orthonormal matrix, $\mathbf{V}^T = \mathbf{V}^{-1}$ and $\mathbf{\Omega} = \mathbf{V}\Sigma^2\mathbf{V}^T$ can be considered as eigen-decomposition of $\mathbf{\Omega}$. In particular, each right singular vector $V^{(\ell)}$ is an eigenvector of the covariance matrix $\mathbf{\Omega}$ with corresponding eigenvalue σ_ℓ^2, for $\ell = 1, 2, \ldots, N_s$. We can obtain the POD basis, which is the first K columns of $\mathbf{U}$ by computing $\mathbf{U} = \mathbf{X}\mathbf{V}\Sigma^{-1}$. That is, each POD basis vector $\Phi_{POD}^{(k)} \in \mathbb{R}^N$, $k = 1, 2, \ldots, K$, for $K \leq N_s$, is given by

$$\Phi_{POD}^{(k)} = \frac{1}{\sigma_k}\sum_{j=1}^{N_s} V^{(k)}(j)\,\mathbf{X}(j), \tag{6}$$

where $V^{(k)}(j)$ is the jth entry of the and kth eigenvector $V^{(k)}$ of the covariance matrix $\mathbf{\Omega}$. Equivalently, the POD basis matrix $\mathbf{\Phi}_{POD} = [\Phi_{POD}^{(1)}, \ldots, \Phi_{POD}^{(K)}] \in \mathbb{R}^{N \times K}$ can be computed from $\mathbf{\Phi}_{POD} = \mathbf{X}\mathbf{V}_K\Sigma_K^{-1}$, where $\mathbf{V}_K = [V^{(1)}, \ldots, V^{(K)}]$ and $\Sigma_K = \text{diag}(\sigma_1, \ldots, \sigma_K)$. The steps for constructing a POD basis matrix by using the method of snapshots are summarized in Algorithm 1.

Algorithm 1 Method of snapshots for constructing POD basis

Input: Snapshots $\mathbf{X}(1), \ldots, \mathbf{X}(N_s) \in \mathbb{R}^N$, $N_s \leq N$ and POD dimension K.
Output: POD basis matrix $\mathbf{\Phi}_{POD} = [\Phi_{POD}^{(1)}, \ldots, \Phi_{POD}^{(K)}] \in \mathbb{R}^{N \times K}$.

1: Create matrix $\mathbf{X} = [\mathbf{X}(1), \ldots, \mathbf{X}(N_s)] \in \mathbb{R}^{N \times N_s}$, and let $r = rank(\mathbf{X})$.
   Form covariance matrix $\mathbf{\Omega} = \mathbf{X}^T\mathbf{X}$.
2: Compute eigen-decomposition $\mathbf{\Omega} = \mathbf{V}\mathbf{D}\mathbf{V}^T$, where $\mathbf{D} = \text{diag}(\sigma_1^2, \ldots, \sigma_r^2)$.
3: Compute POD basis by using (6) or
   using the matrix form: $\mathbf{\Phi}_{POD} = \mathbf{X}\mathbf{V}_K\mathbf{D}_K^{-1/2}$,
   where $\mathbf{V}_K = \mathbf{V}(:, 1:K)$ and $\mathbf{D}_K = \mathbf{D}(1:K, 1:K)$ in MATLAB notation.

The two methods described below illustrate how downsampling can be used in conjunction with POD to create a reduced-order method of storing and reconstructing medical tomography images.

2.3 Downsampled POD Method

We begin with a matrix of n_r sinograms, $\mathbf{X} = [\mathbf{X}(1), \mathbf{X}(2), \ldots, \mathbf{X}(n_r)] \in \mathbb{R}^{N \times n_r}$, where $N = n_\theta \times n_t$. We recall that n_r is the number of slices of the three-dimensional tomographic dataset, n_θ is the number of projections of each slice, and n_t is the number of spatial samples of each projection of each slice. We can make significant gains in efficiency by downsampling the data by taking a reduced resolution and number of layers, and number of projections to approximate the sinograms, $\mathbf{X}$ with a down-sampled matrix $\hat{\mathbf{X}} = [\hat{\mathbf{X}}(1), \hat{\mathbf{X}}(2), \ldots, \hat{\mathbf{X}}(N_r)] \in \mathbb{R}^{\hat{N} \times N_r}$ where $\hat{N} = N_\theta \times N_t$ with dimensions $N_r \ll n_r$, $N_\theta \ll n_\theta$, and $N_t \ll n_t$.

As described in the previous section, we then find and sort the eigenvalues $\hat{\sigma}_k^2$ and corresponding eigenvectors, $\hat{V}^{(k)}$ for $k = 1, 2, \ldots, N_r$ of the covariance matrix $\hat{\Omega} \in \mathbb{R}^{N_r \times N_r}$. By using (6), each POD mode is given by

$$\hat{\Phi}^{(k)}_{POD} = \frac{1}{\hat{\sigma}_k} \sum_{j=1}^{N_r} \hat{V}^{(k)}(j)\hat{\mathbf{X}}(j), \tag{7}$$

for $k = 1, 2, \ldots, N_{POD}$, where $N_{POD} \in \{1, 2, \ldots, N_r\}$ is the desired number of POD modes.

Next, we find the projection of each sinogram onto each POD mode to determine the weights needed to reconstruct the images by taking $\hat{\alpha}^{(j)} = \hat{\mathbf{\Phi}}^T_{POD}\hat{\mathbf{X}}(j) \in \mathbb{R}^{N_{POD}}$ for layers $j = 1, \ldots, N_r$, where $\hat{\mathbf{\Phi}}_{POD} = [\hat{\Phi}^{(1)}_{POD}, \hat{\Phi}^{(2)}_{POD}, \ldots, \hat{\Phi}^{(N_{POD})}_{POD}] \in \mathbb{R}^{\hat{N} \times N_{POD}}$ is the POD basis matrix. Let $\hat{\mathbf{Y}} = [\hat{\mathbf{Y}}(1), \ldots, \hat{\mathbf{Y}}(N_r)]$ be the matrix of tomograms corresponding to our matrix of sinograms, $\hat{\mathbf{X}}$. Then,

$$\hat{\mathbf{Y}}(j) \approx \text{iRadon}\left(\sum_{k=1}^{N_{POD}} \hat{\alpha}_k^{(j)} \hat{\Phi}^{(k)}_{POD} \right), \tag{8}$$

where iRadon represents inversion using the FBP algorithm. By using linearity of this inversion iRadon, (8) is equivalent to

$$\hat{\mathbf{Y}}(j) \approx \sum_{k=1}^{N_{POD}} \hat{\alpha}_k^{(j)} \text{iRadon}\left(\hat{\Phi}^{(k)}_{POD} \right). \tag{9}$$

Let us define $\hat{\Psi}^{(k)}_{POD} = \text{iRadon}\left(\hat{\Phi}^{(k)}_{POD}\right)$. Thus, if we take $N_{POD} \ll N_r$, we only need to store the truncated POD modes, $\{\hat{\Psi}^{(k)}_{POD}\}_{k=1}^{N_{POD}}$, and corresponding truncated $\hat{\alpha}^{(j)}$ values to reconstruct the tomograms $\hat{\mathbf{Y}}(j)$, $j = 1, \ldots, N_r$.

2.4 Hybrid-POD Method

In the second method, we only down sample $\mathbf{X}$ by layers, so that we are using $N_r < n_r$, but the resolution and number of angles remains the same as the full data set. Take $\mathbf{X}_h = [\mathbf{X}_h(1), \ldots, \mathbf{X}_h(N_r)] \in \mathbb{R}^{N \times N_r}$ to be the matrix containing data set down-sampled along the layers only. In this section, we use the same weights from the eigenvectors $\hat{V}^{(j)}$ and eigenvalues $\hat{\sigma}_k^2$ of the covariance matrix $\hat{\Omega}$ together with the coefficient $\hat{\alpha}^{(j)}$, $j = 1, \ldots, N_r$, from the down-sampled method in Sect. 2.3 to construct full-resolution approximate POD modes $\Phi^{(k)}_{hPOD}$ and to reconstruct the corresponding tomograms $\mathbf{Y}_h(1), \ldots, \mathbf{Y}_h(N_r)$. In other words, each POD mode from the hybrid approach can be defined as

$$\Phi^{(k)}_{hPOD} = \frac{1}{\hat{\sigma}_k} \sum_{j=1}^{N_r} \hat{V}^{(k)}(j) \mathbf{X}_h(j) \tag{10}$$

for $k = 1, 2, \ldots, N_{POD}$. The sinograms are then approximated by

$$\mathbf{X}_h(j) \approx \sum_{k=1}^{N_{POD}} \hat{\alpha}_k^{(j)} \Phi^{(k)}_{hPOD}, \qquad j = 1, \ldots, N_r. \tag{11}$$

By using the linearity of the inverse Radon transform, as done in the previous section, we can define $\Psi^{(k)}_{hPOD} = \text{iRadon}\left(\Phi^{(k)}_{hPOD}\right)$ and obtain the approximate tomograms of the form

$$\mathbf{Y}_h(j) \approx \sum_{k=1}^{N_{POD}} \hat{\alpha}_k^{(j)} \Psi^{(k)}_{hPOD}, \qquad j = 1, \ldots, N_r. \tag{12}$$

In practice, we can compute the approximated tomograms in (12) directly without forming the approximated sinograms in (11). The inverse Radon transform of these POD modes has to be computed only once and they can be reused for approximating the reconstructions in all layers. This therefore can significantly reduce the computational cost in practice. Note that, the tomographic reconstruction using this hybrid-POD approach requires less computational cost than the traditional POD reconstruction approach as shown in Table 1. In particular, the hybrid approach has less computational complexity than the traditional POD approach when forming the covariance matrix for constructing POD modes and when computing the coefficients for the POD approximation. The accuracy and efficiency of the hybrid approach are demonstrated in Sect. 3.3.

2.5 Implementation Details

To test our model, we collect ray attenuation data through a test tube of fish eggs from $n_\theta = 200$ angles. The resolution of each sinogram is $n_t = 2048$ pixels so that the dimensions of the sinogram data set is 2048 (resolution) $\times$ 256 (layers) $\times$ 200 (angles). The resolution downsampling is fixed at every 16 rows, beginning with row 16/2 = 8, i.e., we use rows 8, 24, 40, ..., 2024, and 2040. The angles are not down sampled. Thus, for our experiment $N_t = 128$ and $N_\theta = n_\theta = 200$. The number of snapshots used is varied by varying the downsampling of the number of layers. For example, with 64 layers (i.e., $N_r = 64$) we skip every $n_r/N_r = 256/64 =$ 4 layers, beginning with layer 4/2 = 16. Thus, our down-sampled dataset is size 128 (resolution) $\times$ N_r (layers) $\times$ 200 (angles) for $N_r = 8, 16, 32, 64, 128$, and 256 (where $N_r = 256$ means all layers are used in the computation of the POD modes). To transform each 2048 (resolution) by 200 (angles) sinogram to a 1448 by 1448 tomogram, we use a Ram-Lak filters in our backscattered Inverse Radon transform.

3 Results

3.1 Test Tube with Fish Eggs

In this study, we consider a test tube containing fish eggs and water. The setup can be found in Fig. 2. The test tube is X-rayed at 2048 uniformly spaced layers, resulting in 2048 different sinograms (2048 layers). Ray attenuation is measured for 200 angles. Sinogram resolution is 2048 pixels.

Fig. 2 The above image shows high frequency energy passing through the test tube and various fish eggs

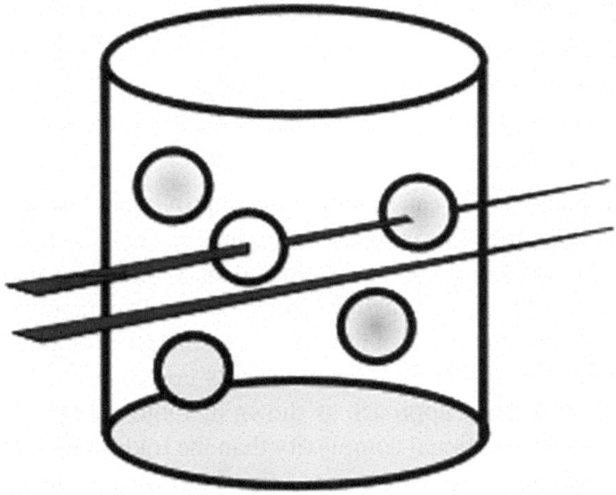

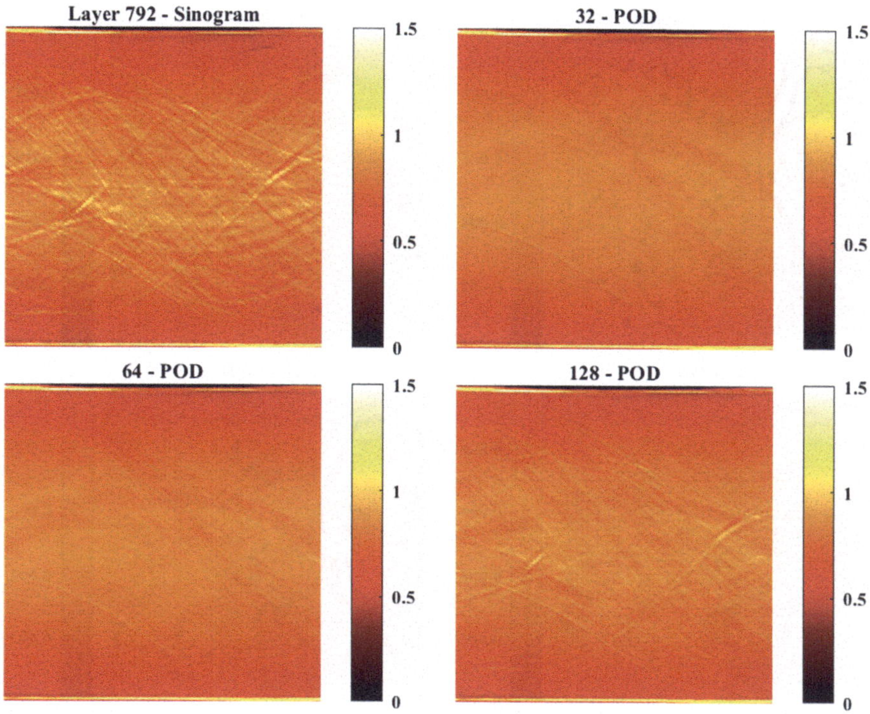

Fig. 3 The above figure shows the sinograms with the original in the top left and then as reconstructed with increasing amounts of modes used

3.2 *Down-Sampling Results*

Our objective in this subsection is to assess how well the POD technique performs in reconstructing the down-sampled data and its computational savings. For down-sampling we apply two different rates to down sample: skip every 8 vertical layers giving $k \in \{4, 12, 20, \ldots, 2044\}$ and skip every 16 spatial pixels giving $j \in \{8, 24, 40, \ldots, 2040\}$. We leave the angular resolution unchanged. Consequently, this changes our data-cube shape from (2048, 2048, 200) to (256, 128, 200). We employ the POD method with 256 snapshots of sinograms and each of which is of dimensions 128 by 200. As a result, there are 256 POD sinograms that one can use to reconstruct any layers. For demonstration purpose, we showed the results for the 792th layer. The sinogram for the 792th layer along with its reconstructions using 32, 64, and 128 POD modes are shown in Fig. 3. Due to the principal components, only a small number of primary POD modes are needed. That is, rather than performing the Radon transformation on all 256 modes, one only needs to perform it on the first few POD modes, which yields the computational savings. The corresponding reconstructed images are shown in Fig. 4. As seen in both figures, the larger number of POD modes are used, the more resemblance the reconstructed image gets. In

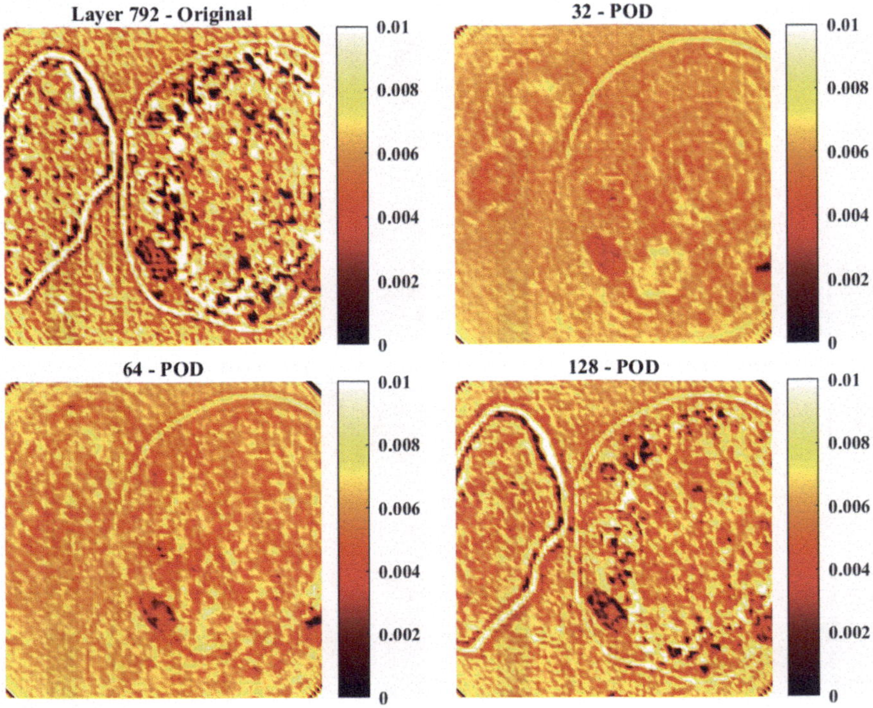

Fig. 4 The above figure shows the constructed image with the original in the top left and then as reconstructed with increasing amounts of modes used

addition, we see increasing detail (and high frequency noise) as the number of modes used increases.

We then examined the Root-Mean-Square Error (RMSE) as the number of modes utilized was varied. The RMSE can be seen in Fig. 5, where the x-axis shows the layer numbers and y-axis shows number of POD modes used and the color bar shows the RMSE. It generally shows that as the number of modes increases the total error decreases. Note that for the first 25 layers did not contain any fish eggs and thus the errors seem low and almost all POD modes. On the other hand, layers between 175 and 210 contain multiple eggs and thus higher number of POD modes are needed to reduce the RMSE. Furthermore, the actual image in Fig. 4 has the scale of order 10^{-2} and with 50 POD modes, one can reconstruct any layer with RMSE less than 5×10^{-4}. Computational savings are shown in Fig. 6. It should be pointed out that the total time for reconstructing all 256 layers takes 1445 s and by using 64 POD modes, it is amount to 70 s or 5% of the time.

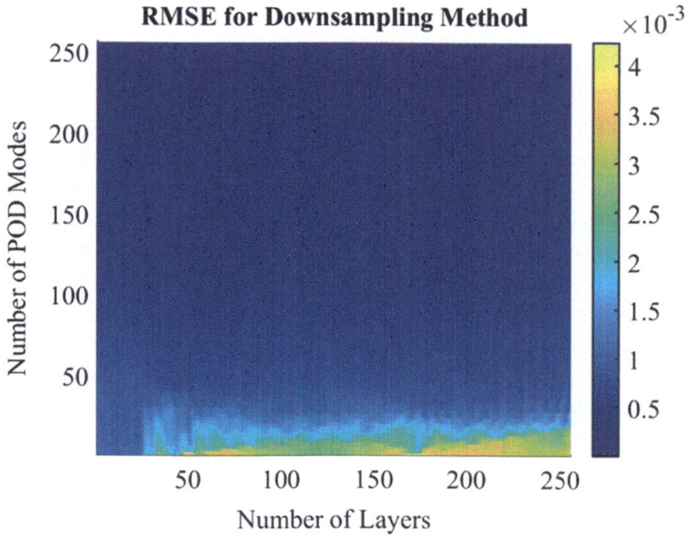

Fig. 5 The above figure shows the RMSE of the constructed image as a function of layer number and the number of POD modes used

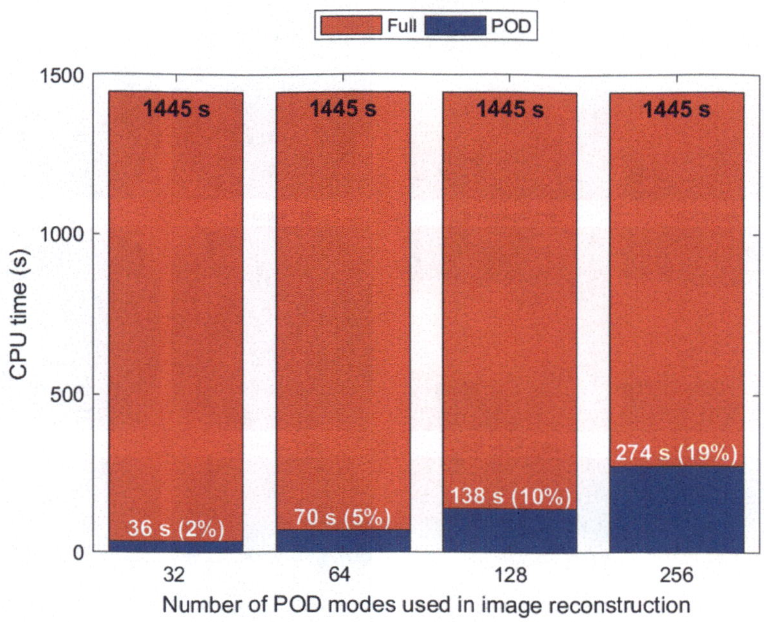

Fig. 6 The computation time with the down-sampling method was significantly shorter due to the much smaller number of layers and decreased image resolution

3.3 Hybrid-POD Method

To fully demonstrating the hybrid-POD technique, we use a finer tomographic data set. Particularly, this set of data contains $N_r = 2048$ layers and each of which has spatial and angular resolutions of $N_t = 2048$ and $N_\theta = 1025$, respectively. The following results demonstrate the accuracy and efficiency of the hybrid-POD method presented in Sect. 2.4. Figures 7, 8, 9, and 10 provide the comparison of the original sinograms and tomograms (in layers 500, 1000, 1500, 2000) with the reconstructed ones from the hybrid-POD approach with $N_{POD} = 128, 256, 512, 1024, 2048$. It should be pointed out that the first 865 layers contain only water. As seen from Fig. 7 that, when there is no object detected in layer 500, the POD-hybrid approach seems to reconstruct the images accurately for all cases of different POD modes. For layers 1000, 1500, 2000 that contain fish eggs, the details of sinogram and tomogram images can be detected more accurately as more POD modes are used, as shown in Figs. 8, 9, and 10. In these cases, the hybrid-POD reconstructions are visually indistinguishable when the number of POD modes $N_{POD} \geq 256$ is used. To measure the accuracy of this approach, we consider the RMSE and its average

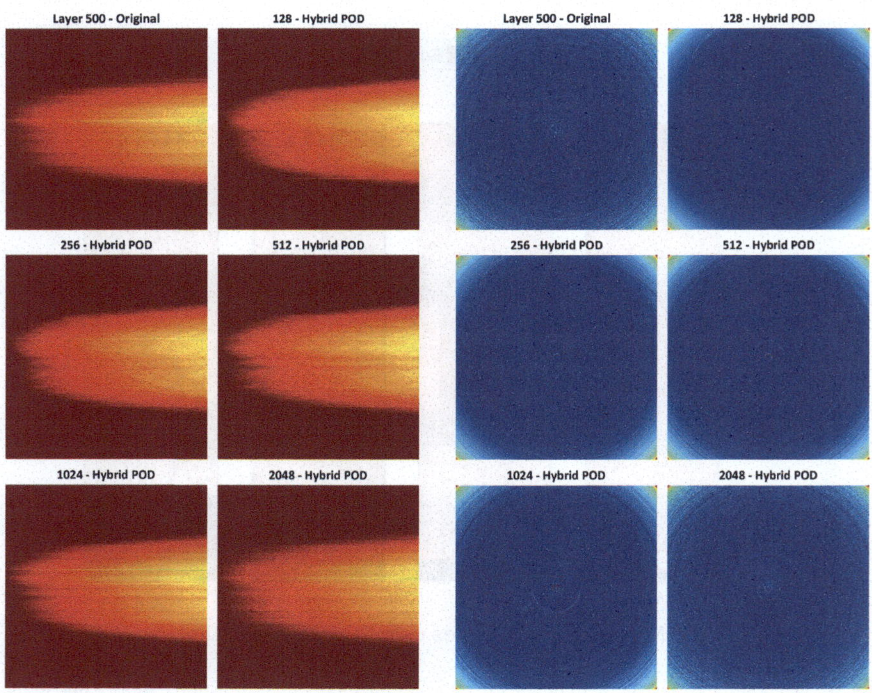

Fig. 7 The comparison of the original sinogram in layer 500 (water with no inclusions) with its sinogram approximations (left plots) and the corresponding tomogram with its approximations (right plots) from hybrid-POD approach using $N_t = 2048$ (resolution) $N_\theta = 1025$ (angles) $N_r = 2048$ (layers) with different dimension of POD basis $N_{POD} = 128, 256, 512, 1024, 2048$

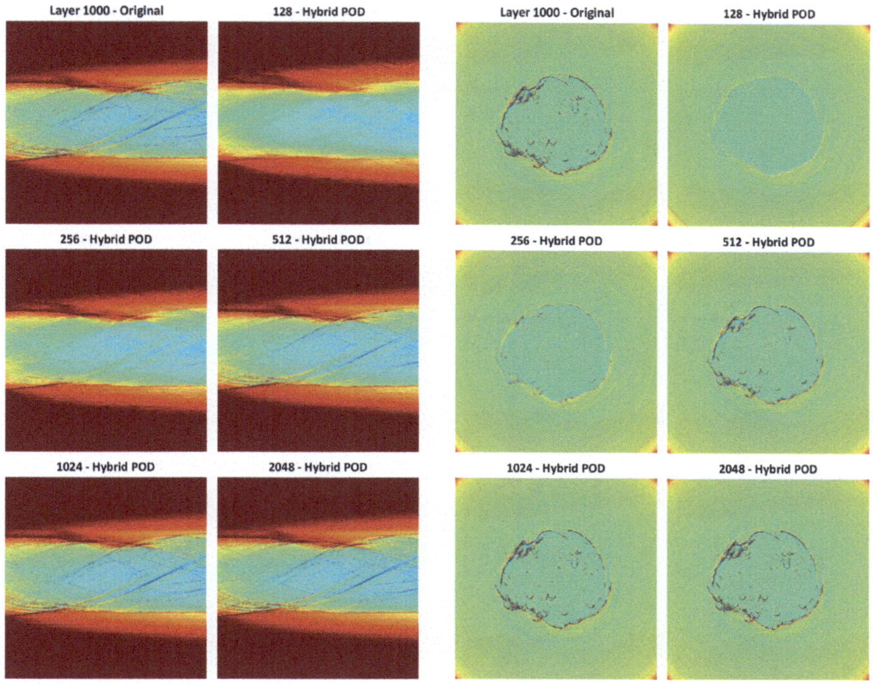

Fig. 8 The comparison of the original sinogram in layer 1000 with its approximations (left plots) and the corresponding tomogram with its approximations (right plots) from hybrid-POD approach using $N_t = 2048$ (resolution) $N_\theta = 1025$ (angles) $N_r = 2048$ (layers) with different dimension of POD basis $N_{POD} = 128, 256, 512, 1024, 2048$

for the reconstructed sinograms and the corresponding tomograms in Figs. 11 and 12, respectively, with different POD modes. As seen in most POD applications, the plots of average relative RMSE in these figures show that when more POD modes are used, the reconstruction becomes more accurate. Note that, in top plot of Fig. 12, the relative RMSEs of the reconstructed tomograms seem to be quite high around the first 865 layers when compared with the RMSE of the remaining layers as clearly shown in the bottom plot of Fig. 12 that separately calculates the average relative RMSE of the first 865 layers and the remaining layers. This might result from the fact that these first 865 layers contain just water with no fish eggs or other objects.

The computational saving from using the hybrid-POD approach for reconstructing tomograms when compared to the direct approach that performs the inverse Radon transform on the sinograms are given in Fig. 13. The computational time considered in Fig. 13 for the POD-hybrid approach includes the CPU time for constructing the coefficients from the down-sample snapshots and CPU time for performing the inverse Radon transform of the hybrid-POD modes. As expected, when dimension of the hybrid-POD basis increases, it requires more computational time to perform the reconstruction. However, all of these POD-hybrid cases can still provide significant

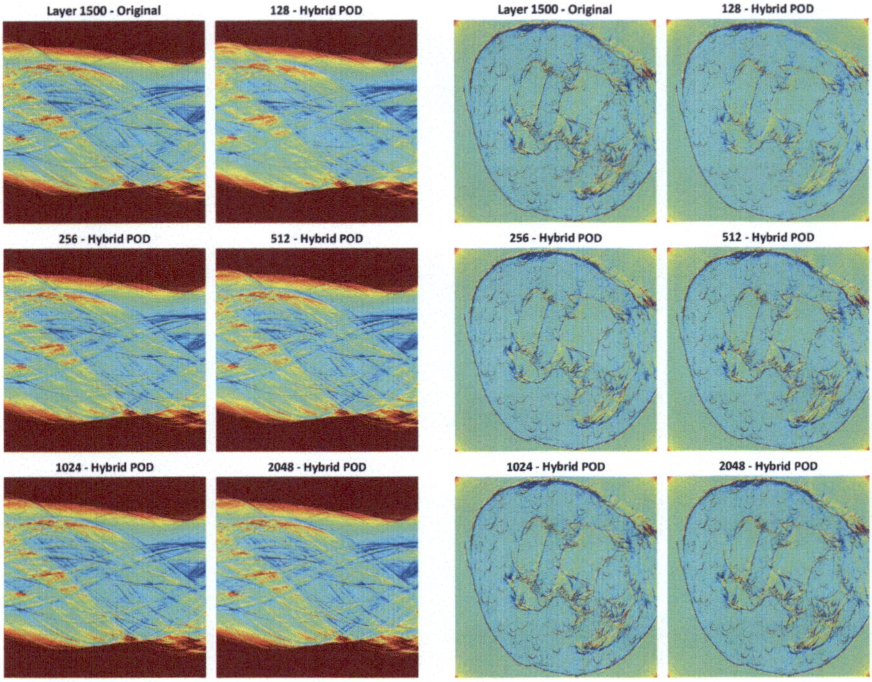

Fig. 9 The comparison of the original sinogram in layer 1500 with its approximations (left plots) and the corresponding tomogram with its approximations (right plots) from hybrid-POD approach using $N_t = 2048$ (resolution) $N_\theta = 1025$ (angles) $N_r = 2048$ (layers) with different dimension of POD basis $N_{POD} = 128, 256, 512, 1024, 2048$

saving. In particular, the POD-hybrid approach uses 4.7, 9.3, 18.3, 36.4, and 72.6% of the CPU time required by the direct tomographic reconstruction when the number of POD modes is $N_{POD} = 128$, 256, 512, 1024, and 2048, respectively.

4 Discussion

There are two main aspects of computational saving in this work. The first one comes from using POD basis to approximate the sinograms. This allows us to obtain tomograms by performing inverse Radon transform on the low-dimensional POD basis, instead of the high-dimensional sinograms. This advantage is clearly shown in Sect. 3.2 when constructing the down-sample case. The reconstruction time is reduced to 5% of the direct reconstruction when 64 POD modes, instead of 256 sinograms, are used in the inverse Radon transform.

Another aspect of computational saving provided in this work is based on applying the POD-hybrid approach described in Sect. 2.4. This approach can substantially

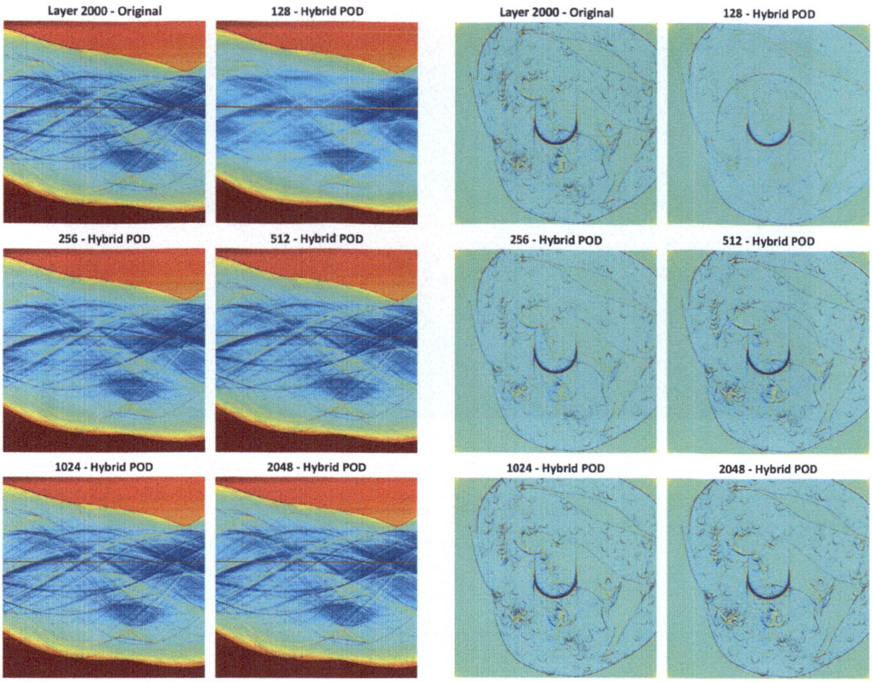

Fig. 10 The comparison of the original sinogram in layer 2000 with its approximations (left plots) and the corresponding tomogram with its approximations (right plots) from POD-hybrid approach using $N_t = 2048$ (resolution) $N_\theta = 1025$ (angles) $N_r = 2048$ (layers) with different dimension of POD basis $N_{POD} = 128, 256, 512, 1024, 2048$

decrease the computational cost when we want to reconstruct tomography images for full-resolution sinograms by using POD method, instead of using down-sampled sinograms. The comparison of complexity for direct reconstruction using the traditional POD approach and for the hybrid-POD approach is shown in Table 1. Notice that, the hybrid approach can clearly decrease the computational cost in Step 1 for forming the covariance matrix and Step 5 for computing the coefficients in the POD approximation. Notice that, in Step 6, the computational cost of performing inverse Radon transform, which is not included in the table, is the same for both the traditional POD and the hybrid-POD approaches because $\Phi^{(k)}_{POD}$ and $\Phi^{(k)}_{hPOD}$ have the same size. The flops count for the inverse Radon transform is discussed in Sect. 2.1. Note that, from Table 1, the additional cost for computing the down-sampled POD modes in Step 4 is negligible when $\hat{N} \ll N$ and $N_s \ll N$, where $\hat{N}$ and N are the dimensions of the down-sampled sinograms and of the full-resolution sinograms, respectively. This numerical saving reflects in the overall CPU time for tomographic reconstruction demonstrated in Sect. 3.3. In particular, the POD-hybrid approach with $N_{POD} = 512$ uses only 12.2% of CPU time for the direct reconstruction (77.8% computational saving) with average relative RMSE of order $\mathcal{O}(10^{-2})$.

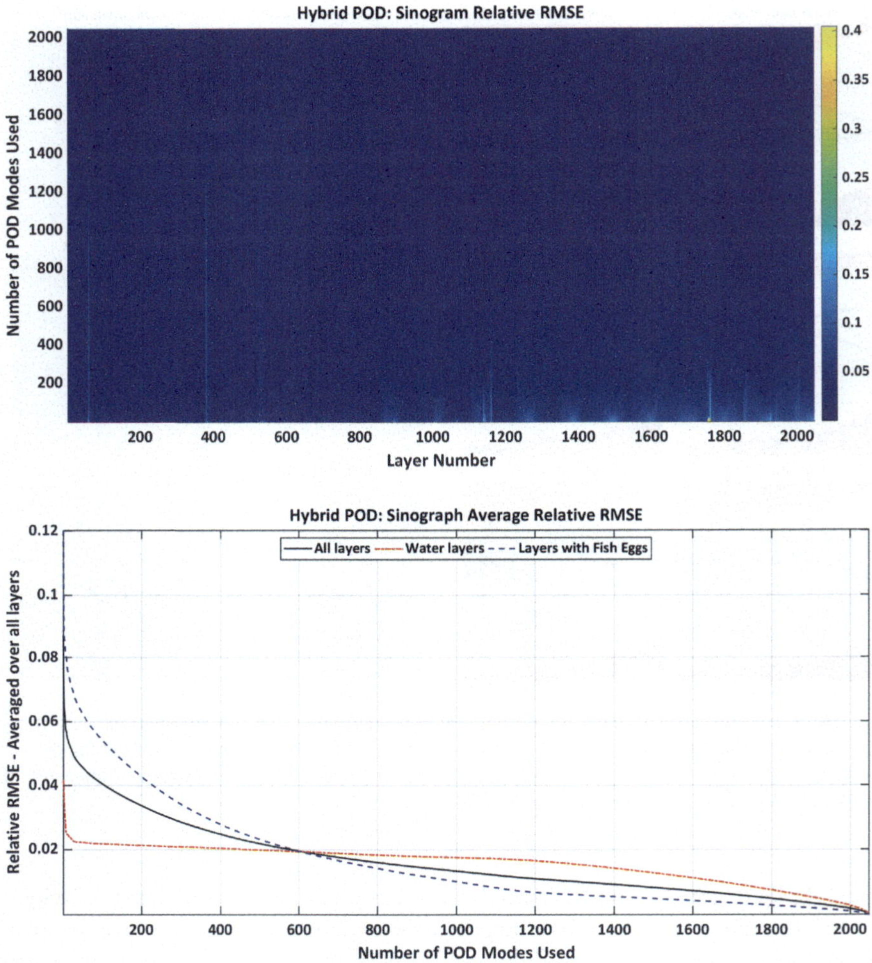

Fig. 11 The relative RMSE of sinograms for each layer (top). The corresponding average relative RMSE of sinograms averaging over (i) all 2048 layers, (ii) layers 1 to 856 that contain water, and (iii) layers 866 to 2048 that contain fish eggs (bottom)

Besides computational cost saving, memory storage can also be substantially reduced by using the down-sampled approach and the hybrid-POD approach. In particular, in order to reconstruct N_s tomograms $\mathbf{Y}(1), \ldots, \mathbf{Y}(N_s)$, we only need to store the K-dimensional POD basis matrix of size $N \times K$ and the coefficient matrix of size $K \times N_s$, which generally require less storage than storing the full-resolution sinograms in $\mathbf{X} = [\mathbf{X}(1), \ldots, \mathbf{X}(N_s)]$ of size $N \times N_s$, especially for large N or for $K < N_s \ll N$.

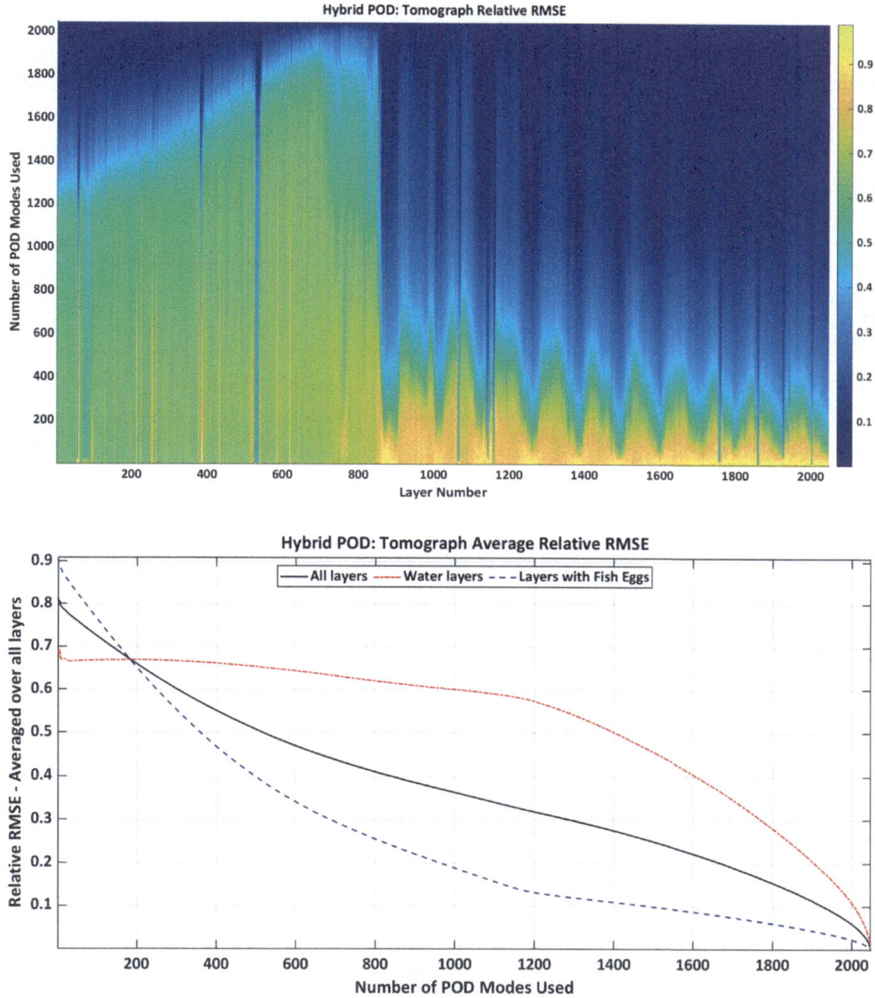

Fig. 12 The relative RMSE of tomograms for each layer (top). The corresponding average relative RMSE of tomograms averaging over (i) all 2048 layers, (ii) layers 1 to 856 that contain water, and (iii) layers 866 to 2048 that contain fish eggs (bottom)

Table 1 Comparison of computational complexity for constructing N_s approximated tomograms $Y(j)$, $j = 1, \ldots, N_S$, by using the traditional POD method and the proposed hybrid-POD method of dimension $N_{POD} = K$. Note that $\Phi^{(k)}_{POD}$ and $\Phi^{(k)}_{hPOD}$ are the kth column of $\mathbf{\Phi}_{POD}$ and $\mathbf{\Phi}_{POD}$, respectively. Note that n_p-by-n_p is size of output each tomogram

Computation	**Traditional POD Reconstruction** Input: Sinograms $\mathbf{X} = [\mathbf{X}(1), \ldots, \mathbf{X}(N_s)] \in \mathbb{R}^{N \times N_s}$		**Hybrid POD Reconstruction** Input: Sinograms $\mathbf{X} = [\mathbf{X}(1), \ldots, \mathbf{X}(N_s)] \in \mathbb{R}^{N \times N_s}$ $\hat{\mathbf{X}} = [\hat{\mathbf{X}}(1), \ldots, \hat{\mathbf{X}}(N_s)] \in \mathbb{R}^{\hat{N} \times N_s}$	
1.Covariance matrix	$\Omega := \mathbf{X}^T\mathbf{X}$	$O(NN_s^2)$	$\hat{\Omega} := \hat{\mathbf{X}}^T\hat{\mathbf{X}}$	$O(\hat{N}N_s^2)$
2.Eigen-decomposition of $N_s \times N_s$ covariance matrix	$\Omega = \mathbf{V}\Sigma^2\mathbf{V}^T$	$O(N_s^3)$	$\hat{\Omega} = \hat{\mathbf{V}}\hat{\Sigma}^2\hat{\mathbf{V}}^T$	$O(N_s^3)$
3.POD basis of dimension K	$_{POD} = \mathbf{X}\mathbf{V}_K\Sigma_K^{-1}$	$O(NN_sK)$	$_{hPOD} = \mathbf{X}\hat{\mathbf{V}}_K\hat{\Sigma}_K^{-1}$	$O(NN_sK)$
4.Down-sampled POD basis of dimension K	-	-	$\widehat{\ }_{POD} = \hat{\mathbf{X}}\hat{\mathbf{V}}_K\hat{\Sigma}_K^{-1}$	$O(\hat{N}N_sK)$
5.Coefficient $j = 1, \ldots, N_s$	$\alpha^{(j)} = {}^T_{POD}\mathbf{X}(j)$	$O(NN_sK)$	$\hat{\alpha}^{(j)} = \widehat{\ }^T_{POD}\hat{\mathbf{X}}(j)$	$O(\hat{N}N_sK)$
6.Approximate tomograms $\mathbf{Y} = [Y(1), \ldots, Y(N_s)]$	Sinograms: $X(j) \approx \sum_{k=1}^{K} \alpha_k^{(j)}\Phi^{(k)}_{POD}$ Tomograms: $Y(j) \approx \sum_{k=1}^{K} \alpha_k^{(j)}\Psi^{(k)}_{POD}$ where $\Psi^{(k)}_{POD} = \text{iRadon}\left(\Phi^{(k)}_{POD}\right)$ $j = 1, \ldots, N_s$	$O(n_p^2 N_s K)$	Sinograms: $X(j) \approx \sum_{k=1}^{K} \hat{\alpha}_k^{(j)}\Phi^{(k)}_{hPOD}$ Tomograms: $Y(j) \approx \sum_{k=1}^{K} \hat{\alpha}_k^{(j)}\Psi^{(k)}_{hPOD}$ where $\Psi^{(k)}_{hPOD} = \text{iRadon}\left(\Phi^{(k)}_{hPOD}\right)$ $j = 1, \ldots, N_s$	$O(n_p^2 N_s K)$

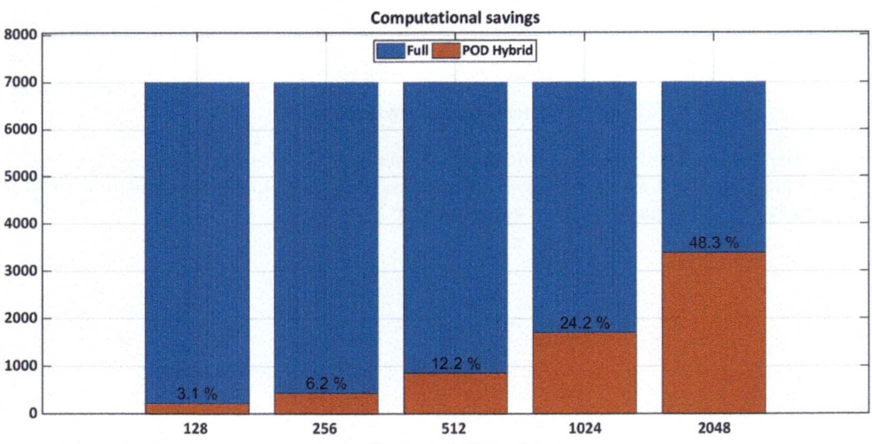

Fig. 13 The computational time for reconstructing the tomograms by using the POD-hybrid approach with number of POD modes $N_{POD} = 128, 256, 512, 1024, 2048$, when compared with the CPU time of the direct reconstruction

5 Conclusion

This work applies POD approximation on tomographic reconstruction to reduce computational time. We first consider the down-sample approach, which can reduce the computational complexity by performing inverse Radon transform on the low-dimensional POD basis, instead of the high-dimensional sinograms. It was shown in Sect. 3.2 that this approach can reduce the reconstruction time by 95% while the RMSE is of order $\mathcal{O}(10^{-3})$. In the case of reconstructing a high-resolution image, we introduce the POD-hybrid approach, which uses some information from down-sample snapshots to construct the weights and the hybrid-POD basis. This approach is shown in Sect. 3.3 to efficiently construct the tomograms with high accuracy, e.g., the hybrid-POD approach with $N_{POD} = 512$ gives 77.8% computational saving with average relative RMSE of order $\mathcal{O}(10^{-2})$. The complexity reduction approaches presented in this work can be readily extended and applied to other general tomographic reconstruction.

Acknowledgements The authors would like to thank the Special Research Center on Optimization and Control at Karl Franzens UniversitVat Graz for their kind hospitality during the 4th Workshop on Model Reduction of Complex Dynamical Systems. Elias S. Helou research was funded by FAPESP grant 2013/07375-0 and CNPq grant 310893/2019-4. Data was acquired at the Brazilian Synchrotron Light Laboratory following proposal IMX-20160215. Charles H. Lee would like to thank Dr. Kari Knutson Miller, former Associate Vice President of International Programs and Global Engagement at California State University Fullerton and Professor Jose Alberto Cuminato at the Institute of Mathematics and Computer Sciences, University of São Paulo for their warm encouragement and strong support for international research collaborations.

References

1. Abbasi, N., Lee, C.H.: Feature extraction techniques on DNA microarray data for cancer detection. In: Proceedings of WACBE World Congress on Bioengineering 2007. Bangkok, Thailand (2007)
2. Andersson, F.: Fast inversion of the Radon transform using log-polar coordinates and partial back-projections. SIAM J. Appl. Math. **65**(3), 818–837 (2005). https://doi.org/10.1137/S0036139903436005, http://link.aip.org/link/?SMM/65/818/1
3. Basu, S., Bresler, Y.: $O(N^2 \log_2 N)$ filtered backprojection reconstruction algorithm for tomography. IEEE Trans. Image Process. **9**(10), 1760–1773 (2000). https://doi.org/10.1109/83.869187
4. Cai, W., Ma, L.: Hyperspectral tomography based on proper orthogonal decomposition as motivated by imaging diagnostics of unsteady reactive flows. Appl. Opt. **49**, 601–610 (2010). https://doi.org/10.1364/AO.49.000601, http://ao.osa.org/abstract.cfm?URI=ao-49-4-601
5. Carnecky, R., Brunner, T., Born, S., Waser, J., Heine, C., Peikert, R.: Vortex detection in 4d MRI data: using the proper orthogonal decomposition for improved noise-robustness. In: EuroVis (Short Papers) (2014)
6. Edholm, P.R., Herman, G.T.: Linograms in image reconstruction from projections. IEEE Trans. Med. Imaging **6**(4), 301–307 (1987). https://doi.org/10.1109/TMI.1987.4307847
7. Edholm, P.R., Herman, G.T., Roberts, D.A.: Image reconstruction from linograms: implementation and evaluation. IEEE Trans. Med. Imaging **7**(3), 239–246 (1988). https://doi.org/10.1109/42.7788

8. George, A., Bresler, Y.: Fast tomographic reconstruction via rotation-based hierarchical backprojection. SIAM J. Appl. Math. **68**(2), 574–597 (2007). https://doi.org/10.1137/060668614
9. Herman, G.T.: Image Reconstruction from Projections: The Fundamentals of Computerized Tomography. Academic, New York (1980)
10. Kak, A.C., Slaney, M.: Principles of Computerized Tomographic Imaging. IEEE Press (1988)
11. Lee, C.A.B., Lee, C.H.: Extended principal orthogonal decomposition method for cancer screening. Int. J. Biosci. Biochem. Bioinform. **2**(2), 136–141 (2012). https://doi.org/10.7763/IJBBB.2012.V2.87
12. Lee, C.H., Tran, K.: Adaptive algorithms for maximizing overall stock return. Decis. Econ. Financ. **33** (2010). https://doi.org/10.1007/s10203-009-0096-5
13. Lee, C.H., Cheung, K.M., Vilnrotter, V.A.: Fast eigen-based signal combining algorithms for large antenna arrays. In: 2003 IEEE Aerospace Conference Proceedings (Cat. No.03TH8652), vol. 2, pp. 1123–1129 (2003). https://doi.org/10.1109/AERO.2003.1235526
14. Lipponen, A., Seppänen, A., Kaipio, J.: Reduced-order model for electrical impedance tomography based on proper orthogonal decomposition (2012). arXiv:1207.0914
15. Lipponen, A., Seppanen, A., Kaipio, J.P.: Electrical impedance tomography imaging with reduced-order model based on proper orthogonal decomposition. J. Electron. Imaging **22**(2), 023008 (2013)
16. Ly, H.V., Tran, H.T.: Modeling and control of physical processes using proper orthogonal decomposition. Math. Comput. Model. **33**(1), 223–236 (2001). https://doi.org/10.1016/S0895-7177(00)00240-5. Computation and Control VI Proceedings of the Sixth Bozeman Conference
17. Ly, H.V., Tran, H.T.: Proper orthogonal decomposition for flow calculations and optimal control in a horizontal CVD reactor. Q. Appl. Math. **60**(4), 631–656 (2002). https://doi.org/10.1090/qam/1939004
18. McGregor, R., Szczerba, D., von Siebenthal, M., Muralidhar, K., Székely, G.: Exploring the use of proper orthogonal decomposition for enhancing blood flow images via computational fluid dynamics. In: Metaxas, D., Axel, L., Fichtinger, G., Székely, G. (eds.) Medical Image Computing and Computer-Assisted Intervention - MICCAI 2008, pp. 782–789. Springer, Berlin (2008)
19. Miqueles, E.X., Koshev, N., Helou, E.S.: A backprojection slice theorem for tomographic reconstruction. IEEE Trans. Image Process. **27**(2), 894–906 (2017). https://doi.org/10.1109/TIP.2017.2766785
20. Natterer, F.: The Mathematics of Computerized Tomography. Wiley, New York (1986)
21. Penna, B., Tillo, T., Magli, E., Olmo, G.: A new low complexity KLT for lossy hyperspectral data compression. In: 2006 IEEE International Symposium on Geoscience and Remote Sensing, pp. 3525–3528 (2006). https://doi.org/10.1109/IGARSS.2006.904
22. Peterson, D., Lee, C.H.: Disease detection technique using the principal orthogonal decomposition on DNA microarray data. In: Proceedings of the 6th Nordic Signal Processing Symposium, 2004. NORSIG 2004, pp. 157–160 (2004)
23. Peterson, D., Lee, C.H.: A DNA-based pattern recognition technique for cancer detection. In: The 26th Annual International Conference of the IEEE Engineering in Medicine and Biology Society, vol. 2, pp. 2956–2959 (2004). https://doi.org/10.1109/IEMBS.2004.1403839
24. Sirovich, L.: Turbulence and the dynamics of coherent structures. i. Coherent structures. Q. Appl. Math. **45**(3), 561–571 (1987)
25. Volkwein, S.: Proper Orthogonal Decomposition: Theory and Reduced-Order Modelling. Lecture Notes, University of Konstanz. http://www.math.uni-konstanz.de/numerik/personen/volkwein/teaching/POD-Book.pdf

Efficient Krylov Subspace Techniques for Model Order Reduction of Automotive Structures in Vibroacoustic Applications

Harikrishnan K. Sreekumar, Rupert Ullmann, Stefan Sicklinger, and Sabine C. Langer

Abstract The paper presents a practical approach to deploy Krylov-based model order reduction techniques for industrial vibroacoustic problems. The numerical analysis of frequency-domain transfer functions provides valuable insights into the model's behavior even at an early stage of product development. Model order reduction is a promising approach to yield faster computations while analyzing large-scale vibroacoustic models, where an expensive evaluation is performed for a significantly large number of frequency points. However, reducing a full-order model including damping mechanisms requires special attention so as to efficiently gain from the reduction process with respect to accuracy and performance. Therefore, one main focus of this contribution is to identify methods and strategies to efficiently reduce vibroacoustic models incorporating different damping mechanisms. The identified and optimized model reduction approach is finally applied to an industrial problem, where a real automotive structure is reduced and evaluated for the coupled system response using a combined substructuring and model reduction scheme. Moreover, through the example, the efficiency of the Krylov-based techniques is demonstrated with respect to localized damping.

Keywords Krylov subspace method · Vibroacoustics · Finite element method · Coupled system response

H. K. Sreekumar (✉) · S. C. Langer
Technische Universität Braunschweig, Institute for Acoustics, Langer Kamp 19, 38106 Braunschweig, Germany
e-mail: hk.sreekumar@tu-braunschweig.de

R. Ullmann
BMW Group, Research, New Technologies, Innovations, Parkring 19, 85748 Garching, Germany

S. Sicklinger
Technische Universität München, Arcisstraße 21, 80333 Munich, Germany

P. Benner et al. (eds.), *Model Reduction of Complex Dynamical Systems*, International Series of Numerical Mathematics 171,
https://doi.org/10.1007/978-3-030-72983-7_12

1 Introduction

Acoustic comfort for passengers is one crucial objective for automotive manufacturers and engineers. Especially for electric and hybrid vehicles, an early assessment of product design has the potential to reduce the amount of structure-borne sound radiated into the passenger cabin. The application of numerical tools for vibroacoustic simulations is a cost-effective option to assist engineers in evaluating necessary acoustic measures. Understanding and quantifying the amount of vibration flowing through a structure provide an insight into the parts that have the potential for design optimization.

The finite element method (FEM) is a popular numerical tool for approximating the mechanical response of structural models with complex design characteristics [1, 2]. With FEM discretization, the propagating structure-borne sound through a structure can be approximated at respective nodes. However, analysis at higher frequencies using FEM yield very fine meshes so as to accurately capture the essential features of a shorter wave. As a result, system matrices of huge dimensions are inevitable to resolve the wave propagation sufficiently. In majority of the cases, high-performance clusters are needed to perform such complex analyses. Therefore, computing becomes highly expensive in terms of solver time as well as memory requirements. Consequently, the application of FEM is limited to available computational resources. Another alternative is to use other suitable numerical tools like the statistical energy analysis for high-frequency ranges [3]. But such a method cannot perform the evaluation for yielding local information, because the method is based on the statistical average of acoustic parameters without spatial discretization. So as to benefit from the powerful FEM capabilities, it is advisable to consider different strategies to reduce computational requirements without compromising on accuracy. Hence, model order reduction (MOR) techniques, where the aim is to reduce the large-scale dimension of a system to significantly lower order with acceptable accuracy, is a promising alternative to enhance computational efficiency.

The MOR techniques have already seen numerous applications and advances in the field of structural dynamics [4–15]. The mathematical background for MOR follows to the book [16] and dissertations [17, 18]. In general, the classical second-order equations of motion are subjected to MOR techniques for a reduction in system dimension. Structure-preserving methods, presented in [19–21], were identified to deliver accurate reduced models without destroying the characteristics of the second-order system. Among the various techniques, the Krylov-based MOR (KMOR) have shown to be efficient when applied to large-scale systems by performing reduction using the structure-preserving second-order Krylov subspaces [15, 22]. In addition, the stability-preserving approaches were developed to address the question of retaining the stability in reduced models. However, stability of the system is not of high importance for analysis in the frequency domain as compared to the analysis in time domain [13]. Unlike the frequency domain analysis that performs solving for some specific frequency points, the time-domain analysis performs the integration of the transient response over time. As a result, for an unstable reduced system, the resulting error in transient response will eventually explode. As the paper demand steady-state dynamic analysis in the frequency domain, less focus is laid on describing the stability of the considered automotive structures.

In the context of vibroacoustics, MOR research has been intensively performed with Krylov subspace methods for instance in [4, 7, 13]. The objective of the contribution is to present a robust Arnoldi KMOR algorithm for the second-order equations of motion with the aim of reducing the large dimensions of the assembly-oriented automotive structures. The focus can be outlined to (a) considering models that include various damping mechanisms, (b) efficient parallelized algorithm that can handle very large scale models where model reduction is performed for MIMO systems with a large number of inputs and outputs, and (c) fast evaluation of coupled system response within a substructuring framework that incorporates the individual reduced-order models (ROM). The following paragraphs brief on the necessity of considering the above-mentioned facts within the paper.

Damping is an important phenomenon in vibroacoustic applications that accounts for the dissipation of vibration energy [23, 24]. Especially at resonance frequencies where the amplitude of vibration is unpredictably high, damping is required to reduce the energy of structure-borne sound. Equivalently, identifying resonant frequencies enable design engineers to incorporate additional damping measures. Hence, with KMOR, reproducing the system dynamics accurately, with respect to resonant modes, in the reduced system is crucial for vibroacoustic applications. One way is to closely analyze the various numerical damping mechanisms and modeling them properly for KMOR. Based on the type of damping model, the KMOR scheme can be optimized for maximum efficiency. The paper discusses in detail the two popular numerical damping models in engineering, namely the Rayleigh damping and the structural damping model. The former was addressed for model reduction in [25]. However, the type of damping model approximates the viscous damping phenomenon to be frequency-independent. On the other hand, the structural damping model considers damping phenomenon to be frequency-dependent. As a result, the structural damping model is most preferred for simulating linear viscoelastic materials. The paper, therefore, investigates the KMOR procedure to efficiently include structural damping mechanisms that can be highly non-proportional in practical applications.

A practical approach to simulate the coupled system response using KMOR techniques is to couple the transfer functions of the comprising subsystems, obtained from the ROMs, in a substructuring framework. As a result, the framework provides flexibility in analysis and faster computations. The method of frequency-based substructuring with Lagrange multipliers in [26, 27] is used in this work. Evaluating coupled system response using substructuring with the classical component mode synthesis (CMS) was presented in [28]. But in the presence of localized damping, the CMS method yields expensive computation and less accurate reduced model. To justify the statement, the paper also presents a comparison of yielded error by deploying the KMOR and CMS method for models in the presence of localized damping.

Finally, the development of an efficient KMOR algorithm with respect to computational time and resources is given equal importance as deriving a reduced model accurately. The paper presents an algorithm developed in C++ that can efficiently handle very large-scale models performing extensive orthogonalization and deflation. Parallelized mathematical routines support the presented KMOR algorithm to efficiently distribute the workload among multiple processors within a shared memory multiprocessor architecture. The performance advantage with model reduction, in comparison to the conventional direct solving, is also presented for the complex automotive assembly.

2 Krylov-Based Model Order Reduction

2.1 *Problem Definition*

The governing equations for vibration analysis are the classical second-order equations of motion obtained from the theory of structural mechanics. The finite element discretization, by deducing the weak form of the governing equations, leads to the assembled system of linear equations, which when represented in the frequency domain can be expressed as

$$\left(-\omega^2\mathbf{M} + i\omega\mathbf{D} + \mathbf{K}\right)\mathbf{z}(\omega) = \mathbf{f}(\omega), \tag{1}$$

where ω is the angular frequency, $\mathbf{M} \in \mathbb{C}^{n\times n}$ is the mass matrix, $\mathbf{D} \in \mathbb{C}^{n\times n}$ is the damping matrix, $\mathbf{K} \in \mathbb{C}^{n\times n}$ is the stiffness matrix, $\mathbf{z} \in \mathbb{C}^n$ is the state or displacement vector, and $\mathbf{f} \in \mathbb{C}^n$ is the load vector.

In the expression above, it is worth noting that the system matrices are considered to span the complex space. There are often cases in structural mechanics where the system matrices belong to the complex space. In this paper, only the stiffness matrix is expected to be complex in the presence of one of the below-mentioned damping model. Therefore, for presenting the general case, the consideration of complex system matrices is continued throughout this contribution.

2.1.1 Damping Models

An undamped model is practically not realizable for evaluating the actual response of a vibrating structure. An overview of various numerical damping models, that enable simulations to include damping, can be found in example [24, 29]. Based on the type of damping model, the KMOR scheme can be optimized for maximum efficiency. The popular damping models in the field of engineering are the

1. Rayleigh damping or the proportional viscous damping and
2. structural damping or hysteric damping.

The Rayleigh damping model is one of the simplest ways to approximate viscous damping, where damping is approximated as frequency-independent. The nature of viscous damping represented as "proportional" means that the damping term is proportional to the stiffness and mass matrices. The same can be expressed as $\mathbf{D} = \alpha\mathbf{M} + \beta\mathbf{K}$, where α and β are proportionality constants. Of the various damping models, Rayleigh damping models can be easily deployed and also accurately predicted using modal techniques as they preserve the normal modes of the undamped case [29]. Or in other words, MOR can be performed for the undamped case and further be extended to the case with Rayleigh damping. Even though the model can be used to understand the damping nature within a system, it cannot be practically used to represent the frequency-dependent damping nature of many industrial problems.

On the other hand, the structural damping model can be used to accurately model the frequency-dependent viscous damping. However, the damping model requires special attention for yielding accurate ROMs due to the facts listed below. With structural damping, the measure of damping in structures is included using a relative measure called the damping loss factor η. The loss factor is proportional to the damping capacity which is the ratio of the dissipated energy to the total energy [24, 30]. Experimental methods to identify the damping loss factor from measured vibration data can be referred to [23, 31]. The damping model mathematically introduces imaginary terms to the stiffness matrix [32], expressed as $\underline{\mathbf{K}} = (1 + i\eta)\mathbf{K} \in \mathbb{C}^{n\times n}$. The imaginary term results in a numerical damping effect within the system by introducing a phase shift between the damping force and the path of vibration. An example is the introduction of a complex Young's modulus $\underline{E} = (1 + i\eta)E$ as the material parameter. For easiness, the complex stiffness matrix is represented further simply as $\mathbf{K}$.

The proportional damping models of both Rayleigh damping and structural damping were considered for model reduction in [33] by exploiting the spectral structure. As a result, an efficient reduction can be performed with real arithmetic saving computational costs in comparison to an equivalent reduction in a complex space. However, to include non-proportional damping, which can be efficiently included with structural damping, complex arithmetic is required. As a non-proportionally damped system is inevitable for practical cases, the structural damping model is preferred in the following discussion over Rayleigh damping or a proportional case of structural damping. Also, the structural damping model is used in the presented industrial example of automotive structures for introducing highly localized damping. In comparison with other viscous damping models that can be efficiently reduced with second-order Krylov subspaces [6, 15, 25], the current paper identifies the structural damping model to yield faster model reduction by using an equivalent first-order subspace. Hence, the main focus of the paper is to deal with the non-proportional structural damping model efficiently for model reduction.

2.2 Reduction Framework

In system theory, the basic linear system of equations in (1) can be rewritten to form the full-order model (FOM) system denoted as Σ with m_1 inputs and m_2 outputs, represented as

$$\Sigma : \begin{cases} (-\omega^2\mathbf{M} + i\omega\mathbf{D} + \mathbf{K})\mathbf{z}(\omega) = \mathbf{G}\mathbf{u}(\omega) \\ \mathbf{y}(\omega) = \mathbf{L}^H\mathbf{z}(\omega) \end{cases} \tag{2}$$

where the superscript H denotes the complex conjugate. The dynamic system excited with an input excitation signal $\mathbf{u}(\omega) \in \mathbb{C}^{m_1}$ results an output system response $\mathbf{y}(\omega) \in \mathbb{C}^{m_2}$. The multiple-input multiple-output (MIMO) configuration can be described with the rectangular matrices $\mathbf{G} \in \mathbb{C}^{n\times m_1}$ and $\mathbf{L} \in \mathbb{C}^{n\times m_2}$, which are the input and output system matrices respectively. Finally, the corresponding transfer function $\mathbf{H} \in \mathbb{C}^{m_2\times m_1}$ for the FOM can be expressed as

$$\mathbf{H}(\omega) = \mathbf{L}^H\left(-\omega^2\mathbf{M} + i\omega\mathbf{D} + \mathbf{K}\right)^{-1}\mathbf{G}. \tag{3}$$

Structure of system matrices: Analysis of structural models yields, in majority cases, symmetric system matrices. Even when the models consist of an unsymmetric connector or distributed coupling elements, the symmetric configuration can be enforced to yield symmetric matrices. However, with the introduction of structural damping, the stiffness matrix becomes complex non-Hermitian symmetric (otherwise termed as complex symmetric) and loses its symmetric structure in the complex space. Or in mathematical terms, structural damping yields stiffness matrix with real part $\Re(\mathbf{K}) = \Re(\mathbf{K})^T$ and imaginary part $\Im(\mathbf{K}) = \Im(\mathbf{K})^T$ where superscript T denotes the real matrix transpose.

The objective of deploying KMOR is to approximate the expensive FOM system, Σ in (2) with corresponding less-expensive ROM system, Σ_R, of smaller dimensions. The statement follows:

$$\Sigma_R : \begin{cases} (-\omega^2\mathbf{M}_R + i\omega\mathbf{D}_R + \mathbf{K}_R)\mathbf{z}_R(\omega) = \mathbf{G}_R\mathbf{u}(\omega) \\ \mathbf{y}(\omega) = \mathbf{L}_R^H\mathbf{z}_R(\omega), \end{cases} \tag{4}$$

where $\mathbf{K}_R, \mathbf{D}_R, \mathbf{M}_R \in \mathbb{C}^{r\times r}$, $\mathbf{z}_R \in \mathbb{C}^r$, $\mathbf{G}_R \in \mathbb{C}^{r\times m_1}$, and $\mathbf{L}_R \in \mathbb{C}^{r\times m_2}$ are the ROMs of the corresponding system matrices with reduced dimension $r \ll n$.

Finally, the transfer function corresponding to the reduced system Σ_R can be expressed as

$$\mathbf{H}_R(\omega) = \mathbf{L}_R^H\left(-\omega^2\mathbf{M}_R + i\omega\mathbf{D}_R + \mathbf{K}_R\right)^{-1}\mathbf{G}_R. \tag{5}$$

With KMOR, the system matrices in the reduced space, in (4), are a result of orthogonal projection using the two projection basis $\mathbf{V}$ and $\mathbf{W}$, such as $[\ ]_R = \mathbf{W}^H[\ ]\mathbf{V}$ for $[\ ] := \{\mathbf{M}, \mathbf{K}, \mathbf{D}\}$, $\mathbf{z} = \mathbf{V}\mathbf{z}_R$, $\mathbf{L}_R = \mathbf{L}^H\mathbf{V}$, and $\mathbf{G}_R = \mathbf{W}^H\mathbf{G}$.

The moment-matching criterion represents the base for approximating FOM within the respective ROM [16–18]. With the interpolatory method of KMOR, the FOM of large-scale dimension is projected into a space of very small order where computations can be performed faster. This is done by projecting the FOM matrices with the two projection bases that are essentially the computed moments. The projection bases are iteratively constructed by computing moments using the Arnoldi algorithm [17]. As a result, with moment matching, the approximated transfer function in the reduced space enables $\Sigma(\omega) \approx \Sigma_R(\omega)$ or $\mathbf{H}(\omega) \approx \mathbf{H}_R(\omega)$ for the desired inputs and outputs.

2.2.1 Deflation and Deflation Tolerance

The convergence of a method is affected by rank deficiency caused by the linearly dependent computed moments. The problem is more pronounced for large-scale systems with many inputs and outputs like the models presented in this paper. One classical approach to deal with the problem is deflation. Other robust methods like the recycling of subspaces presented in [34–37], to accelerate the convergence of a reduction algorithm, are also promising approaches and is a point for future investigations. However, the paper limits the focus to the application of an efficient deflation strategy.

In a block-wise construction of projection bases, deflation can be efficiently performed using the rank-revealing QR (RRQR) decomposition presented in [38]. The rank-revealing $\mathbf{R}$ matrix exposes smaller singular values that help to deflate the corresponding vector entries of the orthogonalized matrix $\mathbf{Q}$. The method is already deployed in [18] for the block-wise framework to compute state-space projection bases. With the presented deflation with RRQR decomposition, singular values corresponding to the linearly dependent columns can be identified and eliminated. As compared to exact arithmetic, in numerical codes linearly dependent columns will be never exactly yielding zero norms. A suitable tolerance σ_{tol} is thereby chosen to safely define the criterion to eliminate the linearly dependent column entries.

The deflation tolerances, used for deflating the redundant moments of the various models presented in this paper, are chosen after performing a sensitivity analysis. The individual models were subjected to different values of deflation tolerance. The study starts from the case of zero deflation which corresponds to a very low deflation tolerance $\sigma_{\text{tol}} = 10^{-25}$. The sensitivity of the RRQR deflation evaluated for increasing σ_{tol}. The sensitivity for the number of deflated moments by performing the placement of repeated or closely spaced expansion points by keeping other KMOR parameters as constant. The method is observed to yield 100% deflation for repeated expansion points. In the second case, expansion points are placed close to each other at constant frequency intervals. An optimal deflation tolerance is then chosen which yields a promising percentage of deflated moments, for instance, 25% for the presented beam example. A similar analysis is performed for the increasing order of Taylor series expansion. The above strategy by observing the sensitivity of deflation has shown to obtain promising values for deflation tolerance. However, the time required to find such an optimal value is not included in the performance comparison. Hence, investigations are proposed for cheap estimation of optimal deflation tolerances.

2.2.2 Choice of Expansion Points

The moments are computed for a small number of interpolating frequency points, also popularly known as the expansion points. The moments computed at the various expansion points are interpolated within the desired frequency range using rational interpolation or multi-point Padé approximation [17]. A suitable choice of expansion points yields accurate and smaller ROM. As the contribution focuses mainly on optimizing KMOR procedures for damping models, a study on various methods to place expansion points is not considered within the actual scope. Hence, the method of Greedy search algorithm, refer to [39], is chosen for the presented examples as a method to place the expansion points based on the error distribution over the entire frequency range.

A discussion on choosing real and complex interpolation points was presented in [17, 40]. In order to avoid confusion from explanations in the cited literature, here the usage of real and imaginary interpolation points are with respect to angular frequency ω in (2); not the Laplace parameter $s = i\omega$. As already mentioned about the requisites for the reduction of vibroacoustic models, resonant modes are to be detected in the respective ROM. Therefore, real interpolations are necessary for yielding local approximation accurately within the desired frequency range [17]. Imaginary interpolation points, which in principle yield broader approximation, require further investigation for vibroacoustic applications. In addition, the choice of imaginary expansion point can also be used sometimes for models with viscous damping to yield moment matching in real space. The work [10] also presents a similar approach. On the other hand, for the structural damping model, this is not possible due to the inherent complex stiffness matrix.

2.2.3 Selection of Krylov Subspace

The reduction of the second-order dynamic system presented in (2) is carried out by performing a structure-preserving approach. As compared to the conventional state-space reduction, structure-preserving approach essentially preserves the second-order characteristics of the governing equation in the ROM [19–21]. According to theory for the general case, the structure-preserved ROMs are achieved by using projection bases spanning the second-order Krylov subspaces expanded for all expansion points:

$$\mathbf{V} \subset \bigcup_{j=1}^{n_{\mathrm{EP}}} \mathcal{K}_{q_1,2}^{\mathrm{input}} \left(-\tilde{\mathbf{K}}_j^{-1} \tilde{\mathbf{D}}_j, -\tilde{\mathbf{K}}_j^{-1} \tilde{\mathbf{M}}_j, -\tilde{\mathbf{K}}_j^{-1} \mathbf{G} \right) \tag{6}$$

$$\mathbf{W} \subset \bigcup_{j=1}^{n_{\mathrm{EP}}} \mathcal{K}_{q_2,2}^{\mathrm{output}} \left(-\tilde{\mathbf{K}}_j^{-H} \tilde{\mathbf{D}}_j^{H}, -\tilde{\mathbf{K}}^{-H} \tilde{\mathbf{M}}_j^{H}, -\tilde{\mathbf{K}}_j^{-H} \mathbf{L} \right) \tag{7}$$

where the subspace matrices at an expansion point ω_j^{EP} take the form:

$$\tilde{\mathbf{K}}_j = -(\omega_j^{\mathrm{EP}})^2\mathbf{M} + i(\omega_j^{\mathrm{EP}})\mathbf{D} + \mathbf{K},\ \tilde{\mathbf{D}}_j = \mathbf{D} + 2i(\omega_j^{\mathrm{EP}})\mathbf{M},\ \tilde{\mathbf{M}}_j = \mathbf{M}. \tag{8}$$

The second-order Krylov subspace spanning the input subspace and output subspace, denoted as $\mathcal{K}_{q_1,2}^{\mathrm{input}}$ and $\mathcal{K}_{q_2,2}^{\mathrm{output}}$, matches q_1 and q_2 number of moments, respectively. For n_{EP} number of expansion points and in case of zero deflation, the projection bases take the form $\mathbf{V} \in \mathbb{C}^{n\times r_1}$ and $\mathbf{W} \in \mathbb{C}^{n\times r_2}$, where $r_1 = m_1 \times q_1 \times n_{\mathrm{EP}}$ and $r_2 = m_2 \times q_2 \times n_{\mathrm{EP}}$ are the reduced dimensions.

In the current paper, the ROMs are achieved by using a single projection basis. Or in other words, the Galerkin projection or the one-sided projection scheme is followed. According to the nature of the numerical damping model, the one-sided projection scheme and the two-sided projection scheme can provide different numbers of matched moments, as discussed below. The former is chosen over the latter, because of (a) the significantly large computational time required to compute the second projection basis—including matrix inversion and deflation, (b) the effort required by following the latter scheme to finally yield square system matrices by controlling the same level of deflation for both the subspaces—this is the case of ensuring if $r_1 = r_2 = r$, and (c) the negligible gain in accuracy when ROMs yielded from the latter scheme are used for applications presented in this paper. While the first disadvantage is straightforward when the corresponding reduction algorithm is considered, the second disadvantage is more pronounced in practical applications considered in this paper which are dealing with models having a large number of inputs and outputs. A detailed explanation of the last disadvantage is provided in the following discussion of the systems with structural damping. Even though the paper considers one-sided projection, the expression for the respective second projection basis is also provided in the case of some damping models. This is to highlight the case of ideal reduction where the two-sided projection scheme can yield full moment matching.

According to the theorems presented in [14, 20], the mapping of FOM to the reduced space using second-order Krylov subspaces matches $q_1 + q_2$ moments of FOM and ROM in the real space. Also it was proven, the consideration of symmetric MIMO configuration, expressed as $\mathbf{G} = \mathbf{L}$, for real symmetric system matrices match $2q$ number of moments ($q_1 = q_2 = q$) delivering ROM matrices with dimension $r \times r$ for $r = m \times q \times n_{\mathrm{EP}}$. In such cases, one-sided projection $\mathbf{V} = \mathbf{W}$ using a single-input Krylov subspace is sufficient for matching the same $2q$ number of moments. However, for vibroacoustic application with structural damping, the moment-matching criterion is to be defined similarly for the complex space. This was proved by Li and Bai [41] for the second-order systems using the structure-preserving approach. An equivalent theorem and its corollary are restated below for moment matching in the complex domain by extending the representation to the second-order Krylov subspaces. Therefore, the proof in [41] applies equally.

Theorem 1 *When the system matrices* $\mathbf{K} \in \mathbb{C}^{n\times n}, \mathbf{D} \in \mathbb{C}^{n\times n}, \mathbf{M} \in \mathbb{C}^{n\times n}$, $\mathbf{G} \in \mathbb{C}^{n\times m_1}$, *and* $\mathbf{L} \in \mathbb{C}^{n\times m_2}$ *are subjected to reduction using second-order Krylov subspaces expressed in* (6) *and* (7), *then* $q_1 + q_2$ *number of moments of FOM transfer function match with corresponding moments of ROM transfer function.*

Corollary 1 *When* $\mathbf{K} = \mathbf{K}^H \in \mathbb{C}^{n\times n}$, $\mathbf{D} = \mathbf{D}^H \in \mathbb{C}^{n\times n}$, $\mathbf{M} = \mathbf{M}^H \in \mathbb{C}^{n\times n}$ *are complex Hermitian and* $\mathbf{L} = \mathbf{G} \in \mathbb{C}^{n\times m}$, *then* $2q$ *number of moments of FOM transfer function match with corresponding moments of ROM transfer function when one-sided projection is performed with second-order Krylov subspace expressed in* (6) *such that* $\mathbf{V} = \mathbf{W}$.

The stated theorem and corollary can be used to describe the moment matching behavior for the various damping models. The different cases of damping models with a discussion on the required subspace and the nature of projection bases are listed below:

1. **Undamped systems**
   For the theoretical undamped systems when the damping matrix $\mathbf{D} = \mathbf{0}$, it has been shown in [6, 15] that the resulting second-order input and output Krylov subspace ($\mathbf{V} \equiv \mathbf{W}$) span an equivalent first-order subspace:

$$\mathbf{V} \subset \bigcup_{j=1}^{n_{\text{EP}}} \mathcal{K}_{q,2}^{\text{input}}\left(-\tilde{\mathbf{K}}_j^{-1}\tilde{\mathbf{D}}, -\tilde{\mathbf{K}}_j^{-1}\tilde{\mathbf{M}}_j, -\tilde{\mathbf{K}}_j^{-1}\mathbf{G}\right) = \bigcup_{j=1}^{n_{\text{EP}}} \mathcal{K}_{q,1}^{\text{input}}\left(-\tilde{\mathbf{K}}_j^{-1}\tilde{\mathbf{M}}, -\tilde{\mathbf{K}}_j^{-1}\mathbf{G}\right) \tag{9}$$

2. **Systems with structural damping**
   As already explained, the structural damping is introduced with the help of complex stiffness matrix. Hence, the discussion on the choice of Krylov subspace is, therefore, an extended discussion of the undamped case with zero viscous damping. Or mathematically, the damping matrix $\mathbf{D} = \mathbf{0}$ retains when structural damping is introduced. As evident from the undamped case, the resulting equivalent first-order subspace is valid for the structural damping model. Therefore, the projection bases:

$$\mathbf{V} \subset \bigcup_{j=1}^{n_{\text{EP}}} \mathcal{K}_{q,1}^{\text{input}}\left(-\tilde{\mathbf{K}}_j^{-1}\tilde{\mathbf{M}}, -\tilde{\mathbf{K}}_j^{-1}\mathbf{G}\right), \text{ and} \tag{10}$$

$$\mathbf{W} \subset \bigcup_{j=1}^{n_{\text{EP}}} \mathcal{K}_{q,1}^{\text{output}}\left(-\tilde{\mathbf{K}}_j^{-H}\tilde{\mathbf{M}}^H, -\tilde{\mathbf{K}}_j^{-H}\mathbf{L}\right) \tag{11}$$

   can be used to reduce the system. A difficulty arises in moment matching using Corollary 1 because the stiffness matrix in the presence of structural damping is complex symmetric (not Hermitian). Hence, an ideal reduction for such a system is to perform two-sided projection as indicated in general Theorem 1 using projection bases expressed in (10) and (11). However, for the application models presented in this paper, one-sided projection is still preferred because (a) the mass matrix is symmetric in all cases, and also (b) as a trade-off between accuracy and performance. Reduction of the application models using the one-sided projection has shown to yield efficient ROMs, even though the entire number of moments

$q_1 + q_2$ according to Theorem 1 have not matched. More discussion about the algorithm and the reduction process follows in the coming sections.
Performing structure-preserving KMOR approach on such a system with structural damping is found to yield ROMs with characteristics of the respective FOM. Or in other words, the reduced stiffness matrix becomes, as expected, unsymmetric due to the lost symmetric characteristics of the respective full-order stiffness matrix (with the introduction of structural damping), whereas the symmetric characteristic of full-order mass matrix remains preserved as Hermitian.
Moreover, analyzing the nature of the stiffness matrix with structural damping, $\Im(\mathbf{K})$ can be either proportional or non-proportional to $\Re(\mathbf{K})$. In both cases, it was found that the imaginary part of the stiffness matrix has to be included for reduction. Unlike the Rayleigh damping model, discussed as the next damping model, the reduced model of the undamped system cannot be reused by multiplying the constants of proportionality. This applies still in the case when the imaginary part is proportional to the stiffness matrix. Therefore, structural damping yields computation of moments in complex domain, i.e., $\mathbf{V} \in \mathbb{C}^{n \times r}$.

3. **Systems with Rayleigh damping**
As already discussed, for the case of Rayleigh damping, the proportionality constants in the damping model are considered as parameters. In such cases, the projection can be performed for the undamped case and the resulting ROM is reused. This can be expressed as $\mathbf{D}_R = \alpha \mathbf{M}_R + \beta \mathbf{K}_R$. The reason being that the system with Rayleigh damping retains the same normal modes of the undamped system. A mathematical proof is presented in [6].
4. **Systems with other damping models**
Though the presented damping models are some among the various other damping models, a general system with viscous damping has to be reduced using the second-order Krylov subspace following the Theorem 1.

The above discussion encourages the introduction of damping into a model with the help of a structural damping model for simulating the frequency-dependent nature of damping, thereby enabling faster computation by using an equivalent first-order subspace. Therefore, the realization of the projection bases for application to vibroacoustics can be performed much faster as compared to the second-order Krylov subspace. Moreover, with a structural damping model, localized damping measures can be included within the reduced model.

3 Numerical Implementation

The paper provides enormous importance to the development of an efficient numerical scheme that can handle the KMOR procedure for reducing the industrial automotive structure. In the previous sections, it was discussed that the structural damping model is suitable to handle non-proportionally damped systems where KMOR is performed by using an equivalent first-order Krylov subspace. However, the introduction of structural damping demands computation of moments in the complex space and cannot be practically avoided.

Table 1 Software specifications

Programming platform	Math library	Sparse solver
C++	Intel® MKL 2018 Special Release	Intel® MKL PARDISO

As a result of the study performed on vibroacoustic models, a block rational Arnoldi first-order Krylov (RAFOK) algorithm is presented as pseudocode in Algorithm 1. The algorithm performs KMOR using a one-sided projection according to (10) so as to deal with a system modeled with structural damping. The algorithm is highly optimized to deliver maximum computational efficiency and can handle large-scale FOMs. Table 1 presents a short overview of the software platform, on which the RAFOK algorithm is deployed. The algorithm is coded in C++ with mathematical routines enabled with threaded Intel® Math Kernel Library (MKL) for BLAS, sparse BLAS, and sparse solver routines [42].

Algorithm 1 Block-RAFOK for reducing systems with structural damping

Input: System matrices $\mathbf{M}, \mathbf{K}$, Starting vectors $\mathbf{G} = \mathbf{L} \in \mathbb{C}^{n \times m}$, Expansion Points $\boldsymbol{\omega}_{\text{EP}}$, Order of Series Expansion q, Deflation Tolerance σ_{tol}
Output: Projection Bases $\mathbf{V}, \mathbf{W} \in \mathbb{C}^{n \times r}$
Function $[\mathbf{V}, \mathbf{W}] =$ `BlockRAFOK(` $\mathbf{M}, \mathbf{K}, \mathbf{G}, \boldsymbol{\omega}_{\text{EP}}, q, \sigma_{\text{tol}}$ `):`
  `// Generation of Projection Basis`
  Initialize $\mathbf{V} = [\ ]$, $n_{\text{EP}} = \text{length}(\boldsymbol{\omega}_{\text{EP}})$
  for $k = 1$ **to** n_{EP} **do** `/* For every expansion point */`
    $\tilde{\mathbf{K}} = -(\omega_{\text{EP}}^{(k)})^2\mathbf{M} + \mathbf{K}$
    $\tilde{\mathbf{V}}_0 = -\tilde{\mathbf{K}}^{-1}\mathbf{G}$ `/* Compute starting Krylov basis (see Eq. 10) */`
    if $k == 1$ **then**
      $\check{\mathbf{V}} =$ `PerformDeflationBlockWiseRRQR(`$\tilde{\mathbf{V}}_0, \sigma_{\text{tol}}$`)` `/* Check for deflation */`
    else
      $\bar{\mathbf{V}} =$ `PerformIterativeGramSchmidt(`$\tilde{\mathbf{V}}_0, \mathbf{V}$`)` `/* Orthogonalize w.r.t already orthogonalized basis */`
      $\check{\mathbf{V}} =$ `PerformDeflationBlockWiseRRQR(`$\bar{\mathbf{V}}, \sigma_{\text{tol}}$`)`
      $\mathbf{V} = [\mathbf{V}\ \check{\mathbf{V}}]$ `/* Append Krylov modes */`
    end
    for $i = 2$ **to** q **do** `/* For every series expansion */`
      $\tilde{\mathbf{V}} = -\tilde{\mathbf{K}}^{-1}\mathbf{M}\check{\mathbf{V}}$ `/* Compute new Krylov basis (see Eq. 10) */`
      $\bar{\mathbf{V}} =$ `PerformIterativeGramSchmidt(`$\tilde{\mathbf{V}}, \mathbf{V}$`)`
      $\check{\mathbf{V}} =$ `PerformDeflationBlockWiseRRQR(`$\bar{\mathbf{V}}, \sigma_{\text{tol}}$`)`
      $\mathbf{V} = [\mathbf{V}\ \check{\mathbf{V}}]$ `/* Append Krylov modes */`
    end
  end
  $\mathbf{W} = \mathbf{V}$
end

The algorithm represents an iterative block Arnoldi algorithm where the moments are computed iteratively as an entire block to finally yield the orthogonal projection basis. For MIMO systems, there is a high chance that the computed moments in every iteration might overlap or linearly dependent on the calculated blocks. The overlapping effect is highly pronounced with increasing system dimensions. Therefore, an efficient orthogonalization procedure with extensive deflation is required to identify and deflate the redundant moments. However, orthogonalization is not an essential condition for moment matching or KMOR. They are required from a numerical point of view in order to gain faster convergence and to avoid rank deficiency in finite precision [17]. In the algorithm, a classical Gram-Schmidt orthogonalization is performed for the whole computed block by incorporating reorthogonalization [43] so as to compensate for the phenomenon of loss of orthogonality with the proposed Arnoldi algorithm. Reorthogonalization is performed for the computed moments until when the length of the orthogonalized basis or respective norm is accurate to its predecessor or orthogonalized moments in the previous iteration. According to [43], the algorithm is more likely to terminate successfully.

The most expensive procedure in RAFOK occurs at factorization Steps 5 and 14. But for each expansion point, the factorization in Step 5 is saved and therefore reused in Step 14. Consideration of structural damping yield factorization to be performed on complex non-Hermitian symmetric matrix. The solver settings to *complex symmetric* yield relatively faster computation with reduced memory requirement, when compared to full representation. The block-wise orthogonalization and deflation steps are also found to be computationally fast as compared to their vector-wise alternative. The linear dependency of computed moments is analyzed for the presented block-wise implementation by performing an RRQR decomposition at Steps 10 and 16. A suitable tolerance σ_{tol} is chosen according to the approach presented in Sect. 2.2.1.

4 Results

The work presents the outcome of model reduction using the presented RAFOK algorithm on two different examples. The first example represents a generic model of a beam discretized using FEM. However, in the field of engineering, the question of applicability of model reduction for very large-scale practical applications is often raised. Therefore, a second model is presented, which represents a real-world automotive model of the BMW i8 rear axle system with high modeling complexity. The models are subjected to steady-state dynamic analysis in the frequency domain for the frequency range between 20 and 1000 Hz. The motivation to choose the specified minimum frequency is the acoustic hearing threshold of humans, whereas the maximum frequency limit is arbitrarily chosen and can be less or high depending on the characteristics of the analyzed mechanical system; for example, the occurrence of important resonances within the frequency range.

The discussion of each application example also includes performance comparison of reduction algorithm with the commercial ABAQUS™ direct solver for the

Table 2 Hardware specifications

Processor	# cores	Base frequency	RAM	Platform
CPU 1: 2 × Intel® Xeon® Silver 4114[a]	20	2.20 GHz	32 GB	Windows 10, x86-64
CPU 2: 2 × Intel® Xeon® Gold 5122[b]	8	3.60 GHz	64 GB	RedHat OS 6, x86-64

[a]Processor used for RAFOK execution and majority of ABAQUS™ direct solve
[b]High-end machine to perform ABAQUS™ direct solve for large-scale rear axle carrier substructure (with 1050897 DoF)

evaluation of all transfer functions (with respect to all possible permutations of considered inputs and outputs). The wall-clock for the KMOR procedure is categorized into the classical offline and online phase [39]. The offline phase includes the KMOR procedure where the presented Krylov subspace is computed so as to yield the respective projection bases and ROMs. Thus, the algorithm receives the FOMs which are already pre-computed using standard FEM software. The wall-clock for the offline phase, therefore, excludes the generation of system matrices. On the other hand, the online phase conducts the frequency-domain analysis using the generated ROMs, see (4). This phase executes the inversion of the assembled matrix using dense solvers for solving the system of linear equations.

An overview of the two types of hardware configurations used for analyses in this contribution can be seen in Table 2. The hardware used for some direct solving with ABAQUS™ is different from that of the hardware used for KMOR execution due to memory restrictions to handle the large-scale model. However, the performance comparison using two different hardware configurations can still indicate significant time savings on fair grounds.

For determining the accuracy of the transfer function $\mathbf{H}_R(\omega)$ in (5) evaluated from ROMs, the original transfer function $\mathbf{H}(\omega)$ in (3) obtained from FOM is used. To measure the overall error of all transfer functions over frequency, a relative error measure can be defined as the maximum relative error of all transfer functions at each frequency represented as:

$$\boldsymbol{\epsilon}_{\text{rel,max}}(\omega) = \max \frac{|\mathbf{H}(\omega) - \mathbf{H}_R(\omega)|}{|\mathbf{H}(\omega)|}, \tag{12}$$

where $\boldsymbol{\epsilon}_{\text{rel,max}}(\omega)$ forms a vector with length equal the number of frequency points. Finally, the L^2 norm can be computed to yield the overall error norm over all frequencies expressed as:

$$\epsilon_{\text{rel,max}} = \left\| \boldsymbol{\epsilon}_{\text{rel,max}}(\omega) \right\|_{L^2}. \tag{13}$$

The maximum relative error measure (12) is used to determine the final accuracy of ROMs within this contribution. However, they cannot be practically realized within an iterative reduction framework so as to evaluate the accuracy of yielded ROM. The reason being the computation of $\mathbf{H}(\omega)$ for the above error estimate is equally expensive as monolithic solving of the respective FOM. As the paper directs the scope toward obtaining an accurate ROM, consideration of cheaper error estimates is not included in the current investigation. Alternate error estimates for MOR are discussed more in [13, 40]. Also, it is crucial to note that the time required for evaluating error measures using the original transfer function is not included within the KMOR wall-clock. This is to enable a better comparison of the efficiency of KMOR approaches for different damping scenarios independent from the type of error estimate.

4.1 Generic System

A generic model of a rectangular beam, shown in Fig. 1, serves as a benchmark model. Discretization is performed with FEM, where the individual elements are assigned to three-dimensional continuum elements with isotropic material properties for steel. The beam is subjected to a steady-state dynamic analysis in the desired frequency range between 20 and 1000 Hz performed for 981 frequency steps. A discretized model respecting proper wave resolution yields a FOM dimension of 20412 degrees of freedom (DoFs). Structural damping is enforced with a constant damping loss factor $\eta = 0.001$. A symmetric MIMO system is formed by including two random nodes with translational DoFs at the two ends of the beam, illustrated in Fig. 1. That means a total of 6 DoFs are considered to be the inputs and outputs defining the transfer function.

The RAFOK algorithm is performed on the model using the parameters described in Table 3. An optimal value for expansion order q and deflation tolerance σ_{tol} is chosen by performing a convergence study. The final ROM yields a dimension of 72 DoFs in the computed Krylov subspace. The accuracy of the resulting transfer function is plotted over frequency in Fig. 3. It is evident that the behavior of the system can be accurately reproduced from ROM for all 981 frequency steps. In order

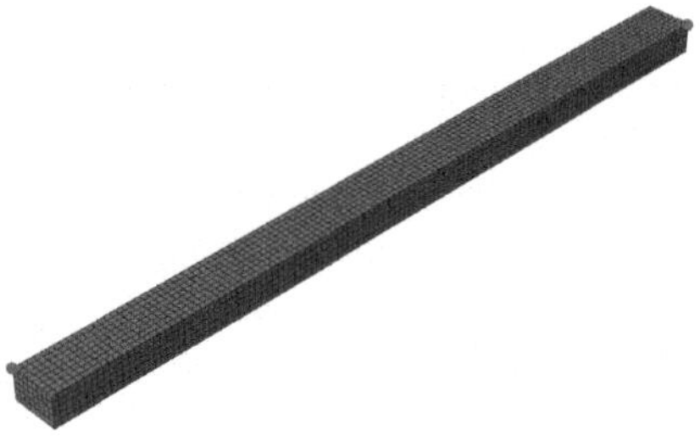

Fig. 1 Generic FE model of simple beam

Table 3 Reduction parameters for the beam model

n	r	m	ω_{EP} (in Hz)	q	σ_{tol}	Overall error norm[a]
20412	72	6	{500 268 970}	4	10^{-11}	3×10^{-9}

[a]Norm of all maximum relative error over frequency from (13)

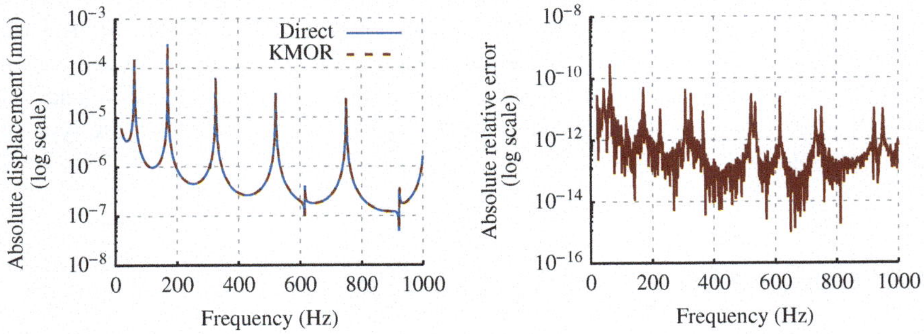

Fig. 2 Plot of a random transfer function $\mathbf{H}^{(3,4)}$ and corresponding relative error

Table 4 Performance comparison for the beam model

Case	Wall-clock time (in minutes)
ABAQUS™ direct solve	13.8[a]
KMOR offline phase	0.04[a]
KMOR online phase	0.002[b]

[a]15 threads on CPU 1
[b]Single thread on CPU 1

to visualize the model complexity which is less in this generic model (as compared to the automotive structure), a random entry of the transfer function $\mathbf{H}^{(3,4)}$ along with the corresponding relative error is plotted in Fig. 2. The expression $\mathbf{H}^{(3,4)}$ denotes the transfer function corresponding to the fourth input and third output signals. As the chosen optimal expansion points and order of series expansion yielded a full ROM dimension, the moments are within the limit of deflation tolerance. Or in other words, no moments were deflated for the optimal KMOR parameter setting in Table 3. The algorithm terminated at this point where sufficient accuracy has been achieved. However, a further iteration with higher tolerance would have yielded linearly dependent moments, whose contribution toward an accurate ROM is very negligible.

In terms of performance, the various clock times are tabulated in Table 4 with respective hardware configurations and the number of CPU threads used for the corresponding execution. A 99% savings in time is evident by comparing KMOR offline phase with the respective direct solve. As expected, the online phase evaluating the system response using yielded ROMs account for a very negligible CPU time.

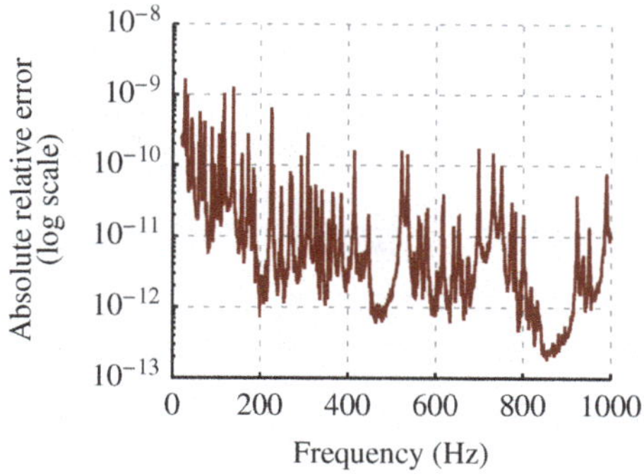

Fig. 3 Overall relative error for all transfer functions

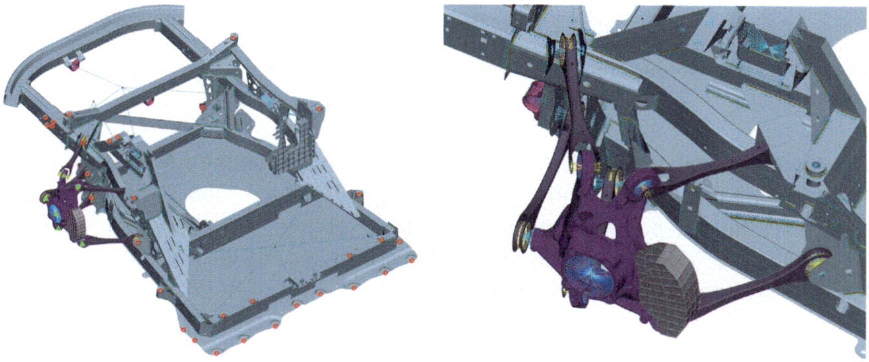

Fig. 4 Left: Complete view of the BMW i8 rear axle assembly with symbols △ and ∘ representing the input/output locations of the wheel carrier and the rear axle carrier, respectively. Right: Magnified view

4.2 Coupled System

The rear axle assembly of the BMW i8 automobile is chosen as the practical example for studying the behavior of systems reproduced by the deployment of their individual ROMs. The model has been chosen to represent a vibroacoustic problem so as to simulate the vibrations that are being transmitted from the wheel carrier to the rear axle carrier. Different views on the FE model of the rear axle assembly are shown in Fig. 4. An equivalent network representation of the whole rear axle assembly with the coupled individual parts has been illustrated in Fig. 5.

The assembly represents various parts or subsystems with localized damping measures. Among the various subsystems in the assembly, the rear axle carrier is considered to be challenging for MOR due to its modeling complexity. Especially the rear axle carrier includes (a) springs connected to lumped masses, (b) various material models with structural damping, and (c) element types assigned to plate, beam, solid, connector, and distributed coupling elements. Moreover, the part demands

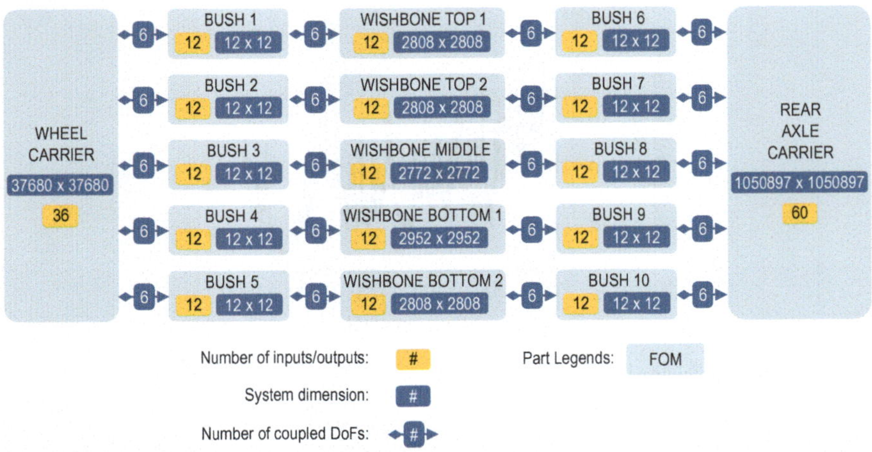

Fig. 5 Monolithic substructuring model of BMW i8 rear axle assembly

high computational power due to its large dimension accounting for approximately one million DoFs. Consequently, CPU 2 configuration (see Table 2) having higher RAM is required to perform conventional direct solving. Therefore, more focus is given to the rear axle carrier to evaluate the performance of the presented RAFOK algorithm with respect to computation time and accuracy of ROM.

The network representation describes the coupling connection between the various subsystems that comprise the whole assembly. The Lagrange multiplier frequency-based substructuring formulation from [26, 27] was deployed to couple the different subsystems of the rear axle assembly. The framework couples the individual transfer functions at the coupling interface between the respective sub-parts so as to determine the combined transfer function of the whole assembly. With KMOR in a substructuring framework, the idea is to replace the individual subsystem's original transfer functions with approximated transfer functions evaluated from their respective ROMs. In the network representation in Fig. 6, the models that have undergone KMOR with RAFOK algorithm are highlighted with their optimal ROM dimension. The bushes are modeled with simple spring and damper elements which do not require a reduction in model order. A summary of respective KMOR parameters and resulting ROM accuracy, in terms of overall error norm, is presented in Table 5 for all substructures of the rear axle assembly.

An efficient deflation procedure is necessary to maintain the stability and accuracy of the yielded ROM. In Table 5, the expected dimension of reduced model r_{exp} is compared to the yielded reduced dimension r of ROM after performing a successful deflation. As already mentioned, the rear axle carrier subsystem is highly challenging for MOR procedure. A random entry of transfer functions is plotted for the rear axle in Fig. 7 along with the corresponding relative error over frequency. From the figure, the modal complexity with respect to dynamics is evident. The accuracy of all the computed transfer functions from their ROMs is presented in Fig. 8. But it can be

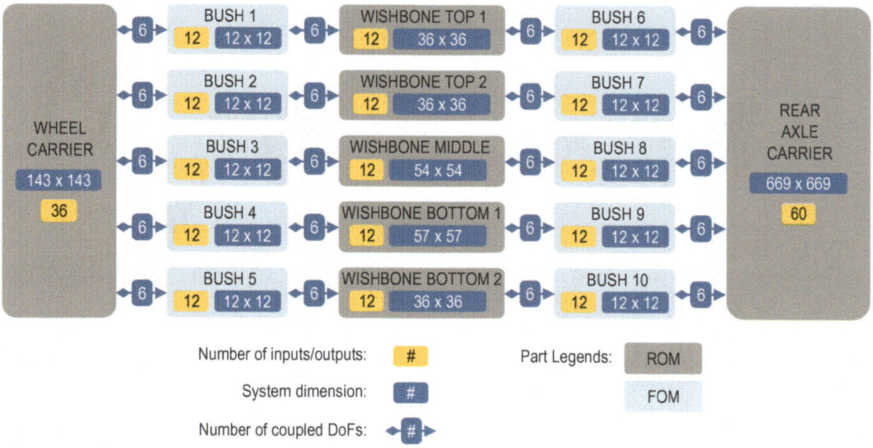

Fig. 6 Hybrid KMOR-substructuring model of BMW i8 rear axle assembly

Table 5 Summary of reduction parameters for different BMW i8 rear axle assembly subsystems

Part	n	r_{exp}	r	r_{CMS} [a]	m	ω_{EP} (in Hz)	q	σ_{tol}	Error norm[b]
Rear axle carrier	1050897	720	669	928	60	{500 129 763 950}	3	10^{-9}	5×10^{-4}
Wheel carrier	37680	288	143	108	36	{500 843}	4	10^{-11}	9×10^{-5}
Wishbone: Top 1	2808	36	36	42	12	{500}	3	10^{-11}	7×10^{-5}
Wishbone: Top 2	2808	36	36	59	12	{500}	3	10^{-11}	6×10^{-4}
Wishbone: Middle	2772	72	54	51	12	{500 44}	3	10^{-11}	5×10^{-5}
Wishbone: Bottom 1	2952	72	57	54	12	{500 37}	3	10^{-11}	1×10^{-5}
Wishbone: Bottom 2	2808	36	36	42	12	{500}	3	10^{-11}	2×10^{-5}

[a]Dimension of ROMs yielded from Craig-Bampton CMS reduction
[b]Overall error norm of all maximum relative error over frequency from (13)

noted that a saturation level has reached for the higher frequency range. A possible solution to reduce the error level even further would be to use the expensive two-sided projection. However, an acceptable accuracy level is achieved for the entire frequency range of interest. In Table 6, the corresponding performance of the KMOR algorithm, compared with the monolithic solving routine, is recorded for the rear axle carrier. Similar plots corresponding to other substructures are omitted in this paper for brevity.

Finally, the coupled system response in terms of power flow through the whole system can be evaluated by computing the nodal force $\mathbf{f}$ and nodal velocity $\mathbf{v}$ at an

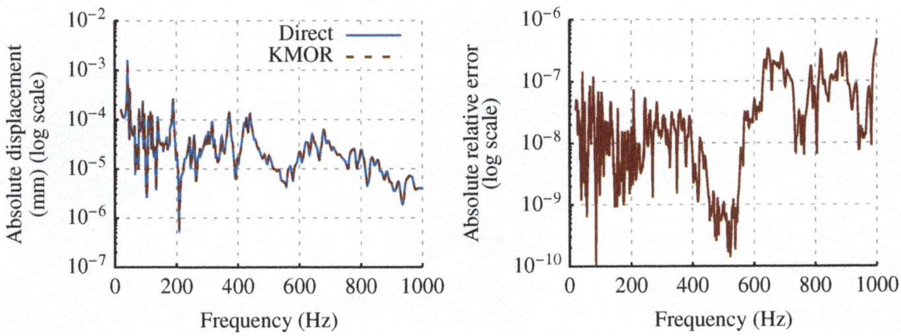

Fig. 7 Plot of a random transfer function of rear axle carrier $\mathbf{H}^{(25,27)}$ and corresponding relative error

Table 6 Performance comparison for the rear axle carrier model

Case	Wall-clock time (in minutes)
ABAQUS™ direct solve	561[a]
KMOR offline phase	13.16[b]
KMOR online phase	0.375[c]

[a] 8 threads on CPU 2
[b] 15 threads on CPU 1
[c] Single thread on CPU 1

arbitrary cutting plane of the model, expressed in [44, 45]:

$$\bar{\mathbf{P}} = \frac{1}{2}\Re\left(\mathbf{f}^H.\mathbf{v}\right). \tag{14}$$

The computed power flow facilitates suitable adaptation to design parameters so as to minimize power flow through the structure. In this paper, the power flow into the rear axle carrier for an excitation at the wheel carrier is presented, refer Fig. 9. Also, the corresponding error measure plotted over frequency is plotted. An additional comparison is made for the resulting power flow with a conventional Craig-Bampton CMS reduction feature of ABAQUS™. The number of modes included for reduction is suitably chosen to finally yield ROMs of comparable dimensions (similar to dimension obtained with RAFOK). The corresponding dimensions of the ROMs obtained from the CMS procedure, denoted as r_{CMS}, are presented in Table 5. As expected for models with non-proportional damping, the CMS technique yield less accurate ROMs for the rear axle assembly with highly localized damping. It is observable that the KMOR approach yields a numerical response with an error significantly smaller than the CMS case. Hence, it is evident that KMOR performs efficiently as compared to CMS for systems with localized damping.

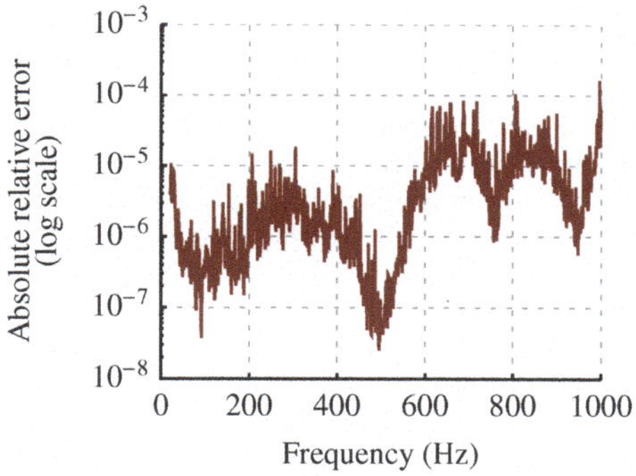

Fig. 8 Overall relative error norm of rear axle carrier for all transfer functions

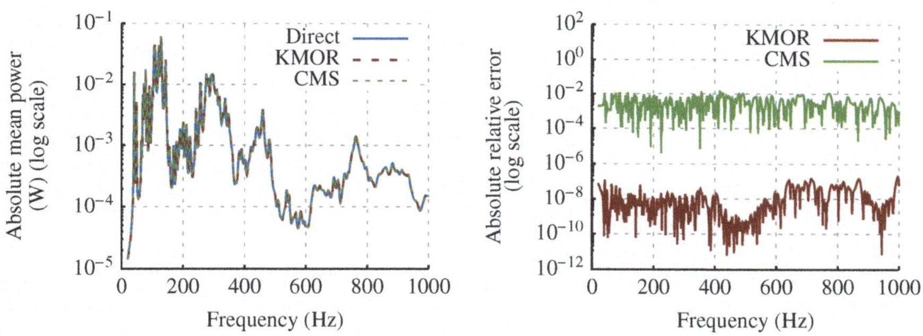

Fig. 9 Plot of mean power flow into the rear axle carrier and corresponding relative error

5 Conclusions and Remarks

In summary, the contribution deals with an efficient way to perform KMOR for large-scale MIMO models, shown up to one million DoFs. An efficient block RAFOK algorithm with the objective of maximum utilization of computational resources and approaches to optimize the algorithm for vibroacoustic applications with structural damping was presented. Finally, the developed algorithm is deployed to reduce a real automotive rear axle assembly incorporating highly localized damping measures. The RRQR deflation strategy has shown to work efficiently to deliver accurate reduced models. Furthermore, the comparison of coupled system response obtained from the reduced models in terms of power flow justifies the KMOR algorithm as a suitable model reduction technique for vibroacoustic models with non-proportional damping.

The scope of the work has considered structural models with a final aim of reducing the presented real automotive structure efficiently for vibration analysis. The outcome serves as a base for further extension to other vibroacoustic problems with fluid-structure coupling. Also, the development of other MOR aspects like the parametric MOR, adaptive methods, efficient error estimators, and deflation tolerances, for the current framework is open for future research.

References

1. Hughes, T.J.R.: The Finite Element Method: Linear Static and Dynamic Finite Element Analysis, 1st edn. Dover Publications Inc., New York (2000)
2. Johnson, C.: Numerical solution of partial differential equations by the finite element method. Acta Appl. Math. **18**, 184–186 (1990). https://doi.org/10.1007/BF0004656
3. Atalla, N., Sgard, F.: Finite Element and Boundary Methods in Structural Acoustics and Vibration. CRC Press, Boca Raton (2015)
4. Davidsson, P.: Structure-Acoustic Analysis: Finite Element Modelling and Reduction Methods. Lund University, Dissertation (2004)
5. Lampe, J., Voss, H.: Second order Arnoldi reduction: application to some engineering problems. http://tubdok.tub.tuhh.de/handle/11420/58 (2005).https://doi.org/10.15480/882.56
6. Eid, R., Salimbahrami, B., Lohmann, B., Rudnyi, E.B., Korvink, J.G.: Parametric order reduction of proportionally damped second-order systems. Sens. Mater. **19**, 149–164 (2007)
7. Srinivasan Puri, R., Morrey, D., Bell, A.J., Durodola, J.F., Rudnyi, E.B., Korvink, J.G.: Reduced order fully coupled structural-acoustic analysis via implicit moment matching. Appl. Math. Model. **33**(11), 4097–4119 (2009). https://doi.org/10.1016/j.apm.2009.02.016
8. Bonin, T., Faßbender, H., Soppa, A., Zaeh, M.: A global Arnoldi method for the model reduction of second-order structural dynamical systems. Linear Algebr. Appl. (2010)
9. Faßbender, H., Soppa, A.: Machine tool simulation based on reduced order FE models. Math. Comput. Simul. **82**(3), 404–413 (2011). https://doi.org/10.1016/j.matcom.2010.10.020
10. Bernstein, D.: Entwurf einer fehlerüberwachten Modellreduktion basierend auf Krylov-Unterraumverfahren und Anwendung auf ein strukturmechanisches Modell. Diploma thesis, Technical University of Dresden (2014)
11. Lein, C., Beitelschmidt, M., Bernstein, D.: Improvement of Krylov-Subspace-Reduced Models by Iterative Mode-Truncation. IFAC-PapersOnLine, vol. 48, issue 1, pp. 178–183 (2015). ISSN 2405-8963
12. Sanchez, R.R., Buchschmid, M., Müller, G.: Model order reduction in structural dynamics. In: Proceedings ECCOMAS Congress (2016)
13. Van de Walle, A.: The power of model order reduction in vibroacoustics. KU Leuven, Dissertation (2018)
14. Lin, Y., Bao, L., Wei, Y.: Model-order reduction of large-scale second-order MIMO dynamical systems via a block second-order Arnoldi method. Int. J. Comput. Math. **84**(7), 1003–1019 (2007). https://doi.org/10.1080/00207160701253836
15. Koustsovasilis, P.: Model Order Reduction in Structural Mechanics. Coupling the Rigid and Elastic Multi Body Dynamics. Technical University of Dresden, Dissertation (2009)
16. Antoulas, A.C.: Approximation of large-scale dynamical systems: society for industrial and applied mathematics. (2005). https://doi.org/10.1137/1.9780898718713
17. Grimme, E.J.: Krylov Projection Methods for Model Reduction. University of Illinois at Urbana-Champaign, Dissertation (1997)
18. Gugercin, S.: Projection Methods for Model Reduction of Large-Scale Dynamical Systems. Rice University, Dissertation (2003)
19. Bai, Z., Su, Y.: Dimension reduction of large-scale second-order dynamical systems via a second-order Arnoldi method. SIAM J. Sci. Comput. **26**(5), 1692–1709 (2005). https://doi.org/10.1137/040605552
20. Salimbahrami, B.: Structure Preserving Order Reduction of Large Scale Second Order Models. Technical University of Munich, Dissertation (2005)
21. Salimbahrami, B., Lohmann, B.: Order reduction of large scale second-order systems using Krylov subspace methods. Linear Algebr. Appl. **415**(2–3), 385–405 (2006). https://doi.org/10.1016/j.laa.2004.12.013

22. Wolf, T., Panzer, H., Lohmann, B.: Gramian-based error bound in model reduction by Krylov subspace methods. In: IFAC Proceedings, vol. 44, no. 1, pp. 3587–3592 (2011). https://doi.org/10.3182/20110828-6-IT-1002.02809
23. Cremer, L., Heckl, M.: Structure-Borne Sound: Structural Vibrations and Sound Radiation at Audio Frequencies, 2nd edn. Springer, Berlin (1988). https://doi.org/10.1007/978-3-662-10121-6
24. Kollmann, F.G., Schösser, T.F., Angert, R.: Praktische Maschinenakustik. Springer, Berlin (2006)
25. Salimbahrami, B., Eid, R., Lohmann, B.: Model Reduction by Second Order Krylov Subspaces: Extensions, Stability and Proportional Damping. IEEE, New Jersey (2006)
26. de Klerk, D., Rixen, D.J.: The frequency based substructuring (FBS) method reformulated according to the dual domain decomposition method. In: A Conference & Exposition on Structural Dynamics. Editor/sn St. Louis, Missouri: IMAC, pp. 1–14 (2006)
27. de Klerk, D., Rixen, D.J., Voormeeren, S.N.: General framework for dynamic substructuring: history, review and classification of techniques. AIAA J. **46**(5), 1169–1181 (2008). https://doi.org/10.2514/1.33274
28. Bampton, M.C.C., Craig, J.R.R.R.: Coupling of substructures for dynamic analyses. AIAA J. **6**(7), 1313–1319 (1968). https://doi.org/10.2514/3.4741
29. Adhikari, S.: Damping Models for Structural Vibration. University of Cambridge, Dissertation (2000)
30. Vér, I.L., Beranek, L.L.: Noise and Vibration Control Engineering: Principles and Applications, 2nd edn. Wiley, Hoboken (2006)
31. Carfagni, M., Lenzi, E., Pierin, M.: The loss factor as a measure of mechanical damping. In: Proceedings of SPIE – the International Society for Optical Engineering, pp. 580–584 (1998)
32. Myklestad, N.O.: The concept of complex damping. J. Appl. Mech. **19**, 284 (1952)
33. Meerbergen, K.: Fast frequency response computation for Rayleigh damping. Int. J. Numer. Methods Eng. **73**(1), 96–106 (2008). https://doi.org/10.1002/nme.2058
34. Parks, M.L., Sturler, E., de Mackey, G., Johnson, D.D., Maiti, S.: Recycling Krylov subspaces for sequences of linear systems. SIAM J. Sci. Comput. **28**(5), 1651–1674 (2006). https://doi.org/10.1137/040607277
35. Ahuja, K., de Sturler, E., Gugercin, S., Chang, E.R.: Recycling BiCG with an application to model reduction. SIAM J. Sci. Comput. **34**(4), A1925–A1949 (2012). https://doi.org/10.1137/100801500
36. Feng, L., Benner, P., Korvink, J.G.: Subspace recycling accelerates the parametric macromodeling of MEMS. Int. J. Numer. Methods Eng. **94**, 84–110 (2013). https://doi.org/10.1002/nme.4449
37. Ahuja, K., Benner, P., de Sturler, E., Feng, L.: Recycling BiCGSTAB with an application to parametric model order reduction. SIAM J. Sci. Comput. **37**(5), S429–S446 (2015). https://doi.org/10.1137/140972433
38. Chan, T.F.: Rank revealing QR factorizations. Linear Algebr. Appl. **88–89**, 67–82 (1987). ISSN 0024-3795
39. Benner, P., Gugercin, S., Willcox, K.: A survey of projection-based model reduction methods for parametric dynamical systems. SIAM Rev. **57**(4), 483–531 (2015). https://doi.org/10.1137/130932715
40. Panzer, H.K.F.: Model Order Reduction by Krylov Subspace Methods with Global Error Bounds and Automatic Choice of Parameters. Technical University of Munich, Dissertation (2014)
41. Li, R., Bai, Z.: Structure-preserving model reduction using a Krylov subspace projection formulation. Commun. Math. Sci. **3**(2), 179–199 (2005)
42. Intel Corporation: Intel(R) Math Kernel Library Reference Manual (2020)
43. Daniel, J.W., Gragg, W.B., Kaufman, L., Stewart, G.W.: Reorthogonalization and stable algorithms for updating the Gram-Schmidt QR factorization. Math. Comput. **30**(136), 772 (1976). https://doi.org/10.1090/S0025-5718-1976-0431641-8

44. Ullmann, R., Sicklinger, S., Buchschmid, M., Müller, G.: Power-based approach for assessment of structure-borne sound in mechanical networks of vehicle structures. Procedia Eng. **199**, 1386–1391 (2017). https://doi.org/10.1016/j.proeng.2017.09.371
45. Sicklinger, S., Ullmann, R.: Structural power as an acoustic design criteria for the early phase of product design. In: Conference proceedings: International Conference on Noise and Vibration Engineering (2018)

Model-Based Adaptive MOR Framework for Unsteady Flows Around Lifting Bodies

Gaetano Pascarella and Marco Fossati

Abstract The problem of performing accurate reconstructions of vortex-dominated unsteady flows by means of reduced basis methods is studied. When faced with the necessity of reconstructing the flow field over a specified time window, a method that aims at automatically and adaptively selecting the most accurate reduction technique among a collection of models is presented. The rationale behind the development of such an adaptive framework is to try to cope with the potential loss of important dynamic information that accompanies classical methods, e.g., proper orthogonal decomposition, where snapshots are treated as statistically independent observation of the dynamical system at study. The adaptive framework will be assessed with respect to two different ways of estimating the reconstruction error by the various methods. One method, referred to as direct error, will employ additional snapshots and will compare explicitly the reduced solution with the reference data. The second method will instead consider a finite volume discretization of the equations and evaluate the error in terms of the unsteady residual of the reduced solution. A backward differencing formula will be used to ensure second-order accuracy in the estimation of the residual. Emphasis will be put on the comparative assessment of the two error estimation methods with respect to the identification of the most suitable reduced method to be used for the reconstruction at a specific instant of time. Problems of relevance to aircraft aerodynamics will be considered such as the impulsive start of 2D airfoils in high-lift configurations.

1 Introduction

The study of the unsteady aerodynamics of lifting surfaces is a central problem in fluid mechanics. The vortices that are generated by lifting bodies are linked to the

---

G. Pascarella · M. Fossati (✉)
Aerospace Centre of Excellence, University of Strathclyde, 75 Montrose street, Glasgow G1 1XJ, UK
e-mail: marco.fossati@strath.ac.uk

G. Pascarella
e-mail: gaetano.pascarella@strath.ac.uk

P. Benner et al. (eds.), *Model Reduction of Complex Dynamical Systems*, International Series of Numerical Mathematics 171,
https://doi.org/10.1007/978-3-030-72983-7_13

efficiency of the body in generating lift and their study and understanding are at the basis of the design of next-generation aircraft. Reduced-order models (ROMs) have been recently developed to allow for accurate evaluations of aerodynamic performance during parametric studies or design processes without requiring heavy computational costs such as high-fidelity computational fluid dynamics simulations. Comprehensive reviews of applications of model reduction to fluid dynamics problems, with a focus also on modal analysis and feature extraction, can be found at [1–3] and references therein. Different methods have been proposed and adopted in the literature to deal specifically with unsteady problems, considering both intrusive and non-intrusive approaches, the difference between the two lying in the use of the governing equations underlying the physical phenomenon under consideration. Hereafter, non-intrusive and equation-free will be used interchangeably. In addition to the classical and widely used proper orthogonal decomposition (POD) [4–6], mostly used in an intrusive manner [7–9], methods such as spectral POD [10–12], dynamic mode decomposition (DMD) [13, 14], and recursive DMD [15] have been introduced in the attempt to obtain models capable to take into account the temporal dynamics of the flow field that is not always accurately represented by the classical POD [16, 17]. Even if these methods have been conceived in the effort to extract pure dynamic information and more exact coherent structures from fluid flows, they are eligible for reduced-order modeling, since they still allow to describe the flow dynamics as a combination of the evolution of a few flow primitives [18, 19]. They can be used in principle in both an intrusive and non-intrusive manner. A recent work [20, 21] has been presented in the literature where all these methods are combined together, in a non-intrusive manner, to define an adaptive approach capable to automatically select the best-in-class method to realize as accurate as possible reconstructions of the unsteady flows. The adaptive ROM has proven to be effective in providing a fairly accurate reconstruction of complex unsteady flows with an accuracy that has shown to be superior to an approach using only pure POD [21]. In the present work, this adaptive method is revisited on the basis of a comparative study of two different approaches to the computation of the reconstruction error. A non-intrusive and an intrusive formulation of the error are presented and compared with respect to their ability in providing an estimation of the reconstruction error and ultimately in their performance in selecting the most appropriate ROMs over a specific time window. A critical analysis is presented that will take into account the requirements of the different error formulae and the trade-off between their computational cost and accuracy in the reconstruction. Section 2 introduces the class of methods that will be considered in the adaptive approach, Sect. 3 will briefly outline the adaptive framework while Sects. 3.2 and 3.3 will discuss the error formulations and their sensitivities. Eventually, Sect. 4 will show the performance of the different methods with respect to two unsteady flows over two airfoils.

2 Linear Reduced Basis Methods

A class of reduced basis methods is considered for the definition of the adaptive approach. The methods considered have been proposed and investigated in the recent literature, they are all implemented considering a non-intrusive modeling, and consist of the classical POD, a recent variant called SPOD, DMD, and RDMD. With the exception of POD, which extracts features without considering any temporal correlation among the training data, all the other methods have been developed with the aim of extracting features which are more capable to unveil the correct underlying dynamics, addressing this point through consideration of some temporal correlation among the available snapshots. The literature on these methods is quite vast and detailed and the interested reader could refer to it. In the following, the key elements of these methods will be reported. POD is a classical data compression method and is based on the following optimality condition:

$$\max_{\phi_i \in \mathbb{R}^n} \langle \mathbf{U}, \phi_i \rangle \quad \text{with} \quad \|\phi_i\|_2 = 1 \quad i = 1, 2, \ldots, N_s \tag{1}$$

which allows to extract the closest set of basis functions $\phi_i^k(\mathbf{x})$ to the initial dataset $\mathbf{U}$. In Eq. 1, $\langle \cdot, \cdot \rangle$ represents the average over time, while $\mathbf{U}$ is a $N_p \times N_s$ matrix with the N_s snapshots, $\mathbf{u}(\mathbf{x}, t)$, arranged in columns, N_p indicates the number of grid points in the high-fidelity mesh. The scalar product in the optimization problem (1) is simply the euclidean scalar product between two vectors of $\mathbb{R}^n$, namely $< \mathbf{x}, \mathbf{y} >= \sum_{i=1}^{n} x_i y_i$, and it is also the inner product used for all the norms introduced hereafter. Using this definitions and the method of snapshots defined in [5], it can be shown that extracting POD feature is equivalent to solving an eigenvalue problem for the matrix $\mathbf{R} = \mathbf{U}\mathbf{U}^T$, where ^{T} denotes transposition. The POD feature can then be obtained projecting the eigenvectors of matrix $\mathbf{R}$ into the high-dimensional space through the snapshots. The fundamental features are eventually linearly combined together to provide the reconstructed field at a desired instant of time [4]. The reconstruction formula based on an non-intrusive way of computing the coefficients is

$$\begin{aligned} \mathbf{u}(\mathbf{x}, \bar{t}) &\simeq \textstyle\sum_{i=1}^{M \leq N_s} a_i^k(\bar{t}) \boldsymbol{\phi}_i^k(\mathbf{x}) \\ a_i^k(\bar{t}) &= p(\bar{t}) + \textstyle\sum_{j=1}^{N_s} w_j f\left(| \, \bar{t} - t_j \, | \right). \end{aligned} \tag{2}$$

In the present work, the coefficients of the linear expansion $a_i^k(\bar{t})$ needed to compute the solution for any new time, $\bar{t}$, are obtained by radial basis function interpolation [22], see second row in Eq. 2. In this equation, $p(\bar{t})$ is a polynomial of low degree and the kernel function f is a real-valued function on $[0; \infty)$. Specifically, the kernel functions used in the present work are gaussian functions, $f\left(| \, \bar{t} - t_j \, | \right) = e^{-\frac{(\bar{t}-t_j)^2}{2\sigma}}$, and the t_j, centers of the RBF, are the time instants corresponding to the components of the POD eigenvectors extracted from matrix $\mathbf{R}$. SPOD is a variant of the POD, not to be confused with the POD in the frequency domain [23], where the basis functions ϕ_i^k are obtained by using a modified correlation matrix of the snapshots,

$\hat{\mathbf{R}} = \mathcal{F}(\mathbf{R})$, that introduces a filter along the diagonals of the matrix, establishing a connection between snapshots at subsequent instants of time [10]. In particular, the generic filtered element of the correlation matrix $\hat{\mathbf{R}}$ is

$$\hat{R_{i,j}} = \sum_{k=-N_f}^{N_f} g_k R_{i+k,j+k}, \tag{3}$$

where the filter width N_f defines the time window over which the temporal correlation between snapshots is considered. The filter g_k can be a user-defined function. In the present work a constant function is considered, namely $g_k = \frac{1}{N_f+1}$, which allows convergence to pure discrete Fourier transform (DFT) when the filter spans over all the correlation matrix $\mathbf{R}$ [10]. The reconstruction formula of SPOD is identical to the one of POD. DMD is a method that takes directly into account the temporal correlation of the snapshots by considering a linear regression in time over all the set of snapshots [13]. Each state is assumed to propagate in time by a constant matrix, namely $\mathbf{U}^{n+1} = \mathbf{A}\mathbf{U}^n$, where $\mathbf{U}^n$, $\mathbf{U}^{n+1}$ are the matrix of snapshots stacked from time t_0 to time t_n and from time t_1 to time t_{n+1}, respectively, and $\mathbf{A}$ is the matrix containing the dynamics information. Practically, DMD aims at extracting eigenvectors and eigenvalues of this matrix, which define spatial structures (eigenvectors) and their associated frequencies, growth/decay rate (eigenvalues). Since the matrix $\mathbf{A}$ for fluid flow problems can be of very high dimension a singular value decomposition is used to express the dynamics in a lower space. Therefore, the eigenvalues and eigenvectors of the corresponding reduced dynamic matrix $\tilde{\mathbf{A}}$ are computed and the dynamic eigenvectors are then projected onto the high-dimensional space to recover the actual DMD flow structures, following the algorithm for the extraction of exact DMD modes reported in [14]. The reconstruction formula takes the form

$$\begin{gathered} \mathbf{u}(\mathbf{x}, \bar{t}) \simeq \textstyle\sum_{i=1}^{M \leq N_s} \alpha_i^{\text{DMD}} \boldsymbol{\phi}_i^{\text{DMD}}(\mathbf{x}) e^{\omega_i \bar{t}} \\ \min_{\alpha \in \mathbb{C}^M} ||\mathbf{U^{n+1}} - \Phi \mathbf{D}_\alpha \mathbf{V}||_2 \end{gathered} \tag{4}$$

where ω_i are the DMD eigenvalues extracted from matrix $\tilde{\mathbf{A}}$, $\boldsymbol{\phi}_i^{\text{DMD}}$ are the exact DMD modes, α_i are the modes coefficients, i.e., DMD modes amplitudes, which can be computed in several ways, mainly based on cosidering the entire set of snapshots [24] or only the initial snapshot [25]. The method developed in the present work considers all the set of available snapshots and uses the solution of the optimization problem reported in [24], which is in the second row of Eq. 4. $\mathbf{D}_\alpha$ is a diagonal matrix containing the DMD coefficients α_i, M is DMD rank, i.e., the number of DMD modes extracted, $\mathbf{V}$ is the Vandermonde matrix built with DMD eigenvalues and Φ is the matrix of the DMD modes. For complex systems, the DMD modes may not provide a sufficiently sound physical interpretation, since the evolution of each mode is limited to a damped/growing or pure periodic oscillation. The recursive dynamic mode decomposition has been recently proposed combining together features from pure DMD and POD. The main aim of the method is trying to bridge the pure frequency extraction of DMD with the optimality property of POD, as SPOD does for POD and DFT. In order to do so, flow features are obtained in an iterative manner where at each step of the recursion process, the DMD mode which is closest to the set of initial data is selected [15] according to the following minimization:

$$\min_{i\in\{1,2,\ldots,N_s-1\}} = \|\mathbf{U}_r - \phi_{i,r}\mathbf{a}_{i,r}\|_2, \tag{5}$$

where $\mathbf{U}_r$ is the dataset on which the DMD extraction is applied at the rth step of recursion, calculated by subtracting the contribution of the first $r-1$ modes from the set of initial snapshots, $\phi_{i,r}\mathbf{a}_{i,r}$ is the reconstruction computed considering only the ith DMD mode at the rth step of recursion. The reconstruction formula of the initial dataset will be

$$\mathbf{u}(\mathbf{x},t_j) = \sum_{i=1}^{M\leq N_s} \gamma_i(t_j)\boldsymbol{\psi}_i(\mathbf{x}) + \mathbf{r}_M \quad j = 1, 2, \ldots, N_s \tag{6}$$

where $\mathbf{r}_M$ bears the residual of the approximation at the Mth step of recursion, the first part is identical to POD, therefore $\gamma_i(\bar{t})$ are the RDMD coefficients, while $\boldsymbol{\psi}_i(\mathbf{x})$ are the RDMD modes and to reconstruct the flow field at a general time instant $\bar{t}$ the same RBF interpolation as POD, reported in the second row of Eq. 2, is used.

3 Adaptive Approach

The idea of the adaptive approach is to be able to automatically select the method and the corresponding number of modes that, for a specific instant of time, will provide the lowest error in the reconstructed field. The ROM construction and error assessment is part of an off-line phase where all methods are evaluated to identify the best method for any specific instant of time. The output of the assessment process takes the form of a convex envelope of all the errors for each method and the associated best in class for a specific instant of time. Figure 1 reports the proposed adaptive approach, where $\mathbf{W}$ is the same as matrix $\mathbf{U}$ defined in Sect. 2, containing the training snapshots. During the offline phase, the high-fidelity training set is generated using a CFD solver, basis functions $\boldsymbol{\phi}_i^k(\mathbf{x})$ are extracted for all the methods reported in Sect. 2, and finally, each ROM is assessed using a specific definition of error (see Sect. 3.2). The output of the offline phase is a database containing the basis functions per each method and the hierarchy of methods for each instant of time of the window under investigation, together with the best choice of the rank for each set of basis functions. During the online phase, the solution is requested for a specific instant of time and the database is explored to select the appropriate basis functions and associated data required for the reconstruction [20], in terms also of the appropriate number of modes to be used. It is worth noticing that, even if the off-line phase requires many computations in terms of extraction of basis functions and construction of the error database (left box

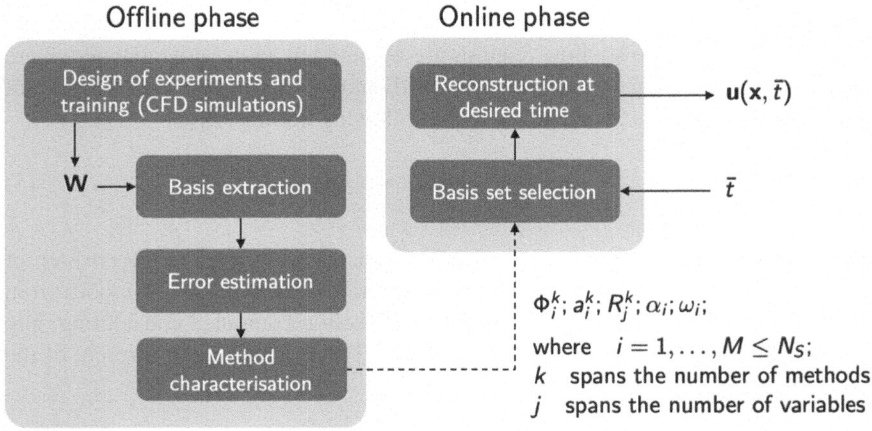

Fig. 1 Schematic of the adaptive approach distinguished in the offline and online phases

in Fig. 1), the online phase is not affected in terms of computational cost with respect to a single ROM, while providing a better accuracy in the flow field reconstruction over the investigated time window (see Sect. 4). Besides that, the time required for the off-line phase is not considered as a crucial parameter in the present study, being this process performed once and for all.

3.1 Physical Problem: Navier-Stokes Equations

The above adaptive framework is here applied to address the problem of dimensionality reduction of the high-dimensional system arising from the discretisation of the set of the classical two-dimensional Navier-Stokes equations of fluid mechanics

$$\frac{\partial \mathbf{u}}{\partial t} + \nabla \cdot \mathcal{F}(\mathbf{u}) = 0. \tag{7}$$

The system of Eq. 7 is a set of four homogeneous PDEs for an advection-diffusion problem representing the dynamics of a viscous fluid. One equation is hyperbolic, the other three are parabolic. $\mathbf{u}$ is the vector representing mass, momentum, and total energy per unit volume for a fluid system. The system of equations is complemented by two thermodynamics equations of state for a perfect ideal gas. The type of boundary and initial conditions depends on the specific problem at hand and it will be made explicit later in Sect. 4. The system of Eq. 7 serves two purposes: (1) obtain high-fidelity solutions for the training of the reduced model (2) construct an error database, when the adaptive framework is equipped with a particular definition of the error which requires the evaluation of the residual produced by the reduced-order solutions when plugged into the finite volume (FV) discretization of Eq. 7. This will be better clarified in the next section, where two definitions of the error will be introduced. The computation of new solutions using the adaptive framework during the online phase is instead completely equation-free and based exclusively on the Eqs. 2 and 4.

3.2 Error Estimation

Central to any adaptation techniques is the estimation of the error of the approach. When dealing with complex unsteady problems involving the full set of Navier-Stokes equations, ROMs lack rigorous a priori and a posteriori error bounds (i.e., certified and reliable ROMs). Lots of efforts have been put in literature to define such bounds, but the study has been usually limited to a POD basis [26] and introducing simplifications to the initial set of the governing equations [27, 28]. In this case, two different methods have been considered, one referred to as direct error and one referred to as residual error. The direct error approach computes the error of the ROM by comparing directly the reconstructed solution with a reference high-fidelity solution. In the present approach, the reference high-fidelity solution does not belong to the set of snapshots used to define the reduced model, therefore, in order to be viable, the direct error estimation requires a higher number of snapshots, some of which will be used to build the reduced model and some others that will be used only for error evaluation, see Fig. 2 top. Approaches exist, where instead a direct error estimation is done by using all the snapshots available without making a distinction between snapshots to be used for error estimation and snapshots to be used for the ROM. The so-called leave-one-out approach (LOO) iteratively excludes one snapshot from the set and uses the remaining $N_s - 1$ to build a "depleted" reduced model used to reconstruct the solution at the conditions corresponding to the excluded snapshot [29]. This approach allows exploiting at most the set of snapshots available, but it is characterized by an estimation that is not based on the actual set of snapshots that will be used to build the full ROM and, therefore, might provide not accurate estimation of the direct error. The formula used here for the direct error is the following norm of the error at every nodal point in the mesh:

$$\epsilon_{\mathrm{D}} = \frac{1}{\sqrt{N_p}} ||\mathbf{u}_{\mathrm{ref}}(\mathbf{x}, \bar{t}) - \mathbf{u}_{\mathrm{ROM}}(\mathbf{x}, \bar{t})||_2 \tag{8}$$

where $\mathbf{x}$ is the vector with the nodal points of the mesh. The residual error is introduced to be able to provide an error estimation approach, that similarly to the LOO approach does not need a discrimination between snapshots for error analysis and model construction, but at the same time, allows to always consider the full set of available snapshots for the estimation, see Fig. 2 bottom. The residual error is computed by plugging the reconstructed solution from the ROM into the same discrete form of the conservation equations that have been used to solve the high-fidelity counterpart of the problem. This process is still part of the offline phase, namely the error estimation step in Fig. 1. Specifically, for the present work, a FV edge-based approximation is used for the discretisation of the conservation equations, together with a backward differencing formula (BDF), first term in the Eq. 9, for the treatment of the unsteady term, leading to the following residual error:

$$\epsilon_{\mathrm{R}}(\bar{t}) = \frac{1}{\sqrt{N_p}} \Bigg\| \Omega_m \frac{3\mathbf{u}_m^n - 4\mathbf{u}_m^{n-1} + \mathbf{u}_m^{n-2}}{2\Delta t_{\mathrm{res}}} + \\ \sum_{l \in \mathcal{L}_{m\neq}} \boldsymbol{\varphi}_{lm}(\mathbf{u}_m^n, \mathbf{u}_l^n, \boldsymbol{\eta}_{lm}) + \sum_{e \in \mathcal{E}^{\partial}} \boldsymbol{\psi}_m^e(\mathbf{u}_m^n, \mathbf{u}_b^e, \boldsymbol{\nu}_b^e) \Bigg\|_2 . \tag{9}$$

In the residual formula, Δt_{res} is a user-defined time step used to account for time accuracy in the BDF formula and it is a parameter that will influence the value of the error as explained later in Sect. 3.3. The superscripts n, $n-1$ and $n-2$ refer to three instants of time used for the evaluation of the error at time n. $n-1$ and $n-2$, respectively, indicate the solution at time $t^n - \Delta t_{\mathrm{res}}$ and at time $t^n - 2\Delta t_{\mathrm{res}}$. In the present approach, also solutions at t^{n-1} and t^{n-2} are obtained by means of the ROM while constructing the residual error database during the offline phase. The second row in the definition of the residual error refers to the spatial discretization and accounts here for a domain term and a boundary term according to the classical FV edge-based discretization [30]. In particular, $\boldsymbol{\varphi}_{lm}(\mathbf{u}_m^n, \mathbf{u}_l^n, \boldsymbol{\eta}_{lm})$ represents the FV discretisation of convective and viscous fluxes in the internal points of the domain, $\mathbf{u}_m^n$, $\mathbf{u}_l^n$ are, respectively, the conserved quantities at time n at node m and l and $\boldsymbol{\eta}_{lm}$ represents the integrated normal along the edge connecting nodes m and l. $\boldsymbol{\psi}_m^e(\mathbf{u}_m^n, \mathbf{u}_b^e, \boldsymbol{\nu}_b^e)$ is the FV discretization of such fluxes on the boundaries, where $\mathbf{u}_b^e$ is the set of conserved quantities obtained by the imposition of the boundary conditions and $\boldsymbol{\nu}_b^e$ is the integrated normal at the boundary of the domain. Ω_m is the cell volume of the mth cell. The above formulae for the direct and residual errors will provide evaluations corresponding to a specific choice of snapshots, number of modes for the reconstruction, and, for the case of the residual error, Δt_{res}. While the present work is not focusing on the choice of snapshots, a critical analysis on the choice of the number of modes and Δt_{res} is considered. More details about the impact of these elements on the error used for selecting the ROM are reported in the following Sect. 3.3.

3.3 Sensitivity

The error estimation depends on a number of parameters. In the case of direct error, the sensitivity of the error is analyzed with respect to the choice of the number of modes in the reconstruction and the method used for the reconstruction. In this analysis, the choice of modes is based on their ranking according to the relative energy content of each mode [4, 12], the output of such analysis being the optimal number of modes to be used for each set of basis function at a specific instant of time, which guarantees the lowest error. In the case of residual error, an additional

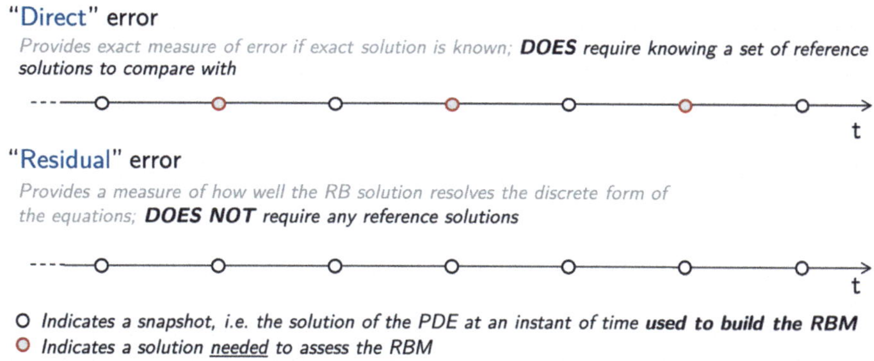

Fig. 2 Conceptual difference between direct and residual errors

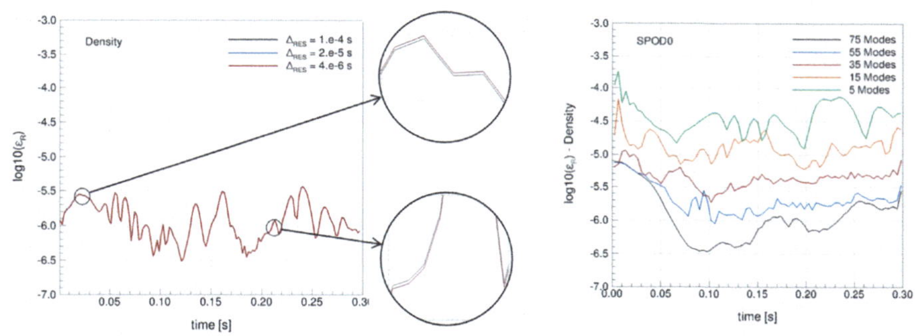

Fig. 3 Sensitivity w.r.t. the choice of Δt_{res} (left), number of modes, and choice of method for density (right)

parameter that is considered here is the choice of the time step used to evaluate the BDF formula. The latter is quite a relevant parameter since it will affect the error measure as a consequence of the necessity of having two additional solutions to evaluate the BDF formula. This has been achieved here by using the same ROM to reconstruct the three solutions, i.e., the one at the instant of time of interest and two previous instants of time. Figure 3 left reports the example of the analysis done to assess the sensitivity of the residual error on the choice of Δt_{res}. In all the analyses presented later, an iterative process has been put into place to reach a state where the changes in residual error as Δt_{res} is reduced is below a specific tolerance set for all cases to 10^{-8}. The right plot of the same figure shows instead the analysis done to assess the impact of the choice of the number of modes in the evaluation of the error. The latter has been done both for residual and direct errors. On the basis of these considerations, an algorithm to compute the error associated to a specific method has been proposed, that on one side automatically identifies the maximum Δt_{res} allowing for independence of the residual error from the choice of Δt_{res}, while on the other allows considering the optimal number of modes to be used for a specific method during the reconstruction as the number of modes guaranteeing the lowest error. The pseudo-algorithm for error estimation is

Pseudo-algorithm

```
for t = 1, N_t

    if ∃ u_REF(t) then
     ε_d(t) = min ϵ_d(t;N_m,N_met)      ∀ N_m,N_met
    end

    if ∄ u_REF(t) .or. Resid then

     Δt_r = Δt_r0
     ε_r (t;Δt_r) = min ϵ_r (t;Δt_r,N_m,N_met)      ∀ N_m,N_met
     ε_r,p = 0
     Δe_r(t;Δ_r) = abs(ε_r - ε_r,p)

     while Δe_r(t;Δ_r) .gt. threshold
      Δt_r = Δt_r/K
      ε_r,p = ε_r(t;Δt_r )
      ε_r(t;Δt_r) = min ϵ_r (t;Δt_r,N_m,N_met)      ∀ N_m,N_met
      Δe_r(t;Δ_r) = abs(ε_r - ε_r,p)
       if Δe_r(t;Δ_r) .lt. threshold
        exit
       end
     end

    end

end
```

The pseudo-algorithm is performed offline (Error estimation step in the left box in Fig. 1). The expected outcome for the direct error ϵ_D is the method N_{met} and its corresponding number of modes N_m, which guarantees the lowest ϵ_D. Equivalently, for the residual error, the result will be the method N_{met} and its corresponding number of modes N_m which guarantees the lowest ϵ_R, with the only difference that a preliminary sensitivity analysis is performed, as stated above, to compute the best Δt_{res} for the evaluation of the residual. The selection of the best values for all these parameters goes through a manual exploration of the parameter space, where a set of N_m, Δt_{res} and all the methods N_{met} are considered. Convergence of the pseudo-algorithm to a globally optimal solution is expected as the number of points in the parameter space is increased.

4 Demonstration on Lifting Surfaces

A series of 2D test cases are considered to assess the performance of the adaptive method using the two different ways of computing the reconstruction error reported in Sect. 3.2. These are both impulsive start flows around a NACA0012 airfoil and a high-lift configuration airfoil known as the 30P30N configuration [31]. These flows exhibit two different interesting flow patterns: both cases are non-periodic flows, with the first featuring an irregular detachment of vortices typical of the path to stall. The 30P30N case instead will feature the dynamics of starting vortices detaching from the three elements of the airfoil that progressively will merge into a single vortex being transported downstream. After the four vortices have merged, a stationary flow field is established in proximity of the airfoil and the dynamics is the one typicals of an advection-dominated problem. All the high-fidelity simulations used in the present work have been obtained using the open-source CFD code SU2 (https://su2code.github.io) [32].

4.1 *Stalled NACA0012 Airfoil*

The conditions of the simulation for the NACA0012 airfoil are reported in Table 1, the corresponding details of the reduced basis method are reported in Table 2. As regards the numerical setup for the high-fidelity simulation, the laminar Navier-Stokes equations have been solved, using a second-order FV discretization for the fluxes (MUSCL approach) and a second-order dual-time stepping scheme to deal with the unsteady part. In particular, the convective fluxes have been discretized using the Roe scheme. As initial condition, the entire domain is initialized to free stream quantities, while the boundary conditions on the body and at the domain borders are no-slip (for momentum equations) adiabatic (for energy equation) and free stream quantities, respectively. The time required for computing a single time step using high-fidelity CFD is approximately 15 min on 1 core, while the time required to compute a single time step with ROM is in the order of a tenth of a second on 1 core. Figure 4 illustrates the vorticity field at two different instants of time. A structured grid was used with 60,600 nodes and 60,196 quadrilateral elements. The training set of snapshots was made of 75 solutions saved each 4×10^{-3} s. The sensitivity analysis on the time step used for the residual evaluation led to a $\Delta \mathbf{t}_{\text{res}}$ of 10^{-5} s. Figure 5 reports the sensitivity of the reconstruction error for the density with respect to the choice of the number of modes and with respect to the different methods. While not reported in the figure, the same analysis has been performed for all the other conserved quantities. The way plots have been represented reflects the steps of the pseudo-algorithm reported in Sect. 3.2. Indeed for each plot in Fig. 5, the minimum envelop of the curves related to different modes is taken as the first step of the pseudo-algorithm. A general trend is observed where the error tends to reduce as the number of modes employed for the reconstruction is increased. This is not always observed in

Table 1 NACA0012 simulation parameters

Mach	α (deg)	Reynolds	T_∞ (K)	Time (s)	Δt (s)	CFL
0.1	15	10,000	288.15	0.3	10^{-3}	5

Table 2 NACA0012 ROM setting

N_S	Δt_{NS} (s)	N. modes	Δt_{RES} (s)	$N_{DOF\text{-}CFD}$	$N_{DOF\text{-}ROM}$	ROM (s)
75	$4\ 10^{-3}$	Error-based	10^{-5}	242,600	$O(10^2)$	1.8

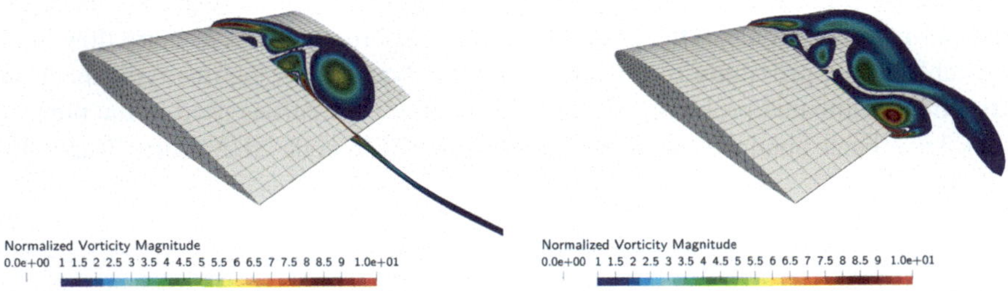

Fig. 4 NACA0012 impulsive start. Unsteady flow at two instants of time. Normalized vorticity magnitude = ω/Vc being V the magnitude of velocity vector and c the airfoil chord

the case of DMD, where time windows exist for which the reconstructed flow with as few as five modes is the one that globally has the lowest direct error. This is supposed to happen as a consequence of the specific flow dynamics that has a very specific frequency captured by the few DMD modes. In Fig. 5 the first column represents the sensitivity analysis on the number of modes performed using the direct error, while the second column reports the same sensitivity performed with the residual error, as they have been defined in Sect. 3.2. It can also be observed that the direct error evaluation tends to be in general lower than the residual error when comparing the same ROM. This difference is obviously related to the fact that the two error definitions lie in different vector spaces.

Figure 6 represents the second and ultimate step of the pseudo-algorithm. Indeed all the minimum envelopes obtained from Fig. 5 for each method in the adaptive framework are combined together to obtain the minimum envelopes reported in the first row of Fig. 6. The procedure is repeated for each conservative variable, obtaining the remaining rows reported in Fig. 6. Therefore, these final envelopes represent the minimum error among all the considered methods and all the number of modes used in the reconstruction. Also for this final step, results considering both error definitions (residual error on the left, direct error on the right), are reported. Table 3 reports the percentage values of the choice of the best method over the time window explored for each one of the conserved quantities, which better clarifies the contribution of each of these methods to the minimum envelopes represented in Fig. 6. It can be observed that for this type of flow, POD and RDMD are the most used methods. Some instants

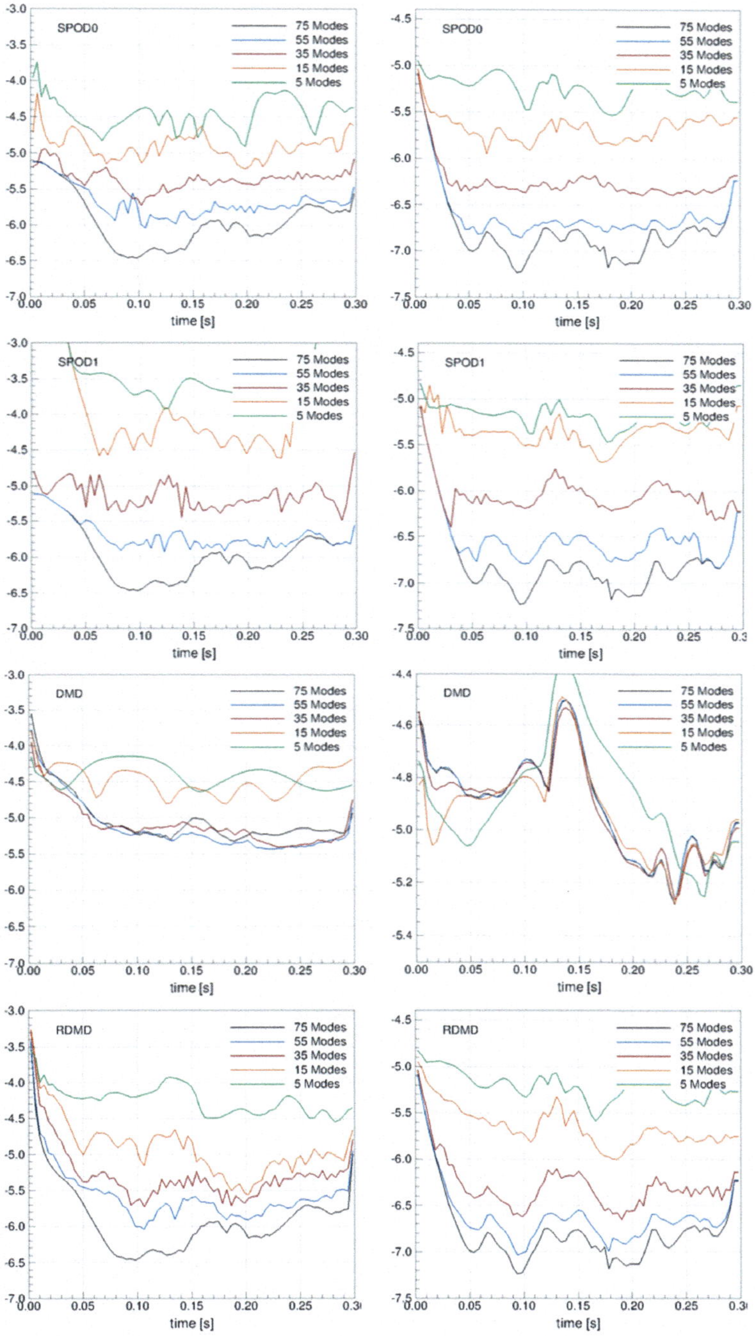

Fig. 5 NACA0012 sensitivity w.r.t. number of modes and choice of method for density. Residual error = $\log_{10}(\varepsilon_r(\rho))$ (left column), Direct error = $\log_{10}(\varepsilon_d(\rho))$ (right column)

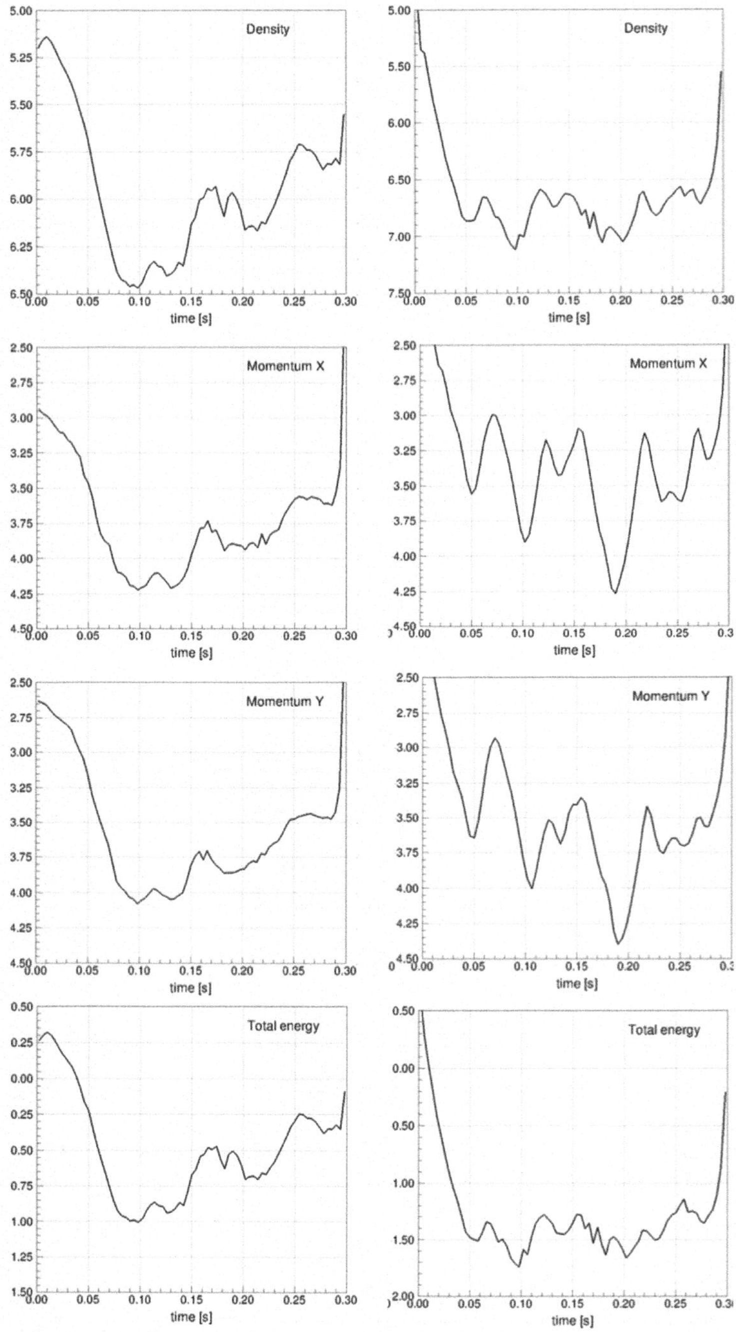

Fig. 6 NACA0012 minimum envelope of errors. Residual error = $\log_{10}(\varepsilon_r(\rho, \rho u, \rho v, \rho e))$ (left column), Direct error = $\log_{10}(\varepsilon_d(\rho, \rho u, \rho v, \rho e))$ (right column)

Table 3 NACA0012 choice of ROM for each conserved quantity according to the residual or direct error

	POD (%)	SPOD1 (%)	DMD (%)	RDMD (%)
ρ_D	48	4	0	48
ρ_R	55	9	0	36
ρu_D	54	1	0	45
ρu_R	55	4	0	41
ρv_D	48	8	0	44
ρv_R	55	4	1	40
ρe_D	32	4	0	64
ρe_R	56	7	0	37

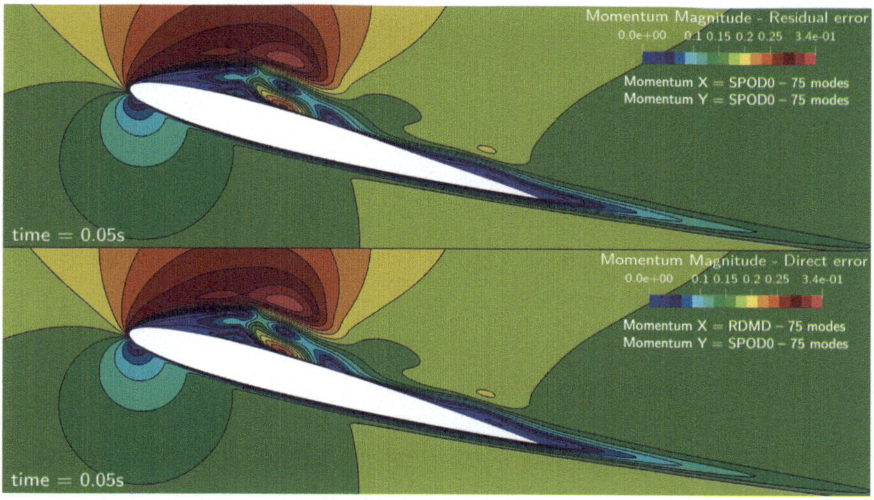

Fig. 7 NACA0012 momentum magnitude contours at 0.05 s. Residual error (top), Direct error (bottom)

of time are best reconstructed using SPOD with a filter of 10 and only very few instants of time are best reconstructed with DMD. Finally, Figs. 7 and 8 report the reconstruction of the flow field by means of the adaptive approach for two different instants of time. A comparison with a reference high-fidelity solution not used in the definition of the ROM is presented. The momentum magnitude is shown as obtained by reconstructing independently the two components ρu and ρv. Colored contours refer to the high-fidelity solution, while solid black lines the reconstructed field. The figure also reports the number of modes and the method used for the reconstruction of the two components of the momentum. Overall, the agreement is good and minor differences are observed between the reconstruction based on the direct error and the one based on the residual error.

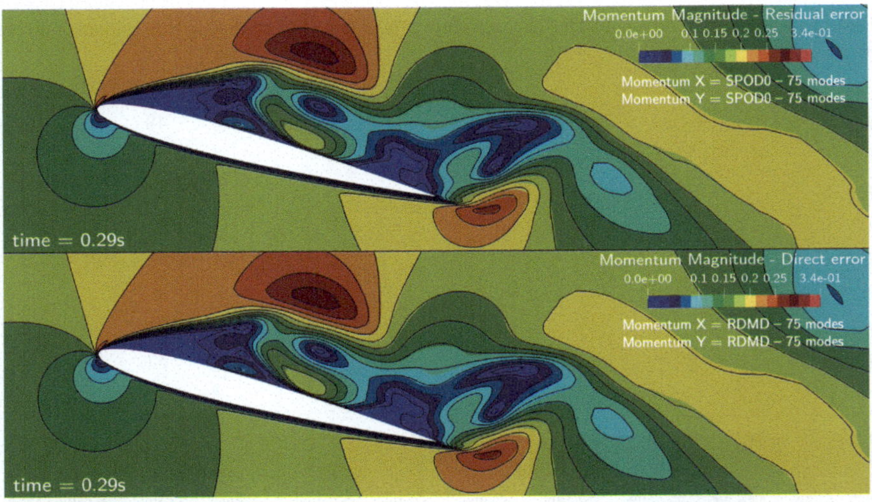

Fig. 8 NACA0012 momentum magnitude contours at 0.29 s. Residual error (top), Direct error (bottom)

Table 4 30P30N simulation parameters

Mach	α (deg)	Reynolds	T_∞ (K)	Time (s)	Δt (s)	CFL
0.2	19	9×10^6	288.15	0.06	10^{-4}	0.4

4.2 High-Lift 30P30N Airfoil

The conditions of the simulation for the 30P30N airfoil are reported in Table 4, the corresponding details of the reduced basis method are reported in Table 5. As regards the numerical setup of the high-fidelity simulation, the unsteady Reynolds averaged Navier-Stokes (URANS) have been solved, being this problem turbulent, using a second-order FV discretization for the fluxes (MUSCL approach) and a second-order dual-time stepping scheme to deal with the unsteady part. In particular, the convective fluxes have been discretized using the Roe scheme. The turbulent model used in the URANS context was SST [33]. Initial and boundary conditions are the same as the ones specified for the previous test case. The time required for computing a single time step using high-fidelity CFD is approximately 60 min on 1 core, while the time required to compute a single time step with ROM is in the order of a tenth of a second on 1 core. Figure 9 illustrates the vorticity field at two different instants of time. A hybrid grid, with quadrilaterals in the boundary layer and unstructured triangles in the rest of the domain, was used with 327,733 nodes and 551,040 quadrilateral elements.

The training set of snapshots was made of 100 solutions saved each 4×10^{-4} s. The sensitivity analysis on the time step used for the residual evaluation led to a

Table 5 30P30N ROM setting

N_S	Δt_{NS} (s)	N. modes	Δt_{RES} (s)	$N_{DOF\text{-}CFD}$	$N_{DOF\text{-}ROM}$	ROM (s)
100	6×10^{-4}	Error-based	5×10^{-6}	1,966,398	$O(10^2)$	24

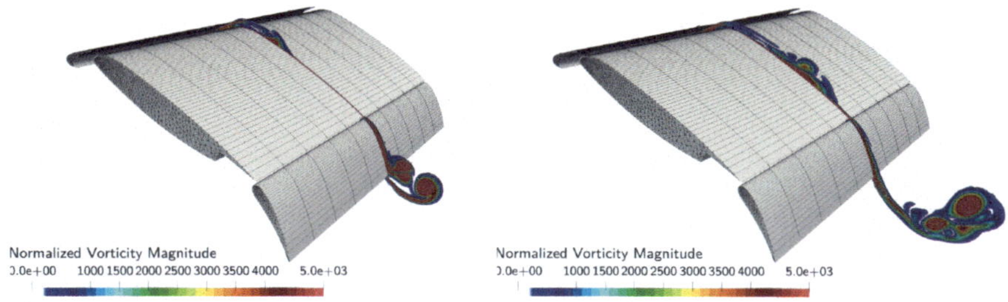

Fig. 9 30P30N impulsive start. Unsteady flow at two instants of time. Normalized vorticity magnitude = ω/Vc being V the magnitude of velocity vector and c the airfoil chord

$\Delta \mathbf{t}_{res}$ of 5×10^{-6} s. As for the previous test case, Fig. 10 reports the sensitivity of the reconstruction error for the density with respect to the choice of the number of modes and with respect to the different methods. Differently from the NACA0012 case, no general trend can be observed with respect to the reduction of reconstruction error as the number of modes increases. This may be related to the different unsteady dynamics of this flow that reaches an advection-dominated status as the vortices coalesce and then get transported downstream. The strong oscillations appearing for the DMD residual error as the number of modes increases (third row on the left column of Fig. 10) might be due to the addition of higher frequency modes as the rank in the DMD algorithm increases, which might not be representative of the actual dynamics and introduce spurious oscillation in the time and space derivatives in the formula 9 for the evaluation of the residual. An investigation of the terms in Eq. 9 that primarily contributes to this spurious oscillation is out of the scope of the present work. Similarly to the NACA0012 case, the direct error evaluation tends to be in general lower than the residual error when comparing the same ROM.

Also similarly to the previous test case, Fig. 11 reports the error curves as a result of the application of the pseudo-algorithm in Sect. 3.3. The steep increase in the direct error envelope at the very end of the investigated time window which can be noticed from this figure (plots on the right columns), might be related to the starting vortex diffusion as it is convected downstream. This same behavior is not present in the residual error envelope, being the diffusion contemplated in the FV discretization of the Navier–Stokes equations. Table 6 reports the percentage values of the choice of the best method over the time window explored for each one of the conserved quantities. Also for this flow, it can be observed that POD is the most used method, but differently from the previous case SPOD with a filter of 10 is the second best choice over the specified time window. Finally, Figs. 12 and 13 report the reconstruction of the flow field by means of the adaptive approach for two

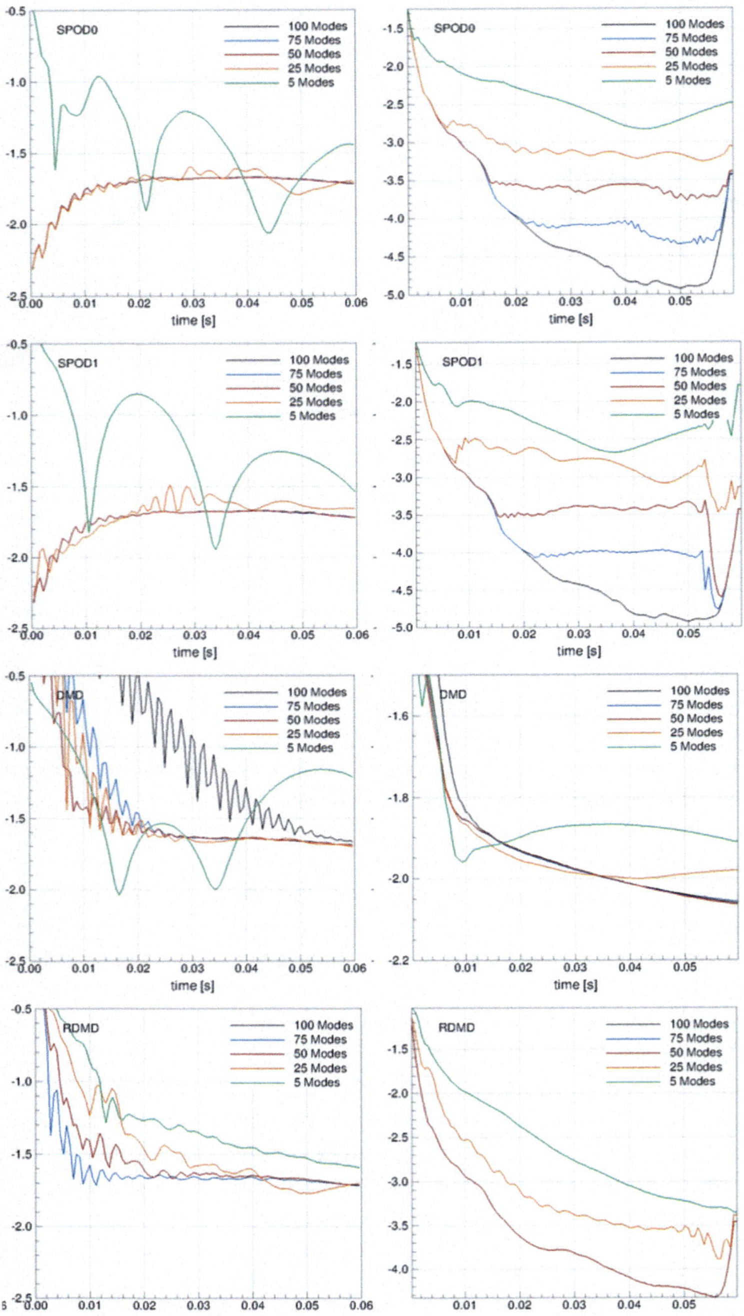

Fig. 10 30P30N sensitivity w.r.t. number of modes and choice of method for density. Residual error = $\log_{10}(\varepsilon_r(\rho))$ (left column), Direct error = $\log_{10}(\varepsilon_d(\rho))$ (right column)

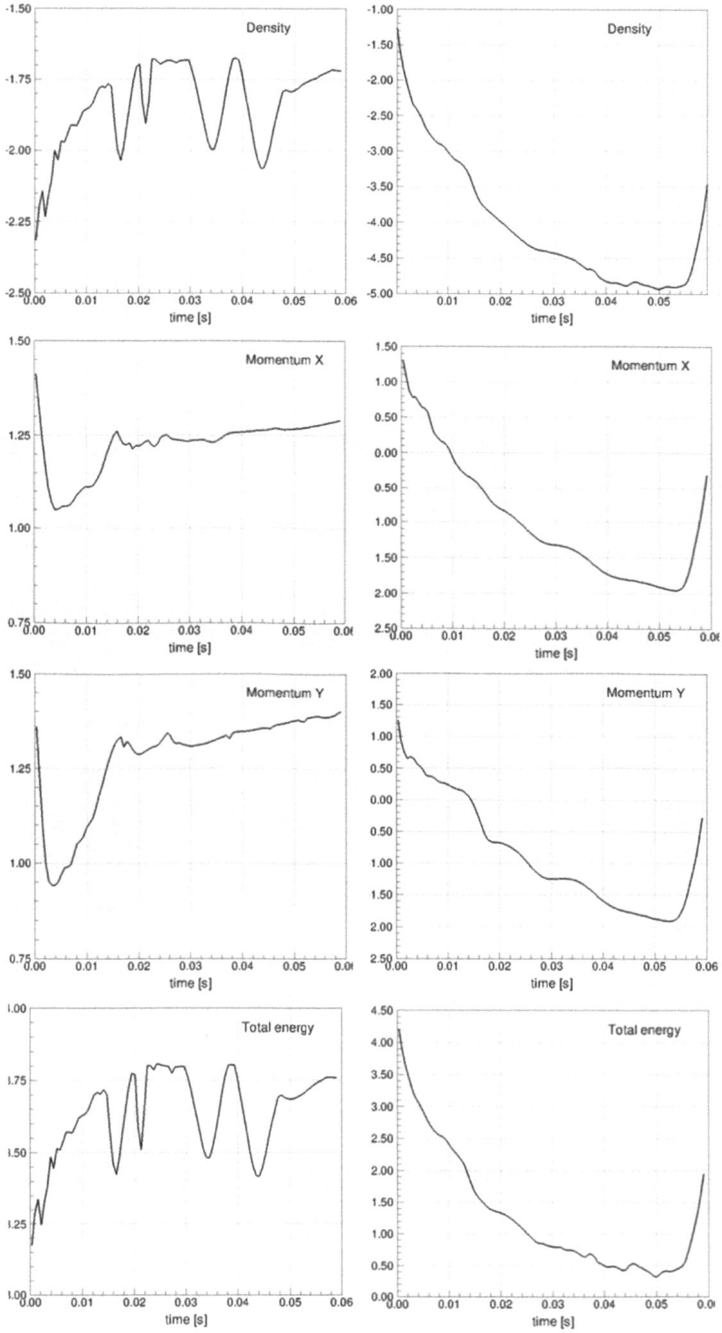

Fig. 11 30P30N minimum envelope of errors. Residual error = $\log_{10}(\varepsilon_r(\rho, \rho u, \rho v, \rho e))$ (left column), Direct error = $\log_{10}(\varepsilon_d(\rho, \rho u, \rho v, \rho e))$ (right column)

Table 6 30P30N choice of ROM for each conserved quantity according to the residual or direct error

	POD (%)	SPOD1 (%)	DMD (%)	RDMD (%)
ρ_D	81	11	0	8
ρ_R	38	26	29	7
ρu_D	82	10	0	8
ρu_R	65	26	8	1
ρv_D	90	9	0	1
ρv_R	36	43	17	4
ρe_D	69	19	0	12
ρe_R	38	28	29	5

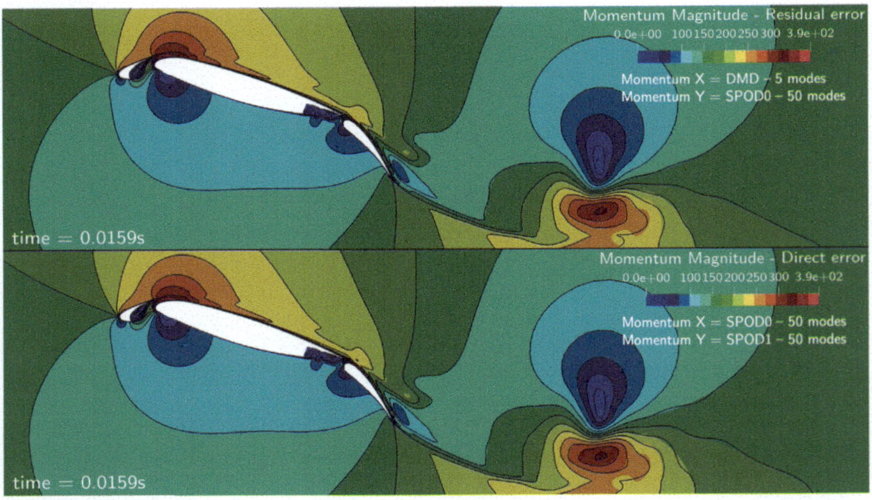

Fig. 12 30P30N momentum magnitude contours at 0.0159 s. Residual error (top), Direct error (bottom)

different instants of time. A comparison with a reference high-fidelity solution that is not used in the definition of the ROM is presented. The momentum magnitude is shown as obtained by reconstructing independently the two components ρu and ρv. Colored contours refer to the high-fidelity solution while solid black lines refer to the reconstructed field. The figure also reports the number of modes and the method used for the reconstruction of the two components of the momentum. Despite the different choice of methods as opposite to what happens in the previous test case (see Figs. 7 and 8), overall, the agreement is good and minor differences are observed between the reconstruction based on the direct error and the one based on the residual error.

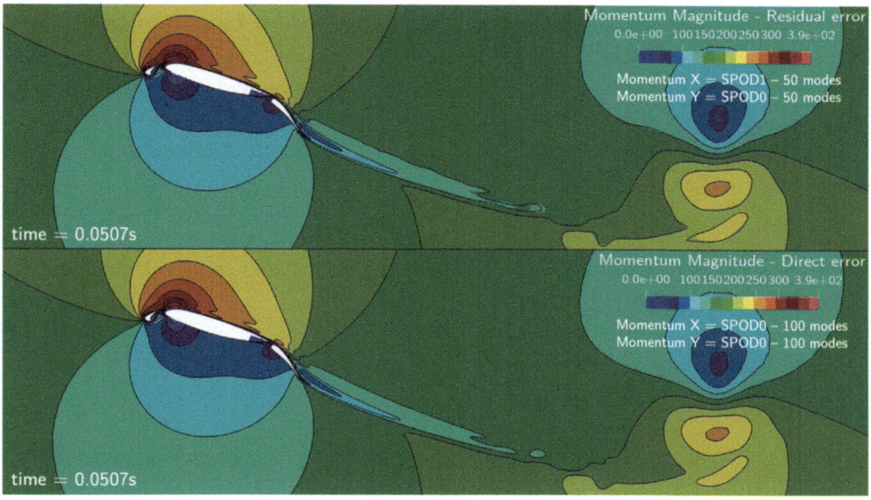

Fig. 13 30P30N momentum magnitude contours at 0.0507 s. Residual error (top), Direct error (bottom)

5 Final Remarks and Outlook

The choice of the error estimator is non-trivial and sometimes driven by engineering/practical considerations and a combination of direct and residual errors can be considered to find the optimal trade-off between the ability to obtain a consistent estimation of the reconstruction error and the number of snapshots that need to be excluded for the ROM due to the evaluation of the direct error. It is worth noticing that the two definitions of the error introduced, which define the two different adaptive frameworks, can be compared only in a heuristic way since they lie in two different spaces. Overall, the trends in terms of choice of method between direct and residual errors are consistent but differences are observed when looking at specific time windows, i.e., for the 30P30N test case, the choice reported in Figs. 12 and 13 are different for the two definitions of the error. The direct error tends to be a more reliable estimation of the error since no pollution is expected, nevertheless, it requires a bigger database of snapshots to be able to use some of them only for error estimation and the rest for the ROM construction. Despite the difference in the method selection, the reconstructed solutions show a good agreement with a reference CFD solution. The present analysis was based on conservative quantities, i.e., error estimation and reconstruction were performed for mass, momentum, and total energy. In case the ROM is required to obtain a primitive or another derived quantity, these can be obtained from the conservative ones. An alternative approach is under evaluation, for the adaptive framework equipped with the residual error, considering an estimation of the residual of such non-conserved quantities as a function of the residuals of the conserved quantities. An analysis of the influence of the

SPOD filter is underway including the possibility to deal with non-uniform time steps (e.g., skew-normal SPOD). Smoothness and regularity of reconstructed solution when transitioning from one method to the other is under consideration (e.g., when is it actually needed from a "practical" viewpoint?). Finally, a zonal approach is under development that aims at adaptivity "within" each computational domain and not only in the time domain.

Acknowledgements The authors wish to thank Dr. G. Barrenechea from Strathclyde University for his suggestions and comments. The simulations were done on the Archie-WeST supercomputer (https://www.archie-west.ac.uk).

References

1. Rowley, C.W., Dawson, S.T.M.: Model reduction for flow analysis and control. Annu. Rev. Fluid Mech. **49**, 387–417 (2017)
2. Taira, K., et al.: Modal analysis of fluid flows: an overview. AIAA J. **55**(12), 4013–4041 (2017)
3. Taira, K., et al.: Modal analysis of fluid flows: applications and outlook. AIAA J. **58**(3), 998–1022 (2020)
4. Holmes, P., et al.: Turbulence, Coherent Structures, Dynamical Systems and Symmetry. Cambridge University Press, Cambridge (2012)
5. Sirovich, L.: Turbulence and the dynamics of coherent structures. I. Coherent Struct. Q. Appl. Math. **45**, 561–571 (1987)
6. Cazemier, W., Verstappen, R.W.C.P., Veldman, A.E.P.: Proper orthogonal decomposition and low-dimensional models for driven cavity flows. Phys. Fluids **10**, 1685–1699 (1998)
7. Rowley, C.W., Colonius, T., Murray, R.M.: Model reduction for compressible flows using POD and Galerkin projection. Phys. D: Nonlinear Phenom. **189**(1–2), 115–129 (2004)
8. Carlberg, K., Bou-Mosleh, C., Farhat, C.: Efficient non-linear model reduction via a least-squares Petrov-Galerkin projection and compressive tensor approximations. Int. J. Numer. Methods Eng. **86**(2), 155–181 (2011)
9. Carlberg, K., Barone, M., Antil, H.: Galerkin v. least-squares Petrov-Galerkin projection in nonlinear model reduction. J. Comput. Phys. **330**, 693–734 (2017)
10. Sieber, M., Paschereit, C.O., Oberleithner, K.: Spectral proper orthogonal decomposition. J. Fluid Mech. **792**, 798–828 (2016)
11. Sieber, M., Paschereit, C.O., Oberleithner, K.: On the nature of spectral proper orthogonal decomposition and related modal decompositions (2017). arXiv preprint arXiv:1712.08054
12. Pascarella, G., Barrenechea, G.R., Fossati, M.: Impact of POD modes energy redistribution on flow reconstruction for unsteady flows of impulsively started airfoils and wings. Int. J. Comput. Fluid Dyn. (2019) (Special Issue on Advances in Reduced Order Methods in CFD)
13. Schmid, P.J.: Dynamic mode decomposition of numerical and experimental data. J. Fluid Mech. **656**, 5–28 (2010)
14. Tu, J.H., et al.: On dynamic mode decomposition: theory and applications (2013). arXiv preprint arXiv:1312.0041
15. Noack, B.R., et al.: Recursive dynamic mode decomposition of transient and post-transient wake flows. J. Fluid Mech. **809**, 843–872 (2016)
16. Rowley, C.W.: Model reduction for fluids, using balanced proper orthogonal decomposition. Int. J. Bifurc. Chaos **15**(03), 997–1013 (2005)
17. Noack, B.R.: From snapshots to modal expansions-bridging low residuals and pure frequencies. J. Fluid Mech. **802**, 1–4 (2016)
18. Alla, A., Kutz, J.N.: Nonlinear model order reduction via dynamic mode decomposition. SIAM J. Sci. Comput. **39**(5), B778–B796 (2017)

19. Tissot, G., et al.: Model reduction using dynamic mode decomposition. C. R. Méc. **342**(6–7), 410–416 (2014)
20. Pascarella, G., Barrenechea, G.R., Fossati, M.: Adaptive reduced basis methods for the reconstruction of unsteady vortex-dominated flows. Comput. Fluids **190**, 382–397 (2019)
21. Pascarella, G., Barrenechea, G.R., Fossati, M.: Model-based adaptive reduced basis methods for unsteady aerodynamic studies. AIAA 2019 Aviation and Aeronautics Forum and Exposition (2019)
22. Carr, J.C., et al.: Reconstruction and representation of 3D objects with radial basis functions. In: Proceedings of the 28th Annual Conference on Computer Graphics and Interactive Techniques, pp. 67–76. ACM (2001)
23. Towne, A., Schmidt, O.T., Colonius, T.: Spectral proper orthogonal decomposition and its relationship to dynamic mode decomposition and resolvent analysis (2017). arXiv preprint arXiv:1708.04393
24. Jovanović, M.R., Schmid, P.J., Nichols, W.: Sparsity-promoting dynamic mode decomposition. Phys. Fluids **26**(2), 024103 (2014)
25. Kutz, J.N., et al.: Dynamic Mode Decomposition: Data-Driven Modeling of Complex Systems. Society for Industrial and Applied Mathematics, Philadelphia (2016)
26. Kunisch, K., Volkwein, S.: Galerkin proper orthogonal decomposition methods for a general equation in fluid dynamics. SIAM J. Numer. Anal. **40**(2), 492–515 (2002)
27. Rozza, G., Huynh, D.B.P., Patera, A.T.: Reduced basis approximation and a posteriori error estimation for affinely parametrized elliptic coercive partial differential equations. Arch. Comput. Methods Eng. **15** (2007)
28. Nguyen, N.C., Rozza, G., Patera. A.T.: Reduced basis approximation and a posteriori error estimation for the time-dependent viscous Burgers-equation. Calcolo **46**, 157–185 (2009)
29. Fossati, M.: Evaluation of aerodynamic loads via reduced-order methodology. AIAA J. **53**, 1685–1699 (2015)
30. Selmin, V.: The node-centred finite volume approach: bridge between finite differences and finite elements. Comput. Methods Appl. Mech. Eng. **102**, 107–138 (1993)
31. Rumsey, C.L., Gatski, T.B.: Recent turbulence model advances applied to multielement airfoil computations. J. Aircraft **38**, 904–910 (2001)
32. Palacios, F., et al.: Stanford University Unstructured (SU2): an open-source integrated computational environment for multi-physics simulation and design. In: AIAA Paper 2013-0287 (2013)
33. Menter, F.L.: Improved two-equation k-omega turbulence models for aerodynamic flows. NASA Technical Memorandum 103975 (1992)

Reduced Basis Methods for Quasilinear Elliptic PDEs with Applications to Permanent Magnet Synchronous Motors

Michael Hinze and Denis Korolev

Abstract In this paper, we propose a certified reduced basis (RB) method for *quasilinear* elliptic problems together with its application to *nonlinear magnetostatics* equations, where the later model permanent magnet synchronous motors (PMSM). The parametrization enters through the geometry of the domain and thus, combined with the nonlinearity, drives our reduction problem. We provide a residual-based a-posteriori error bound which, together with the Greedy approach, allows to construct reduced basis spaces of small dimensions. We use the empirical interpolation method (EIM) to guarantee the efficient *offline-online* computational procedure. The reduced basis solution is then obtained with the surrogate of Newton's method. The numerical results indicate that the proposed reduced basis method provides a significant computational gain, compared to a finite element method.

1 Introduction

A crucial task in the design of electric motors is the creation of proper magnetic circuits. In permanent magnet electric motors, the latter is created by electromagnets and permanent magnets. The corresponding mathematical model is governed by a quasilinear elliptic PDE (magnetostatic approximation of Maxwell equations) which describes the magnetic field generated by the sources. One of the engineering design goals consists of improving the performance of the motor through modifying the size and/or location of the permanent magnets. This problem can be viewed as a parameter optimization problem [2, 4, 9, 10], where the parameters determine the geometry of the computational domain. The underlying optimization problem then requires repeated solutions of the nonlinear (in general) elliptic problem on the parametrized domain. Therefore, there is an increasing demand for fast and reliable

M. Hinze · D. Korolev (✉)
University of Koblenz-Landau, Mathematical Institute, Koblenz, Germany
e-mail: korolev@uni-koblenz.de

M. Hinze
e-mail: hinze@uni-koblenz.de

P. Benner et al. (eds.), *Model Reduction of Complex Dynamical Systems*,
International Series of Numerical Mathematics 171,
https://doi.org/10.1007/978-3-030-72983-7_14

reduced models as surrogates in the optimization problem. To achieve this goal, we use the reduced basis method [7, 11]. The extension of reduced basis techniques to nonlinear problems is a non-trivial task and the crucial ingredients of the method then highly dependent on the underlying problem. Efficient implementation of the greedy procedure requires a posteriori error bounds, which, to the best of our knowledge, are not yet available for the problem we consider. In [1] the reduced basis method is applied to approximate the micro-problems in a homogenization procedure for quasilinear elliptic PDEs with non-monotone nonlinearity. However, we note that this is different from our approach, where we use the reduced basis method for the approximation of the solution of a quasilinear PDE. In our case, the monotonicity of the problem allows the a posteriori control of the global reduced basis approximation error. We provide the corresponding error bound for quasilinear elliptic equations, which is based on a monotonicity argument and can be viewed as a generalization of the classical error bound for linear elliptic problems [12], where the coercivity constant is now substituted by the monotonicity constant of the spatial differential operator. The computational efficiency of the reduced basis method is based on the so-called offline-online decomposition. The offline phase corresponds to the construction of the surrogate model and depends on high-dimensional simulations, and thus is expensive. The online phase, where the surrogate model is operated, is usually decoupled from high-dimensional simulations and thus in general is inexpensive. This splitting is feasible if all the quantities in the problem admit e.g. the affine decomposition, which essentially means that all parameter dependencies can be separated from the spatial variables. The recovery of the affine decomposition in the presence of nonlinearities represents an additional challenge and it is usually treated with the empirical interpolation method (EIM) [3, 6]. The EIM algorithm requires additional data, i.e., the basis for interpolation is constructed from nonlinearity snapshots in the "truth" space. For the efficient numerical solution of the reduced basis problem with Newton's method, we extend the computational machinery, proposed in [6] for semilinear PDEs. It leads to a reduced numerical scheme with full affine decomposition and thus to a considerable acceleration in the online phase, compared to the original finite element simulations.

2 The Quasilinear Parametric Elliptic PDE

2.1 Abstract Formulation

We start by introducing the model for a permanent magnet synchronous machine. We consider a three-phase six-pole permanent magnet synchronous machine (PMSM) with one buried permanent magnet per pole. We parametrize the problem through the size of the magnet by introducing a three-dimensional parameter $p = (p_1, p_2, p_3)$ which characterizes magnet's width p_1, magnet's height p_2, and the perpendicular distance from the magnet to the rotor p_3 in mm. In Fig. 1, the geometry of the problem is shown. PMSM then can be described with sufficient accuracy by the magnetostatic approximation of Maxwell's equations

$$-\nabla \cdot (\nu(x, |\nabla u(p)|)\nabla u(p)) = J_e - \frac{\partial}{\partial x_2} H_{pm,1}(p) + \frac{\partial}{\partial x_1} H_{pm,2}(p) \text{ in } \Omega(p) \quad (1)$$

with boundary conditions

$$u|_{BC} = u|_{DA} = 0 \quad \text{and} \quad u|_{AB} = -\, u|_{CD} \,.$$

Here AB, BC, CD, DA represent parts of the boundary $\partial\Omega$ and marked in Fig. 1. We assume that $\Omega(p)$ represents the cross section of the electric motor which is located in the $x_1 - x_2$ plane of $\mathbb{R}^3$ and the solution u is the x_3-component of the magnetic vector potential. The x_3-component of the current density is represented by J_e, and $H_{pm,1}(p)$ and $H_{pm,2}(p)$ are components of the permanent magnet magnetic field. The nonlinear magnetic reluctivity function

$$\nu(x, \eta) = \begin{cases} \nu_1(\eta), & \text{for } x \in \Omega^1(p) \\ \nu_2(x), & \text{for } x \in \Omega^2(p), \end{cases} \quad (2)$$

represents ferromagnetic properties of the material. Here we split the domain $\Omega(p)$ into two non-overlapping subdomains $\Omega^1(p)$ (ferromagnetic steel) and $\Omega^2(p)$ (air, magnet, coils) such that $\nu_1 \in C^1(\Omega^1(p))$ and ν_2 is piecewise constant on $\Omega^2(p)$ (i.e., constant for each material). In practice, we reconstruct ν_1 from the real $B - H$ measurements of PMSM by using cubic spline interpolation. The scheme preserves desired physical properties of the reluctivity function (see, e.g. [8] for the details of the interpolation scheme) and provides the fast-growing nonlinearity of exponential type. We use physical constants for ν_2. Then the reluctivity function satisfies

$$0 < \nu_{\text{LB}} \le \nu(x, \eta) \le \nu_0, \quad \forall x \in \Omega(p), \quad (3)$$

where ν_{LB} can be chosen independently of the parameter p (see Sect. 3.4 for details).

We continue with an abstract formulation of a two-dimensional nonlinear magnetostatic field problem with geometric parametrization, where the parameter set is given by $\mathcal{D} \subset \mathbb{R}^3$ and describes the geometry of the permanent magnet. The regular, bounded, and p-dependent domain $\Omega(p) \subset \mathbb{R}^2$ gives rise to a p-dependent real and separable Hilbert space $X(p) := X(\Omega(p))$ and the corresponding dual space $X'(p) := X'(\Omega(p))$. The function space $X(p)$ is such that

$$X(p) := \{v|\ v \in L^2(p), \nabla v \in (L^2(p))^2,\ u|_{BC} = u|_{DA} = 0,\ u|_{AB} = -\, u|_{CD}\}$$

with $H_0^1(p) \subset X(p) \subset H^1(p)$, where $H^1(p) := \{v|\ v \in L^2(p), \nabla v \in (L^2(p))^2\}$, $H_0^1(p) := \{v|\ \ v \in H^1(p),\ v|_{\partial\Omega} = 0\}$. The inner product on $X(p)$ is defined by $(w, v)_{X(p)} = \int_{\Omega(p)} \nabla w \cdot \nabla v \, dx$ and the induced norm is given by $\|v\|_{X(p)} =$

$(v, v)^{1/2}_{X(p)}$, which is indeed a norm due to Poincare-Friedrichs inequality. Then the abstract problem reads as follows: for $p \in \mathcal{D}$, find $u(p) \in X(p)$ such that

$$a[u(p)](u(p), v; p) = f(v, p), \quad \forall v \in X(p), \tag{4}$$

where we have

$$a[u](w, v; p) = \int_{\Omega(p)} \nu(x, |\nabla u|)\nabla w \cdot \nabla v \, dx, \tag{5}$$

$$f(v; p) = \int_{\Omega(p)} \left(J_e v - H_{pm,2}\frac{\partial v}{\partial x_1} + H_{pm,1}\frac{\partial v}{\partial x_2} \right) dx. \tag{6}$$

The quasilinear form $a[\cdot](\cdot, \cdot; p)$ is strongly monotone on $X(p)$ with monotonicity constant $\nu_{\text{LB}} > 0$, i.e.,

$$a[v](v, v-w; p) - a[w](w, v-w; p) \geq \nu_{\text{LB}} \|v - w\|^2_{X(p)} \quad \forall v, w \in X(p), \tag{7}$$

and Lipschitz continuous on $X(p)$ with Lipschitz constant $3\nu_0 > 0$, i.e.,

$$|a[u](u, v; p) - a[w](w, v; p)| \leq 3\nu_0 \|u - w\|_{X(p)} \|v\|_{X(p)} \quad \forall u, w, v \in X(p). \tag{8}$$

The conditions (7), (8) are established, e.g. in [8]. Then problem (4) admits a unique solution (see [14], Th 25.B). Moreover, those properties will be needed for the error estimates.

In order to avoid domain re-meshing caused by the change of the parameters, we transfer the domain $\Omega(p)$ to a fixed domain $\hat{\Omega} := \Omega(\hat{p})$, where $\hat{p}$ is the reference parameter with $\hat{x} := x(\hat{p})$ as a spatial coordinate on $\hat{\Omega}$ (see e.g., [12]). Further we assume that $\hat{\Omega} = \hat{\Omega}^1 \cup \hat{\Omega}^2$ and this can be decomposed into $L = L_1 + L_2$ (in our case $L = 12$) non-overlapping triangles (see Fig. 1) so that $\hat{\Omega} = \cup_{d=1}^{L} \hat{\Omega}_d$ and in particular $\hat{\Omega}^1 = \cup_{d=1}^{L_1} \hat{\Omega}^1_d$ and $\hat{\Omega}^2 = \cup_{d=1}^{L_2} \hat{\Omega}^2_d$. The transformation $\mathcal{T}(p)$ on each triangle is affine, whereas piecewise affine and continuous over the whole domain according to

$$\begin{aligned} \mathcal{T}(p)|_{\hat{\Omega}_d} : \hat{\Omega}_d &\to \Omega(p) \\ \hat{x} &\mapsto C_d(p)\hat{x} + z_d(p), \end{aligned} \tag{9}$$

for $d = 1, \ldots, L$, where $C_d(p) \in \mathbb{R}^{2\times 2}$ and $z_d(p) \in \mathbb{R}^2$. According to (9), the Jacobian matrix $J_{\mathcal{T}}(p)$ of the transformation $\mathcal{T}(p)$ is constant on each region of the given parametrization, i.e., we have $J_{\mathcal{T}}(p)|_{\hat{\Omega}_d} = C_d(p)$.

Now we state the problem (4) on the reference domain $\hat{\Omega}$ with the corresponding Hilbert space $\hat{X} := X(\hat{p})$ equipped with the inner product $(\hat{w}, \hat{v})_{\hat{X}} = \int_{\hat{\Omega}} \nabla\hat{w} \cdot \nabla\hat{v} d\hat{x}$ and the induced norm $\|\hat{v}\|_{\hat{X}} = (\hat{v}, \hat{v})^{1/2}_{\hat{X}}$. It reads as follows: for $p \in \mathcal{D}$, find $\hat{u}(p) \in \hat{X}$ so that

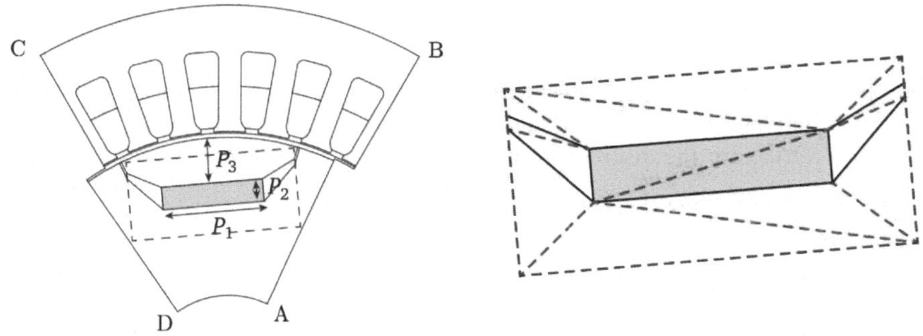

Fig. 1 The cross section of one pole of the machine with the magnet depicted in gray and the region of the geometric parametrization indicated by the dashed box. The dashed lines indicate the triangulation into L triangles. Figure is adapted from [4]

$$a[\hat{u}(p)](\hat{u}(p), \hat{v}; p) = f(\hat{v}, p), \quad \forall \hat{v} \in \hat{X}, \tag{10}$$

where the quasilinear form in (5) is now transformed with the change of variables formula into

$$a[\hat{u}](\hat{w}, \hat{v}; p) = \int_{\hat{\Omega}} \nu(\hat{x}, |J_{\mathcal{T}}^{-T}(p)\nabla\hat{u}|)[J_{\mathcal{T}}^{-T}(p)\nabla\hat{w}] \cdot [J_{\mathcal{T}}^{-T}(p)\nabla\hat{v}]|\det J_{\mathcal{T}}(p)|\, d\hat{x}. \tag{11}$$

Similarly, the linear form in (5) is transformed into

$$f(\hat{v}; p) = \int_{\hat{\Omega}} [f \circ \mathcal{T}(p)]\hat{v}|\det J_{\mathcal{T}}(p)|\, d\hat{x}. \tag{12}$$

Since $\hat{\Omega} = \hat{\Omega}^1 \cup \hat{\Omega}^2$, we have the decomposition

$$a[\hat{w}](\hat{w}, \hat{v}; p) := a^{\nu_1}[\hat{w}](\hat{w}, \hat{v}; p) + a^{\nu_2}(\hat{w}, \hat{v}; p), \tag{13}$$

where a^{ν_1} is the restriction of (11) to $\hat{\Omega}^1$ with nonlinear reluctivity function ν_1, and a^{ν_2} is the restriction of (11) to $\hat{\Omega}^1$ with piecewise constant reluctivity function ν_2. Application of Newton's method requires the computation of the derivative of a^{ν_1}, which is given by

$$a'[u](w, v; p) = \int_{\Omega^1(p)} \frac{\nu_1'(|\nabla u|)}{|\nabla u|}(\nabla u \cdot \nabla w)(\nabla u \cdot \nabla v)dx + a^{\nu_1}(w, v; p) \tag{14}$$

and transformed as in (11) to the reference domain $\hat{\Omega}^1$ with the change of variables formula.

We then introduce a high-dimensional finite element discretization ("truth" approximation) of our problem in the space $\hat{X}_{\mathcal{N}} = \text{span}\{\phi_1, \dots, \phi_{\mathcal{N}}\} \subset \hat{X}$ of piecewise linear and continuous finite element functions. The finite element approximation is obtained by a standard Galerkin projection: given the ansatz $\hat{u}_{\mathcal{N}}(p) = \sum_{j=1}^{\mathcal{N}} \hat{u}_{\mathcal{N}\, j}(p)\phi_j$ for the discrete solution and testing against the basis elements in $\hat{X}_{\mathcal{N}}$ leads to the system

$$\sum_{j=1}^{\mathcal{N}} A_{ij}^{\mathcal{N}}(p)\hat{u}_{\mathcal{N}\, j}(p) = F_{\mathcal{N}\, i}(p), \quad 1 \le i \le \mathcal{N}, \tag{15}$$

of nonlinear algebraic equations, where $F_{\mathcal{N}}(p) \in \mathbb{R}^{\mathcal{N}}$, $F_{\mathcal{N}\, j}(p) = f(\phi_j; p)$, $1 \le j \le \mathcal{N}$ and $A^{\mathcal{N}}(p) \in \mathbb{R}^{\mathcal{N}\times\mathcal{N}}$, $A_{ij}^{\mathcal{N}}(p) = a[\hat{u}_{\mathcal{N}}(p)](\phi_j, \phi_i; p)$, $1 \le i, j \le \mathcal{N}$. We then apply a Newton iterative scheme: given a current iterate $\hat{\bar{u}}_{\mathcal{N}\, j}(p)$, $1 \le j \le \mathcal{N}$, we find an increment $\delta\hat{u}_{\mathcal{N}\, j}(p)$, $1 \le \ j \le \mathcal{N}$, such that

$$\sum_{j=1}^{\mathcal{N}} \bar{D}_{ij}^{\mathcal{N}}(p)\delta\hat{u}_{\mathcal{N}\, j}(p) = F_{\mathcal{N}\, i}(p) - \sum_{j=1}^{\mathcal{N}} \bar{A}_{ij}^{\mathcal{N}}(p)\hat{\bar{u}}_{\mathcal{N}\, j}(p), \quad 1 \le i \le \mathcal{N}, \tag{16}$$

where $\bar{D}^{\mathcal{N}}(p) \in \mathbb{R}^{\mathcal{N}\times\mathcal{N}}$, $\bar{D}_{ij}^{\mathcal{N}}(p) = a'[\hat{\bar{u}}_{\mathcal{N}}](\phi_j, \phi_i; p)$ and $\bar{A}^{\mathcal{N}}(p) \in \mathbb{R}^{\mathcal{N}\times\mathcal{N}}$, $\bar{A}_{ij}^{\mathcal{N}}(p) = a[\hat{\bar{u}}_{\mathcal{N}}](\phi_j, \phi_i; p)$, $1 \le i, j \le \mathcal{N}$ are computed at each Newton's iteration.

From here onward by the "truth" solution $\hat{u}(p)$, we understand its finite element approximation $\hat{u}_{\mathcal{N}}(p)$, assuming that the given finite element approximation is good enough.

3 Reduced Basis Approximation

3.1 An EIM-RB Method

To perform the reduced basis approximation, we first introduce a subset $\mathcal{D}_{train} \subset \mathcal{D}$ from which a sample $\mathcal{D}_N^u = \{\bar{p}_1 \in \mathcal{D}, \dots, \bar{p}_N \in \mathcal{D}\}$ with associated reduced basis space $\hat{W}_N^u = \text{span}\{\zeta_n := \hat{u}(\bar{p}_n),\ \ 1 \le n \le N\}$ of dimension N are built with the help of a weak Greedy algorithm. This algorithm constructs iteratively nested (Lagrangian) spaces $\hat{W}_n^u$, $1 \le n \le N$ using an a posteriori error estimator $\Delta_u(Y; p)$, which predicts the expected approximation error for a given parameter p in the space $\hat{W}_n^u = Y$. We want the expected approximation error to be less than the prescribed tolerance ε_{RB}. We initiate the algorithm with an arbitrary chosen parameter $\bar{p}_1$ with the corresponding snapshot $\hat{u}(\bar{p}_1)$ for the basis enrichment. Next we proceed as stated in the following Algorithm 1.

Algorithm 1 RB-Greedy algorithm

Input: Tolerance ε_{RB}, max. number of iterations $N_{\max}$, parameter set $\mathcal{D}_{train} \subset \mathcal{D}$
Output: RB spaces $\{\hat{W}_n^u\}_{n=1}^N$

while $m \leq N_{\max}$ and $\varepsilon_n := \max_{p \in \mathcal{D}_{train}} \Delta_u\,(\hat{W}_n^u, p) > \varepsilon_{RB}$ **do**
  $\bar{p}_n \leftarrow \arg\max_{p \in \mathcal{D}_{train}} \Delta_u\,(\hat{W}_{n-1}^u, p)$
  $\mathcal{D}_n^u \leftarrow \mathcal{D}_{n-1}^u \cup \{\bar{p}_n\}$
  $\hat{W}_n^u \leftarrow \hat{W}_{n-1}^u \bigoplus \mathrm{span}\{\zeta_n \equiv \hat{u}(\bar{p}_n)\}$
  $n \leftarrow n+1$
end while

We note that the basis functions ζ_n are also orthonormalized relative to the $(\cdot,\cdot)_{\hat{X}}$ inner product with a Gram-Schmidt procedure to generate a well-conditioned system of equations.

The empirical interpolation method (EIM) [3] is used to ensure the availability of offline/online decomposition in the presence of the nonlinearity. For the EIM nonlinearity approximation, we construct a sample $\mathcal{D}_M^\nu = \{p_1^\nu \in \mathcal{D}, \ldots, p_M^\nu \in \mathcal{D}\}$ and associated approximation spaces $W_M^\nu = \mathrm{span}\{\xi_m := \nu_1(\hat{u}(p_m^\nu); \hat{x}; p_m^\nu), 1 \leq m \leq M\} = \mathrm{span}\{q_1, \ldots, q_M\}$ together with a set of interpolation points $T_M = \{\hat{x}_1^M, \ldots, \hat{x}_M^M\}$. Then we build an affine approximation $\nu_1^M(\hat{u}(p); \hat{x}; p)$ of $\nu_1(\hat{u}(p); \hat{x}; p)$ as

$$
\begin{aligned}
\nu_1(\hat{u}(p); \hat{x}; p) :&= \nu_1(|J_{\mathcal{T}}^{-T}(\hat{x}, p)\nabla\hat{u}(\hat{x}, p)|) \approx \sum_{m=1}^{M} \varphi_m(p) q_m(\hat{x}) \\
&= \sum_{m=1}^{M} (B_M^{-1}\nu_p)_m q_m(\hat{x}) := \nu_1^M(\hat{u}(p); \hat{x}; p),
\end{aligned} \tag{17}
$$

where $\nu_p := \{\nu_1(\hat{u}(p); \hat{x}_m^M; p)\}_{m=1}^M \in \mathbb{R}^M$ and $B_M \in \mathbb{R}^{M \times M}$ with $(B_M)_{ij} = q_j(\hat{x}_i^M)$ is the interpolation matrix. The EIM algorithm is initiated with an arbitrary chosen sample point $p_1^\nu \in \mathcal{D}$ and then associated quantities are computed as follows:

$$
\xi_1 = \nu_1(\hat{u}(p_1^\nu); \hat{x}; p_1^\nu), \quad \hat{x}_1^M = \arg\sup_{\hat{x} \in \hat{\Omega}} |\xi_1(\hat{x})|, \quad q_1 = \frac{\xi_1}{\xi_1(\hat{x}_1^M)}. \tag{18}
$$

The next parameters in the sample S_M^ν are selected according to the following Algorithm 2:

Algorithm 2 EIM algorithm

Input: Tolerance ϵ_{EIM}, max. number of iterations $M_{\max}$, parameter set $\mathcal{D}_{train} \subset \mathcal{D}$
Output: Approximation spaces $\{W_m^{\nu}\}_{m=1}^{M}$, interpolation points $\{T_m\}_{m=1}^{M}$

while $m \le M_{\max}$ and $\delta_m^{max} > \epsilon_{EIM}$ **do**
  $[\delta_m^{max}, p_m^{\nu}] \leftarrow \arg\max\limits_{p\in\mathcal{D}_{train}} \inf\limits_{z\in W_{m-1}^{\nu}} \|\nu_1(\hat{u}(p); .; p) - z\|_{L^\infty(\hat{\Omega})}$
  $\mathcal{D}_m^{\nu} \leftarrow \mathcal{D}_{m-1}^{\nu} \cup \{p_m^{\nu}\}$
  $r_m(\hat{x}) = \nu_1(\hat{u}(p_m^{\nu}); \hat{x}; p_m^{\nu}) - \nu_1^m(\hat{u}(p_m^{\nu}); \hat{x}; p_m^{\nu})$
  $\hat{x}_m^M = \arg\sup\limits_{\hat{x}\in\hat{\Omega}} |r_m(\hat{x})|, \quad q_m = r_m / r_m(\hat{x}_m^M)$
  $m \leftarrow m + 1$
end while

The EIM approximation of ν_1 results in the EIM-approximation $a_M[\cdot](\cdot, \cdot; p)$ of the quasilinear form $a[\cdot](\cdot, \cdot; p)$ and then the reduced basis approximation is obtained by a standard Galerkin projection: given $p \in \mathcal{D}$, find $\hat{u}_{N,M}(p) \in \hat{W}_N^u$ such that

$$a_M[\hat{u}_{N,M}(p)](\hat{u}_{N,M}(p), \hat{v}_N; p) = f(\hat{v}_N; p), \quad \forall \hat{v}_N \in \hat{W}_N^u \tag{19}$$

holds. Since $\hat{\Omega} = \hat{\Omega}^1 \cup \hat{\Omega}^2$, we have the decomposition

$$a_M[\hat{w}](\hat{w}, \hat{v}; p) := a_M^{\nu_1}[\hat{w}](\hat{w}, \hat{v}; p) + a^{\nu_2}(\hat{w}, \hat{v}; p), \tag{20}$$

where $a_M^{\nu_1}[\cdot](\cdot, \cdot; p)$ is the EIM approximation of $a^{\nu_1}[\cdot](\cdot, \cdot; p)$ with nonlinear reluctivity $\nu_1(p)$ replaced by its EIM counterpart $\nu_1^M(p)$.

3.2 *Error Estimation*

We define $W_N^u(p) := \{w_N \mid w_N = \hat{w}_N \circ \mathcal{T}^{-1}, \ \hat{w}_N \in \hat{W}_N^u\}$ as a push-forward reduced basis space over the parametrized domain $\Omega(p)$ for error estimation purposes, where $\mathcal{T}^{-1}$ is the inverse of the geometric transformation (9). First, we study the convergence of $\hat{u}_{N,M}(p) \to \hat{u}(p)$.

Proposition 3.1 (A-priori Error Bound) *Assume that the EIM approximation error of the nonlinearity satisfies* $\sup_{p\in\mathcal{D}} \|\nu_1(p) - \nu_1^M(p)\|_{L^\infty} \le \epsilon_M$. *Assume further that* $a(\cdot; \cdot, \cdot; p)$ *is Lipschitz continuous on* $X(p)$ *with Lipschitz constant* $3\nu_0 > 0$ *and that the EIM approximation* $a_M(\cdot; \cdot, \cdot; p)$ *of* $a(\cdot; \cdot, \cdot; p)$ *is strongly monotone with monotonicity constant* $\tilde{\nu}_{LB} := \nu_{LB} - \epsilon_a > 0$. *Then we have*

$$\|\hat{u}(p) - \hat{u}_{N,M}(p)\|_{\hat{X}} \le \sqrt{\frac{C_2(p)}{C_1(p)}} \inf_{\hat{w}_N \in \hat{W}_N^u} \left\{ \left(1 + \frac{3\nu_0}{\tilde{\nu}_{LB}}\right) \|\hat{u}(p) - \hat{w}_N\|_{\hat{X}} + \frac{\epsilon_M}{\tilde{\nu}_{LB}} \|\hat{w}_N\|_{\hat{X}} \right\}$$

with the geometric constants

$$C_1(p) := \min_{1\le d\le L}\{\lambda_{min}(C_d(p)^{-1}C_d(p)^{-T})|\det C_d(p)|\} \tag{21}$$

and

$$C_2(p) := \max_{1\le d\le L}\{\lambda_{max}(C_d(p)^{-1}C_d(p)^{-T})|\det C_d(p)|\} \tag{22}$$

Proof Set $u := u(p) \in X(p)$, $u_{N,M} := u_{N,M}(p) \in W_N^u(p)$ and let $w_N \in W_N^u(p)$ be arbitrary. Set $\sigma_{N,M} := u_{N,M} - w_N$. First, we note that

$$a_M[u_{N,M}](u_{N,M}, w_N) - a[u](u, w_N) = 0, \ \forall w_N \in W_N^u(p) \subset X(p). \tag{23}$$

Then we use (23), the strong monotonicity condition and Lipschitz continuity to obtain the bound

$$\begin{aligned}
\tilde{\nu}_{\rm LB}\|\sigma_{N,M}\|^2_{X(p)} &\le a_M[u_{N,M}](u_{N,M}, \sigma_{N,M}) - a_M[w_N](w_N, \sigma_{N,M}) \\
&= a[u](u, \sigma_{N,M}) - a[w_N](w_N, \sigma_{N,M}) \\
&+ a[w_N](w_N, \sigma_{N,M}) - a_M[w_N](w_N, \sigma_{N,M}) \\
&\le 3\nu_0\|u - w_N\|_{X(p)}\|\sigma_{N,M}\|_{X(p)} \\
&+ \sup_{p\in\mathcal{D}}\|\nu_1(p) - \nu_1^M(p)\|_{L^\infty}\|w_N\|_{X(p)}\|\sigma_{N,M}\|_{X(p)}.
\end{aligned}$$

Dividing both sides by $\tilde{\nu}_{\rm LB}\|\sigma_{N,M}\|_{X(p)}$ and using the triangle inequality

$$\|u - u_{N,M}\|_{X(p)} \le \|u - w_N\|_{X(p)} + \|\sigma_{N,M}\|_{X(p)},$$

we obtain the estimate

$$\|u(p) - u_{N,M}(p)\|_{X(p)} \le \left(1 + \frac{3\nu_0}{\tilde{\nu}_{\rm LB}}\right)\|u(p) - w_N\|_{X(p)} + \frac{\epsilon_M}{\tilde{\nu}_{\rm LB}}\|w_N\|_{X(p)}. \tag{24}$$

Inspecting the geometric dependence with the lower bound,

$$\begin{aligned}
\|v\|^2_{X(p)} &= \sum_{d=1}^{L}\sum_{i,j=1}^{2}[C_d(p)^{-1}C_d(p)^{-T}]_{ij}|\det C_d(p)|\int_{\hat{\Omega}_d}\frac{\partial\hat{v}}{\partial\hat{x}_i}\frac{\partial\hat{v}}{\partial\hat{x}_j}\,d\hat{x} \qquad (25)\\
&\geqslant \min_{1\le d\le L}\{\lambda_{min}(C_d(p)^{-1}C_d(p)^{-T})|\det C_d(p)|\}\ \|\hat{v}\|^2_{\hat{X}} = C_1(p)\|\hat{v}\|^2_{\hat{X}},
\end{aligned}$$

applied to the left-hand side of (24), together with the similarly established upper bound

$$\|v\|^2_{X(p)} \le \max_{1\le d\le L}\{\lambda_{max}(C_d(p)^{-1}C_d(p)^{-T})|\det C_d(p)|\}\ \|\hat{v}\|^2_{\hat{X}} = C_2(p)\|\hat{v}\|^2_{\hat{X}} \tag{26}$$

applied to the right-hand side of (24), the desired result follows after a short calculation. □

For efficient implementation of the reduced basis methodology and the verification of the error, it is necessary to provide an a posteriori error bound, which can be quickly evaluated. For this, we establish an error bound based on the residual. We denote by $r_M(\cdot; p) \in \hat{X}'$ the residual (formed on the reference domain) of the problem, defined naturally as

$$r_M(\hat{v}; p) = f(\hat{v}; p) - a_M[\hat{u}_{N,M}](\hat{u}_{N,M}, \hat{v}; p). \tag{27}$$

We have the following:

Proposition 3.2 (A posteriori Error Bound) *Let* $\nu_{LB} > 0$ *be the lower bound of the monotonicity constant. Then, the RB-EIM error* $\hat{e}_{N,M}(p) := \hat{u}(p) - \hat{u}_{N,M}(p)$ *can be bounded by*

$$\|\hat{e}_{N,M}(p)\|_{\hat{X}} \leq \frac{\|r_M(\cdot; p)\|_{\hat{X}'}}{\nu_{LB}\, C_1(p)} + \frac{C_2(p)\delta_M(p)}{\nu_{LB} C_1(p)} \|\hat{u}_{N,M}(p)\|_{\hat{X}} := \Delta_{N,M}(p) \tag{28}$$

with the geometric constants (21), (22) *and the EIM approximation error*

$$\delta_M(p) = \sup_{\hat{x} \in \hat{\Omega}} |\nu_1(|J_{\mathcal{T}}^{-T}(\hat{x}, p)\nabla\hat{u}_{N,M}(\hat{x}; p)|) - \nu_1^M(|J_{\mathcal{T}}^{-T}(\hat{x}, p)\nabla\hat{u}_{N,M}(\hat{x}; p)|)| \tag{29}$$

of the nonlinearity

Proof Since in the case $e_{N,M} = 0$, there is nothing to show, we assume that $e_{N,M} \neq 0$. We then use strong monotonicity condition (7) and the definition of the residual (27) to estimate

$$\begin{aligned}
\nu_{\text{LB}} \|e_{N,M}\|^2_{X(p)} &\leq a[u](u, e_{N,M}) - a[u_{N,M}](u_{N,M}, e_{N,M}) \\
&= f(e_{N,M}) - a_M[u_{N,M}](u_{N,M}, e_{N,M}) \\
&\quad + a_M[u_{N,M}](u_{N,M}, e_{N,M}) - a[u_{N,M}](u_{N,M}, e_{N,M}) \\
&:= r_M(e_{N,M}) + a_M[u_{N,M}](u_{N,M}, e_{N,M}) - a[u_{N,M}](u_{N,M}, e_{N,M}) \\
&= r_M(\hat{e}_{N,M}) + a_M[u_{N,M}](u_{N,M}, e_{N,M}) - a[u_{N,M}](u_{N,M}, e_{N,M}) \\
&\leq \|r_M\|_{\hat{X}'} \|\hat{e}_{N,M}\|_{\hat{X}} + \delta_M(p) \|u_{N,M}\|_{X(p)} \|e_{N,M}\|_{X(p)}
\end{aligned}$$

Now, the final result follows from the estimate (25) and (26), applied to $\|e_{N,M}\|^2_{X(p)}$ and the right-hand side of the inequality, corresponding,ly. □

We address the computational realization of the estimator (28) in the next section. Next we denote by $r(\cdot; p) \in \hat{X}'$ the residual of the original problem (without EIM reduction), defined as

$$r(\hat{v}; p) = f(\hat{v}; p) - a[\hat{u}_N](\hat{u}_N, \hat{v}; p) \tag{30}$$

and let $\hat{e}_N(p) := \hat{u}(p) - \hat{u}_N(p)$ be the error of the reduced basis approximation. Along the lines of Proposition 3.2 one can prove the error bound

$$\|\hat{e}_N(p)\|_{\hat{X}} \leq \frac{\|r(\cdot; p)\|_{\hat{X}'}}{\nu_{\mathrm{LB}}\, C_1(p)} := \Delta_N(p). \tag{31}$$

We use (31) to investigate the factor of overestimation in the reduced basis approximation.

Proposition 3.3 (Effectivity bound for RB-approximation) *Let* $\eta_N(p) = \frac{\Delta_N(p)}{\|\hat{e}_N\|_{\hat{X}}}$. *Then*

$$\eta_N(p) \leq \frac{3\nu_0}{\nu_{LB}} \sqrt{C_1(p) C_2(p)} \tag{32}$$

Proof Let $\hat{v}_r \in \hat{X}$ denote the Riesz representative of $r(\cdot; p)$. Then we have

$$\langle \hat{v}_r, \hat{v} \rangle_{\hat{X}} = r(\hat{v}; p), \;\; \hat{v} \in \hat{X}, \quad \|\hat{v}_r\|_{\hat{X}} = \|r(\cdot; p)\|_{\hat{X}'}.$$

Now let $v_r := \hat{v}_r \circ \mathcal{T}^{-1} \in X(p)$. Then, using Lipshitz continuity of (8), we have

$$\begin{aligned} \|v_r\|^2_{X(p)} = \langle v_r, v_r \rangle_{X(p)} = r(v_r; \mu) &= a[u](u, v_r; p) - a[u_N](u_N, v_r; p) \\ &\leq 3\nu_0 \|e_N\|_{X(p)} \|v_r\|_{X(p)}. \end{aligned}$$

With the estimates (25) and (26), applied to both sides of this inequality, we obtain

$$\frac{\|\hat{v}_r\|_{\hat{X}}}{\|\hat{e}_N\|_{\hat{X}}} \leq 3\nu_0 \sqrt{\frac{C_2(p)}{C_1(p)}}.$$

With (31), we then conclude

$$\eta_N(p) = \frac{\Delta_N(p)}{\|\hat{e}_N\|_{\hat{X}}} = \frac{\|\hat{v}_r\|_{\hat{X}}}{\nu_{\mathrm{LB}}\, C_1(p) \|\hat{e}_N\|_{\hat{X}}} \leq \frac{3\nu_0}{\nu_{\mathrm{LB}}} \sqrt{C_1(p) C_2(p)}.$$

and obtain the effectivity bound. □

This bound is further used to explain the gap between the true error and the estimator.

3.3 Computational Procedure

The computational process in the reduced basis modelling can be split into the offline and the online phase. The computations in the offline phase depend on the dimension $\mathcal{N}$ of the finite element space and are expensive, but should be performed only once. The computations in the online phase are independent of $\mathcal{N}$, with computational complexity which depends only on the dimension N of the reduced basis approximation space and the dimension M of the EIM approximation space. The key concept utilized here is parameter separability (or affine decomposition) of all the forms involved in the problem. With EIM, we can achieve an affine decomposition of the quasilinear form

$$a_M^{\nu_1}[\hat{u}_{N,M}(p)](\hat{w},\hat{v};p) = \sum_{m=1}^{M}\sum_{d=1}^{L_1}\sum_{i,j=1}^{2} \varphi_m(p)\Phi_{d,L_1}^{i,j}(p)a_{m,d}^{i,j}(\hat{w},\hat{v}), \tag{33}$$

$$a^{\nu_2}(\hat{w},\hat{v};p) = \sum_{d=1}^{L_2}\sum_{i,j=1}^{2} \Phi_{d,L_2}^{i,j}(p)a_d^{i,j}(\hat{w},\hat{v}),$$

such that $\Phi_{d,L_1}^{i,j} : \mathcal{D} \to \mathbb{R}$ for $d = 1,\ldots,L_1, i, j = 1,2$ and $\Phi_{d,L_2}^{i,j} : \mathcal{D} \to \mathbb{R}$ for $d = 1,\ldots,L_2, i, j = 1,2$ are functions depending on p and on the parameter-independent forms

$$a_{m,d}^{i,j}(\hat{w},\hat{v}) = \int_{\hat{\Omega}_d^1} q_m \frac{\partial \hat{w}}{\partial \hat{x}_i}\frac{\partial \hat{v}}{\partial \hat{x}_j}\, d\hat{x},\ 1 \le d \le L_1,\ 1 \le i, j \le 2,$$

$$a_d^{i,j}(\hat{w},\hat{v}) = \int_{\hat{\Omega}_d^2} \frac{\partial \hat{w}}{\partial \hat{x}_i}\frac{\partial \hat{v}}{\partial \hat{x}_j}\, d\hat{x},\ 1 \le d \le L_2,\ 1 \le i, j \le 2.$$

For notational convenience, we set $c_m(\hat{w},\hat{v};p) := \sum_{d=1}^{L_1}\sum_{i,j=1}^{2} \Phi_{d,L_1}^{i,j}(p)a_{m,d}^{i,j}(\hat{w},\hat{v})$ so that

$$a_M^{\nu_1}[\hat{u}_{N,M}(p)](\hat{w},\hat{v};p) = \sum_{m=1}^{M} \varphi_m(p)c_m(\hat{w},\hat{v};p).$$

Similarly, the affine decomposition of f has the form

$$f(\hat{v};p) = \int_{\hat{\Omega}} J_e \hat{v}\, d\hat{x} - \sum_{d=1}^{L}\sum_{i=1}^{2} |\det C_d(p)| C_d(p)^{-T}_{1\,i} \int_{\hat{\Omega}_d} H_{pm,1}\frac{\partial \hat{v}}{\partial \hat{x}_i} d\hat{x}$$

$$+ \sum_{d=1}^{L}\sum_{i=1}^{2} |\det C_d(p)| C_d(p)^{-T}_{2\,i} \int_{\hat{\Omega}_d} H_{pm,2}\frac{\partial \hat{v}}{\partial \hat{x}_i} d\hat{x} = \sum_{q=1}^{Q_f} \Phi_q^f(p) f_q(\hat{v}),$$

where $\Phi_q^f : \mathcal{D} \to \mathbb{R}$ for $q = 1, \ldots, Q_f$ are parameter-dependent functions and parameter-independent forms $f_q(\hat{v})$.

We now give the details of the numerical scheme for the nonlinear part, defined on the domain $\hat{\Omega}^1$. The second term in (33) is linear and can be treated similarly. We expand our reduced basis solution as $\hat{u}_{N,M}(p) = \sum_{j=1}^{N} \hat{u}_{N,M\ j}\zeta_j$ and test against the basis elements in $\hat{W}_N^u$ to obtain the algebraic equations

$$\sum_{j=1}^{N}\sum_{m=1}^{M} \varphi_m(p) C_{i\ m}^{j(N,M)}(p)\hat{u}_{N,M\ j}(p) = F_{N\ i}(p), \quad 1 \le i \le N, \tag{34}$$

where $C^{j(N,M)}(p) \in \mathbb{R}^{N\times M}$, $C_{i\ m}^{j(N,M)}(p) = c_m(\zeta_j, \zeta_i; p)$, $1 \le i \le N$, $1 \le m \le M$, $1 \le j \le N$, and $F_{N\ i}(p) = f(\zeta_i; p)$. Since $\varphi_M(p) = \{\varphi_{M\ k}(p)\}_{k=1}^{M} \in \mathbb{R}^M$ is given by

$$\begin{aligned}\sum_{k=1}^{M} B_{m\ k}^{M}\varphi_{M\ k}(p) &= \nu_1(\hat{u}_{N,M}(\hat{x}_m^M; p); \hat{x}_m^M; p), \quad 1 \le m \le M \\ &= \nu_1\left(\sum_{n=1}^{N} \hat{u}_{N,M\ n}(p)\zeta_n(\hat{x}_m^M); \hat{x}_m^M; p\right), \quad 1 \le m \le M.\end{aligned} \tag{35}$$

We then insert (35) into (34) to get the following nonlinear algebraic equation system:

$$\sum_{j=1}^{N}\sum_{m=1}^{M} D_{i\ m}^{j(N,M)}(p)\nu_1\left(\sum_{n=1}^{N} \hat{u}_{N,M\ n}(p)\zeta_n(\hat{x}_m^M); \hat{x}_m^M; p\right) \hat{u}_{N,M\ j}(p) = F_{N\ i}(p), \tag{36}$$

where $1 \le i \le N$ and $D^{j(N,M)}(p) = C^{j(N,M)}(p)(B^M)^{-1} \in \mathbb{R}^{N\times M}$.

To solve (36) for $\hat{u}_{N,M\ j}(p)$, $1 \le j \le N$, we apply a Newton's iterative scheme: given the current iterate $\hat{\hat{u}}_{N,M\ j}(p)$, $1 \le j \le N$, compute an increment $\delta\hat{u}_{N,M\ j}(p)$, $1 \le j \le N$, from

$$\sum_{j=1}^{N}[\bar{A}_{ij}^{N}(p) + \bar{E}_{ij}^{N}(p)]\delta\hat{u}_{N,M\ j}(p) = R_{N\ i}(p), \quad 1 \le i \le N, \tag{37}$$

and update $\hat{\hat{u}}_{N,M\ j}(p) := \hat{\hat{u}}_{N,M\ j}(p) + \delta\hat{u}_{N,M\ j}(p)$, where the residual $R_N(p) \in \mathbb{R}^N$ for the Newton's scheme must be calculated at every Newton iteration according to

$$R_{N\ i}(p) = F_{N\ i}(p) - \sum_{j=1}^{N}\sum_{m=1}^{M} D_{i\ m}^{j(N,M)}(p)\nu_1(\hat{\hat{u}}_{N,M}(\hat{x}_m^M; p); \hat{x}_m^M; p)\hat{\hat{u}}_{N,M\ j}(p). \tag{38}$$

Furthermore, $\bar{A}^N(p) \in \mathbb{R}^{N\times N}$, $\bar{A}^N_{ij}(p) = a^{\nu_1}_M[\hat{\bar{u}}_{N,M}(p)](\zeta_j, \zeta_i; p)$ and $\bar{E}^N(p) \in \mathbb{R}^{N\times N}$ with

$$\bar{E}^N_{ij}(p) = \sum_{s=1}^{N} \hat{\bar{u}}_{N,M\,s}(p) \sum_{m=1}^{M} D^{s(N,M)}_{i\,m}(p) \frac{g^j_m(p)\partial_1 \nu_1(\hat{\bar{u}}_{N,M}(\hat{x}^M_m; p); \hat{x}^M_m; p)}{|J^{-T}_{\mathcal{T}}(\hat{x}^M_m, p)\nabla\hat{\bar{u}}_{N,M}(\hat{x}^M_m; p)|}, \quad (39)$$

where $1 \leq i, j \leq N$ and

$$g^j_m(p) = [J^{-T}_{\mathcal{T}}(\hat{x}^M_m, p)\nabla\hat{\bar{u}}_{N,M}(\hat{x}^M_m; p)] \cdot [J^{-T}_{\mathcal{T}}(\hat{x}^M_m, p)\nabla\zeta_j(\hat{x}^M_m)]$$

for $1 \leq m \leq M$. In (39), $\partial_1\nu_1$ denotes the partial derivative of ν_1 with respect to its first argument

Although (39) looks quite involved, it possesses an affine decomposition and allows efficient assembling in the online phase. Indeed, the matrix $D^{j(N,M)}(p)$ is parameter-separable, since $C^{j(N,M)}(p)$ is parameter-separable and the evaluation of $g^j \in \mathbb{R}^M$ in (39) requires the evaluation of the reduced basis functions only on the set of interpolation points T_M. Therefore, these quantities can be computed and stored in the offline phase and can be assembled in the online phase independently of $\mathcal{N}$. The operation count associated with each Newton's update is then as follows: the assembling of the residual $R_N(p)$ in (38) is achieved at cost $O(MN^2)$ together with the EIM system solve at cost $O(M^2)$. The Jacobian $\bar{A}^N(p) + \bar{E}^N(p)$ in (36) is assembled at cost $O(MN^3)$, where the dominant cost is for the assembling of $\bar{E}^N(p)$. It is then inverted at cost $O(N^3)$. The operation count in the online phase is thus $O(MN^3)$ per Newton iteration. However, we observe in our numerical experiment that it is sufficient to use $\bar{A}^N(p)$ and drop $\bar{E}^N(p)$ term in (37), which results in $O(MN^2 + N^3)$ operations per Newton iteration.

Next, we address the computation of the a posteriori error bound (28). It requires the computation of the dual norm of the residual (27). Since the right-hand side $f(\cdot; p)$ and $a_M[\cdot](\cdot, \cdot; p)$ are parameter-separable, the residual $r_M(\cdot; p)$ is also parameter-separable and admits an affine decomposition together with its Riesz representative $\hat{v}_r(p) \in \hat{X}$ according to

$$r_M(\hat{v}; p) = \sum_{q=1}^{Q_r} \Phi^r_q(p) r_{M\,q}(\hat{v}), \quad \hat{v}_r(p) = \sum_{q=1}^{Q_r} \Phi^r_q(\mu)\hat{v}_{r\,q}, \quad (40)$$

where $r_M(\hat{v}; p) = (\hat{v}_r(p), \hat{v})_{\hat{X}}$ for all $\hat{v} \in \hat{X}$ and $Q_r = Q_f + N(M + 4ML_1 + 4L_2)$. Since the dual norm of the residual is equal to the norm of its Riesz representative, we have

$$\|r_M(\cdot; p)\|_{\hat{X}'} = \|\hat{v}_r(p)\|_{\hat{X}} = (\mathbf{\Phi}^r(p)^T G_r \mathbf{\Phi}^r(p))^{1/2}, \quad (41)$$

where $\mathbf{\Phi}^r(p) = \{\Phi^r_q(p)\}_{q=1}^{Q_r} \in \mathbb{R}^{Q_r}$ and $G_r \in \mathbb{R}^{Q_r\times Q_r}$ with $(G_r)_{ij} = (v_{r\,i}, v_{r\,j})_{\hat{X}}$ and the dual norm (41) is then computed at cost $O(Q_r^2)$. The evaluation of the norm

$\|\hat{u}_{N,M}(p)\|_{\hat{X}}$ is at cost $O(N^2)$. Once ν_{LB} is available, the constants $C_1(p)$ and $C_2(p)$ in (28) are computed directly.

The EIM approximation error (29) is computed on the discretized domain $\hat{\Omega}_h \subset \hat{\Omega}$: the nonlinearity depends on the gradient and it is evaluated on the triangle barycenters $\hat{x}_{b_j}$, $1 \leq j \leq \mathcal{N}_T$, where $\mathcal{N}_T$ is the total number of triangles in the iron material region for a given finite element triangulation. The EIM procedure results in the set of triangle barycenter points $T_M = \{\hat{x}^M_{b_1}, \ldots, \hat{x}^M_{b_M}\}$, where $M << \mathcal{N}_T$. In the offline phase, we evaluate the gradients $\{\nabla\zeta_n\}_{n=1}^N$ for each basis element $\{\zeta_n\}_{n=1}^N$ of the reduced basis space $\hat{W}^u_N$ on the interpolation barycenters T_M. We thus store offline $\{\nabla\zeta_n|_{T_M}(\hat{x}^M_{b_j})\}_{j=1}^M \in \mathbb{R}^{M\times 2}$ for $1 \leq n \leq N$ and then efficiently evaluate the nonlinearity on T_M with the ansatz $\nabla\hat{u}_{N,M}(p)\big|_{T_M} = \sum_{j=1}^N \hat{u}_{N,M\ j}(p)\ \nabla\zeta_j\big|_{T_M}$ online. The operation count for the EIM approximation $\nu_1^M(\hat{u}_{N,M}(\hat{x};\, p);\, p)$ is then $O(M^2 + \mathcal{N}_T M)$, and the evaluation of ν_1 at M points. We note that (29) requires the knowledge of $\nu_1(\hat{u}_{N,M}(p);\, \hat{x};\, p)$ and thus one full evaluation of the nonlinearity. In order to increase the online computational efficiency, an one-point estimator $\hat{\varepsilon}_M(p)$ can be used (see, e.g., [6]). It requires the evaluation of the nonlinearity at only one point, but $\hat{\varepsilon}_M(p) \leq \delta_M(p)$ in general, thus this lower bound estimator must be effective, i.e., $\frac{\hat{\varepsilon}_M(p)}{\delta_M(p)}$ should close to 1. In our case, the nonlinearity is of the exponential type and the effectivity of the bound is of the order 10^2 in practice.

3.4 Numerical Results

First, we introduce a parameter set $\mathcal{D} = [18, 19] \times [4, 5] \times [7, 8]$. The nonlinear reluctivity function $\nu_1(p)$ is reconstructed from the real $B - H$ measurements using cubic spline interpolation. Finite element simulations are based on a mesh composed of 121012 triangles and 60285 nodes (excluding Dirichlet boundary nodes). Piecewise linear, continuous finite element functions are chosen for the finite element approximation. We solve the finite element problem with Newton's method. We iterate unless the norm of the residual is less than the tolerance level, which we set to 10^{-4}. The tolerance level 10^{-5} is used for the RB Newton's method.

We generate the RB-EIM model as follows: we start from $\mathcal{D}^{EIM(1)}_{train} \subset \mathcal{D}$ (a regular $6 \times 6 \times 6$ grid over $\mathcal{D}$ of size 216) and compute finite element solutions for each parameter in $\mathcal{D}^{EIM(1)}_{train}$ to approximate the nonlinearity with the EIM within the prescribed tolerance $\epsilon_{EIM} = 5 \cdot 10^{-1}$. Since the norm $\|\hat{u}_{N,M}(p)\|_{\hat{X}}$ is of the order 10^{-2}, we hope to further balance the contributions of the reduced basis and EI nonlinearity approximation in the estimator on the test set. Next we run the RB-Greedy procedure with the prescribed tolerance $\epsilon_{RB} = 10^{-2}$ for the estimator (28) on $\mathcal{D}_{train} \subset \mathcal{D}$, where $\mathcal{D}_{train}$ is a regular $10 \times 10 \times 10$ grid over $\mathcal{D}$ of size 1000. We set $\nu_{LB} = 110$, since

$$\nu_{LB} \leq \min_{\hat{x}\in\hat{\Omega}} \nu_1(|J_T^{-T}(\hat{x},\, p)\nabla\hat{u}_{N,M}(\hat{x};\, p)|) \simeq 110 \tag{42}$$

for all $p \in \mathcal{D}_{train}$ in our setting. This is a robust heuristic procedure, since for small N, the reduced basis solution $\hat{u}_{N,M}(\hat{x}; p)$ is a good approximation to $\hat{u}(\hat{x}; p)$ in the regions with low magnetic flux density $|\nabla \hat{u}(\cdot; p)|$. The size of the magnet (change in the parameter p) influences only the high values of the magnetic flux density $|\nabla \hat{u}_{N,M}(\cdot; p)|$ in the magnetic circuit and does not have an impact on the minimum of the reluctivity function. We note that the evaluation of $\delta_M(p)$ (29) requires one full evaluation of the nonlinearity, thus it is available for the computation in (42) for the a posteriori error estimation.

Once the reduced basis model is constructed ($N_{max} = 12$, $M_{max} = 50$), we use it to improve the quality of the nonlinearity approximation: we generate the reduced basis solutions over $\mathcal{D}_{train}^{EIM(2)} := \mathcal{D}_{train}$ and use them to construct the improved EIM approximation space W_M^{ν} of dimension $M_{max} = 50$. With the new approximation of the nonlinearity, we run the RB-Greedy procedure over $\mathcal{D}_{train}$ again with the prescribed tolerance $\epsilon_{RB} = 10^{-2}$, which results in the reduced basis space $\hat{W}_N^u$ of dimension $N_{max} = 10$.

Next, we introduce a parameter test sample $\mathcal{D}_{test} \subset \mathcal{D}$ of size 343 ($7 \times 7 \times 7$ grid with uniformly random sampling on each interval) and verify the convergence with N of $\max \Delta_{N,M} = \max\limits_{p \in \mathcal{D}_{test}} \Delta_{N,M}(p)$ for different values of M (see Fig. 2a). We see that with $N = 8$ and $M = 50$ the estimator is below the prescribed tolerance $\epsilon_{RB} = 10^{-2}$ on the test set. One observes that there is an increase in the estimator for $N \geq 8$ and for $M < 50$ due to the poor quality of the EIM approximation. Moreover, we can naturally split the estimator into two parts: the reduced basis and the nonlinearity approximation error estimation contributions

$$\Delta_{N,M}^{RB}(p) := \frac{\|r_M(\cdot; p)\|_{\hat{X}'}}{\nu_{\mathrm{LB}}\, C_1(p)}, \quad \Delta_{N,M}^{EI}(p) := \frac{C_2(p)\delta_M(p)}{\nu_{\mathrm{LB}}\, C_1(p)} \|\hat{u}_{N,M}(p)\|_{\hat{X}}. \tag{43}$$

We then set

$$\Delta_{N,M}^{RB} := \max_{p \in \mathcal{D}_{test}} \Delta_{N,M}^{RB}(p), \quad \Delta_{N,M}^{EI} := \max_{p \in \mathcal{D}_{test}} \Delta_{N,M}^{EI}(p). \tag{44}$$

The strategy is to balance two contributions in (44) for the specified tolerance level ε_{RB}, e.g., by choosing $N = 8$ and $M = 50$, see Fig. 2b. In Fig. 2b we can also see the improvement from the described above additional EIM step.

In Table 1 we present, as a function of N and M, the maximum error bound $\max\limits_{p \in \mathcal{D}_{test}} \Delta_{N,M}(p)$ as well as the mean $\bar{\eta}_{N,M}$ and $\max \eta_{N,M}$ of the effectivity $\eta_{N,M}(p) := \frac{\Delta_{N,M}(p)}{\|\hat{e}_{N,M}\|_{\hat{X}}}$. The effectivities require the knowledge of "truth" solution, therefore we compute the finite element solutions for all the parameters in the test set. We observe that the values of $\bar{\eta}_{N,M}$ and $\max \eta_{N,M}$ are quite large, which partially can be explained by the estimate (32) for the effectivity $\eta_N(p)$ of the reduced basis approximation. In our example, we have

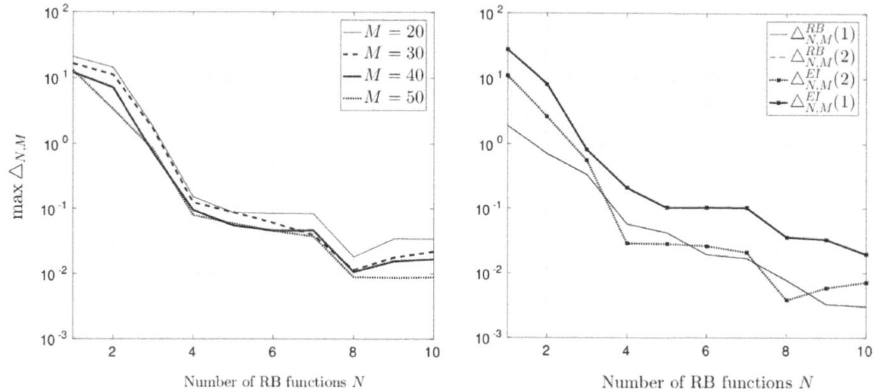

(a) (b)

Fig. 2 Convergence with N of $\max \triangle_{N,M}$ for different values of M on the test set (**a**). Convergence with N of $\triangle^{RB}_{N,M}$ and $\triangle^{EI}_{N,M}$ contributions for $M = 50$ on the test set. The number in the label bracket indicates the EIM step (**b**)

Table 1 Performance of RB-EIM model on the test set

N	M	$\max \triangle_{N,M}$	$\bar{\triangle}_{N,M}(p)$	$\bar{\eta}_{N,M}$	$\max \eta_{N,M}$
4	30	1.24 E-01	4.74 E-02	7.41 E02	1.46 E03
6	40	4.59 E-02	2.37 E-02	3.98 E02	7.18 E02
8	45	9.30 E-03	5.10 E-03	2.46 E02	6.24 E02
8	50	8.90 E-03	5.51 E-03	2.49 E02	6.32 E02
10	50	8.90 E-03	5.30 E-03	8.48 E02	4.65 E03

$$\max_{\hat{x}\in\hat{\Omega}} \nu_1(|J_{\mathcal{T}}^{-T}(\hat{x}, p)\nabla \hat{u}_{N,M}(\hat{x}; p)|) \leq \nu_0$$

on $\mathcal{D}_{test}$, where $\nu_0 \approx 7.95 \times 10^5$ is the reluctivity of air. Therefore, the upper bound constant for $\eta_N(p)$ is of order 10^3 in practice.

In Fig. 3 we plot the reduced basis solutions, i.e., the magnetic equipotential lines for several parameters and the corresponding reluctivity functions, evaluated fully with splines and with EIM. Next, we compare the average CPU time required for both the finite element method, which takes $\approx$150 s to obtain the solution, and the RB method ($N_{max} = 10$, $M_{max} = 50$), which takes $\approx$0.27/0.95 s without/with the error bound evaluation and results in the speedup factors of 555 and 158, respectively.[1] The computation of the error bound significantly increases the total CPU time since the complexity of the error bound evaluation scales quadratically with Q_r, where Q_r is large and requires one full evaluation of the nonlinearity. The offline phase requires the knowledge of the "truth" finite-element solutions for the first EIM approximation

[1] All the computations are performed in MATLAB on Intel Xeon(R) CPU E5-1650 v3, 3.5 GHz x 12 cores, 64 GB RAM.

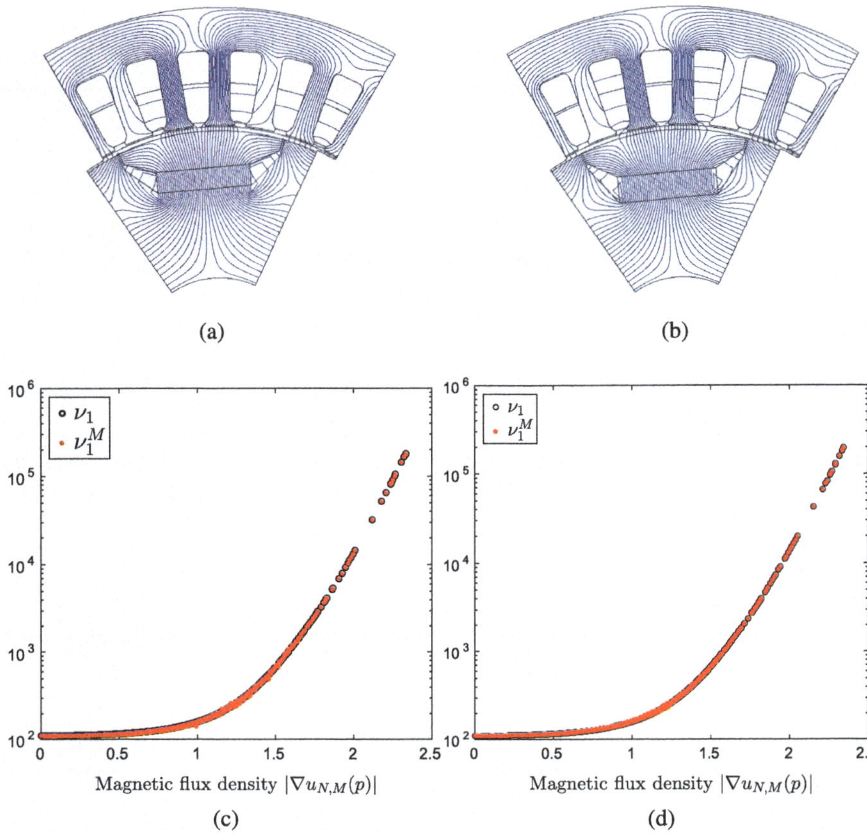

Fig. 3 Magnetic equipotential lines, computed with reduced basis method (10 RB functions, 50 EIM basis functions) for parameter value **a** $p = (18, 4, 7)$, **b** $p = (19, 5, 8)$. Reluctivity function $\nu_1(p)$, computed with full spline approximation and its EIM counterpart $\nu_1^M(p)$ for parameter value **c** $p = (18, 4, 7)$, **d** $p = (19, 5, 8)$

step. Since 216 finite element solutions were generated in the consecutive order, it takes $\approx$9 h, but it can be done in parallel to reduce the computational time. The Greedy algorithm execution takes $\approx$4 h, and since we run it twice, it takes $\approx$8 h for our implementation. We note that our implementation may not be optimal, therefore the offline time is only a rough estimate.

We also note that in the presented numerical example the relatively small parameter domain $\mathcal{D}$ was chosen. In the author's opinion, it is possible to enlarge the parameter domain with the increasing cost of the nonlinearity approximation by combining few additional EIM steps as described above and exploiting divide-and-conquer principles and hp-adaptivity in the Greedy procedure (see, e.g., [5, 13]).

4 Conclusion

In this paper, we propose the reduced basis method for quasilinear elliptic PDEs with application to the nonlinear magnetostatic problem. The geometric parametrization for the PDE is introduced in the setting of magnet design for the permanent magnet electric motor. We present a new a -posteriori error bound for the class of problems we consider and use it for the weak Greedy algorithm and corresponding reduced basis construction. The affine decomposition of the quasilinear form was achieved with the help of EIM. Numerical results confirm a significant speed-up factor which supports the validity of the proposed approach.

Acknowledgements Both authors acknowledge the support of the collaborative research project PASIROM funded by the German Federal Ministry of Education and Research (BMBF) under grant no. 05M2018.

References

1. Abdulle, A., Bai, Y., Vilmart, G.: Reduced basis finite element heterogeneous multiscale method for quasilinear elliptic homogenization problems. Discret. Contin. Dyn. Syst. S **8**(1), 91–118 (2015)
2. Alla, A., Hinze, M., Kolvenbach, P., et al.: A certified model reduction approach for robust parameter optimization with PDE constraints. Adv. Comput. Math. **45**, 1221–1250 (2019)
3. Barrault, M., Maday, Y., Nguyen, N.C., Patera, A.T.: An empirical interpolation method: application to efficient reduced-basis discretization of partial differential equations. C.R. Acad. Sci. Paris Ser. I **339**(9), 667–672 (2004)
4. Bontinck, Z., Lass, O., Schöps, S., et al.: Robust optimisation formulations for the design of an electric machine. IET Sci. Meas. Technol. **12**(8), 939–948 (2018)
5. Eftang, J.L., Stamm, B.: Parameter multi-domain ' hp' empirical interpolation. Int. J. Numer. Methods Eng. **90**, 412–428 (2012)
6. Grepl, M.A., Maday, Y., Nguyen, N.C., Patera, A.T.: Efficient reduced-basis treatment of non-affine and nonlinear partial differential equations. ESAIM: Math. Model. Numer. Anal. **41**(3), 575–605 (2007)
7. Haasdonk, B.: Reduced basis methods for parametrized PDEs - a tutorial introduction for stationary and instationary problems. In: Benner, P., Cohen, A., Ohlberger, M., Willcox, K. (eds.) Chapter in Model Reduction and Approximation: Theory and Algorithms, pp. 65–136. SIAM, Philadelphia (2017)
8. Heise, B.: Analysis of a fully discrete finite element method for a nonlinear magnetic field problem. SIAM J. Numer. Anal. **31**(3), 745–759 (1994)
9. Ion, I.G., Bontinck, Z., Loukrezis, D., et al.: Robust shape optimization of electric devices based on deterministic optimization methods and finite-element analysis with affine parametrization and design elements. Electr. Eng. **100**, 2635–2647 (2018)
10. Lass, O., Ulbrich, S.: Model order reduction techniques with a posteriori error control for nonlinear robust optimization governed by partial differential equations. SIAM J. Sci. Comput. **39**, S112–S139 (2017)
11. Quarteroni, A., Manzoni, A., Negri, F.: Reduced Basis Methods for Partial Differential Equations: An Introduction, vol. 92. Springer International Publishing, Switzerland (2016)
12. Rozza, G., Huynh, D.B.P., Patera, A.T.: Reduced basis approximation and a posteriori error estimation for affinely parametrized elliptic coercive partial differential equations. Arch. Comput. Methods Eng. **15**(3), 229–275 (2008)

13. Sen, S.: Reduced basis approximation and a posteriori error estimation for many-parameter heat conduction problems. Numer. Heat Transf. Part B **54**, 369–389 (2008)
14. Zeidler, E.: Nonlinear Functional Analysis and Its Applications II/B: Nonlinear Monotone Operators. Springer Science + Business Media, New York (1990)

Structure-Preserving Reduced- Order Modeling of Non-Traditional Shallow Water Equation

Süleyman Yildiz, Murat Uzunca, and Bülent Karasözen

Abstract An energy- preserving reduced -order model (ROM) is developed for the non-traditional shallow water equation (NTSWE) with full Coriolis force. The NTSWE in the noncanonical Hamiltonian/Poisson form is discretized in space by finite differences. The resulting system of ordinary differential equations is integrated in time by the energy preserving average vector field (AVF) method. The Poisson structure of the discretized NTSWE exhibits a skew-symmetric matrix depending on the state variables. An energy- preserving, computationally efficient reduced order model (ROM) is constructed by proper orthogonal decomposition with Galerkin projection. The nonlinearities are computed for the ROM efficiently by discrete empirical interpolation method. Preservation of the discrete energy and the discrete enstrophy are shown for the full- order model, and for the ROM which ensures the long- term stability of the solutions. The accuracy and computational efficiency of the ROMs are shown by two numerical test problems.

Keywords Shallow water equation · Model order reduction · Hamiltonian mechanics · Finite difference methods · Implicit time integrator

1 Introduction

The shallow water equation (SWE) consists of a set of two-dimensional partial differential equations (PDEs) describing a thin inviscid fluid layer flowing over the

S. Yildiz (✉)
Institute of Applied Mathematics, Middle East Technical University, Ankara, Turkey
e-mail: yildiz.suleyman@metu.edu.tr

M. Uzunca
Department of Mathematics, Sinop University, Sinop, Turkey
e-mail: muzunca@sinop.edu.tr

B. Karasözen
Institute of Applied Mathematics & Department of Mathematics, Middle East Technical University, Ankara, Turkey
e-mail: bulent@metu.edu.tr

P. Benner et al. (eds.), *Model Reduction of Complex Dynamical Systems*,
International Series of Numerical Mathematics 171,
https://doi.org/10.1007/978-3-030-72983-7_15

topography in a rotating frame. SWE is a hyperbolic PDEs describing geophysical wave phenomena, e.g., the Kelvin and Rossby waves in the atmosphere and the oceans. SWEs are frequently used in large-scale geophysical flow prediction [5, 16], investigation of baroclinic instability [8, 43], and planetary flows [44]. Energy and enstrophy are the most important conserved quantities of the SWEs, whereas the energy cascades to large scales, whilst while enstrophy cascades to small scales [4, 42].

Real-time simulation of SWEs requires a large amount of computer memory and computing time. The reduced- order models (ROMs) have emerged as a powerful approach to reduce the computational cost of evaluating large systems of PDEs like the SWE by constructing a low-dimensional linear reduced subspace, that approximately represents the solution to the system of PDEs with a significantly reduced computational cost. The solutions of the high fidelity full- order model (FOM), generated by space-time discretization of PDEs are projected usually on low- dimensional reduced spaces using the proper orthogonal decomposition (POD), which is the widely used reduced order modeling technique. Applying POD Galerkin projection, the dominant POD modes of the PDEs are extracted from the snapshots of the FOM solutions. The computation of the FOM solutions is performed in the offline stage, whereas the reduced system from the low-dimensional subspace is solved in the online stage. The primary challenge in producing the low- dimensional models of the high dimensional discretized PDEs is the efficient evaluation of the nonlinearities. The computational cost is reduced by sampling the nonlinear terms and interpolating, known as hyper-reduction techniques [2, 3, 9, 11, 32, 47].

The naive application of POD or DEIM may not preserve the geometric structures, like the symplecticness, energy preservation, and passivity of Hamiltonian, Lagrangian, and port- Hamiltonian PDEs. The stability of reduced models over long-time integration and the structure-preserving properties has been recently investigated in the context of Lagrangian systems [10, 25], and for port-Hamiltonian systems [13]. For linear and nonlinear Hamiltonian systems, the symplectic model reduction technique, proper symplectic decomposition (PSD) is constructed for Hamiltonian systems like linear wave equation, sine-Gordon equation, nonlinear Schrödinger equation to ensure long- term stability of the reduced model [1, 34]. Recently, the average vector field (AVF) method is used as a time integrator to construct ROMsreduced order models for Hamiltonian systems like Korteweg-de Vries equation [23, 31] and nonlinear Schrödinger equation [24]. ROMsReduced order models for the SWEs are constructed in conservative form using POD-DEIM [27, 28], in the β-plane by POD-DEIM and tensorial POD [38, 39], by dynamic mode decomposition [6, 7], the f-plane using POD [19]. In these articles, the preservation of the energy and other conservative quantities in the reduced space isare not discussed.

In this paper, we have constructed structure-preserving ROMs for the non-traditional shallow water equation (NTSWE) [17, 40, 42] with the full Coriolis force. Replacing the first- order derivatives that appear in the NSTWE, a skew-gradient system, i.e., a non-canonical Hamiltonian system of ordinary differential equations (ODEs) is obtained. Time discretization of this system of non-canonical Hamiltonian system of ODEs by the AVF [15] leads to FOM, which preserves the discrete Hamiltonian and Casimirs. The skew-symmetric structure of the full- order skew-gradient system is preserved using the reduced- order technique in [23, 24, 31]. The full- order and reduced- order NTSWE have state- dependent skew-symmetric matrices, which does not allow separation of online and offline computations of the nonlinear terms. Following [31], we have shown that the complexity of the ROM can be reduced for the POD and for the discrete empirical interpolation method (DEIM) [11]. The numerical results for two different representative examples of the NTSWE confirm the structure- preserving features like preserving the Hamiltonian (energy) and enstropy. The efficiency of the ROMs are is demonstrated by achieved speed-ups with the POD and DEIM over the FOM solutions.

The paper is organized as follows. In Sect. 2, the NTSWE is described in the Hamiltonian form. The structure- preserving FOM in space and time is developed in Sect. 3. The ROM with POD and DEIM are constructed in Sect. 4. In Sect. 5, numerical results for two NTSWE examples are presented. The paper ends with some conclusions.

2 Shallow Water Equation

Most of the models of the ocean and atmosphere include only the contribution to the Coriolis force from the component of the planetary rotation vector that is locally normal to geopotential surfaces when the vertical length scales are much smaller than the horizontal length scales. This approach is known as traditional approximation. However, many atmospheric and oceanographic phenomena are substantially influenced by the non-traditional component of the Coriolis force [41], such as deep convection [30], Ekman spirals [26], and internal waves [22]. The nondimensional NTSWE [17, 40, 42] has the same structural form as the traditional SWE [37] by distinguishing between the canonical velocities $\tilde{u}(x, y, t)$ and $\tilde{v}(x, y, t)$, and particle velocities $u(x, y, t)$ and $v(x, y, t)$

$$
\begin{aligned}
\frac{\partial \tilde{u}}{\partial t} &= hqv - \frac{\partial \Phi}{\partial x}, \\
\frac{\partial \tilde{v}}{\partial t} &= -hqu + \frac{\partial \Phi}{\partial y}, \\
\frac{\partial h}{\partial t} &= -\frac{\partial}{\partial x}(hu) - \frac{\partial}{\partial y}(hv),
\end{aligned}
\tag{1}
$$

where x and y denote horizontal distances within a constant geopotential surface, and $h(x, y, t)$ is the height field. The one-layer NTSWE (1) describes an inviscid fluid flowing over bottom topography at $z = h_b(x, y)$ in a frame rotating with angular velocity vector $\mathbf{\Omega} = (\Omega^{(x)}, \Omega^{(y)}, \Omega^{(z)})$. The orientation of the x and y axes are considered arbitrary with respect to North. In traditional rotating and non-rotating SWEs, only the particle velocity components appear. The canonical velocity components are related to the canonical momentum per mass or to the depth average of particle velocities as

$$\tilde{u} = u + 2\Omega^{(y)}\left(h_b + \frac{1}{2}h\right), \quad \tilde{v} = v - 2\Omega^{(x)}\left(h_b + \frac{1}{2}h\right). \tag{2}$$

The Bernoulli potential Φ and potential vorticity q are given by

$$\Phi = \frac{1}{2}(u^2 + v^2) + g(h_b + h) + h(\Omega^{(x)}v - \Omega^{(y)}u),$$
$$q = \frac{1}{h}(2\Omega^{(z)} + \tilde{v}_x - \tilde{u}_y).$$

The traditional SWE and NTSWE differ only by a function of the space alone, so their time derivatives are identical. The non-rotating, traditional SWE [36] and NTSWE (1) have the same Hamiltonian structure and Poisson bracket [17, 40, 42]

$$\frac{\partial \tilde{z}}{\partial t} = \mathscr{J}(\tilde{z})\frac{\delta \mathscr{H}}{\delta \tilde{z}} = \begin{pmatrix} 0 & q & -\partial_x \\ -q & 0 & -\partial_y \\ -\partial_x & -\partial_y & 0 \end{pmatrix}\begin{pmatrix} hu \\ hv \\ \Phi \end{pmatrix}, \tag{3}$$

where $z = (u, v, h)$ and $\tilde{z} = (\tilde{u}, \tilde{v}, h)$, $\mathscr{J}$ is the symplectic (Poisson) matrix, and $\delta\mathscr{H}$ denotes the variational derivative of the Hamiltonian. The Hamiltonian or the energy of (1) is given in terms of particle velocity components by

$$\mathscr{H}(z) = \iint \left\{\frac{1}{2}h(u^2 + v^2) + gh\left(h_b + \frac{1}{2}h\right)\right\}d\mathbf{x}, \tag{4}$$

over a periodic domain. We remark that the Hamiltonian (4) is treated as a function of the canonical velocity components $\tilde{u}$ and $\tilde{v}$ and the layer thickness using the relations (2).

The non-canonical Hamiltonian form of NTSWE (3) is determined by the skew-adjoint Poisson bracket of two functionals $\mathscr{A}$ and $\mathscr{B}$ [29, 37] as

$$\{\mathscr{A}, \mathscr{B}\} = \iint \left(q\frac{\delta(\mathscr{A}, \mathscr{B})}{\delta(\tilde{u}, \tilde{v})} - \frac{\delta\mathscr{A}}{\delta\tilde{\boldsymbol{v}}}\cdot\nabla\frac{\delta\mathscr{B}}{\delta h} + \frac{\delta\mathscr{B}}{\delta\tilde{\boldsymbol{v}}}\cdot\nabla\frac{\delta\mathscr{A}}{\delta h}\right)d\mathbf{x}, \tag{5}$$

where $\tilde{\boldsymbol{v}} = (\tilde{u}, \tilde{v})$. The functional Jacobian is given by

$$\frac{\delta(\mathscr{A}, \mathscr{B})}{\delta(\tilde{u}, \tilde{v})} = \frac{\delta\mathscr{A}}{\delta\tilde{u}}\frac{\delta\mathscr{B}}{\delta\tilde{v}} - \frac{\delta\mathscr{B}}{\delta\tilde{u}}\frac{\delta\mathscr{A}}{\delta\tilde{v}}.$$

The Poisson bracket (5) is related to the skew-symmetric symplectic matrix $\mathscr{J}$ as $\{\mathscr{A}, \mathscr{B}\} = \{\mathscr{A}, \mathscr{J}\mathscr{B}\}$. Although the matrix $\mathscr{J}$ in (3) is not skew-symmetric, the skew-symmetry of the Poisson bracket appears after integrations by parts [29], and the Poisson bracket satisfies the Jacobi identity

$$\{\mathscr{A}, \{\mathscr{B}, \mathscr{D}\}\} + \{\mathscr{B}, \{\mathscr{D}, \mathscr{A}\}\} + \{\mathscr{D}, \{\mathscr{A}, \mathscr{B}\}\} = 0,$$

for any three functionals $\mathscr{A}$, $\mathscr{B}$, and $\mathscr{D}$. The preservation of the Hamiltonian follows from the antisymmetry of the Poisson bracket (5)

$$\frac{d\mathscr{H}}{dt} = \{\mathscr{H}, \mathscr{H}\} = 0.$$

Besides the Hamiltonian, there are other conserved quantities in form of Casimirs

$$\mathscr{C} = \iint hG(q)d\mathbf{x},$$

where G is an arbitrary function of the potential vorticity q. The Casimirs are additional constants of motion which commute with any functional $\mathscr{A}$, i.e., the Poisson bracket vanishes. Important special cases are the potential enstrophy

$$\mathscr{Z} = \frac{1}{2}\int\int hq^2 d\mathbf{x} = \frac{1}{2}\int\int \frac{1}{h}\left(\Omega^{(z)} + \frac{\partial \tilde{v}}{\partial x} - \frac{\partial \tilde{u}}{\partial y}\right)^2 d\mathbf{x},$$

the mass $\mathscr{M} = \iint hd\mathbf{x}$, and the vorticity $\mathscr{V} = \iint hqd\mathbf{x}$.

3 Full- Order Model

The skew-symmetry of the Poisson form is preserved for the thermal Shallow water equation using finite elements [18], for the rotational SWE with discontinuous Galerkin method [21] and finite volume method [35].

The NTSWE (1) is discretized by finite differences on a uniform grid in the spatial domain $(a, b) \times (c, d)$ with the nodes $\mathbf{x}_{ij} = (x_i, y_j)^T$, where $x_i = a + (i - 1)\Delta x$ and $y_j = c + (j-1)\Delta y$, $i = 1, \ldots, N_x + 1$, $j = 1, \ldots, N_y + 1$, and then discretized in space canonical and particle velocity components and height are given by

$$\begin{aligned}
\mathbf{u}(t) &= (u_{11}(t), \ldots, u_{1N_y}(t), u_{21}(t), \ldots, u_{2N_y}(t), \ldots, u_{N_xN_y}(t))^T,\\
\mathbf{v}(t) &= (v_{11}(t), \ldots, v_{1N_y}(t), v_{21}(t), \ldots, v_{2N_y}(t), \ldots, v_{N_xN_y}(t))^T,\\
\tilde{\mathbf{u}}(t) &= (\tilde{u}_{11}(t), \ldots, \tilde{u}_{1N_y}(t), \tilde{u}_{21}(t), \ldots, \tilde{u}_{2N_y}(t), \ldots, \tilde{u}_{N_xN_y}(t))^T,\\
\tilde{\mathbf{v}}(t) &= (\tilde{v}_{11}(t), \ldots, \tilde{v}_{1N_y}(t), \tilde{v}_{21}(t), \ldots, \tilde{v}_{2N_y}(t), \ldots, \tilde{v}_{N_xN_y}(t))^T,\\
\mathbf{h}(t) &= (h_{11}(t), \ldots, h_{1N_y}(t), h_{21}(t), \ldots, h_{2N_y}(t), \ldots, h_{N_xN_y}(t))^T.
\end{aligned}$$

where for $w = u, v, \tilde{u}, \tilde{v}, h$, $w_{ij}(t)$ denotes the approximation of $w(\mathbf{x}, t)$ at the grid nodes $\mathbf{x}_{ij}$ at time t, $i = 1, \ldots, N_x$, $j = 1, \ldots, N_y$. We note that the degree of freedom is given by $N = N_x N_y$ because of the periodic boundary conditions, i.e., the most right and the most top grid nodes are not included. Throughout the paper, we do not explicitly represent the time dependency of the semi-discrete solutions for simplicity, and we write $\mathbf{u}$, $\mathbf{v}$, $\tilde{\mathbf{u}}$, $\tilde{\mathbf{v}}$ and $\mathbf{h}$. The semi-discrete vector for the solution vectors are defined by $\mathbf{z} = (\mathbf{u}, \mathbf{v}, \mathbf{h}) \in \mathbb{R}^{3N}$ and $\tilde{\mathbf{z}} = (\tilde{\mathbf{u}}, \tilde{\mathbf{v}}, \mathbf{h}) \in \mathbb{R}^{3N}$.

For the approximation of the first- order partial derivative terms, we use one-dimensional central finite differences to the first- order derivative terms in either x- and or y- direction, and we extend them to two dimensions utilizing the Kronecker product. For a positive integer s, let $\widetilde{D}_s$ denotes the matrix related to the central finite differences to the first- order ordinary differential operator under periodic boundary conditions

$$\widetilde{D}_s = \begin{pmatrix} 0 & 1 & & & -1 \\ -1 & 0 & 1 & & \\ & \ddots & \ddots & \ddots & \\ & & -1 & 0 & 1 \\ 1 & & & 1 & 0 \end{pmatrix} \in \mathbb{R}^{s \times s}.$$

Then, on the two- dimensional mesh, the central finite difference matrices corresponding to the first- order partial derivative operators ∂_x and ∂_y are given, respectively, by

$$D_x = \frac{1}{2\Delta x} \widetilde{D}_{N_x} \otimes I_{N_y} \in \mathbb{R}^{N \times N}, \quad D_y = \frac{1}{2\Delta y} I_{N_x} \otimes \widetilde{D}_{N_y} \in \mathbb{R}^{N \times N},$$

where $\otimes$ denotes the Kronecker product, and I_{N_x} and I_{N_y} are the identity matrices of size N_x and N_y, respectively.

Central difference approximation of the first- order differential operators ensures that the discretized ODE is in skew-symmetric form, which is necessary for the preservation of the Hamiltonian and other conserved quantities. The skew-symmetry is also preserved using finite elements for the thermal shallow water equation [18], the rotational and linear SWEs with discontinuous Galerkin method [21, 46] and finite volume method [35]. Discretization of SWE as a hyperbolic system with nonlinear conservation laws using cell-centered finite volume methods [20] and the well-balanced schemes central upwind and finite volume Galerkin method [14] leads to stable and accurate solutions of the rotational SWE with waves and geostrophic jets.

We further partition the time interval $[0, T]$ into N_t uniform intervals with the step size $\Delta t = T/N_t$ as $0 = t_0 < t_1 < \cdots < t_{N_t} = T$, and $t_k = k\Delta t$, $k = 0, 1, \ldots, N_t$. Then, we denote by $\tilde{\mathbf{u}}^k = \tilde{\mathbf{u}}(t_k)$, $\tilde{\mathbf{v}}^k = \tilde{\mathbf{v}}(t_k)$ and $\mathbf{h}^k = \mathbf{h}(t_k)$ the full discrete solution vectors at time t_k. Similar setting is used for the other components, as well.

The full discrete form of the energy and the enstrophy at a time instance t_k are given as

$$H^k = \sum_{i=1}^{N} \left\{ \frac{1}{2}\mathbf{h}_i^k \left((\mathbf{u}_i^k)^2 + (\mathbf{v}_i^k)^2\right) + g\mathbf{h}_i^k \left((\mathbf{h}_b)_i + \frac{1}{2}\mathbf{h}_i^k\right) \right\} \Delta x \Delta y, \tag{6}$$

$$Z^k = \frac{1}{2} \sum_{i=1}^{N} \frac{\left((D_x \tilde{\mathbf{v}}^k)_i - (D_y \tilde{\mathbf{u}}^k)_i + \Omega^{(z)}\right)^2}{\mathbf{h}_i^k} \Delta x \Delta y.$$

The semi-discrete formulation of the NTSWE (1) leads to a $3N$- dimensional system of Hamiltonian ODEs in skew-gradient form

$$\frac{d\tilde{\mathbf{z}}}{dt} = J(\tilde{\mathbf{z}}) \nabla_{\mathbf{z}} H(\mathbf{z}) = \begin{pmatrix} 0 & \mathbf{q}^d & -D_x \\ -\mathbf{q}^d & 0 & -D_y \\ -D_x & -D_y & 0 \end{pmatrix} \begin{pmatrix} \mathbf{u} \circ \mathbf{h} \\ \mathbf{v} \circ \mathbf{h} \\ \mathbf{\Phi} \end{pmatrix}, \tag{7}$$

with the discrete Bernoulli potential

$$\mathbf{\Phi} = \frac{1}{2}(\mathbf{u} \circ \mathbf{u} + \mathbf{v} \circ \mathbf{v}) + g(\mathbf{h} + \mathbf{h}_b) + \mathbf{h}\left(\Omega^{(x)}\mathbf{v} - \Omega^{(y)}\mathbf{u}\right),$$

where $\circ$ denotes element-wise or Hadamard product. The matrix $\mathbf{q}^d \in \mathbb{R}^{N \times N}$ is the diagonal matrix with the diagonal elements $\mathbf{q}_{ii}^d = \mathbf{q}_i$ where $\mathbf{q}$ is the semi-discrete vector of the potential vorticity q, $i = 1, \ldots, N$.

To confirm the energy- conserving property of the space discretised discretized equations, we apply the energy- conserving time integrator, the average vector field method (AVF). The AVF method preserves higher order polynomial Hamiltonians [15], including the cubic Hamiltonian $\mathscr{H}$ of the NTSWE (1). Quadratic Casimir's functions like mass and circulation are preserved exactly by AVF method. But higher -order polynomial Casimirs like the enstrophy (cubic) can not be preserved. Practical implementation of the AVF method requires the evaluation of the integral on the right-hand side of (8). Since the Hamiltonian $\mathscr{H}$ and the discrete form of the Casimirs, potential enstrophy, mass, and circulation are polynomial, they can be exactly integrated with a Gaussian quadrature rule of the appropriate degree. The AVF method is a fully implicit, second- order accurate Poisson integrator. The AVF method is used with finite element discretization of the rotational SWE [4, 45] and for thermal SWE [18] in Poisson form. After time integration of the semi-discrete NTSWE (7) by the AVF integrator, the full discrete problem reads as: for $k = 0, 1, \ldots, N_t - 1$, given $\tilde{\mathbf{z}}^k$ find $\tilde{\mathbf{z}}^{k+1}$ satisfying

$$\tilde{\mathbf{z}}^{k+1} = \tilde{\mathbf{z}}^k + \Delta t J\left(\frac{\tilde{\mathbf{z}}^{k+1} + \tilde{\mathbf{z}}^k}{2}\right) \int_0^1 \nabla_{\mathbf{z}} H(\xi(\mathbf{z}^{k+1} - \mathbf{z}^k) + \mathbf{z}^k) d\xi. \tag{8}$$

4 Reduced- Order Model

In this section, we construct ROMs that preserve the skew-gradient structure of the NTSWE (7) and, consequently, the discrete Hamiltonian (6). Because the NTSWE is a non-canonical Hamiltonian PDE with a state- dependent Poisson structure, a straightforward application of the POD will not preserve the skew-gradient structure of the NTSWE (7) in reduced form. Energy preserving POD reduced systems are constructed for Hamiltonian systems with constant skew-symmetric matrices like the Korteweg- de Vries equation [23, 31] and nonlinear Schrödinger equation (NLSE) [24]. The approach in [23] can be applied to skew-gradient systems with state-dependent skew-symmetric structure as the NTSWE (7). We show that the state-dependent skew-symmetric matrix in (7) can be evaluated efficiently in the online stage independent of the full dimension N. The full and reduced models are computed separately approximating the nonlinear terms by DEIM. The DEIM also preserves the skew-symmetric form as shown for the (NLSE) [24].

The POD basis are computed through the mean subtracted snapshot matrices $S_{\tilde{u}}$, $S_{\tilde{v}}$ and S_h, constructed by the solutions of the full discrete high- fidelity model (8)

$$
\begin{aligned}
S_{\tilde{u}} &= \left(\tilde{\mathbf{u}}^1 - \bar{\tilde{\mathbf{u}}}, \ldots, \tilde{\mathbf{u}}^{N_t} - \bar{\tilde{\mathbf{u}}}\right) \in \mathbb{R}^{N\times N_t}, \\
S_{\tilde{v}} &= \left(\tilde{\mathbf{v}}^1 - \bar{\tilde{\mathbf{v}}}, \ldots, \tilde{\mathbf{v}}^{N_t} - \bar{\tilde{\mathbf{v}}}\right) \in \mathbb{R}^{N\times N_t}, \\
S_h &= \left(\mathbf{h}^1 - \bar{\mathbf{h}}, \ldots, \mathbf{h}^{N_t} - \bar{\mathbf{h}}\right) \in \mathbb{R}^{N\times N_t},
\end{aligned}
$$

where $\bar{\tilde{\mathbf{u}}}, \bar{\tilde{\mathbf{v}}}, \bar{\mathbf{h}} \in \mathbb{R}^N$ denote the time averaged mean of the solutions

$$
\bar{\tilde{\mathbf{u}}} = \frac{1}{N_t}\sum_{k=0}^{N_t} \tilde{\mathbf{u}}^k , \quad \bar{\tilde{\mathbf{v}}} = \frac{1}{N_t}\sum_{k=0}^{N_t} \tilde{\mathbf{v}}^k , \quad \bar{\mathbf{h}} = \frac{1}{N_t}\sum_{k=0}^{N_t} \mathbf{h}^k .
$$

The mean-subtracted ROMs is used frequently in fluid dynamics to stabilize the reduced system, and it guarantees that ROM solutions would satisfy the same boundary conditions for the FOM [6].

The POD modes are computed by applying singular value decomposition (SVD) to the snapshot matrices

$$
S_{\tilde{u}} = W_{\tilde{u}}\Sigma_{\tilde{u}}U_{\tilde{u}}^T , \quad S_{\tilde{v}} = W_{\tilde{v}}\Sigma_{\tilde{v}}U_{\tilde{v}}^T , \quad S_h = W_h\Sigma_h U_h^T ,
$$

where for $i = \tilde{u}, \tilde{v}, h$, $W_i \in \mathbb{R}^{N\times N_t}$ and $U_i \in \mathbb{R}^{N_t\times N_t}$ are orthonormal matrices, and $\Sigma_i \in \mathbb{R}^{N_t\times N_t}$ is the diagonal matrix with its diagonal entries are the singular values $\sigma_{i,1} \geq \sigma_{i,2} \geq \cdots \geq \sigma_{i,N_t} \geq 0$. Then, the matrix $V_{i,n} \in \mathbb{R}^{N\times n}$ of rank n POD modes consists of the first n left singular vectors from W_i corresponding to the n largest singular values, which satisfies the following least- squares error

$$\min_{V_{i,n} \in \mathbb{R}^{N \times n}} ||S_i - V_{i,n} V_{i,n}^T S_i||_F^2 = \sum_{j=n+1}^{N_t} \sigma_{i,j}^2 \, , \quad i = \tilde{u}, \tilde{v}, h,$$

where $\| \cdot \|_F$ is the Frobenius norm. Moreover, we have the reduced approximations

$$\tilde{\mathbf{u}} \approx \overline{\tilde{\mathbf{u}}} + V_{\tilde{\mathbf{u}},n} \tilde{\mathbf{u}}_r, \quad \tilde{\mathbf{v}} \approx \overline{\tilde{\mathbf{v}}} + V_{\tilde{\mathbf{v}},n} \tilde{\mathbf{u}}_r, \quad \mathbf{h} \approx \overline{\mathbf{h}} + V_{h,n} \mathbf{h}_r, \tag{9}$$

where the reduced (coefficient) vectors $\tilde{\mathbf{u}}_r$, $\tilde{\mathbf{u}}_r$ and $\mathbf{h}_r$ are the solutions of the following ROM of (7):

$$\frac{d}{dt} \tilde{\mathbf{z}}_r = V_{z,n}^T J(\tilde{\mathbf{z}}) \nabla_{\mathbf{z}} H(\mathbf{z}), \tag{10}$$

where $\tilde{\mathbf{z}}_r = (\tilde{\mathbf{u}}_r, \tilde{\mathbf{v}}_r, \mathbf{h}_r)$, and the components of the vector $\tilde{\mathbf{z}} = (\tilde{\mathbf{u}}, \tilde{\mathbf{v}}, \mathbf{h})$ are given as in (9). The block diagonal matrix $V_{z,n}$ contains the matrix of POD modes for each solution component given by

$$V_{z,n} = \begin{pmatrix} V_{\tilde{\mathbf{u}},n} & & \\ & V_{\tilde{\mathbf{v}},n} & \\ & & V_{\mathbf{h},n} \end{pmatrix} \in \mathbb{R}^{3N \times 3n}.$$

The ROM (10) is not a skew-gradient system. A reduced skew-gradient system is obtained formally by inserting $V_{z,n} V_{z,n}^T$ between $J(\tilde{\mathbf{z}})$ and $\nabla_{\mathbf{z}} H(\mathbf{z})$ [23], leading to the ROM

$$\frac{d}{dt} \tilde{\mathbf{z}}_r = J_r(\tilde{\mathbf{z}}) \nabla_{\mathbf{z}_r} H(\mathbf{z}), \tag{11}$$

where $J_r(\tilde{\mathbf{z}}) = V_{z,n}^T J(\tilde{\mathbf{z}}) V_{z,n}$ and $\nabla_{\mathbf{z}_r} H(\mathbf{z}) = V_{z,n}^T \nabla_{\mathbf{z}} H(\mathbf{z})$. The reduced- order NTSWE (11) is also solved by the AVF.

The reduced NTSWE (11) can be written explicitly as

$$\frac{d}{dt} \tilde{\mathbf{z}}_r = \begin{pmatrix} 0 & V_{u,n}^T \mathbf{q}^d V_{v,n} & -V_{u,n}^T D_x V_{h,n} \\ -V_{v,n}^T \mathbf{q}^d V_{u,n} & 0 & -V_{v,n}^T D_y V_{h,n} \\ -V_{h,n}^T D_x V_{u,n} & -V_{h,n}^T D_y V_{v,n} & 0 \end{pmatrix} V_{z,n}^T \nabla_{\mathbf{z}} H(\mathbf{z}). \tag{12}$$

The reduced system (12) has constant matrices which can be precomputed in offline stage, whereas the matrices $V_{u,n}^T \mathbf{q}^d V_{v,n}$ and $V_{v,n}^T \mathbf{q}^d V_{u,n}$ should be computed in online stage depending on the full- order system. Exploiting the diagonal structure of $\mathbf{q}^d$, the computational complexity of evaluating the state- dependent skew-symmetric matrix in (12) can be reduced similar to the skew-gradient systems with constant skew-symmetric matrices as in [31]. Let vec$(\cdot)$ denotes vectorization of a matrix. For any $A \in \mathbb{R}^{m \times n}$ and $B \in \mathbb{R}^{n \times p}$

$$\text{vec}(AB) = (I_p \otimes A)\text{vec(B)} = (B^\top \otimes I_m)\text{vec(A)}.$$

Thus, for a diagonal matrix $D \in \mathbb{R}^{n\times n}$ and $V \in \mathbb{R}^{n\times r}$,

$$\begin{aligned}\text{vec}(V^\top DV) &= (I_r \otimes V^\top)\text{vec}(DV)\\ &= (I_r \otimes V^\top)(V^\top \otimes I_n)\text{vec}(D)\\ &= (V \otimes V)^\top \text{vec}(D)\\ &= (V \otimes V)^\top M^\top \tilde{D}\\ &= \begin{pmatrix} V(1,:) \otimes V(1,:)\\ \vdots \\ V(n,:) \otimes V(n,:)\end{pmatrix}^\top \tilde{D},\end{aligned}$$

where $M \in \mathbb{R}^{n\times n^2}$ is a matrix satisfying $M(a \otimes b) = a \circ b$ for any vector $a, b \in \mathbb{R}^n$, and $\tilde{D} = [D_{11}, D_{22}, \ldots, D_{nn}]^T \in \mathbb{R}^n$. Using the above result, the computational complexity of the matrix products $V_{u,n}^T \mathbf{q}^d V_{v,n}$ and $V_{v,n}^T \mathbf{q}^d V_{u,n}$ is reduced from $\mathcal{O}(n \cdot N(n+N))$ to $\mathcal{O}(n^2 \cdot N)$.

Due to nonlinear terms, the computation of the reduced system still scales with the dimension N of the FOM. This can be reduced by applying the hyper-reduction technique such as DEIM [11]. The ROM (11) can be rewritten as a nonlinear ODE system of the form:

$$\frac{d}{dt}\tilde{\mathbf{z}}_r = V_{z,n}^T F(\tilde{\mathbf{z}}) = \begin{pmatrix} V_{u,n}^T F_1(\tilde{\mathbf{z}})\\ V_{v,n}^T F_2(\tilde{\mathbf{z}})\\ V_{h,n}^T F_3(\tilde{\mathbf{z}})\end{pmatrix}. \tag{13}$$

The DEIM is applied by sampling the nonlinearity $F(\cdot)$ and then interpolating with hyper-reduction. To obtain the DEIM basis, we form the snapshot matrices defined by

$$G_i = (F_i^1, F_i^2, \ldots, F_i^{N_t}) \in \mathbb{R}^{N\times N_t}, \quad i = 1, 2, 3,$$

where $F_i^k = F_i(\tilde{\mathbf{z}}^k)$ denotes the ith component of the nonlinearity $F(\tilde{\mathbf{z}})$ in (13) at time t_k computed by using the FOM solution vector $\tilde{\mathbf{z}}$, $k = 1, \ldots, N_t$. Then, we can approximate each $F_i(\tilde{\mathbf{z}})$ in the column space of the snapshot matrices G_i. We first apply POD to the snapshot matrices G_i and find the basis matrices $V_{F_i,m} \in \mathbb{R}^{N\times m}$ whose columns are the basis vectors spanning the column space of the snapshot matrices G_i. We apply the DEIM algorithm [11] to find a projection matrix $P_i \in \mathbb{R}^{N\times m}$

$$F_i(\tilde{\mathbf{z}}) \approx V_{F_i,m}(P_i^T V_{F_i,m})^{-1} P_i^T F_i(\tilde{\mathbf{z}}),$$

and then we get the DEIM approximation to the reduced nonlinearities in (13) as

$$V_{u,n}^T F_1(\tilde{\mathbf{z}}) \approx \mathcal{V}_{u,1}(P_1^T F_1(\tilde{\mathbf{z}})), \quad V_{v,n}^T F_2(\tilde{\mathbf{z}}) \approx \mathcal{V}_{v,2}(P_2^T F_2(\tilde{\mathbf{z}})), \quad V_{h,n}^T F_3(\tilde{\mathbf{z}}) \approx \mathcal{V}_{h,3}(P_3^T F_3(\tilde{\mathbf{z}})),$$

where

$$\mathscr{V}_{u,1} = V_{u,n}^T V_{F_1,m}(P_1^T V_{F_1,m})^{-1}, \; \mathscr{V}_{v,2} = V_{v,n}^T V_{F_2,m}(P_2^T V_{F_2,m})^{-1}, \; \mathscr{V}_{h,3} = V_{h,n}^T V_{F_3,m}(P_3^T V_{F_3,m})^{-1}$$

are all the matrices of size $n \times m$, and they are precomputed in the offline stage. Using the DEIM approximations, the ROM (13) becomes

$$\frac{d}{dt}\tilde{\mathbf{z}}_r = \begin{pmatrix} \mathscr{V}_{u,1} F_{r,1}(\tilde{\mathbf{z}}) \\ \mathscr{V}_{v,2} F_{r,2}(\tilde{\mathbf{z}}) \\ \mathscr{V}_{h,3} F_{r,3}(\tilde{\mathbf{z}}) \end{pmatrix}, \tag{14}$$

where the reduced nonlinearities $F_{r,i}(\tilde{\mathbf{z}}) = P_i^T F_i(\tilde{\mathbf{z}})$ are computed by considering just $m \ll N$ entries of the nonlinearities $F_i(\tilde{\mathbf{z}})$ among N entries, $i = 1, 2, 3$. In addition, being an approximation to the right- hand side of the ROM (13), the ROM (14) with DEIM approximately preserves the skew-gradient structure, but exactly at the interpolation points.

5 Numerical Results

In this section, we present two numerical examples to demonstrate the efficiency of the ROMs. We consider the propagation of the inertia-gravity waves by Coriolis force, known as geostrophic adjustment [42], and the shear instability in the form of roll-up of an unstable shear layer, known as barotropic instability [42]. For numerical simulations, we consider the nondimensional form of the NTSWE (1) with the setting

$$x = R_d\hat{x}, \quad y = R_d\hat{y}, \quad u = c\hat{u}, \quad v = c\hat{v}, \quad h = H\hat{h}, \quad h_b = H\hat{h}_b,$$

$$\left(\Omega^{(x)}, \Omega^{(y)}, \Omega^{(z)}\right) = \Omega\left(\hat{\Omega}^{(x)}, \hat{\Omega}^{(y)}, \hat{\Omega}^{(z)}\right),$$

where a component with a hat denotes a dimensionless variable, and Ω is planetary rotation rate to construct the gravity wave speed c

$$c = \sqrt{gH}, \quad R_d = \frac{c}{2\Omega}, \quad \sigma = \frac{H}{R_d} = \frac{2\Omega H}{c}.$$

The non-traditional parameter is given as $\sigma = H/R_d$, where H represents the layer thickness scale and R_d is Rossby deformation radius, and g denotes the gravitational acceleration [17, 40]. The parameters are taken following [42] as $H = 1000\,\text{m}$, $\Omega \approx 7.3 \times 10^{-5}\,\text{rad s}^{-1}$, $g = 10^{-3}\,\text{ms}^{-2}$. The dimensionless components of the rotation vector at latitude ϕ are taken as

$$\hat{\Omega}^{(x)} = 0, \quad \hat{\Omega}^{(y)} = \cos(\phi), \quad \hat{\Omega}^{(z)} = \sin(\phi),$$

where we set $\phi = \pi/4$ in the numerical experiments. In all examples, the spatial and temporal mesh sizes are taken as $\Delta x = 0.1$ and $\Delta t = 0.1$, respectively.

In order to determine the numbers n and m of the POD and DEIM modes, respectively, we use the so-called relative cumulative energy criteria for a desired number $p = m, n$

$$\min_p \frac{\sum_{j=1}^{p} \sigma_j^2}{\sum_{j=1}^{N_t} \sigma_j^2} > 1 - \kappa, \tag{15}$$

where κ is a user-specified tolerance. In our simulations, we set $\kappa = 10^{-3}$ and $\kappa = 10^{-5}$ to catch at least 99.9% and 99.999% of data information for POD and DEIM modes, respectively. We take the same number of modes for each state variable.

The error between a discrete FOM solution and a discrete reduced approximation (FOM-ROM error) are measured for the components $\mathbf{w} = \tilde{\mathbf{u}}, \tilde{\mathbf{v}}, \mathbf{h}$ using the following time averaged relative errors in L^2-norm

$$\|\mathbf{w} - \widehat{\mathbf{w}}\|_{rel} = \frac{1}{N_t} \sum_{k=1}^{N_t} \frac{\|\mathbf{w}^k - \widehat{\mathbf{w}}^k\|_{L^2}}{\|\mathbf{w}^k\|_{L^2}}, \quad \|\mathbf{w}^k\|_{L^2}^2 = \sum_{i=1}^{N} (\mathbf{w}_i^k)^2 \Delta x \Delta y,$$

where $\widehat{\mathbf{w}} = \overline{\mathbf{w}} + V_{\mathbf{w},n}\mathbf{w}_r$ denotes the reduced approximation to $\mathbf{w}$. All simulations are performed on a machine with Intel® Core^{TM} i7 2.5 GHz 64 bit CPU, 8 GB RAM, Windows 10, using 64 bit MatLab R2014.

5.1 Single-Layer Geostrophic Adjustment

We consider the NTSWE on the periodic spatial domain $[-5, 5]^2$ and on the time interval [0, 150] [42]. The initial conditions are prescribed in form of a motionless layer with an upward bulge of the height field

$$\begin{aligned} h(x, y, 0) &= 1 + \frac{1}{2} \exp\left[-\left(\frac{4x}{5}\right)^2 - \left(\frac{4y}{5}\right)^2 \right], \\ u(x, y, 0) &= 0, \\ v(x, y, 0) &= 0. \end{aligned}$$

The inertia-gravity waves propagate after the collapse of the initial symmetric peak with respect to axes. Nonlinear interactions create shorter waves propagating around the domain and increasingly more complicated patterns are formed.

For this test problem, each snapshot matrix $S_{\tilde{u}}$, $S_{\tilde{v}}$ and S_h has size 10000 × 1500 (same for the nonlinearity snapshots). According to the energy criteria (15), we take $n = 40$ POD modes and $m = 240$ DEIM modes. In Fig. 1, the singular values decay slowly, which is the characteristic of the problems with wave phenomena in fluid

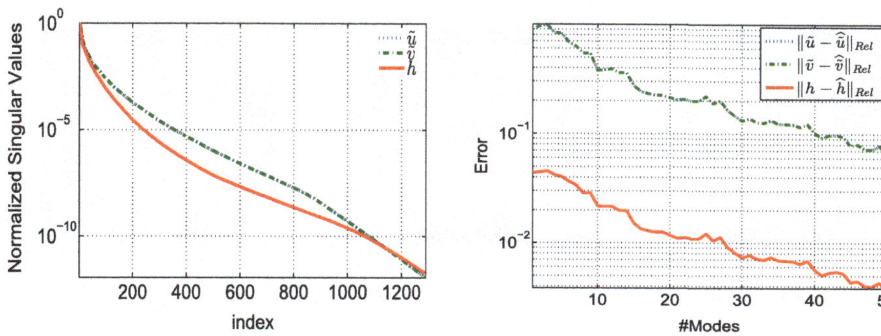

Fig. 1 Normalized singular values for solution snapshots (left) and Relative FOM-ROM errors (right)

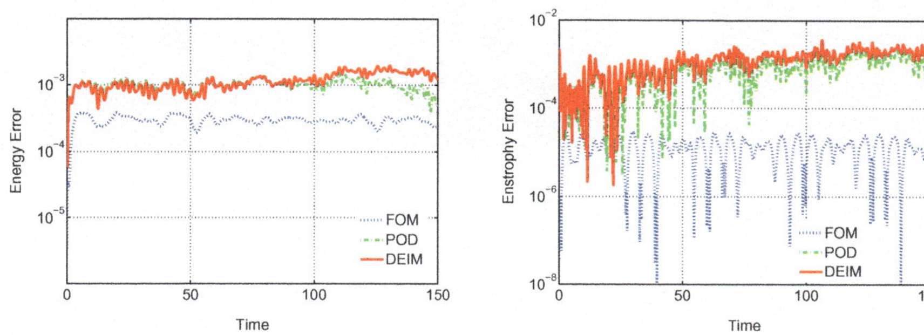

Fig. 2 Energy error $|H^k - H^0|$ (left) and enstrophy error $|Z^k - Z^0|$ (right)

dynamics [33]. Due to the slow decay of the singular values, FOM-ROM errors for all components with a varying number of POD modes in Fig. 1 decrease slowly with small oscillations.

The energy and the enstrophy errors in Fig. 2 show small drifts with bounded oscillations over the time, i.e., they are preserved approximately at the same level of accuracy. In Figs. 3 and 4, the height $\mathbf{h}$ and the potential vorticity $\mathbf{q}$ are shown at the final time. It was show in [12] that a priori error bounds are proportional to the sums of the singular values corresponding to neglected POD basis vectors in the reduced system and in the DEIM approximation of the nonlinear terms. Large number of DEIM points are needed for convergence of Newton method for solving the nonlinear fully discrete form of the reduced system (12). It can be seen from the Figs. 3 and 4 and Tables 1 and 2 that POD, POD-DEIM reduced solutions, and conserved reduced quantities have almost the same level accuracy. The speed-up factors in Table 3 show that the ROM with DEIM increases the computational efficiency further.

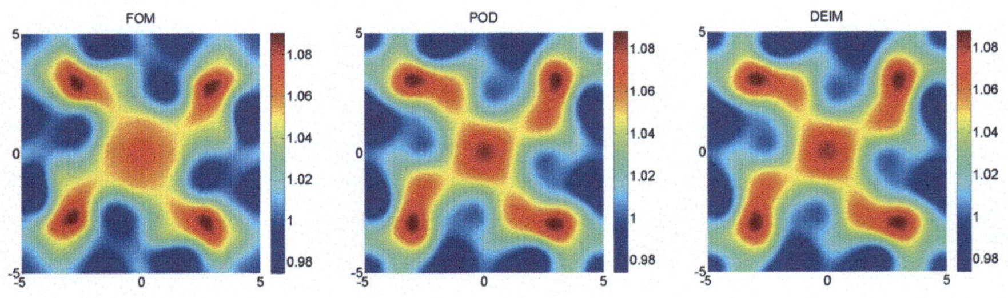

Fig. 3 Full and reduced solutions for the height **h** at the final time

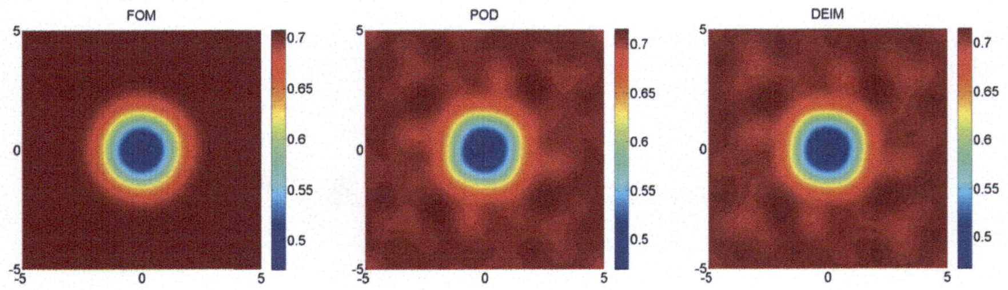

Fig. 4 Full and reduced solutions for the potential vorticity **q** at the final time

5.2 Single-Layer Shear Instability

We consider the NTSWE on the periodic spatial domain $[0, 10]^2$ and on the time interval [0, 150] [42]. The initial conditions are given as

$$
\begin{aligned}
h(x, y, 0) &= 1 + \Delta_h \sin\left\{\frac{2\pi}{L}\left[y - \Delta_y \sin\left(\frac{2\pi x}{L}\right)\right]\right\},\\
u(x, y, 0) &= -\frac{2\pi\Delta_h}{\Omega^z L}\cos\left\{\frac{2\pi}{L}\left[y - \Delta_y \sin\left(\frac{2\pi x}{L}\right)\right]\right\},\\
v(x, y, 0) &= -\frac{4\pi^2\Delta_h\Delta_y}{\Omega^z L^2}\cos\left\{\frac{2\pi}{L}\left[y - \Delta_y \sin\left(\frac{2\pi x}{L}\right)\right]\right\}\cos\left(\frac{2\pi x}{L}\right)
\end{aligned}
$$

where $\Delta_h = 0.2$, $\Delta_y = 0.5$ and the dimensionless spatial domain length $L = 10$, as the case in the first test example. This problem illustrates the roll-up of an unstable shear layer.

Decay of the singular values and FOM-ROM errors in Fig. 5 are similar to the single-layer geostrophic adjustment case in Fig. 5.

Similar to the previous example, each snapshot matrix has size 10000×1500, and the number of POD and DEIM modes are set as $n = 10$ and $m = 345$, respectively, according to the energy criteria (15).

The energy and enstropy errors in Fig. 6 are bounded over time with small oscillations as in the case of the first test example. Similarly, the height **h** and the potential

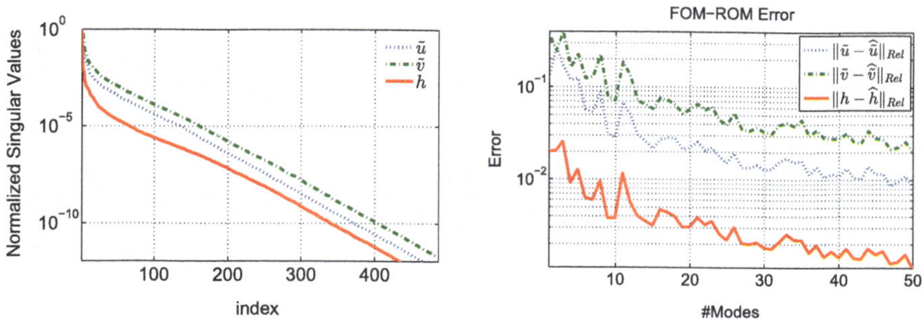

Fig. 5 Normalized singular values for solution snapshots (left) and Relative FOM-ROM errors (right)

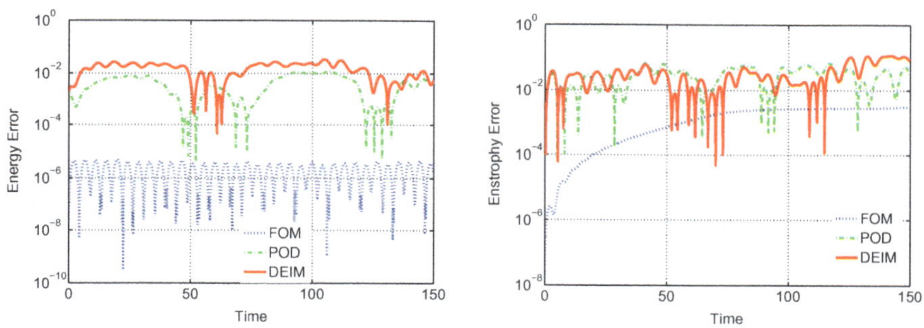

Fig. 6 Energy error $|H^k - H^0|$ (left) and enstrophy error $|Z^k - Z^0|$ (right)

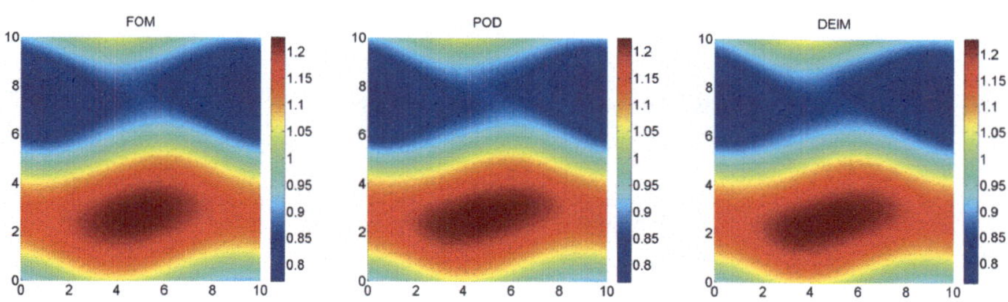

Fig. 7 Full and reduced solutions for the height **h** at the final time

vorticity **q** are well approximated by the ROMs at the final time in Figs. 7 and 8. In Tables 1, 2 and 3, the accuracy and computational efficiency of the reduced approximations are demonstrated.

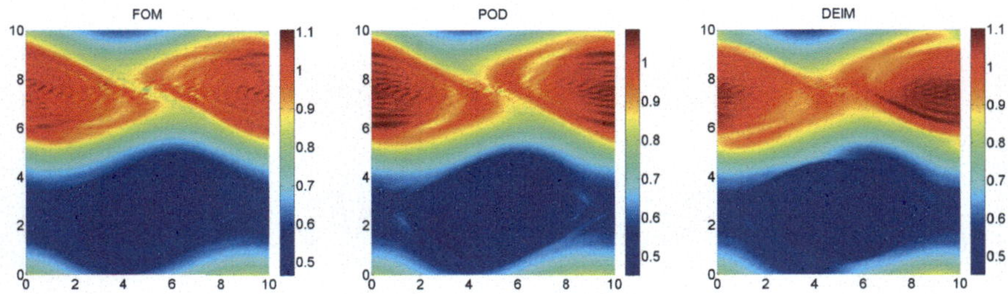

Fig. 8 Full and reduced solutions for the potential vorticity $\mathbf{q}$ at the final time

Table 1 Time averaged relative L^2-errors

		$\|\tilde{\mathbf{u}} - \widehat{\tilde{\mathbf{u}}}\|_{Rel}$	$\|\tilde{\mathbf{v}} - \widehat{\tilde{\mathbf{v}}}\|_{Rel}$	$\|\mathbf{h} - \widehat{\mathbf{h}}\|_{Rel}$
Example 5.1	40 POD modes	1.151e-01	1.151e-01	6.172e-03
	240 DEIM modes	1.156e-01	1.156e-01	6.180e-03
Example 5.2	10 POD modes	3.946e-02	9.088e-02	4.224e-03
	345 DEIM modes	4.859e-02	1.011e-01	5.464e-03

Table 2 Mean absolute FOM-ROM errors of the conserved quantities

		Energy	Enstrophy
Example 5.1	40 POD modes	7.094e-04	9.067e-04
	240 DEIM modes	8.837e-04	1.370e-03
Example 5.2	10 POD modes	4.450e-03	3.068e-02
	345 DEIM modes	1.529e-02	3.589e-02

Table 3 CPU time (in seconds) and speed-up factors

		Example 5.1		Example 5.2	
		CPU time	Speed-up	CPU time	Speed-up
FOM		1051.0		1038.1	
POD	Basis computation	61.6		23.2	
	Online computation	412.7	2.55	167.4	6.2
DEIM	Basis computation (POD & DEIM)	129.2		54.3	
	Online computation	87.1	12.1	47.3	22.0

6 Conclusions

In contrast to the canonical Hamiltonian systems like the NLS and non-canonical Hamiltonian systems with constant Poisson structure, NTSWE possesses state-dependent Poisson structure. In this paper, the Hamiltonian/energy reduced- order modeling approach in [23] is applied by reducing further the computational cost of the ROM in the online stage by exploiting the special structure of the skew-symmetric matrix corresponding to the discretized Poisson structure. The accuracy and computational efficiency of the reduced solutions are demonstrated by numerical examples for the POD and DEIM. Preservation of the energy and enstrophy shows further the stability of the reduced solutions over time.

Acknowledgements This work was supported by 100/2000 Ph.D. Scholarship Program of the Turkish Higher Education Council. The authors would like to thank to the referees which helped to improve the paper.

References

1. Afkham, B.M., Hesthaven, J.S.: Structure preserving model reduction of parametric Hamiltonian systems. SIAM J. Sci. Comput. **39**(6), A2616–A2644 (2017). https://doi.org/10.1137/17M1111991
2. Astrid, P., Weiland, S., Willcox, K., Backx, T.: Missing point estimation in models described by proper orthogonal decomposition. IEEE Trans. Autom. Control **53**(10), 2237–2251 (2008). https://doi.org/10.1109/TAC.2008.2006102
3. Barrault, M., Maday, Y., Nguyen, N.C., Patera, A.T.: An empirical interpolation method: application to efficient reduced-basis discretization of partial differential equations. C. R. Math. **339**(9), 667–672 (2004). https://doi.org/10.1016/j.crma.2004.08.006
4. Bauer, W., Cotter, C.: Energy-enstrophy conserving compatible finite element schemes for the rotating shallow water equations with slip boundary conditions. J. Comput. Phys. **373**, 171–187 (2018). https://doi.org/10.1016/j.jcp.2018.06.071
5. Belanger, E., Vincent, A.: Data assimilation (4D-VAR) to forecast flood in shallow-waters with sediment erosion. J. Hydrol. **300**(1–4), 114–125 (2005)
6. Bistrian, D.A., Navon, I.M.: An improved algorithm for the shallow water equations model reduction: dynamic mode decomposition vs POD. Int. J. Numer. Methods Fluids **78**(9), 552–580 (2015). https://doi.org/10.1002/fld.4029
7. Bistrian, D.A., Navon, I.M.: The method of dynamic mode decomposition in shallow water and a swirling flow problem. Int. J. Numer. Methods Fluids **83**(1), 73–89 (2017)
8. Boss, E., Paldor, N., Thompson, L.: Stability of a potential vorticity front: from quasi-geostrophy to shallow water. J. Fluid Mech. **315**, 65–84 (1996)
9. Carlberg, K., Farhat, C., Cortial, J., Amsallem, D.: The GNAT method for nonlinear model reduction: effective implementation and application to computational fluid dynamics and turbulent flows. J. Comput. Phys. **242**, 623–647 (2013). https://doi.org/10.1016/j.jcp.2013.02.028
10. Carlberg, K., Tuminaro, R., Boggs, P.: Preserving Lagrangian structure in nonlinear model reduction with application to structural dynamics. SIAM J. Sci. Comput. **37**(2), B153–B184 (2015). https://doi.org/10.1137/140959602
11. Chaturantabut, S., Sorensen, D.C.: Nonlinear model reduction via discrete empirical interpolation. SIAM J. Sci. Comput. **32**(5), 2737–2764 (2010). https://doi.org/10.1137/090766498

12. Chaturantabut, S., Sorensen, D.C.: A state space error estimate for POD-DEIM nonlinear model reduction. SIAM J. Numer. Anal. **50**(1), 46–63 (2012). https://doi.org/10.1137/110822724
13. Chaturantabut, S., Beattie, C., Gugercin, S.: Structure-preserving model reduction for nonlinear port-Hamiltonian systems. SIAM J. Sci. Comput. **38**(5), B837–B865 (2016). https://doi.org/10.1137/15M1055085
14. Chertock, A., Dudzinski, M., Kurganov, A., Lukáčová-Medvid'ová, M.: Well-balanced schemes for the shallow water equations with Coriolis forces. Numer. Math. **138**(4), 939–973 (2018). https://doi.org/10.1007/s00211-017-0928-0
15. Cohen, D., Hairer, E.: Linear energy-preserving integrators for Poisson systems. BIT Numer. Math. **51**(1), 91–101 (2011). https://doi.org/10.1007/s10543-011-0310-z
16. Cotter, C.J., Shipton, J.: Mixed finite elements for numerical weather prediction. J. Comput. Phys. **231**(21), 7076–7091 (2012)
17. Dellar, P.J., Salmon, R.: Shallow water equations with a complete Coriolis force and topography. Phys. Fluids **17**(10), 106601 (2005). https://doi.org/10.1063/1.2116747
18. Eldred, C., Dubos, T., Kritsikis, E.: A quasi-Hamiltonian discretization of the thermal shallow water equations. J. Comput. Phys. **379**, 1–31 (2019). https://doi.org/10.1016/j.jcp.2018.10.038
19. Esfahanian, V., Ashrafi, K.: Equation-free/Galerkin-free reduced-order modeling of the shallow water equations based on Proper Orthogonal Decomposition. J. Fluids Eng. **131**(7), 071401–13 (2009). https://doi.org/10.1115/1.3153368
20. Fjordholm, U.S., Mishra, S., Tadmor, E.: Well-balanced and energy stable schemes for the shallow water equations with discontinuous topography. J. Comput. Phys. **230**(14), 5587–5609 (2011). https://doi.org/10.1016/j.jcp.2011.03.042
21. Gassner, G.J., Winters, A.R., Kopriva, D.A.: A well balanced and entropy conservative discontinuous Galerkin spectral element method for the shallow water equations. Appl. Math. Comput. **272**, 291–308 (2016). https://doi.org/10.1016/j.amc.2015.07.014
22. Gerkema, T., Shrira, V.I.: Near-inertial waves in the ocean: beyond the 'traditional approximation'. J. Fluid Mech. **529**, 195–219 (2005)
23. Gong, Y., Wang, Q., Wang, Z.: Structure-preserving Galerkin POD reduced-order modeling of Hamiltonian systems. Comput. Methods Appl. Mech. Eng. **315**, 780–798 (2017). https://doi.org/10.1016/j.cma.2016.11.016
24. Karasözen, B., Uzunca, M.: Energy preserving model order reduction of the nonlinear Schrödinger equation. Adv. Comput. Math. **44**(6), 1769–1796 (2018). https://doi.org/10.1007/s10444-018-9593-9
25. Lall, S., Krysl, P., Marsden, J.E.: Structure-preserving model reduction for mechanical systems. Phys. D **184**(1–4), 304–318 (2003). https://doi.org/10.1016/S0167-2789(03)00227-6
26. Leibovich, S., Lele, S.: The influence of the horizontal component of Earth's angular velocity on the instability of the Ekman layer. J. Fluid Mech. **150**, 41–87 (1985)
27. Lozovskiy, A., Farthing, M., Kees, C., Gildin, E.: POD-based model reduction for stabilized finite element approximations of shallow water flows. J. Comput. Appl. Math. **302**, 50–70 (2016). https://doi.org/10.1016/j.cam.2016.01.029
28. Lozovskiy, A., Farthing, M., Kees, C.: Evaluation of Galerkin and Petrov-Galerkin model reduction for finite element approximations of the shallow water equations. Comput. Methods Appl. Mech. Eng. **318**, 537–571 (2017). https://doi.org/10.1016/j.cma.2017.01.027
29. Lynch, P.: Hamiltonian methods for geophysical fluid dynamics: an introduction (2002)
30. Marshall, J., Schott, F.: Open-ocean convection: observations, theory, and models. Rev. Geophys. **37**(1), 1–64 (1999)
31. Miyatake, Y.: Structure-preserving model reduction for dynamical systems with a first integral. Jpn. J. Ind. Appl. Math. **36**(3), 1021–1037 (2019). https://doi.org/10.1007/s13160-019-00378-y
32. Nguyen, N.C., Patera, A.T., Peraire, J.: A "best points" interpolation method for efficient approximation of parametrized functions. Int. J. Numer. Methods Eng. **73**(4), 521–543 (2008). https://doi.org/10.1002/nme.2086
33. Ohlberger, M., Rave, S.: Reduced basis methods: success, limitations and future challenges. In: Proceedings of the Conference Algoritmy, pp. 1–12 (2016)

34. Peng, L., Mohseni, K.: Symplectic model reduction of Hamiltonian systems. SIAM J. Sci. Comput. **38**(1), A1–A27 (2016). https://doi.org/10.1137/140978922
35. Ranocha, H.: Shallow water equations: split-form, entropy stable, well-balanced, and positivity preserving numerical methods. GEM Int. J. Geomath. **8**(1), 85–133 (2017). https://doi.org/10.1007/s13137-016-0089-9
36. Salmon, R.: Hamiltonian fluid mechanics. Annu. Rev. Fluid Mech. **20**(1), 225–256 (1988). https://doi.org/10.1146/annurev.fl.20.010188.001301
37. Salmon, R.: Poisson-bracket approach to the construction of energy- and potential-enstrophy-conserving algorithms for the shallow-water equations. J. Atmos. Sci. **61**(16), 2016–2036 (2004). https://doi.org/10.1175/1520-0469(2004)0612016:PATTCO2.0.CO;2
38. Ştefănescu, R., Navon, I.M.: POD/DEIM nonlinear model order reduction of an ADI implicit shallow water equations model. J. Comput. Phys. **237**, 95–114 (2013). https://doi.org/10.1016/j.jcp.2012.11.035
39. Ştefănescu, R., Sandu, A., Navon, I.M.: Comparison of pod reduced order strategies for the nonlinear 2D shallow water equations. Int. J. Numer. Methods Fluids **76**(8), 497–521 (2014). https://doi.org/10.1002/fld.3946
40. Stewart, A.L., Dellar, P.J.: Multilayer shallow water equations with complete Coriolis force. part 1. derivation on a non-traditional beta-plane. J. Fluid Mech. **651**, 387-413 (2010). https://doi.org/10.1017/S0022112009993922
41. Stewart, A.L., Dellar, P.J.: Multilayer shallow water equations with complete Coriolis force. part 3. hyperbolicity and stability under shear. J. Fluid Mech. **723**, 289–317 (2013)
42. Stewart, A.L., Dellar, P.J.: An energy and potential enstrophy conserving numerical scheme for the multi-layer shallow water equations with complete Coriolis force. J. Comput. Phys. **313**, 99–120 (2016). https://doi.org/10.1016/j.jcp.2015.12.042
43. Vallis, G.K.: Atmospheric and Oceanic Fluid Dynamics. Cambridge University Press, Cambridge (2017)
44. Warneford, E.S., Dellar, P.J.: Thermal shallow water models of geostrophic turbulence in Jovian atmospheres. Phys. Fluids **26**(1), 016603 (2014)
45. Wimmer, G., Cotter, C., Bauer, W.: Energy conserving upwinded compatible finite element schemes for the rotating shallow water equations. arXiv preprint (2019)
46. Xu, Y., van der Vegt, J.J.W., Bokhove, O.: Discontinuous Hamiltonian finite element method for linear hyperbolic systems. J. Sci. Comput. **35**(2–3), 241–265 (2008). https://doi.org/10.1007/s10915-008-9191-y
47. Zimmermann, R., Willcox, K.: An accelerated greedy missing point estimation procedure. SIAM J. Sci. Comput. **38**(5), A2827–A2850 (2016). https://doi.org/10.1137/15M1042899

Benchmarks and Software of Model Order Reduction

A Non-stationary Thermal-Block Benchmark Model for Parametric Model Order Reduction

Stephan Rave and Jens Saak

Abstract In this contribution, we aim to satisfy the demand for a publicly available benchmark for parametric model order reduction that is scalable both in degrees of freedom as well as parameter dimension.

1 Introduction

Model order reduction (MOR) of parametric problems (PMOR) is accepted to be an important field of research, in particular, due to its relevance for multi-query applications such as uncertainty quantification, inverse problems, or parameter studies in the engineering sciences. Still, publicly available software is often either tailored to a very specific problem or bound to a specific PDE discretization software. The joint feature of the software packages, `emgr` [13], M-M.E.S.S. [23], MORLAB [7] and pyMOR [18], reported in this volume is the attempt to make (P)MOR available in a more general-purpose fashion. Further packages that fall into this category are rbMIT [17], RBmatlab [11, 21], RBniCS [5, 12], redbKIT [16, 19], psssMOR [8]. So far, comparison of PMOR methods is a difficult task [6]. We think that one of the difficulties is the lack of models that can be easily used and fairly compared in all packages. It is the goal of this benchmark to overcome some of the shortcomings of available benchmarks.

The MOR community Wiki [24] already provides a number of parametric benchmark models. However, most of them have not only large dimensions making them difficult to access directly for dense matrix-based packages like [7], but also cumbersome to use during development and testing of new sparse methods. Other bench-

S. Rave
University of Münster, Orleans-Ring 10, 48149 Münster, Germany
e-mail: stephan.rave@uni-muenster.de

J. Saak (✉)
Max Planck Institute for Dynamics of Complex Technical Systems, Sandtorstr. 1, 39106 Magdeburg, Germany
e-mail: saak@mpi-magdeburg.mpg.de

P. Benner et al. (eds.), *Model Reduction of Complex Dynamical Systems*, International Series of Numerical Mathematics 171,
https://doi.org/10.1007/978-3-030-72983-7_16

marks have rather limited parameter dimension, i.e., they feature only scalar or at most two-dimensional parameters. A very common feature among the benchmarks in the Wiki is that essentially all of them are matrix-based, giving easy access for MATLAB®-based solvers, but at the same time making it difficult for packages like pyMOR [3, 15, 18] to show their full flexibility.

Therefore, the new benchmark introduced in this chapter has a few features addressing exactly these problems. The model is of limited dimension in the basic version provided as matrices. On the other hand, it also provides the FEniCS [2, 14]-based procedural setup[1] allowing for easy generation of larger versions or integration into FEniCS-based software packages. The current version features one to four parameters, but the setup can be extended to higher parameter dimensions by tweaking the basic domain description given as plain text input for gmsh [10]. Thus, we provide maximal flexibility with a small, but scalable, benchmark with up to four independent parameters, given in a description that can easily be adapted for many PDE discretization tools. The benchmark we introduce here is a specific version of the so-called thermal-block benchmark. This type of model has been a standard test case in the reduced basis community for many years, e.g., [22]. This specific model setup is also known as the "cookie baking problem" [4] in the numerical linear algebra community. It further presents a flattened 2D version of what is sometimes referred to as the "skyscraper model" in high-performance computing, e.g., [9, p. 216]. We choose the common name used for this type of model in the reduced-basis community.

The remainder of this chapter is organized as follows. The next section provides a basic, abstract description of the model problem. After that, in Sect. 3, we present three variants of our model that will be used in the numerical experiments of the following chapters.

2 Problem Description

We consider a basic parabolic "thermal-block"-type benchmark problem. To this end, consider the computational domain $\Omega := (0, 1)^2$ which we partition into subdomains

$$\begin{aligned}
\Omega_1 &:= \{\xi \in \Omega \mid |\xi - (0.3, 0.3)| < 0.1\}, & \Omega_2 &:= \{\xi \in \Omega \mid |\xi - (0.7, 0.3)| < 0.1\},\\
\Omega_3 &:= \{\xi \in \Omega \mid |\xi - (0.7, 0.7)| < 0.1\}, & \Omega_4 &:= \{\xi \in \Omega \mid |\xi - (0.3, 0.7)| < 0.1\},\\
\Omega_0 &:= \Omega \setminus (\Omega_1 \cup \Omega_2 \cup \Omega_3 \cup \Omega_4),
\end{aligned}$$

with its boundary partitioned into

$$\Gamma_{in} := \{0\} \times (0, 1), \quad \Gamma_D := \{1\} \times (0, 1), \quad \Gamma_N := (0, 1) \times \{0, 1\},$$

cf. Fig. 1. Given a parameter $\mu \in \mathbb{R}^4_{\geq 0}$, let the heat conductivity $\sigma(\xi; \mu)$ given by

[1] Actually, the core feature is the unified form language (UFL) [1] that also other packages, e.g., firedrake [20] use.

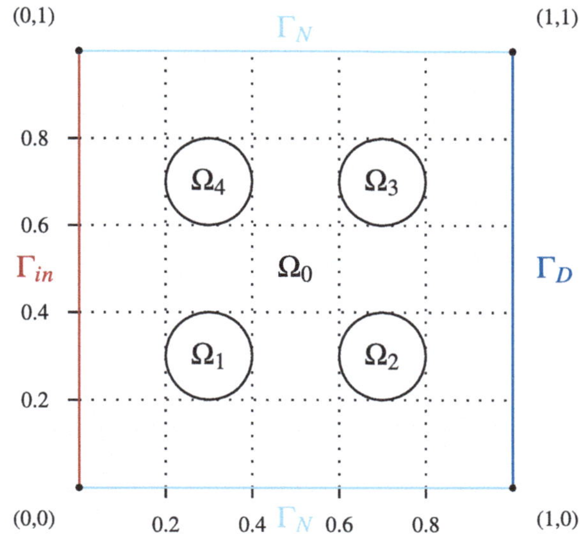

Fig. 1 Computational domain and boundaries

$$\sigma(\xi;\mu) := \begin{cases} 1 & \xi \in \Omega_0 \\ \mu_i & \xi \in \Omega_i, \quad 1 \le i \le 4, \end{cases} \tag{1}$$

and let the temperature $\theta(t,\xi;\mu)$ in the time interval $[0,T]$ for thermal input $u(t)$ at Γ_{in} be given by

$$\begin{aligned} \partial_t\theta(t,\xi;\mu) + \nabla\cdot(-\sigma(\xi;\mu)\nabla\theta(t,\xi;\mu)) &= 0 && t\in(0,T),\ \xi\in\Omega, \\ \sigma(\xi;\mu)\nabla\theta(t,\xi;\mu)\cdot n(\xi) &= u(t) && t\in(0,T),\ \xi\in\Gamma_{in}, \\ \sigma(\xi;\mu)\nabla\theta(t,\xi;\mu)\cdot n(\xi) &= 0 && t\in(0,T),\ \xi\in\Gamma_N, \\ \theta(t,\xi;\mu) &= 0 && t\in(0,T),\ \xi\in\Gamma_D, \\ \theta(0,\xi;\mu) &= 0 && \xi\in\Omega. \end{aligned}$$

More precisely, we let $\theta \in L^2(0,T;V)$ with $\partial_t\theta(\mu) \in L^2(0,T;V')$ be given as the solution of the weak parabolic problem

$$\langle \partial_t\theta(t,\cdot;\mu),\, v\rangle + \int_\Omega \sigma(\mu)\nabla\theta(t,\xi;\mu)\cdot\nabla v d\xi = \int_{\Gamma_{in}} u(t) v ds \quad t\in(0,T),\ v\in V, \tag{2}$$

$$\theta(0,\xi;\mu) = 0, \tag{3}$$

where $V := \{v \in H_0^1(\Omega) \mid v_{\Gamma_D} = 0\}$ denotes the space of Sobolev functions with vanishing trace on Γ_D and V' is its continuous dual.

As outputs $y(t;\mu) \in \mathbb{R}^4$ we consider the average temperatures in the subdomains Ω_i, i.e.

$$y_i(t;\mu) := \frac{1}{|\Omega_i|}\int_{\Omega_i} \theta(t,\xi;\mu)\mathrm{d}\xi, \qquad 1 \le i \le 4. \tag{4}$$

To ease the notation, we drop the explicit dependence on ξ in the following.

In view of the definition (1) of σ as a linear combination of characteristic functions, we can write (2)–(4) as

$$\begin{aligned} \partial_t m(\theta(t;\mu), v) + a_0(\theta(t;\mu), v) + \sum_{i=1}^{4} \mu_i \cdot a_i(\theta(t;\mu), v) &= \varphi(v)\cdot u(t) \in V \\ \theta(0;\mu) &= 0 \\ y_i(t;\mu) &= \psi_i(\theta(t;\mu)), \end{aligned} \tag{5}$$

for $t \in (0, T)$, $v \in V$, $1 \le i \le 4$, with bilinear forms $m, a_i \in \mathrm{Bil}(V, V)$ given by

$$m(w, v) := \int_{\Omega} wv\mathrm{d}\xi \quad \text{and} \quad a_i(w, v) := \int_{\Omega_i} \nabla w \cdot \nabla v \mathrm{d}\xi$$

and linear forms $\varphi, \psi_i \in V'$ given by

$$\varphi(v) := \int_{\Gamma_{in}} v\mathrm{d}s, \quad \text{and} \quad \psi_i(v) := \frac{1}{|\Omega_i|}\int_{\Omega_i} v\mathrm{d}\xi.$$

To arrive at a discrete approximation of (5), we perform a Galerkin projection onto a space $S^1(\mathcal{T}) \cap V$ of linear finite elements w.r.t. a simplicial triangulation of Ω approximating the decomposition into the subdomains $\Omega_0, \ldots, \Omega_4$. Assembling matrices $E \in \mathbb{R}^{n\times n}$ for m, $A_i \in \mathbb{R}^{n\times n}$ for a_i, $B \in \mathbb{R}^{n\times 1}$ for φ, and $C \in \mathbb{R}^{4\times n}$ for ψ_i, all w.r.t. the finite element basis, we arrive at the linear time-invariant system

$$\begin{aligned} E \cdot \partial_t x(t;\mu) &= A_0 \cdot x(t;\mu) + \sum_{i=1}^{4} \mu_i A_i \cdot x(t;\mu) + B \cdot u(t) \\ y(t;\mu) &= C \cdot x(t;\mu). \end{aligned} \tag{6}$$

Here, n denotes the dimension of the finite element space and x is the coefficient vector of the discrete solution state θ w.r.t. the finite element basis.

For the numerical experiments in the following chapters, the mesh $\mathcal{T}$ was generated with gmsh version 3.0.6 with "clscale" set to 0.1, for which the system matrices were assembled using FEniCS 2019.1.

The source code of the model implementation as well as the resulting system matrices are available at

https://doi.org/10.5281/zenodo.3691894

Note that due to the handling of Dirichlet constraints in FEniCS, all matrices were assembled over the full unconstrained space $S^1(\mathcal{T})$. Rows of E, A_i corresponding to degrees of freedom located on Γ_D have zero off-diagonal entries. The corresponding diagnal entries are 1 for E, -1 for A_0, and 0 for $A_1, \ldots, A_4$. Rows of B corresponding to Dirichlet degrees of freedom are set to 0. Consequently, all system matrices $A(\mu) := A_0 + \sum_{i=1}^{4} \mu_i A_i$ have a k-dimensional eigenspace with eigenvalue -1 spanned by the k finite element basis functions associated with Γ_D.

3 Problem Variants

The following chapters test the model introduced in the previous section in three different variants. The simplest case is a basic non-parametric version with all parameters fixed. For the parametric versions, either all four parameters are considered independent or they are all scaled versions of a single scalar parameter. This section introduces all of them with the specific parameter selections and allowed parameter domains.

3.1 Four-Parameter LTI System

This represents exactly the model in (5), or (6), with its full flexibility with respect to the parameters. Note that by construction the model becomes singular in case any of the μ_i becomes zero. Thus, we limit the μ_i from below by 10^{-6}. This will also limit the condition numbers of the linear systems involving the matrices E and A_i ($i = 0, \ldots, 4$) in the PDE solvers as well as MOR routines. At the same time, we do not allow for the subdomain heat conductivities to be drastically larger than the conductivity for Ω_0. So, we limit also from the above, resulting in parameter domains $\mu_i \in [10^{-6}, 10^2]$, ($i = 1, \ldots, 4$).

Figure 2 shows the final heat distribution, at $t = 1$ after 100 steps of implicit Euler with $\mu = [10^2, 10^{-2}, 10^{-3}, 10^{-4}]$, in pyMOR 2019.2.

3.2 Single-Parameter LTI System

In this variation of the model, the parameters are limited in flexibility. We make them all use the same order of magnitude by defining

$$\mu = \tilde{\mu} \cdot \begin{bmatrix} 0.2 \\ 0.4 \\ 0.6 \\ 0.8 \end{bmatrix},$$

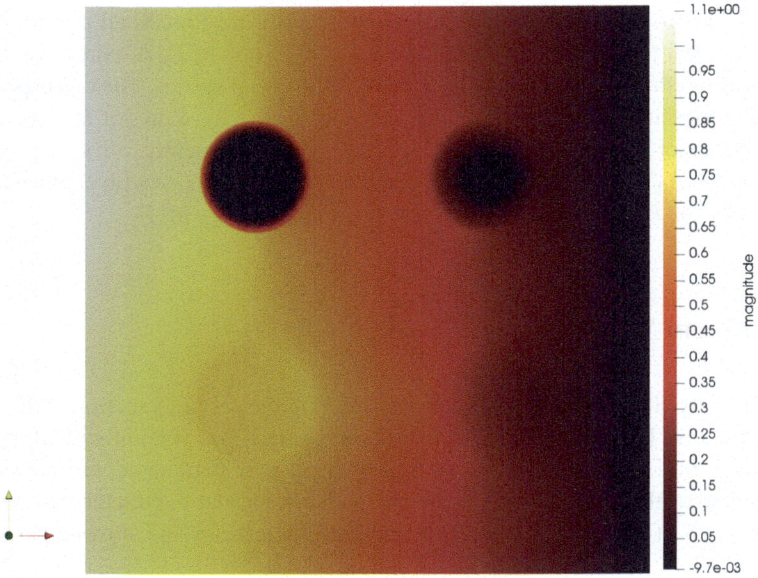

Fig. 2 A sample final heat distribution

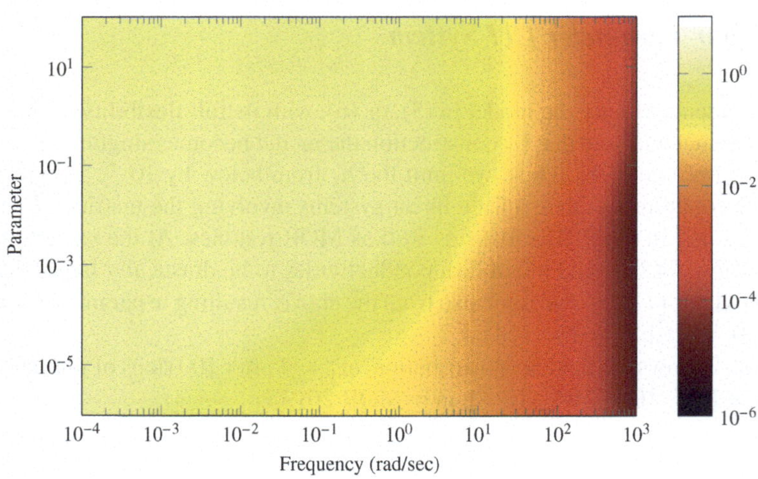

Fig. 3 Sigma magnitude plot for the single-parameter LTI system

for a single scalar parameter $\tilde{\mu} \in [10^{-6}, 10^{2}]$. The transfer function, arising after Laplace transformation of (6) is a rational matrix-valued function of the frequency and the parameters. Its Sigma-magnitude plot, i.e., the maximum singular value of the transfer function matrix, with this restriction on μ, is shown in Fig. 3.

3.3 Non-parametric LTI System

This is the simplest version of the benchmark. We use the setup described in Sect. 3.2 with $\tilde{\mu} = \sqrt{10}$. Note that this value of $\tilde{\mu}$ is rather arbitrary. Depending on the desired application, different values may be insightful. For both time-domain and frequency-domain investigations, variation is strongest in the parameter range $[10^{-5}, 10^{-1}]$. On the other hand, values between 1.25 and 5.0 essentially turn the model into a simple heat equation on the unit square with almost homogeneous heat conductivity $\sigma(t, \xi, \tilde{\mu}) \approx 1$. Hence, $\tilde{\mu} = \sqrt{10} \approx 3.1623$ appears to be a proper choice to get reasonably close to an easy to solve textbook problem, here. Smaller values of μ, especially when approaching $\mu = 0$, can be used to make the problem arbitrarily ill-conditioned.

4 Conclusion

We have specified a flexible, scalable benchmark that can be used both based on pre-generated matrices or based on a procedural inclusion into an existing finite element setting. The new model has been added to the benchmark collection hosted at the MOR Wiki [24, Thermal Block].

Acknowledgements The authors would like to thank Christian Himpe, Petar Mlinarić and Steffen W. R. Werner for helpful comments and discussions during the creation of the model.
Funded by the Deutsche Forschungsgemeinschaft (DFG, German Research Foundation) under Germany's Excellence Strategy EXC 2044-390685587, Mathematics Münster: Dynamics-Geometry-Structure. Funded by German Bundesministerium für Bildung und Forschung (BMBF, Federal Ministry of Education and Research) under grant number 05M18PMA in the programme "Mathematik für Innovationen in Industrie und Dienstleistungen".

References

1. Alnæs, M.S.: UFL: a finite element form language, chap. 17. Springer (2012). https://doi.org/10.1146/10.1145/2566630
2. Alnæs, M.S., Blechta, J., Hake, J., Johansson, A., Kehlet, B., Logg, A., Richardson, C., Ring, J., Rognes, M.E., Wells, G.N.: The FEniCS project version 1.5. Arch. Numer. Softw. **3**(100), 9–23 (2015). https://doi.org/10.1146/10.11588/ans.2015.100.20553
3. Balicki, L., Mlinarić, P., Rave, S., Saak, J.: System-theoretic model order reduction with pyMOR. Proc. Appl. Math. Mech. **19**(1) (2019). https://doi.org/10.1146/10.1002/pamm.201900459
4. Ballani, J., Kressner, D.: Reduced basis methods: from low-rank matrices to low-rank tensors. SIAM J. Sci. Comput. **38**(4), A2045–A2067 (2016). https://doi.org/10.1146/10.1137/15M1042784
5. Ballarin, F., Rozza, G.: RBniCS. https://mathlab.sissa.it/rbnics
6. Baur, U., Benner, P., Haasdonk, B., Himpe, C., Martini, I., Ohlberger, M.: Comparison of methods for parametric model order reduction of time-dependent problems. In: Benner, P.,

Cohen, A., Ohlberger, M., Willcox, K. (eds.) Model Reduction and Approximation: Theory and Algorithms, pp. 377–407. SIAM (2017). https://doi.org/10.1146/10.1137/1.9781611974829.ch9
7. Benner, P., Werner, S.W.R.: MORLAB – Model Order Reduction LABoratory (version 5.0) (2019). https://doi.org/10.1146/10.5281/zenodo.3332716. See also: http://www.mpi-magdeburg.mpg.de/projects/morlab
8. Chair of Automatic Control TUM, Technical University of Munich: psssMOR. https://www.mw.tum.de/rt/forschung/modellordnungsreduktion/software/psssmor/
9. Dolean, V., Jolivet, P., Nataf, F.: An Introduction to Domain Decomposition Methods: Algorithms, Theory, and Parallel Implementation. Society for Industrial and Applied Mathematics (SIAM), Philadelphia, PA (2015). https://doi.org/10.1146/10.1137/1.9781611974065.ch1
10. Geuzaine, C., Remacle, J.F.: Gmsh Reference Manual (2010). http://www.geuz.org/gmsh/doc/texinfo/gmsh.pdf
11. Haasdonk, B.: Reduced basis methods for parametrized PDEs–a tutorial introduction for stationary and instationary problems, chap. 2, pp. 65–136. SIAM Publications (2017). https://doi.org/10.1146/10.1137/1.9781611974829.ch2
12. Hesthaven, J.S., Rozza, G., Stamm, B.: Certified Reduced Basis Methods for Parametrized Partial Differential Equations, 1 edn. Springer Briefs in Mathematics. Springer International Publishing (2016). https://doi.org/10.1146/10.1007/978-3-319-22470-1. http://www.springer.com/us/book/9783319224695
13. Himpe, C.: emgr – EMpirical GRamian framework (version 5.4). http://gramian.de (2018). https://doi.org/10.1146/10.5281/zenodo.1241532
14. Logg, A., Mardal, K.A., Wells, G. (eds.): Automated Solution of Differential Equations by the Finite Element Method. Lecture Notes in Computational Science and Engineering, vol. 84, 1 edn. Springer, Berlin (2012)
15. Milk, R., Rave, S., Schindler, F.: pyMOR - generic algorithms and interfaces for model order reduction. SIAM J. Sci. Comput. **38**(5), S194–S216 (2016). https://doi.org/10.1146/10.1137/15M1026614
16. Negri, F.: redbKIT Version 2.2. http://redbkit.github.io/redbKIT/ (2016)
17. Patera, A., Rozza, G.: Reduced basis approximation and a posteriori error estimation for parametrized partial differential equations. Version 1.0, Copyright MIT (2006). https://orms.mfo.de/project?id=316
18. pyMOR developers and contributors: pyMOR - model order reduction with Python. https://pymor.org
19. Quarteroni, A., Manzoni, A., Negri, F.: Reduced Basis Methods for Partial Differential Equations. La Matematica per il 3+2, vol. 92. Springer International Publishing (2016). https://doi.org/10.1146/10.1007/978-3-319-15431-2. https://www.springer.com/us/book/9783319154305
20. Rathgeber, F., Ham, D.A., Mitchell, L., Lange, M., Luporini, F., McRae, A.T.T., Bercea, G.T., Markall, G.R., Kelly, P.H.J.: Firedrake: automating the finite element method by composing abstractions. ACM Trans. Math. Softw. **43**(3), 24:1–24:27 (2016). https://doi.org/10.1146/10.1145/2998441
21. RBmatlab: https://www.morepas.org/software/rbmatlab/
22. Rozza, G., Huynh, D.B.P., Patera, A.T.: Reduced basis approximation and a posteriori error estimation for affinely parametrized elliptic coercive partial differential equations. Arch. Comput. Methods Eng. **15**(3), 229–275 (2008). https://doi.org/10.1146/10.1007/s11831-008-9019-9
23. Saak, J., Köhler, M., Benner, P.: M-M.E.S.S. – the matrix equations sparse solvers library. https://doi.org/10.1146/10.5281/zenodo.632897. See also: https://www.mpi-magdeburg.mpg.de/projects/mess
24. The MORwiki Community: MORwiki - Model Order Reduction Wiki. http://modelreduction.org

Parametric Model Order Reduction Using pyMOR

Petar Mlinarić, Stephan Rave, and Jens Saak

Abstract pyMOR is a free software library for model order reduction that includes both reduced basis and system-theoretic methods. All methods are implemented in terms of abstract vector and operator interfaces, which allows a direct integration of pyMOR's algorithms with a wide array of external PDE solvers. In this contribution, we give a brief overview of the available methods and experimentally compare them for the parametric instationary thermal-block benchmark defined in [12].

1 Introduction

pyMOR is a free software library for building model order reduction applications with the Python programming language [9, 11]. Originally only implementing reduced basis methods, since version 0.5, released in January 2019, it additionally implements system-theoretic methods such as balanced truncation [10] and IRKA [2]. Here, we focus on version 2019.2, released in December 2019, which added support for parametric system-theoretic methods.

We consider model reduction of the thermal-block model defined in [12], which takes the form

$$
\begin{aligned}
E\dot{x}(t;\mu) &= A(\mu)x(t;\mu) + Bu(t), \quad x(0;\mu) = 0,\\
y(t;\mu) &= Cx(t;\mu),
\end{aligned}
$$

P. Mlinarić (✉) · J. Saak
Max Planck Institute for Dynamics of Complex Technical Systems, Sandtorstr. 1, 39106 Magdeburg, Germany
e-mail: mlinaric@mpi-magdeburg.mpg.de

J. Saak
e-mail: saak@mpi-magdeburg.mpg.de

S. Rave
University of Münster, Orleans-Ring 10, 48149 Münster, Germany
e-mail: stephan.rave@uni-muenster.de

P. Benner et al. (eds.), *Model Reduction of Complex Dynamical Systems*, International Series of Numerical Mathematics 171,
https://doi.org/10.1007/978-3-030-72983-7_17

with system matrices $E, A(\mu) \in \mathbb{R}^{n\times n}$, input matrix $B \in \mathbb{R}^{n\times 1}$, output matrix $C \in \mathbb{R}^{p\times n}$, state $x(t) \in \mathbb{R}^n$, input $u(t) \in \mathbb{R}$, and output $y(t) \in \mathbb{R}^p$, where $\mu \in \mathcal{P} \subset \mathbb{R}^d$ is the parameter. The matrix-valued function A additionally has parameter affine form $A(\mu) = A_0 + \sum_{i=1}^{d} \mu_i A_i$, where $\mu = (\mu_1, \mu_2, \ldots, \mu_d)$. We also consider a non-parametric version, for which we write A instead of $A(\mu)$.

We begin, in Sect. 2, with a brief discussion of pyMOR's software design. In Sect. 3, we give a brief overview of the methods implemented in pyMOR 2019.2. Next, we give numerical results in Sect. 4. A conclusion follows in Sect. 5.

2 Software Design

The central goal of pyMOR's design is to allow an easy integration with external PDE solver libraries. To this end, generic interfaces for vectors and operators have been defined that give pyMOR access to the solver's internal data structures representing vectors, matrices, or nonlinear operators, as well as operations on them, e.g., the computation of inner products or the solution of the linear equation system.

All high-dimensional model reduction operations in pyMOR, for instance POD computation or Petrov-Galerkin projection, are expressed in terms of these interfaces. Compared to a file-based exchange of matrices or solution snapshots, this approach enables the usage of problem adapted solvers implemented in the PDE library or the reduction of very large MPI-distributed problems [9].

3 Overview of Model Order Reduction Methods

The majority of MOR methods implemented in pyMOR are projection-based methods, i.e., they consist of finding basis matrices V and W and defining the reduced-order model as

$$\begin{aligned} \hat{E}\dot{\hat{x}}(t;\mu) &= \hat{A}(\mu)\hat{x}(t;\mu) + \hat{B}u(t), \quad \hat{x}(0;\mu) = 0, \\ \hat{y}(t;\mu) &= \hat{C}\hat{x}(t;\mu), \end{aligned}$$

where $\hat{E} = W^{\mathrm{T}} E V$, $\hat{A}(\mu) = W^{\mathrm{T}} A(\mu) V = \hat{A}_0 + \sum_{i=1}^{d} \mu_i \hat{A}_i$, $\hat{A}_i = W^{\mathrm{T}} A_i V$, $\hat{B} = W^{\mathrm{T}} B$, and $\hat{C} = CV$. If $V = W$, we call it a Galerkin projection and otherwise a Petrov-Galerkin projection.

In the following, we give short descriptions of some projection-based methods with remarks on their implementation in pyMOR.

3.1 Reduced Basis Method

We consider a weak POD-Greedy algorithm [8] to build a basis matrix V for which the maximum state-space approximation error

$$\max_{\mu \in \mathcal{S}_{\text{train}}} \sum_{i=1}^{N} x(t_i; \mu) - V\hat{x}(t_i; \mu)\|^2_{H_0^1(\Omega)}$$

for constant input $u \equiv 1$ over some training set $\mathcal{S}_{\text{train}}$ of parameters is minimized in the Sobolev H_0^1-norm. To this end, in each iteration of the Greedy algorithm the current reduced-order model is solved for all $\mu \in \mathcal{S}_{\text{train}}$ and the parameter $\mu_{\max}$ is selected for which an (online-efficient) estimate of the MOR error is maximized [7]. For this parameter, the matrix of full-order model (FOM) solution snapshots

$$X = \left[x(t_1; \mu_{\max})\ x(t_2; \mu_{\max})\ \cdots\ x(t_N; \mu_{\max}) \right],$$

is computed, and the first left-singular vectors of its H_0^1-orthonormal projection onto the H_0^1-orthogonal complement of V are added to V.

Note that, in the non-parametric case, POD-Greedy reduces to POD, i.e., using the first few left singular vectors of the snapshot matrix X as a Galerkin projection basis.

3.2 System-Theoretic Methods

3.2.1 Balanced Truncation

For non-parametric models, balanced truncation (BT) consists of solving two Lyapunov equations

$$\begin{aligned} APE^{\mathrm{T}} + EPA^{\mathrm{T}} + BB^{\mathrm{T}} &= 0, \\ A^{\mathrm{T}}QE + E^{\mathrm{T}}QA + C^{\mathrm{T}}C &= 0. \end{aligned} \tag{1}$$

Based on the solutions P and Q, it computes V and W of the Petrov-Galerkin projection. pyMOR provides bindings to dense Lyapunov equation solvers in SciPy [16], Slycot [14] (Python wrappers for SLICOT [13]), and Py-M.E.S.S. [6]. For the reduction of large-scale models, there are bindings for low-rank solvers in Py-M.E.S.S.. Since Py-M.E.S.S. does not allow generic vectors, there is also an implementation of the alternating direction implicit iteration in pyMOR [3].

It is known that BT preserves asymptotic stability and has a priori bounds for Hardy $\mathcal{H}_\infty$ and $\mathcal{H}_2$ errors depending on the truncated Hankel singular values (the square roots of the eigenvalues of $E^{\mathrm{T}}QEP$).

For parametric models, there are several possible extensions of BT [4, 15, 17]. We focus on the simplest global basis approach by concatenating several local basis matrices. Let $\mu^{(1)}, \mu^{(2)}, \ldots, \mu^{(\ell)} \in \mathcal{P}$ be parameter samples and $V^{(1)}, V^{(2)}, \ldots, V^{(\ell)}$ and $W^{(1)}, W^{(2)}, \ldots, W^{(\ell)}$ corresponding local basis matrices. To guarantee asymptotic stability, we use Galerkin projection with

$$\left[V^{(1)}\ V^{(2)}\ \ldots\ V^{(\ell)}\ W^{(1)}\ W^{(2)}\ \ldots\ W^{(\ell)}\right]$$

after orthogonalization and rank truncation.

3.2.2 LQG Balanced Truncation

LQG balanced truncation (LQGBT) is a variant of BT related to the linear-quadratic-Gaussian (LQG) optimal control problem. Unlike BT, LQGBT consists of solving Riccati equations:

$$\begin{aligned} APE^{\mathrm{T}} + EPA^{\mathrm{T}} - EPC^{\mathrm{T}}CPE^{\mathrm{T}} + BB^{\mathrm{T}} &= 0, \\ A^{\mathrm{T}}QE + E^{\mathrm{T}}QA - E^{\mathrm{T}}QBB^{\mathrm{T}}QE + C^{\mathrm{T}}C &= 0. \end{aligned}$$

Similar to BT, it guarantees the preservation of asymptotic stability and has an a priori error bound. As for Lyapunov equations, pyMOR provides bindings for external Riccati equation solvers and an implementation of the low-rank RADI method [5].

Additionally, there is bounded-real BT in pyMOR, but it currently relies on a dense solver which does not respect the vector and operator interfaces, so it is not possible to use it with a PDE solver.

3.2.3 Iterative Rational Krylov Algorithm

Iterative rational Krylov algorithm (IRKA) is a locally optimal MOR method in the Hardy $\mathcal{H}_2$ norm. In each step, it computes (tangential) rational Krylov subspaces

$$\begin{aligned} \mathcal{V} &= \operatorname{span}\{(\sigma_1 E - A)^{-1}Bb_1, (\sigma_2 E - A)^{-1}Bb_2, \ldots, (\sigma_r E - A)^{-1}Bb_r\}, \\ \mathcal{W} &= \operatorname{span}\{(\sigma_1 E - A)^{-\mathrm{T}}C^{\mathrm{T}}c_1, (\sigma_2 E - A)^{-\mathrm{T}}C^{\mathrm{T}}c_2, \ldots, (\sigma_r E - A)^{-\mathrm{T}}C^{\mathrm{T}}c_r\}. \end{aligned} \tag{2}$$

The interpolation points $\sigma_1, \sigma_2, \ldots, \sigma_r$ for the next step are chosen as reflected poles $-\lambda_1, -\lambda_2, \ldots, -\lambda_r$ of the projected matrix pencil $\lambda W^{\mathrm{T}}EV - W^{\mathrm{T}}AV$ (vectors $b_1, b_2, \ldots, b_r$ and $c_1, c_2, \ldots, c_r$ are computed based on the eigenvectors). Even if the original model has real poles, the projected poles can be complex. Since the complex number support is limited in PDE solvers, solving complex shifted linear systems $(\sigma E - A)x = b$ needs to be done using an iterative method. Implementing efficient preconditions for such systems is a future research topic for pyMOR. For this reason, we demonstrate IRKA only on the non-parametric example in Sect. 4.1.

In the parametric case, we only use one-sided IRKA (OS-IRKA), where W in (2) is replaced by V, which guarantees real interpolation points for the heat equation example we consider. To generate the global basis matrix, we concatenate the local basis matrices $V^{(i)}$ and do a rank truncation.

3.2.4 Generating Reduced Models

All system-theoretic methods in pyMOR can be called similarly. For instance, BT can be run with

```
bt = BTReductor(fom, mu=mu)
rom = bt.reduce(10)
```

where `fom` is the (parametric) full-order model (an instance of `LTIModel`) and `mu` is the parameter sample. The `reduce` method of `bt` accepts the reduced order as a parameter (among others) and returns the non-parametric reduced-order model `rom` (again an instance of `LTIModel`). The basis matrices are then available as `VectorArrays` in `bt.V` and `bt.W`.

4 Numerical Results

Here, we present results of applying MOR methods discussed in Sect. 3 to parametric models, in particular the thermal-block example. To demonstrate pyMOR's integration with external PDE solvers, we used FEniCS 2019.1.0 ([1]) to define the full-order model.

We use the Hardy $\mathcal{H}_2$ norm to quantify the results, which is defined for non-parametric, asymptotically stable systems

$$\begin{aligned} E\dot{x}(t) &= Ax(t) + Bu(t), \quad x(0) = 0, \\ y(t) &= Cx(t), \end{aligned} \tag{3}$$

as the $\mathcal{L}_2$ norm of the impulse response $h\colon [0, \infty) \to \mathbb{R}^{p \times 1}$ defined by $h(t) = C \exp(tE^{-1}A)E^{-1}B$, assuming E is invertible [2]. This can be computed using

$$h^2_{\mathcal{L}_2([0,\infty);\mathbb{R}^{p\times 1})} = \operatorname{tr}\left(CPC^{\mathrm{T}}\right) = \operatorname{tr}\left(B^{\mathrm{T}}QB\right), \tag{4}$$

where P and Q are as in (1). Note that for a reduced-order model

$$\begin{aligned} \hat{E}\dot{\hat{x}}(t) &= \hat{A}\hat{x}(t) + \hat{B}u(t), \quad \hat{x}(0) = 0, \\ \hat{y}(t) &= \hat{C}\hat{x}(t), \end{aligned}$$

the error system

$$\begin{bmatrix} E & 0 \\ 0 & \hat{E} \end{bmatrix} \begin{bmatrix} \dot{x}(t) \\ \dot{\hat{x}}(t) \end{bmatrix} = \begin{bmatrix} A & 0 \\ 0 & \hat{A} \end{bmatrix} \begin{bmatrix} x(t) \\ \hat{x}(t) \end{bmatrix} + \begin{bmatrix} B \\ \hat{B} \end{bmatrix} u(t),$$
$$y(t) - \hat{y}(t) = \begin{bmatrix} C & -\hat{C} \end{bmatrix} \begin{bmatrix} x(t) \\ \hat{x}(t) \end{bmatrix},$$

is of the same form as the FOM (3), which allows us to compute $\mathcal{H}_2$ errors, i.e., the $\mathcal{H}_2$ norm of the error system, using (4).

We chose to use the $\mathcal{H}_2$ norm because it is independent of the input u. Additionally, it can be computed efficiently using the low-rank Lyapunov equation solver available in pyMOR.

We begin with the non-parametric version in Sect. 4.1, comparing system-theoretic methods with POD. Then, in Sects. 4.2 and 4.3 we compare methods for parametric versions.

The source code of the implementations used to compute the presented results can be obtained from

https://doi.org/10.5281/zenodo.3928528

and is authored by Petar Mlinarić and Stephan Rave.

4.1 Non-parametric Version

Fig. 1 compares BT, LQGBT, IRKA, OS-IRKA, and POD in terms of $\mathcal{H}_2$ error. The POD model was trained using the step response ($u(t) = 1$ for $t \geq 0$). We see that BT, LQGBT, and IRKA give similar results, while OS-IRKA and POD give worse errors. Interestingly, POD is mostly better than OS-IRKA in this example.

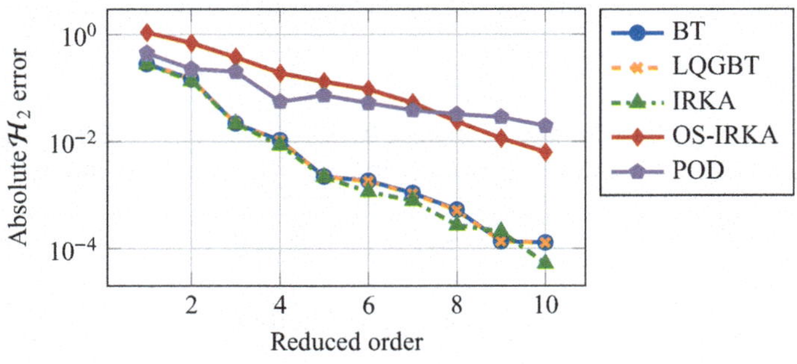

Fig. 1 Comparison of the methods from Sect. 3 for the non-parametric model (Sect. 4.1)

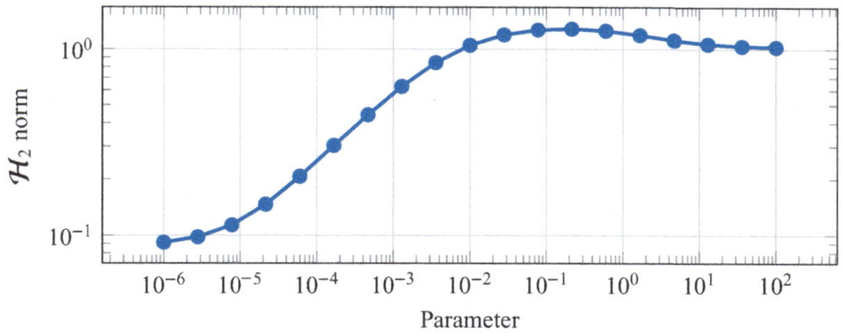

Fig. 2 The $\mathcal{H}_2$ norms of the one-parameter model for different parameter values

4.2 Single-Parameter Version

In this setting, as the training set we chose 10 logarithmically equi-spaced parameter values from 10^{-6} to 10^2. For testing, we added additional 9 in-between points. We used BT and OS-IRKA to get reduced models of order 10 for each parameter value and concatenated their local bases as explained in Sect. 3.2.1. After truncation, BT's global basis was of order 175 and OS-IRKA's was 67. To have a fairer comparison, we further truncated BT's global basis to the same order as OS-IRKA.

Figure 2 shows the $\mathcal{H}_2$ norm of the full-order model for different parameters, from which we see that it only changes by about an order of magnitude over the parameter range. Therefore, we restrict to showing only the absolute $\mathcal{H}_2$ errors in the following plots. In particular, Fig. 3 shows the absolute $\mathcal{H}_2$ error for BT and OS-IRKA. Possibly related to BT being a Petrov-Galerkin projection method, its global basis produces worse results than the local bases. On the other hand, OS-IRKA improves with using the global basis.

Finally, Fig. 4 compares BT and OS-IRKA with RB. For RB, we used the same training set to generate a model of order 67. In this example, OS-IRKA performed best near the boundaries of the parameter set and comparable to other methods in the middle. On the other hand, BT gave the worst results near the boundaries. RB produced an almost flat absolute $\mathcal{H}_2$ error curve, which is not surprising since it tries to minimize the worst error.

4.3 Four-Parameter Version

Here, we randomly sampled 20 points e_i from the uniform distribution over $[-6, 2]^4$ to generate the training set $\mu^{(i)} = 10^{e_i}$ and additional 20 such points for testing. As before, we used BT and OS-IRKA to find reduced models of order 10 at each training parameter point. Here, after truncation, BT's global basis was of order 347 and OS-IRKA's was 128. Figure 5 compares them, where the first 20 parameter values are

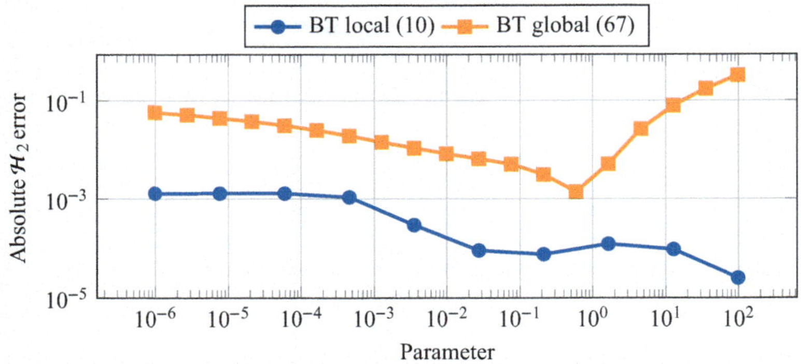

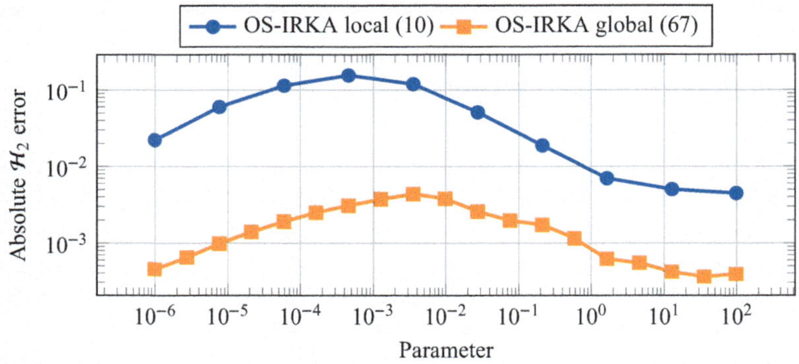

Fig. 3 Comparison of using local and global bases (see Sect. 3.2.1) for balanced truncation (BT) and one-sided iterative rational Krylov algorithm (OS-IRKA) for the one-parameter model

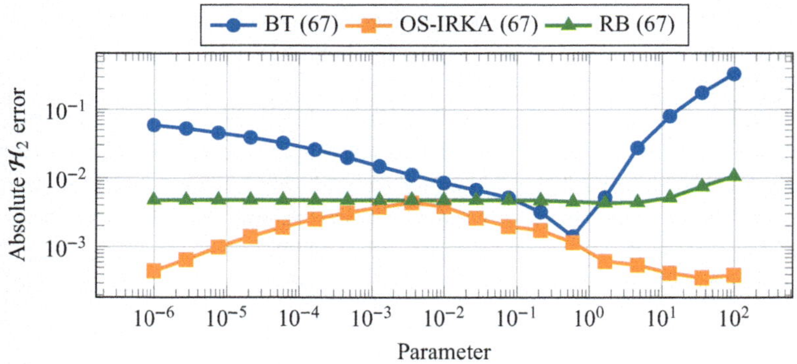

Fig. 4 Comparison of methods for the one-parameter model for fixed reduced order (67)

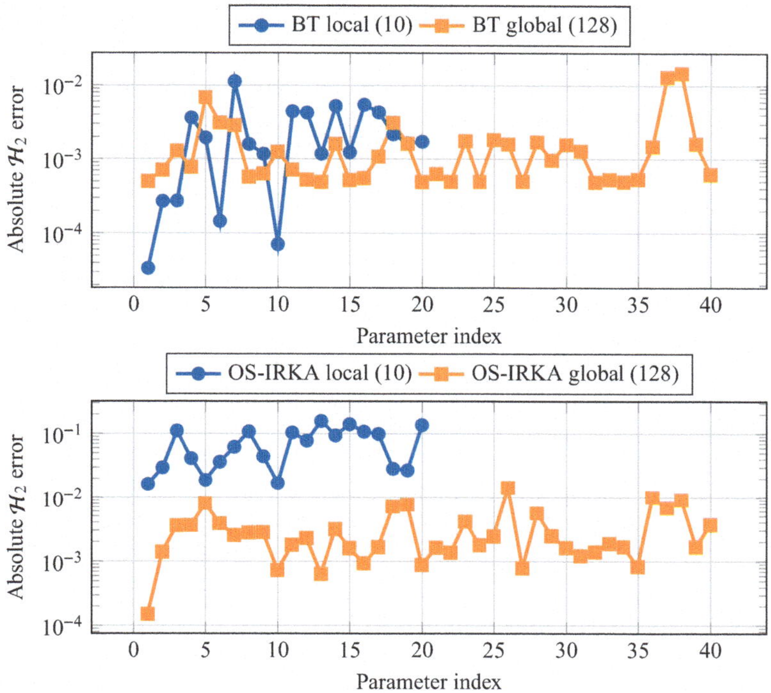

Fig. 5 Comparison of using local and global bases (see Sect. 3.2.1) for balanced truncation (BT) and one-sided iterative rational Krylov algorithm (OS-IRKA) for the four-parameter model. The first 20 parameters are used to construct local bases and global bases are tested on further 20 parameters (cf. Fig. 3)

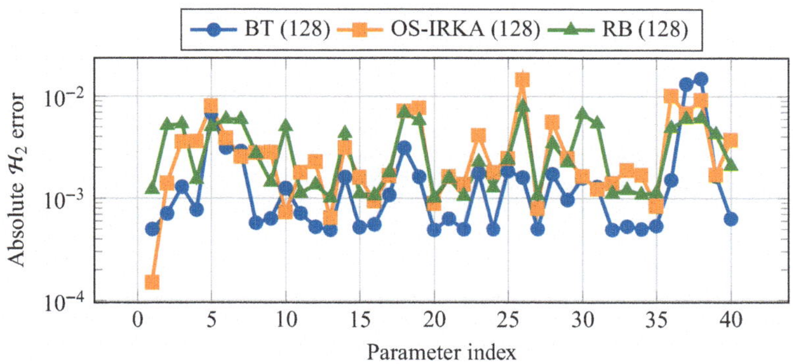

Fig. 6 Comparison of methods for the four-parameter model for fixed reduced order (128)

from the training set and the other for testing. As we had in the previous example, OS-IRKA gives better results on a global basis.

Figure 6 compares the two methods with RB. We see that they give comparable results, although they are rather different methods. On closer inspection, we note that, in this example, BT gives better errors the most and RB shows the smallest maximum error and the least variation in error.

5 Conclusions

We briefly presented pyMOR, a freely available Python package for MOR, built on generic interfaces for easy integration with external PDE solvers. We then described some of the MOR methods implemented in pyMOR, which includes both system-theoretic and reduced basis methods. Lastly, we compared methods on a thermal-block benchmark discretized with FEniCS.

Acknowledgements Funded by the Deutsche Forschungsgemeinschaft (DFG, German Research Foundation) under the project SA 3477/1-1 and Germany's Excellence Strategy EXC 2044–390685587, Mathematics Münster: Dynamics-Geometry-Structure. Funded by German Bundesministerium für Bildung und Forschung (BMBF, Federal Ministry of Education and Research) under grant number 05M18PMA in the programme "Mathematik für Innovationen in Industrie und Dienstleistungen".

References

1. Alnæs, M.S., Blechta, J., Hake, J., Johansson, A., Kehlet, B., Logg, A., Richardson, C., Ring, J., Rognes, M.E., Wells, G.N.: The FEniCS project version 1.5. Arch. Numer. Softw. **3**(100), 9–23 (2015). https://doi.org/10.11588/ans.2015.100.20553
2. Antoulas, A.C., Beattie, C.A., Gugercin, S.: Interpolatory model reduction of large-scale dynamical systems. In: Mohammadpour, J., Grigoriadis, K.M. (eds.) Efficient Modeling and Control of Large-Scale Systems, pp. 3–58. Springer, Berlin (2010). https://doi.org/10.1007/978-1-4419-5757-3_1
3. Balicki, L.: Low-rank alternating direction implicit iteration in pyMOR. GAMM Arch. Stud. **2**(1), 1–13 (2020). https://doi.org/10.14464/gammas.v2i1.420
4. Baur, U., Benner, P.: Modellreduktion für parametrisierte Systeme durch balanciertes Abschneiden und Interpolation (Model reduction for parametric systems using balanced truncation and interpolation). at-Automatisierungstechnik **57**(8), 411–420 (2009). https://doi.org/10.1524/auto.2009.0787
5. Benner, P., Bujanović, Z., Kürschner, P., Saak, J.: RADI: a low-rank ADI-type algorithm for large scale algebraic Riccati equations. Numer. Math. **138**(2), 301–330 (2018). https://doi.org/10.1007/s00211-017-0907-5
6. Benner, P., Köhler, M., Saak, J.: M.E.S.S. – the matrix equations sparse solvers library. https://www.mpi-magdeburg.mpg.de/projects/mess
7. Grepl, M.A., Patera, A.T.: A posteriori error bounds for reduced-basis approximations of parametrized parabolic partial differential equations. ESAIM: M2AN **39**(1), 157–181 (2005). https://doi.org/10.1051/m2an:2005006
8. Haasdonk, B.: Convergence rates of the POD-Greedy method. ESAIM: Math. Model. Numer. Anal. **47**(3), 859–873 (2013)

9. Milk, R., Rave, S., Schindler, F.: pyMOR - generic algorithms and interfaces for model order reduction. SIAM J. Sci. Comput. **38**(5), S194–S216 (2016). https://doi.org/10.1137/15M1026614
10. Moore, B.C.: Principal component analysis in linear systems: controllability, observability, and model reduction. IEEE Trans. Autom. Control **AC–26**(1), 17–32 (1981). https://doi.org/10.1109/TAC.1981.1102568
11. pyMOR developers and contributors: pyMOR - model order reduction with Python. https://pymor.org
12. Rave, S., Saak, J.: A non-stationary thermal-block benchmark model for parametric model order reduction (2020). arXiv preprint arXiv:2003.00846 [math.NA] (Chapter 16 in this volume)
13. SLICOT. http://www.slicot.org
14. Slycot developers and contributors: Slycot. https://github.com/python-control/Slycot
15. Son, N.T., Stykel, T.: Solving parameter-dependent Lyapunov equations using the reduced basis method with application to parametric model order reduction. SIAM J. Matrix Anal. Appl. **38**(2), 478–504 (2017). https://doi.org/10.1137/15M1027097
16. Virtanen, P., Gommers, R., Oliphant, T.E., Haberland, M., Reddy, T., Cournapeau, D., Burovski, E., Peterson, P., Weckesser, W., Bright, J., van der Walt, S.J., Brett, M., Wilson, J., Millman, K.J., Mayorov, N., Nelson, A.R.J., Jones, E., Kern, R., Larson, E., Carey, C.J., Polat, İ., Feng, Y., Moore, E.W., VanderPlas, J., Laxalde, D., Perktold, J., Cimrman, R., Henriksen, I., Quintero, E.A., Harris, C.R., Archibald, A.M., Ribeiro, A.H., Pedregosa, F., van Mulbregt, P.: SciPy 1.0: fundamental algorithms for scientific computing in Python. Nat. Methods (2020). https://doi.org/10.1038/s41592-019-0686-2
17. Wittmuess, P., Tarin, C., Keck, A., Arnold, E., Sawodny, O.: Parametric model order reduction via balanced truncation with Taylor series representation. IEEE Trans. Autom. Control **61**(11), 3438–3451 (2016). https://doi.org/10.1109/TAC.2016.2521361

Matrix Equations, Sparse Solvers: M-M.E.S.S.-2.0.1—Philosophy, Features, and Application for (Parametric) Model Order Reduction

Peter Benner, Martin Köhler, and Jens Saak

Abstract Matrix equations are omnipresent in (numerical) linear algebra and systems theory. Especially in model order reduction (MOR), they play a key role in many balancing-based reduction methods for linear dynamical systems. When these systems arise from spatial discretizations of evolutionary partial differential equations, their coefficient matrices are typically large and sparse. Moreover, the numbers of inputs and outputs of these systems are typically far smaller than the number of spatial degrees of freedom. Then, in many situations, the solutions of the corresponding large-scale matrix equations are observed to have low (numerical) rank. This feature is exploited by M-M.E.S.S. to find successively larger low-rank factorizations approximating the solutions. This contribution describes the basic philosophy behind the implementation and the features of the package, as well as its application in the MOR of large-scale linear time-invariant (LTI) systems and parametric LTI systems.

1 Introduction

The M-M.E.S.S. toolbox [55] for MATLAB® (or package for GNU Octave) in version 2.0.1 focuses on the solution of large-scale symmetric algebraic and differential matrix equations and their application in model order reduction (MOR) and linear-quadratic regulator (LQR) problems. The basis for all considerations and problem formulations are linear dynamical systems of the form

---

P. Benner · M. Köhler · J. Saak (✉)
Max Planck Institute for Dynamics of Complex Technical Systems, Sandtorstr. 1, 39106 Magdeburg, Germany
e-mail: saak@mpi-magdeburg.mpg.de

P. Benner
e-mail: benner@mpi-magdeburg.mpg.de

M. Köhler
e-mail: koehlerm@mpi-magdeburg.mpg.de

P. Benner et al. (eds.), *Model Reduction of Complex Dynamical Systems*, International Series of Numerical Mathematics 171,
https://doi.org/10.1007/978-3-030-72983-7_18

$$\begin{aligned} E\dot{x}(t) &= Ax(t) + Bu(t), \\ y(t) &= Cx(t) + Du(t), \end{aligned} \tag{Σ}$$

where $E, A \in \mathbb{R}^{n\times n}$, $B \in \mathbb{R}^{n\times m}$, $C \in \mathbb{R}^{p\times n}$, $D \in \mathbb{R}^{p\times m}$, and $x(t) \in \mathbb{R}^n$, for all time instances $t \in [0, T]$. We assume that E is invertible, and often in addition that (Σ) is asymptotically stable.

Some of the supported matrix equations have applications in H_∞-control, where the slightly more structured system

$$\begin{aligned} E\dot{x}(t) &= Ax(t) + B_1u(t) + B_2w(t), \\ y(t) &= C_1x(t) + D_{11}u(t) + D_{12}w(t), \\ z(t) &= C_2x(t) + D_{21}u(t) + D_{22}w(t), \end{aligned} \tag{Σ_∞}$$

is considered.

M-M.E.S.S. aims at systems, where $n \in \mathbb{N}$ is too large to store an $n \times n$ matrix in the computer's memory. This will usually be accounted for by the facts, that $p, m \ll n$ and E, A are sparse or have a sparse realization that we can exploit in computations. We present more details about the exploitable structures in Sect. 2.

Similarly, for systems (Σ_∞), the matrices B_1, B_2, C_1, C_2 are considered thin and rectangular and the parts D_{ij}, $i, j \in \{1, 2\}$ correspondingly small.

The contribution of this document is two-fold. On the one hand, we give the first concise introduction to M-M.E.S.S., its general philosophy and current features. On the other hand, we show how the software, that is in core intended for the solution of large-scale matrix equations, can be employed in the implementation of basic parametric MOR (PMOR) methods for systems of the form (Σ).

Before moving on to the historical evolution of the package, we state the equations that can currently be solved by M-M.E.S.S.. The following is a list of all equations for which at least one solver function exists:

Algebraic Lyapunov equations

$$\begin{aligned} 0 &= APE^\mathsf{T} + EPA^\mathsf{T} + BB^\mathsf{T} \\ 0 &= A^\mathsf{T}QE + E^\mathsf{T}QA + C^\mathsf{T}C \end{aligned} \tag{CALE}$$

Algebraic Riccati equations

$$\begin{aligned} 0 &= APE^\mathsf{T} + EPA^\mathsf{T} + BB^\mathsf{T} - EPC^\mathsf{T}CPE^\mathsf{T} \\ 0 &= A^\mathsf{T}QE + E^\mathsf{T}QA + C^\mathsf{T}C - E^\mathsf{T}QBB^\mathsf{T}QE \end{aligned} \tag{CARE}$$

$$\begin{aligned} 0 &= \tilde{A}PE^\mathsf{T} + EP\tilde{A}^\mathsf{T} + \tilde{B}_1\tilde{B}_1^\mathsf{T} - EP\left(\tilde{C}_1^\mathsf{T}\tilde{C}_1 - \tilde{C}_2^\mathsf{T}\tilde{C}_2\right)PE^\mathsf{T} \\ 0 &= \tilde{A}^\mathsf{T}QE + E^\mathsf{T}Q\tilde{A} + \tilde{C}_1^\mathsf{T}\tilde{C}_1 - E^\mathsf{T}Q\left(\tilde{B}_1\tilde{B}_1^\mathsf{T} - \tilde{B}_2\tilde{B}_2^\mathsf{T}\right)QE \end{aligned} \tag{$\mathcal{H}_\infty - ARE$}$$

In the last pair of equations, the matrix $\tilde{A}$ is sparse plus low-rank (splr), i.e. $\tilde{A} = A + UV^{\mathsf{T}}$, where U, V are tall and skinny. Moreover, the matrices $\tilde{B}_1$, $\tilde{B}_2$, $\tilde{C}_1$, $\tilde{C}_2$ are derived from the given system data by scaling and potentially rotation of the matrices B_1, B_2, C_1, C_2.

For finite time horizon linear-quadratic control problems, one needs to solve differential Riccati equations. We restrict to providing only the controller equations here, while the dual "filter-type" equations are supported as well.

Autonomous differential Riccati equations

$$-E^{\mathsf{T}}\dot{Q}(t)E = A^{\mathsf{T}}Q(t)E + E^{\mathsf{T}}Q(t)A + C^{\mathsf{T}}C - E^{\mathsf{T}}Q(t)BB^{\mathsf{T}}Q(t)E \quad \text{(ADRE)}$$

Non-autonomous differential Riccati equations

$$\begin{aligned} -E(t)^{\mathsf{T}}\dot{Q}(t)E(t) &= \left(A(t)+\dot{E}(t)\right)^{\mathsf{T}}Q(t)E(t) + E(t)^{\mathsf{T}}Q(t)\left(A(t)+\dot{E}(t)\right) \\ &\quad + C(t)^{\mathsf{T}}C(t) - E(t)^{\mathsf{T}}Q(t)B(t)B(t)^{\mathsf{T}}Q(t)E(t) \end{aligned} \quad \text{(NDRE)}$$

The last equations are formulated for the time-varying counterpart of (Σ), i.e., the system where all matrices are allowed to depend on time as well. Both DREs contain the case of differential Lyapunov equations. Optimized solvers for those are still work in progress and must at the moment be implemented by setting either B or C (in the dual equation) to zero and thus eliminating the quadratic term. Available solution methods in M-M.E.S.S. are described in Sect. 2.

Classic Lyapunov equation-based balanced truncation is known to preserve asymptotic stability of the original system in the reduced-order model. Other balancing-based methods have been developed to preserve other properties like passivity or contractivity. For these special balancing-type MOR methods, other matrix equations need to be solved that do not have a tailored solver in M-M.E.S.S., yet. Still, they can be reformulated into one of the types above. In order to have a more complete picture of what equations can be solved with the current M-M.E.S.S., we list them here, but get back to them in Sect. 3 and describe their reformulations into the special cases above and why they can still be solved using M-M.E.S.S..

Positive real balancing

$$\begin{aligned} 0 &= APE^{\mathsf{T}} + EPA^{\mathsf{T}} + (EPC^{\mathsf{T}} - B)(D + D^{\mathsf{T}})^{-1}(EPC^{\mathsf{T}} - B)^{\mathsf{T}} \\ 0 &= A^{\mathsf{T}}QE + E^{\mathsf{T}}QA + (E^{\mathsf{T}}QB - C^{\mathsf{T}})(D + D^{\mathsf{T}})^{-1}(E^{\mathsf{T}}QB - C^{\mathsf{T}})^{\mathsf{T}} \end{aligned} \quad \text{(PRARE)}$$

Bounded real balancing

$$\begin{aligned} 0 &= APE^{\mathsf{T}} + EPA^{\mathsf{T}} + BB^{\mathsf{T}} + (EPC^{\mathsf{T}} + BD^{\mathsf{T}})(I - DD^{\mathsf{T}})^{-1}(EPC^{\mathsf{T}} + BD^{\mathsf{T}})^{\mathsf{T}} \\ 0 &= A^{\mathsf{T}}QE + E^{\mathsf{T}}QA + C^{\mathsf{T}}C + (E^{\mathsf{T}}QB + C^{\mathsf{T}}D)(I - D^{\mathsf{T}}D)^{-1}(E^{\mathsf{T}}QB + C^{\mathsf{T}}D)^{\mathsf{T}} \end{aligned} \quad \text{(BRARE)}$$

Linear-quadratic-Gaussian balancing

$$
\begin{aligned}
0 &= APE^{\mathsf{T}} + EPA^{\mathsf{T}} + BB^{\mathsf{T}} - (EPC^{\mathsf{T}} + BD^{\mathsf{T}})(I + DD^{mathsfT})^{-1}(EPC^{\mathsf{T}} + BD^{\mathsf{T}})^{\mathsf{T}} \\
0 &= A^{\mathsf{T}}QE + E^{\mathsf{T}}QA + C^{\mathsf{T}}C - (E^{\mathsf{T}}QB + C^{\mathsf{T}}D)(I + D^{\mathsf{T}}D)^{-1}(E^{\mathsf{T}}QB + C^{\mathsf{T}}D)^{\mathsf{T}}
\end{aligned}
\tag{LQGARE}
$$

1.1 A Brief History of M-M.E.S.S.

Early days, the LyaPack years

The package M-M.E.S.S. originates in the work of Penzl [14, 47, 48] around the year 2000. More precisely, we understand M-M.E.S.S. as a continuation and successor of Penzl's LyaPack-toolbox [49] for MATLAB. While most of the basic ideas from the original package have been preserved, some features have been abandoned and some have been altered to improve efficiency and reliability.

The treatment of generalized state-space systems, i.e., systems (Σ) with nontrivial, i.e., non-identity, E-matrices have been added first. These changes still happened under the LyaPack-label in versions 1.1–1.8 until about 2007.

Transition to M-M.E.S.S. and present

The transition to the relabeled M-M.E.S.S.-1.0 package included a complete reorganization of the process data. Also, LyaPack used string manipulations and `eval`-calls to mimicked function pointers, which we replaced by proper function handles supported in modern MATLAB and GNU Octave. Moreover, the formulation of the low-rank alternating directions implicit (LR-ADI) iteration, which always was the heart and soul of LyaPack, was greatly updated to allow for cheaper evaluation of stopping criteria and an iteration inherent generation of real solution factors, which could only be achieved through post-processing in LyaPack.

The necessity for an a priori selection of shift parameters for convergence acceleration used to be a major point of criticism regarding the ADI-based solvers. The selection of shift generation methods was extended in M-M.E.S.S. and especially a new method that automatically generates the shifts during the iteration [34] was added, which makes the solvers accessible also to non-experts.

Other than that, version 1.0 saw general code modernization to support optimized features in MATLAB and to be 100% GNU Octave compatible.

The two major contributions of version 2.0 were the inclusion of the RADI iteration [7] for (CARE) and several solvers for differential Riccati equations in both the autonomous (ADRE) and non-autonomous (NDRE) cases.

Moreover, over time more system classes, including specially structured differential-algebraic equation (DAE)-based systems and second-order systems, have been added.

Future development plans

The most immediate upcoming feature in the near future is the inclusion of Krylov subspace projection methods for algebraic Lyapunov [59, 61] and Riccati equations [39, 60, 62]. The infrastructure and solvers are under current development and the feature is going to be part of version 3.0. The plans for the more distant future include, inclusion of low-rank solvers for Sylvester equations [12] and non-symmetric AREs [13], as well as the discrete-time counterparts of the existing equations, i.e., Stein equations [12, 32, 34, 38, 51] and discrete-time Riccati equations. Also, more complex sets of equations like Lur'e equations [50] and Lyapunov-plus-positive equations [8, 19, 58] are currently investigated and will be added if solvers can be implemented in a robust and efficient way using the M-M.E.S.S. infrastructure.

1.2 Structure of This Chapter

The following section introduces M-M.E.S.S. and its basic implementation philosophy. It further elaborates on supported system structures beyond the basic form in (Σ) and describes the current basic features of the package. Section 3 is dedicated to the description of MOR methods contained or demonstrated in M-M.E.S.S., while Sect. 4 shows how the existing tools in M-M.E.S.S. can be used to implement basic PMOR methods from the literature. The last section demonstrates how M-M.E.S.S. can be employed in PMOR, solving a selection of the above equations, for the benchmark example introduced in the separate chapter [52] of this volume. Similarly, this benchmark setting is considered in the other software chapters [18, 31, 40], in order to compare the applicability of the individual packages in a standardized setting.

2 M-M.E.S.S.—Philosophy and Features

The M-M.E.S.S. philosophy relies on three simple principles:

Abstract state-spacesystem All routines assume to work on a system of the form (Σ), or (Σ_∞). For a simple spatially discretized parabolic PDE, (Σ) is exactly given by the sparse matrices describing the semi-discretized system. For other systems, (Σ) may be a dense, inaccessible realization, like, e.g. a projection to a hidden manifold for a Stokes-type DAE system.

Implicit reformulation When the system matrices are potentially dense or even inaccessible, or otherwise prohibitive to use, the matrices are never formed explicitly, but only their actions are expressed in terms of the original data. For the aforementioned DAEs, this means, only the given semi-explicit system matrices are employed, but the algorithm runs as if it was formulated on the hidden manifold, i.e., for the implicit ordinary differential equation. This technique is often also called *implicit index-reduction*. For second-order systems, similarly, it is sometimes prohibitive to work with the double-sized phase-space realization in companion form. Again, all operations are executed only using the original second-order matrices, while solutions live in the double-sized space.

operation abstraction The abstraction of operations is realized via the so-called user-supplied function sets (usfs), which we have inherited from LyaPack. In comparison to LyaPack, we have slightly extended this set of functions. At the same time, we have removed the necessity to provide empty functions, which are now automatically replaced by a `do_nothing` function. While making things far more complicated in, e.g. the `default` case (see Table 1), where all matrices are expected to be available, this allows to hide the actual matrix realization from the algorithms. This way, in principle, the algorithms can run matrix-free with respect to A and E as demonstrated in [16].

The basic structure and design, of M-M.E.S.S., was decided when object-oriented features in MATLAB were in their early stages and essentially absent in GNU Octave. Still, some of the design follows object-oriented paradigms. We mimic the object orientation by passing three central data structures through all relevant functions. These three items of type `struct` are

`eqn` This structure essentially holds all relevant information about the underlying system (Σ), or (Σ_∞) and determines which equation in the dual pair we are aiming to solve, by `eqn.type='N'`, or `eqn.type='T'` representing the transposition on the left multiplication by A.

`oper` The operator structure, generated by the function `operatormanager`, holds all function handles for the relevant operations with the system matrices A and E. A list of these operations can be found in Table 2. Most functions in the list are accompanied by two functions, with appendices `_pre` and `_post`, called at the beginning and the end of a function working with them. They are intended for the generation and clean up of helper data, like the pre-factorization of matrices, when a sparse direct solver is used, or the generation of a preconditioner for an iterative solver.

`opts` The actual options structure is a structure of structures, i.e., it has a substructure for each algorithm/function but also holds central information on the top level. For example, `opts.norm` defines the norm that should consistently be used in all operations and hierarchy levels of the potentially cascaded algorithms, while substructures like `opts.adi`, or `opts.shifts` provide the specific control parameters for the LR-ADI algorithm and the shift computation.

Note that for all matrix operations in the usfs, we allow for corresponding `_pre` and `_post` functions. Other functions like `init` or `size` do not support `_pre` and `_post`.

While the function handles in `oper` work on the original A from (Σ), sometimes it is necessary to actually work with low-rank updated versions of A in the form $A + UV^{\mathsf{T}}$. We have seen an example in ($\mathcal{H}_\infty - ARE$), where $\tilde{A}$ is in the very form. Another prominent appearance is the Newton-Kleinman iteration (see [33] for classic iteration and [14] for the low-rank version) for (CARE), wherein iteration j, the step equation (for the second equation in the pair) takes the form

Table 1 Supported system structures via user-supplied function sets (usfs)

usfs	`default`	`so_1 / so_2`	`dae_1`	`dae_2`	`dae_1/2/3_so`
System	Standard/ generalized state-space form	Second-order 1st/2nd companion form	Semi-explicit index-1 DAE	Semi-explicit index-2 Stokes-type DAEs	Semi-explicit second-order index-1/2/3 DAEs using companion form
Demos	FDM [49], Rail [15]	TripleChain [54, 66]	DAE1 (BIPS Power-systems model [24])	DAE2 Stokes [57], Kárman vortex shedding [69]	Constrained TripleChain

$$\left(A - BK_{j-1}\right)^{\mathsf{T}} X_j E + E^{\mathsf{T}} X_j \left(A - BK_{j-1}\right) = \left[C\ K_{j-1}\right]^{\mathsf{T}} \left[C\ K_{j-1}\right].$$

Therefore, most solvers in M-M.E.S.S. assume that this structure can be given. The flag `eqn.haveUV` set to a non-zero value indicates that this is the case. Then the fields `eqn.U` and `eqn.V` need to hold the corresponding dense rectangular matrices of compatible dimensions. Similarly, the field `eqn.haveE` indicates that a non-trivial, i.e., non-identity E matrix is present and needs to be used via the function handles in Table 2.

Note that it is prohibitive to form $A + UV^{\mathsf{T}}$ explicitly, since even for very sparse A it can easily be a dense matrix. Especially, it is prohibitive to use direct solvers based on matrix decompositions on it, since then even if $A + UV^{\mathsf{T}}$ manages to preserve some sparsity, the fill-in will make the triangular factors dense. Therefore, all linear systems with $A + UV^{\mathsf{T}}$ are solved via the Sherman-Morrison-Woodbury matrix-inversion formula (see, e.g. [27, Sect. 2.1.4]) in M-M.E.S.S..

2.1 Available Solver Functions and Underlying Methods

We provide two solvers for the standard cases in (CALE) and (CALE) that are purely matrix-based, intended for large-scale sparse matrix coefficients and classic two-term low-rank factorizations of the constant terms and cores in the quadratic terms. The functions are called `mess_lyap` and `mess_care` and mimic the calls of `lyap` and `care` from MATLAB's control systems toolbox, or the GNU Octave control package, for dense matrices.

Other than that, we have the functions in Table 3 that allow for more flexible tuning, solve a large variety of equations and, especially, benefit from the full potential of the user-supplied functions. In the table, we give references to the most state-of-the-art presentations of the algorithms in the literature, on which our implementations are based.

Table 2 User-supplied function names and their actual operation

Function call	Operation
Y = oper.mul_A(eqn,opts,opA,B,opB)	$Y = A^{\mathrm{opA}} B^{\mathrm{opB}}$
Y = opr.mul_E(eqn,opts,opE,B,opB)	$Y = E^{\mathrm{opE}} B^{\mathrm{opB}}$
Y = oper.mul_ApE(eqn,opts,opA,p,opE,B,opB)	$Y = \left(A^{\mathrm{opA}} + pE^{\mathrm{opE}}\right) B^{\mathrm{opB}}$
X = oper.sol_A(eqn,opts,opA,B,opB)	$A^{\mathrm{opA}} X = B^{\mathrm{opB}}$
X = oper.sol_E(eqn,opts,opE,B,opB)	$E^{\mathrm{opE}} X = B^{\mathrm{opB}}$
X = oper.sol_ApE(eqn,opts,opA,p,opE,B,opB)	$\left(A^{\mathrm{opA}} + pE^{\mathrm{opE}}\right) X = B^{\mathrm{opB}}$
Result = oper.init(eqn,opt,oper,f1,f2)	General initialization and sanity checks
[W,res0] = oper.init_res(eqn,opts,oper,V)	Compute initial residual factor `W` from `V`, and `res0` $= \|W\|$
[eqn,opts,oper] = eval_matrix_functions(eqn,opts,oper,t)	In the time-varying case, fix all the above to time instance `t`
n = oper.size(eqn,opts,oper)	Returns the dimension n in (Σ)

3 Model Order Reduction in M-M.E.S.S.

The basic MOR facilities in M-M.E.S.S. are limited. Still, all building blocks for projection-based MOR using balancing methods, where matrix equations are most obviously applied, are available. For the sake of completeness and to fix our notation, we repeat the basics of projection-based MOR. Given a state-space system of the form (Σ), we search for the two rectangular transformation matrices $V, W \in \mathbb{R}^{n\times r}$ that define the actual oblique projection $T = V(W^{\mathsf{T}}V)^{-1}W^{\mathsf{T}}$, but transform the system into the reduced coordinates directly. The reduced-order model then takes the form

$$\begin{aligned}\hat{E}\dot{\hat{x}}(t) &= \hat{A}\hat{x}(t) + \hat{B}u(t),\\ \hat{y}(t) &= \hat{C}\hat{x}(t) + Du(t),\end{aligned} \tag{ROM}$$

where $\hat{E} = W^{\mathsf{T}}EV$, $\hat{A} = W^{\mathsf{T}}AV \in \mathbb{R}^{r\times r}$, $\hat{B} = W^{\mathsf{T}}B \in \mathbb{R}^{r\times m}$, and $\hat{C} = CV \in \mathbb{R}^{p\times r}$.

The number of actual MOR routines in M-M.E.S.S. is rather limited. In version 2.0.1, we have `mess_balanced_truncation` implementing classic Lyapunov balancing [37, 42, 65], for systems (Σ) realized with sparse E and A [14, 29, 54], and `mess_tangential_irka` implementing the tangential iterative Krylov algorithm (IRKA) [28] for first- and second-order systems. Our Gramian computation methods are integrated in a range of MOR software packages, though. While sss-

Table 3 Solver functions with algorithm and feature descriptions and latest and most feature complete literature references

Solver	Description	Reference
Algebraic Lyapunov equations		
mess_lradi	The low-rank alternating directions implicit (LR-ADI) iteration in residual-based formulation and with automatic shift selection for (CALE)	[34]
Algebraic Riccati equations		
mess_lrnm	An inexact Kleinman-Newton iteration with line search for (CARE)	[69]
mess_lrradi	The RADI iteration for (CARE)	[7]
mess_lrri	A low-rank version of the Riccati iteration [36] for $(\mathcal{H}_\infty - ARE)$	
Differential Riccati equations		
mess_bdf_dre	Low-rank formulation of backward differentiation formulas for large-scale differential Riccati equations (ADRE), (NDRE)	[35]
mess_rosenbrock_dre	Low-rank formulation of Rosenbrock methods for large-scale differential Riccati equations (ADRE)	[35]
mess_splitting_dre	Splitting schemes for large-scale differential Riccati equations (ADRE), (NDRE)	[63, 64]

MOR [20] directly calls M-M.E.S.S.-1.0.1, for other packages like MOREMBS [23] and MORPACK [43], we have contributed tailored versions of our algorithms.

Also, we provide tools like a square root method (SRM) function to compute the transformation matrices V and W from given Gramian factors. This function currently only uses the classic Lyapunov balancing error bound in the adaptive mode. This is subject to change in future versions.

3.1 IRKA and Classic Balanced Truncation

Consider that all matrices in (Σ) are available. As an example we use the Steel Profile benchmark [15, 45], included in M-M.E.S.S., using the version with $n = 1\,357$. Then computing the reduced-order matrices E_r, A_r, B_r, C_r for maximum reduced order 25 using the tangential IRKA [28] is as easy as calling:

```
eqn = getrail(1);
opts.irka.r = 25;
[Er, Ar, Br, Cr] = …
    mess_tangential_irka(eqn.E_,eqn.A_, eqn.B, eqn.C, …
    opts)
```

This will use default values for maximum iteration numbers and stopping criteria, which can be refined via the `opts.irka` structure. For a list of available options see `help mess_tangential_irka`.

Analogously, to compute a (Lyapunov) balanced truncation approximation of maximum order 50 and with an absolute H_∞-error tolerance of 10^{-2} for the same model the simplest call is:

```
eqn = getrail(1);
[Er, Ar, Br, Cr] = ...
    mess_balanced_truncation(eqn.E_, eqn.A_, eqn.B, …
        eqn.C, 50, 1e-2);
```

Note that Lyapunov balancing leaves the D matrix untouched in general, while it is absent in this example anyway. Note, further, that the interface may change slightly in future releases to make it more consistent with that of the IRKA function and to allow for the addition of the other balancing methods.

The balanced truncation approximation can be achieved in a step-by-step procedure, first, by computing the two Gramian factors, then applying them in the SRM to determine V and W, and finally, compressing the large-scale matrices to the reduced-order system matrices. This can all be executed using the procedural building blocks of `mess_balanced_truncation`. The example `bt_mor_rail_tol` in the `DEMOS/Rail` folder, residing in the main installation folder of M-M.E.S.S.-2.0.1, demonstrates this procedure. The step-wise approach can also be used for a number of structured systems like second-order systems, or semi-explicit DAE systems, while `mess_balanced_truncation` only supports generalized systems with invertible E, and all coefficients given explicitly as matrices, at the moment. See Table 4 for an overview of demonstration examples explaining these procedures.

3.2 *Further Variants of Balanced Truncation*

We have shown the Riccati equations defining the Gramians employed in positive-real, bounded-real, and linear-quadratic-Gaussian balanced truncation in equations (PRARE), (BRARE), (LQGARE) in the Introduction. Assuming, we have computed the Gramian factors, the reduced-order models can be derived, along the lines of the demonstration examples from Table 4. This can be done using the same M-M.E.S.S. function at least for a fixed desired reduced order. The error bound based order decision in the SRM needs adaptation to the specific error bound in some cases, though, see e.g., [2, Sect. 7.5] for a comparison of the bounds and procedures.

Table 4 Demonstration examples for balanced truncation of structured systems in M-M.E.S.S.

Example	Description	References
bt_mor_DAE1_tol	Balanced truncation for a semi-explicit power systems model of differential index 1	[24]
bt_mor_DAE2	Balanced truncation for Stokes and Oseen equations of index 2	[17, 30]
BT_TripleChain	First-order and structure-preserving balanced truncation for a model with three coupled mass-spring-damper chains	[53, 54, 66]
BT_sym_TripleChain	As above, but exploiting state-space symmetry of the tailored companion form first-order reformulation	
BT_DAE3_SO	First-order and structure-preserving balanced truncation for a variant of the above system that has a constraint turning it into an index-3 DAE	[56, 67, 68]

Here, we restrict ourselves to presenting how the specially structured Riccati equations can be solved with the existing functionality in M-M.E.S.S..

3.2.1 Positive-Real Balancing

For positive-real systems, by definition $D + D^{\mathsf{T}}$ is positive-definite, when it is invertible. This is always the case when the Riccati equations exist and do not degenerate to a set of Lur'e equations. Then we can decompose $D + D^{\mathsf{T}}$ into Cholesky factors, i.e., $R^{\mathsf{T}} R = D + D^{\mathsf{T}}$. Using these Cholesky factors, we define

$$\tilde{E} = E, \qquad \tilde{A} = A + UV^{\mathsf{T}}, \qquad \tilde{B} = BR^{-1}, \qquad \tilde{C} = R^{-\mathsf{T}}C,$$

with $U = \tilde{B}$ and $V^{\mathsf{T}} = \tilde{C}$, and a straight forward calculation shows that (PRARE) can be rewritten in the form

$$\begin{aligned} 0 &= \tilde{A}P\tilde{E}^{\mathsf{T}} + \tilde{E}P\tilde{A}^{\mathsf{T}} + \tilde{B}\tilde{B}^{\mathsf{T}} + \tilde{E}P\tilde{C}^{\mathsf{T}}\tilde{C}P\tilde{E}^{\mathsf{T}} \\ 0 &= \tilde{A}^{\mathsf{T}}Q\tilde{E} + \tilde{E}^{\mathsf{T}}Q\tilde{A} + \tilde{C}^{\mathsf{T}}\tilde{C} + \tilde{E}^{\mathsf{T}}Q\tilde{B}\tilde{B}^{\mathsf{T}}Q\tilde{E}. \end{aligned}$$

This resembles the Riccati case in ($\mathcal{H}_\infty - ARE$) with a low-rank updated matrix A and only the positive quadratic term present. This case is supported by the `mess_lrri` routine. Note that $D + D^{\mathsf{T}}$ is of small dimension, such that this reformulation is always feasible.

3.2.2 Bounded-Real Balancing

The bounded-real assumptions guarantee that $I - DD^{\mathsf{T}}$ and $I - D^{\mathsf{T}}D$ are symmetric positive definite. Therefore, we can decompose them into Cholesky factors, i.e., $R^{\mathsf{T}}R = I - DD^{\mathsf{T}}$ and $L^{\mathsf{T}}L = I - D^{\mathsf{T}}D$. Now, we define

$$\tilde{E} = E, \qquad \tilde{A} = A + UV^{\mathsf{T}}, \qquad \tilde{B} = BL^{-1}, \qquad \tilde{C} = R^{-1}C,$$

with $U = BD^{\mathsf{T}}$ and $V^{\mathsf{T}} = (I - DD^{\mathsf{T}})^{-1}C$ and another technical, but straight forward, calculation shows that (BRARE) can be rewritten in the form:

$$\begin{aligned} 0 &= \tilde{A}P\tilde{E}^{\mathsf{T}} + \tilde{E}P\tilde{A}^{\mathsf{T}} + \tilde{B}\tilde{B}^{\mathsf{T}} + \tilde{E}P\tilde{C}^{\mathsf{T}}\tilde{C}P\tilde{E}^{\mathsf{T}} \\ 0 &= \tilde{A}^{\mathsf{T}}Q\tilde{E} + \tilde{E}^{\mathsf{T}}Q\tilde{A} + \tilde{C}^{\mathsf{T}}\tilde{C} + \tilde{E}^{\mathsf{T}}Q\tilde{B}\tilde{B}^{\mathsf{T}}Q\tilde{E}. \end{aligned}$$

This, again, falls into the class of equations in $(\mathcal{H}_\infty - ARE)$ with a low-rank updated matrix A and only the positive square term present. As mentioned above, this case is supported by the `mess_lrri` routine. For the same reason as above, this reformulation can always be done.

3.2.3 Linear-Quadratic-Gaussian Balancing

For linear-quadratic-Gaussian balanced truncation, an important special case (see, e.g., [3, 11, 41, 44]) is $D = 0$. In that case (LQGARE) obviously reduces to the standard Riccati equation (CARE) that can be solved using `mess_lrnm` or `mess_lrradi`. The corresponding M-M.E.S.S. workflow is demonstrated in the `lqgbt_mor_FDM` example for a simple heat equation model semi-discretized by the finite difference method.

On the other hand, when $D \neq 0$, it is, by standard assumptions in M-M.E.S.S., real and all eigenvalues of DD^{T} and $D^{\mathsf{T}}D$ are non-negative. Therefore, $I + DD^{\mathsf{T}}$ and $I + D^{\mathsf{T}}D$ are symmetric and positive definite and analogous to the above, we can decompose into Cholesky factorizations $R^{\mathsf{T}}R = I + DD^{\mathsf{T}}$ and $L^{\mathsf{T}}L = I + D^{\mathsf{T}}D$. We now define

$$\tilde{E} = E, \qquad \tilde{A} = A + UV^{\mathsf{T}}, \qquad \tilde{B} = BL^{-1}, \qquad \tilde{C} = R^{-1}C,$$

with $U = -BD^{\mathsf{T}}$ and $V^{\mathsf{T}} = (I + DD^{\mathsf{T}})^{-1}C$. An analogous calculation to the bounded-real case shows that (LQGARE) can be rewritten in the form

$$0 = \tilde{A}P\tilde{E}^{\mathsf{T}} + \tilde{E}P\tilde{A}^{\mathsf{T}} + \tilde{B}\tilde{B}^{\mathsf{T}} - \tilde{E}P\tilde{C}^{\mathsf{T}}\tilde{C}P\tilde{E}^{\mathsf{T}}$$
$$0 = \tilde{A}^{\mathsf{T}}Q\tilde{E} + \tilde{E}^{\mathsf{T}}Q\tilde{A} + \tilde{C}^{\mathsf{T}}\tilde{C} - \tilde{E}^{\mathsf{T}}Q\tilde{B}\tilde{B}^{\mathsf{T}}Q\tilde{E}.$$

Due to the different signs, here, we end up with a standard Riccati equation (CARE), just like in the case $D = 0$. Again the transformation is always feasible in the sense of M-M.E.S.S. applicability.

4 Parametric Model Order Reduction Using M-M.E.S.S.

PMOR aims to preserve symbolic parameters in the original system description also in the reduced-order model. In the most general case, the system

$$\begin{aligned} E(\mu)\dot{x}(\mu,t) &= A(\mu)x(\mu,t) + B(\mu)u(t), \\ y(\mu,t) &= C(\mu)x(\mu,t) + D(\mu)u(t), \end{aligned} \qquad (\Sigma(\mu))$$

is transformed into

$$\begin{aligned} \hat{E}(\mu)\dot{\hat{x}}(\mu,t) &= \hat{A}(\mu)\hat{x}(\mu,t) + \hat{B}(\mu)u(t), \\ \hat{y}(t) &= \hat{C}(\mu)\hat{x}(\mu,t) + D(\mu)u(t). \end{aligned} \qquad (\mathrm{ROM}(\mu))$$

By default, M-M.E.S.S.-2.0.1 does not support PMOR. It is, however, very easy to implement basic PMOR routines building up on the methods from the previous section. The key ingredient, that at the same time establishes the link to the previous section in many methods, is the necessity to evaluate standard MOR problems in certain training points for given parameter values $\mu^{(i)}$ $(i = 1, \dots, k)$, e.g., on a sparse-grid in the parameter domain. While piecewise MOR approaches (e.g., [4]) aim to find constant global (with respect to the parameter) transformation matrices V and W to derive (ROM(μ)), other methods aim to establish it by interpolation of some kind. The literature basically provides three approaches interpolating different system features, see, e.g., [10] for further categorization of PMOR methods:

- matrix interpolation, i.e., function interpolation of the parameter-dependent coefficient matrices, or the transformation matrices, (e.g., [1, 25, 26, 46]),
- interpolation of the transfer functions in the parameter variable [5],
- interpolation of system poles (e.g., [9, 70]).

We demonstrate the basic steps for piecewise and interpolatory methods along with the lines of [4, 5] in the remainder of this section and give numerical illustrations in Sect. 5.

4.1 Piecewise MOR

We have mentioned above that the aim, here, is to find V and W constant, such that $\hat{E}(\mu) = W^{\mathsf{T}} E(\mu) V$, $\hat{A}(\mu) = W^{\mathsf{T}} A(\mu) V$, $\hat{B}(\mu) = W^{\mathsf{T}} B(\mu)$, $\hat{C}(\mu) = C(\mu) V$. The strong point of this method is that it trivially allows the ROMs in the parameters $\mu^{(i)}$ to vary in their reduced order. This is due to the fact that

$$V = \left[V^{(1)} \cdots V^{(k)}\right] \qquad \text{and} \qquad W = \left[W^{(1)} \cdots W^{(k)}\right],$$

with $V^{(i)}$ and $W^{(i)}$ the transformation matrices at parameter sample $\mu^{(i)}$. This concatenation should be followed by a rank truncation to eliminate linear dependencies.

It essentially does not matter how the single transformation matrices have been generated. We follow the presentation in [4], where IRKA is used. In the numerical experiments, we also compare to versions using balanced truncation in the training samples.

4.2 Interpolation of Transfer Functions

The representation of $(\Sigma(\mu))$ in frequency domain after Laplace transformation in the time variable (t), leads to the transfer function

$$H(\mu, s) = C(\mu)(sE(\mu) - A(\mu))^{-1} B(\mu).$$

For fixed μ, the IRKA method seeks to interpolate this function in s-direction, while the well-known balanced truncation method computes an approximation to this function, with a computable error bound. Therefore, it is an obvious task to try to extend these features by interpolation in μ-direction. Baur and Benner meet this goal in [5] for local balanced truncation approximations of $(\Sigma(\mu))$, achieving both stability preservation and an error bound, i.e., the selling points of balanced truncation. Moreover, their method shares the flexibility with respect to the ROM orders since the interpolation is done via the transfer function, that has a fixed dimension independent of the realization of the system. On the other hand, interpolation on matrix manifolds and with respect to system invariants need to fix the dimensions of those objects.

For simplicity we restrict ourselves to the case of scalar parameters. The approach in [5] defines (ROM(μ)) via its transfer function, which is chosen as an interpolant of the form

$$\begin{aligned}\hat{H}(\mu, s) = \sum_{i=1}^{k} \ell_i(\mu) \hat{H}^{(i)}(s) &= \sum_{i=1}^{k} \ell_i(\mu) \hat{C}^{(i)} \left(s\hat{E}^{(i)} - \hat{A}^{(i)}\right)^{-1} \hat{B}^{(i)} \\ &= \sum_{i=1}^{k} \hat{C}^{(i)}(\mu) \left(s\hat{E}^{(i)} - \hat{A}^{(i)}\right)^{-1} \hat{B}^{(i)}\end{aligned}$$

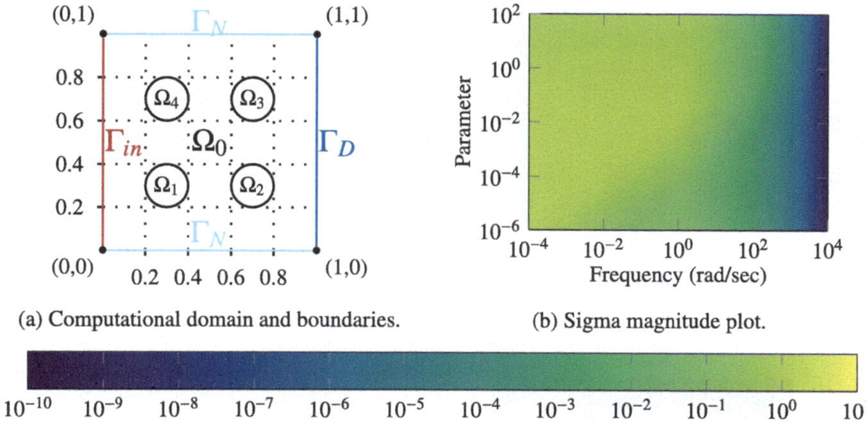

(a) Computational domain and boundaries. (b) Sigma magnitude plot.

Fig. 1 Computational domain and sigma magnitude plot for the thermal block model

with scalar coefficients functions $\ell_i(\mu)$, $\hat{H}^{(i)}(s)$ the transfer function of the ROM at parameter sample $\mu^{(i)}$ and $\hat{C}^{(i)}(\mu) = \ell_i(\mu)\hat{C}^{(i)}$. One can use the last identity to define the matrices for the ROM realization

$$\hat{E} = \begin{bmatrix} \hat{E}^{(1)} & & \\ & \ddots & \\ & & \hat{E}^{(k)} \end{bmatrix}, \quad \hat{A} = \begin{bmatrix} \hat{A}^{(1)} & & \\ & \ddots & \\ & & \hat{A}^{(k)} \end{bmatrix}, \quad \hat{B} = \begin{bmatrix} \hat{B}^{(1)} \\ \vdots \\ \hat{B}^{(k)} \end{bmatrix},$$
$$\hat{C}(\mu) = \left[\hat{C}^{(1)}(\mu) \cdots \hat{C}^{(k)}(\mu)\right],$$

such that

$$\hat{H}(\mu, s) = \hat{C}(\mu)\left(s\hat{E} - \hat{A}\right)^{-1}\hat{B}.$$

Note that the parameter could as well be put into $\hat{B}$. The specific choice of Lagrange polynomials is not necessary. We present experiments with both classic polynomial interpolation and spline interpolation in the next section. Since we are dealing with scalar coefficient functions here, it is advisable for a modern MATLAB implementation to exploit the power of Chebfun [21, 22]. We do this for the polynomial interpolation and the generation of the grid of training parameters, while the splines use our own implementation.

5 Numerical Experiments

The experiments reported here have been executed in MATLAB R2019a on a Lenovo X380 Yoga equipped with an Intel® i7 8770 and 32GB of main memory running 64bit Linux based on Ubuntu 18.04. The experiments use M-M.E.S.S.-2.0.1 [55] and Chebfun version 5.7.0 [21, 22].

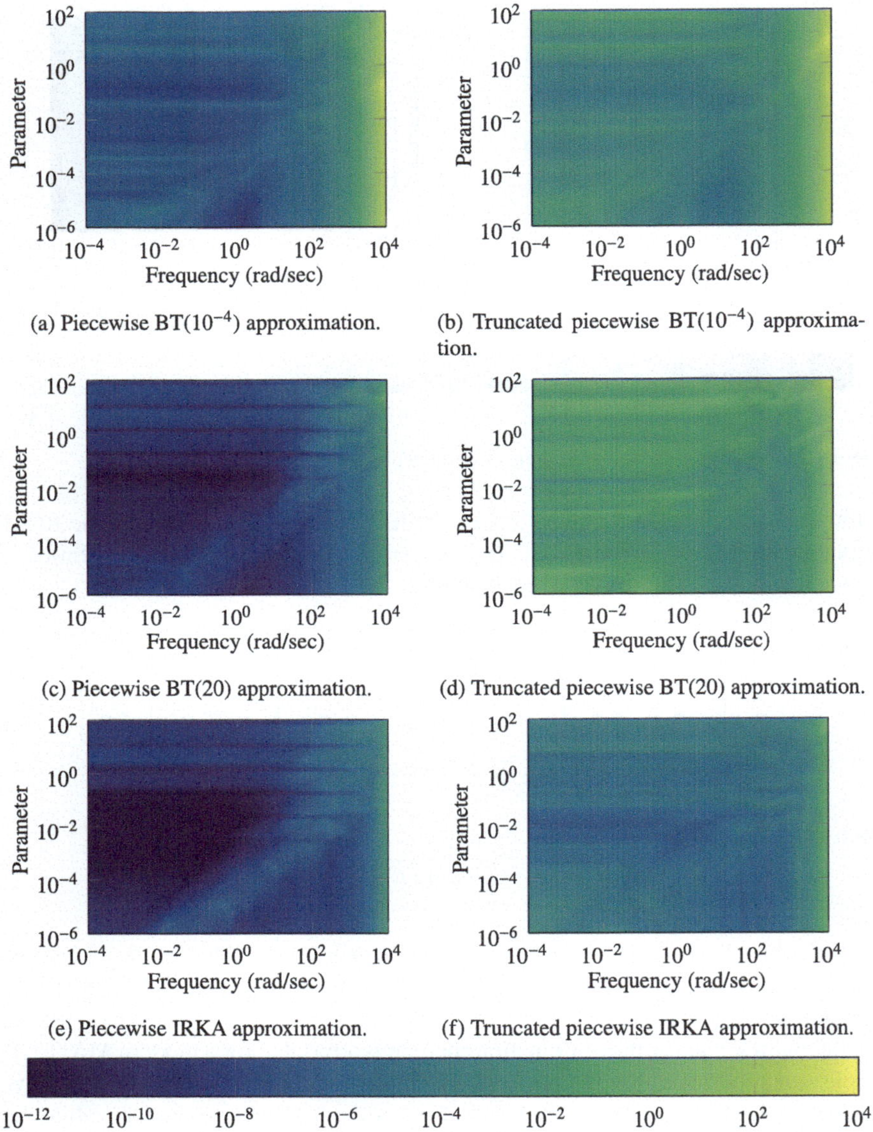

(a) Piecewise BT(10^{-4}) approximation.

(b) Truncated piecewise BT(10^{-4}) approximation.

(c) Piecewise BT(20) approximation.

(d) Truncated piecewise BT(20) approximation.

(e) Piecewise IRKA approximation.

(f) Truncated piecewise IRKA approximation.

Fig. 2 Relative sigma-magnitude errors of different piecewise parametric reduction approaches for the thermal-block model

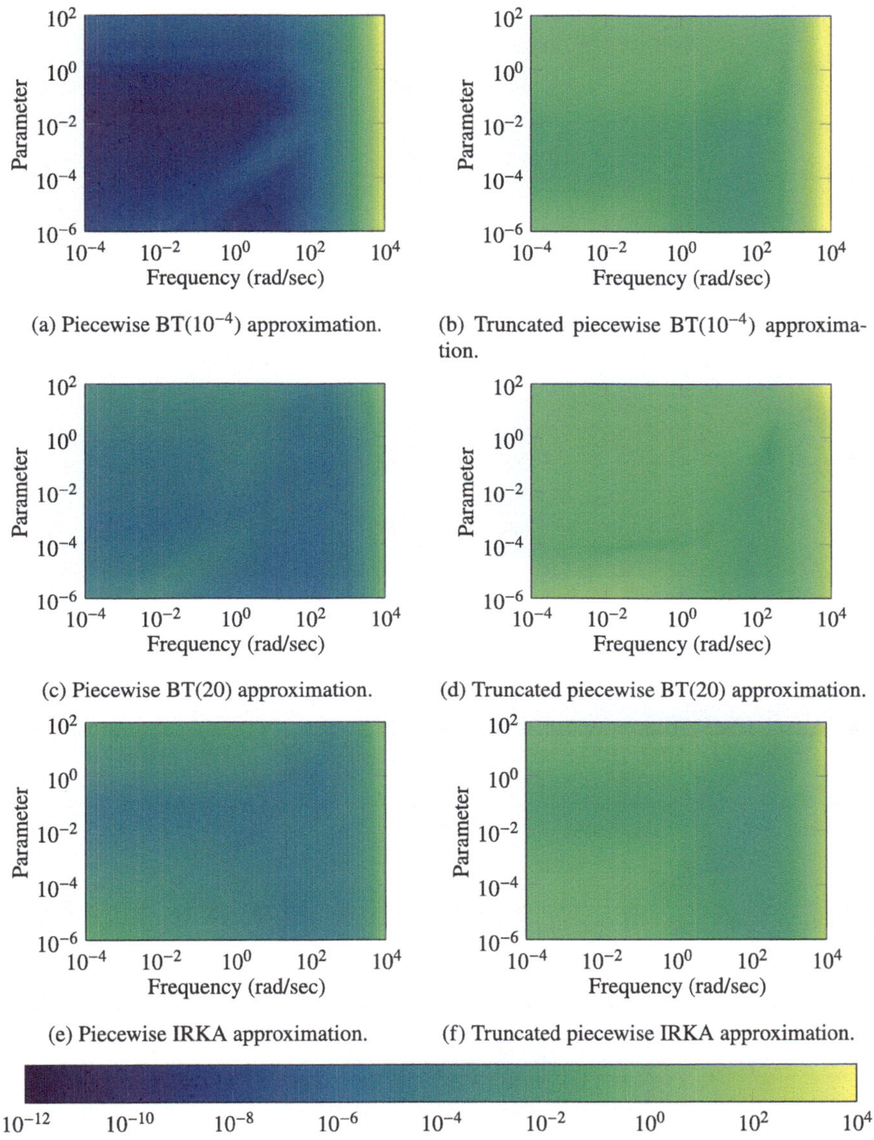

(a) Piecewise BT(10^{-4}) approximation.

(b) Truncated piecewise BT(10^{-4}) approximation.

(c) Piecewise BT(20) approximation.

(d) Truncated piecewise BT(20) approximation.

(e) Piecewise IRKA approximation.

(f) Truncated piecewise IRKA approximation.

Fig. 3 Relative sigma-magnitude errors of different piecewise parametric one-sided reduction approaches for the thermal-block model

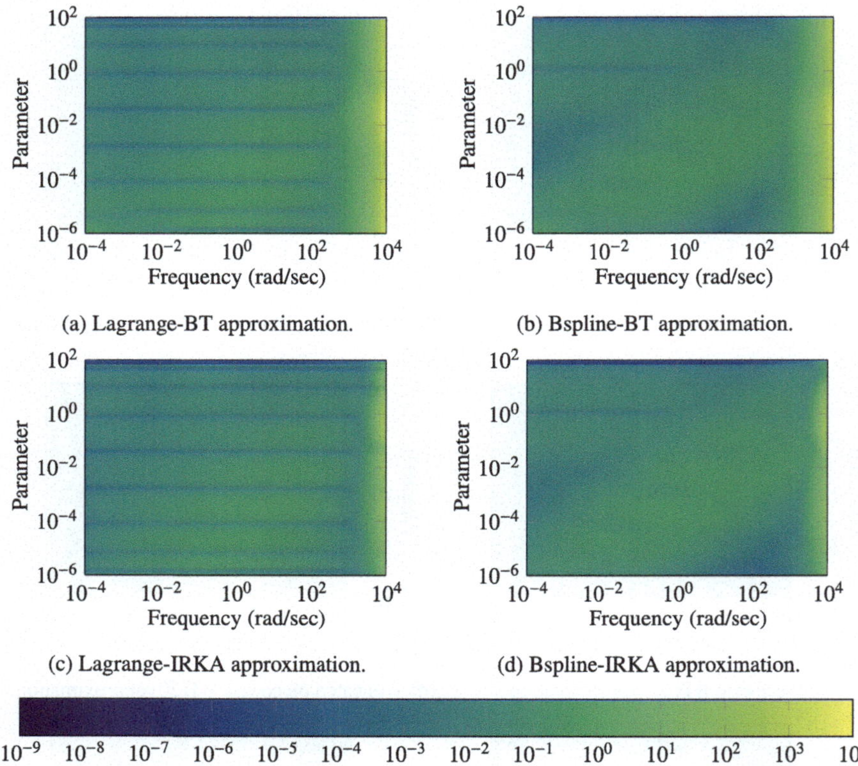

(a) Lagrange-BT approximation.

(b) Bspline-BT approximation.

(c) Lagrange-IRKA approximation.

(d) Bspline-IRKA approximation.

Fig. 4 Relative sigma-magnitude errors of different transfer function interpolation methods for parametric reduction for the thermal-block model

The source code of the implementations used to compute the presented results can be obtained from:

https://doi.org/10.5281/zenodo.3678213

and is authored by Jens Saak and Steffen W. R. Werner.

For easier comparison with the other reported software packages, all experiments use the thermal-block benchmark introduced in Chapter 16 of this volume. It describes a simple heat transfer model on the domain depicted in Fig. 1a. Here, we investigate the one-parameter version of the benchmark. That means, the heat transfer coefficients on the four circular sub-domains are given as 0.2, 0.4, 0.6, and $0.8\,\mu$ for a single scalar parameter $\mu \in [10^{-6}, 10^{2}] = M \subset \mathbb{R}$. The full order model has dimension $n = 7\,488$ and one input but 4 outputs. In Fig. 1b, we present the sigma-magnitude plot of the full-order model (FOM), i.e., we plot $\|H(\mu, s)\|_2 = \sigma_{\max}(H(\mu, s))$ over the full parameter range and the frequency range $[10^{-4}, 10^{4}]$. The plot is based on 100 logarithmically equi-spaced sample points (`logspace`-generated) in each direction. We also use this sampling for all relative sigma-magnitude error plots in the other figures. The error plots analogously show $\|H(\mu, s) - \hat{H}(\mu, s)\|_2 / \|H(\mu, s)\|_2$.

Excluding the 10 000 evaluations for the pre-sampling of the original transfer function, all computations for generation of the ROMs and evaluation of the approximation errors can be executed in less than 8 min.

We compare both IRKA and classic (Lyapunov) balanced truncation (BT) in the piecewise as well as the transfer function interpolation context. For IRKA, we use fixed order $r = 20$ in all training samples, while for BT, we run in two modes. Since we have the BT error-bound that allows for adaptive processing, i.e., automatic choice of the reduced order, we do that with absolute error tolerance 10^{-4}. On the other hand, for a more fair comparison to IRKA, we also run BT for fixed order $r = 20$. We refer to these two modes as BT(10^{-4}) and BT(20).

For the piecewise approaches, we use ten logarithmically equi-spaced (`logspace`-generated) parameter samples in M as the training positions. For the interpolatory approaches, we choose 10 Chebyshev-roots generated by Chebfun. We have mentioned the final rank-truncation after basis concatenation in Sect. 4.1. We use a tolerance equal to `eps` in the standard case. Alternatively, to further compress the final parametric ROM, we truncate with tolerance 10^{-6} and refer to this approach by the name *truncated piecewise*.

For the training, BT can not reuse information from previous samples very easily. On the other hand, IRKA can be initialized with the ROM from the previous parameter sample, which in most cases made it converge after less than five steps (mostly being stopped by the criterion monitoring the relative change of the model in the $\mathcal{H}_2$-norm). For further implementation details, we refer to the scripts in the code package.

Although BT guarantees the local ROMs in the sample points to preserve the asymptotic stability of the original model, and also IRKA preserves stability upon convergence, this feature is in general lost after concatenating the bases to the global one. Still, for a one-sided projection the stability of the global ROM can be preserved. Due to stability and symmetry of the thermal-block model, Bendixson's theorem [6] guarantees this. Therefore, we compare to a one-sided approach that simply combines V and W into one matrix. The comparison can be found in Figs. 2 and 3. And the corresponding ROM orders are given in the first block of Table 5.

For the interpolatory approaches, we compare Lagrange polynomials and variation diminishing B-splines of order 2. Here, we always use BT(10^{-4}) in the BT case, since the results are already hard to distinguish from the IRKA-based ones in this case and we do not expect much improvement from the higher local orders.

It can be seen from Table 5 that the piecewise BT models are, in parts significantly, smaller than the piecewise IRKA models. This comes at the price that the accuracy is not as good in parts of the domain. Nonetheless, e.g., the truncated one-sided BT(10^{-4}) approximation yields a relative error of below 1% on a majority (around 70%) of the investigated frequency parameter domain with a model size that is 3.7 to 5.6 times smaller. There is a significant increase in error for those frequencies, where the transfer function has very small values (see Fig. 1b) that can be considered to be on the noise level (Fig. 3).

Table 5 Reduced orders of the training-sample ROMs and final ROM (numbers in () are after additional truncation with tolerance 10^{-6})

Method	ROMs	Full	One-sided
Piecewise			
BT(10^{-4})	9/12/15/13/12/9/8/9/8/7	102 (52)	200 (36)
BT(20)	20/20/20/20/20/20/20/20/20/20	199 (64)	200 (72)
IRKA	20/20/20/20/20/20/20/20/20/20	200 (132)	200 (132)
Lagrange			
BT(10^{-4})	9/9/12/15/12/9/8/8/7/7	96	–
IRKA	20/20/20/20/20/20/20/20/20/20	200	–
B-spline			
BT(10^{-4})	9/9/12/15/12/9/8/8/7/7	96	–
IRKA	20/20/20/20/20/20/20/20/20/20	200	–

The results are very satisfactory and so are the computation times. This indicates that the implementations can be used for larger and more challenging examples, that we can not report here due to space restrictions.

References

1. Amsallem, D., Farhat, C.: An online method for interpolating linear parametric reduced-order models. SIAM J. Sci. Comput. **33**(5), 2169–2198 (2011). https://doi.org/10.1137/100813051
2. Antoulas, A.C.: Approximation of Large-Scale Dynamical Systems, Advances in Design and Control, vol. 6. SIAM Publications, Philadelphia (2005). https://doi.org/10.1137/1.9780898718713
3. Batten King, B., Hovakimyan, N., Evans, K.A., Buhl, M.: Reduced order controllers for distributed parameter systems: LQG balanced truncation and an adaptive approach. Math. Comput. Model. **43**(9), 1136–1149 (2006). https://doi.org/10.1016/j.mcm.2005.05.031
4. Baur, U., Beattie, C.A., Benner, P., Gugercin, S.: Interpolatory projection methods for parameterized model reduction. SIAM J. Sci. Comput. **33**(5), 2489–2518 (2011). https://doi.org/10.1137/090776925
5. Baur, U., Benner, P.: Modellreduktion für parametrisierte Systeme durch balanciertes Abschneiden und Interpolation (Model reduction for parametric systems using balanced truncation and interpolation). at-Automatisierungstechnik **57**(8), 411–420 (2009). https://doi.org/10.1524/auto.2009.0787
6. Bendixson, I.: Sur les racines d'une équation fondamentale. Acta Math. **25**(1), 359–365 (1902). https://doi.org/10.1007/BF02419030
7. Benner, P., Bujanović, Z., Kürschner, P., Saak, J.: RADI: a low-rank ADI-type algorithm for large scale algebraic Riccati equations. Numer. Math. **138**(2), 301–330 (2018). https://doi.org/10.1007/s00211-017-0907-5

8. Benner, P., Goyal, P.: Balanced truncation model order reduction for quadratic-bilinear systems (2017). arXiv preprint arXiv:1705.00160 [math.OC]
9. Benner, P., Grundel, S., Hornung, N.: Parametric model order reduction with a small $\mathcal{H}_2$-error using radial basis functions. Adv. Comput. Math. **41**(5), 1231–1253 (2015). https://doi.org/10.1007/s10444-015-9410-7
10. Benner, P., Gugercin, S., Willcox, K.: A survey of model reduction methods for parametric systems. SIAM Rev. **57**(4), 483–531 (2015). https://doi.org/10.1137/130932715
11. Benner, P., Heiland, J.: LQG-balanced truncation low-order controller for stabilization of laminar flows. In: King, R. (ed.) Active Flow and Combustion Control 2014. Notes on Numerical Fluid Mechanics and Multidisciplinary Design, vol. 127, pp. 365–379. Springer International Publishing (2015). https://doi.org/10.1007/978-3-319-11967-0_22
12. Benner, P., Kürschner, P.: Computing real low-rank solutions of Sylvester equations by the factored ADI method. Comput. Math. Appl. **67**(9), 1656–1672 (2014). https://doi.org/10.1016/j.camwa.2014.03.004
13. Benner, P., Kürschner, P., Saak, J.: Low-rank Newton-ADI methods for large nonsymmetric algebraic Riccati equations. J. Frankl. Inst. **353**(5), 1147–1167 (2016). https://doi.org/10.1016/j.jfranklin.2015.04.016
14. Benner, P., Li, J.R., Penzl, T.: Numerical solution of large-scale Lyapunov equations, Riccati equations, and linear-quadratic optimal control problems. Numer. Linear Algebr. Appl. **15**(9), 755–777 (2008). https://doi.org/10.1002/nla.622
15. Benner, P., Saak, J.: A semi-discretized heat transfer model for optimal cooling of steel profiles. In: Benner, P., Mehrmann, V., Sorensen, D. (eds.) Dimension Reduction of Large-Scale Systems. Lecture Notes in Computational Science and Engineering, vol. 45, pp. 353–356. Springer, Berlin/Heidelberg (2005). https://doi.org/10.1007/3-540-27909-1_19
16. Benner, P., Saak, J., Schieweck, F., Skrzypacz, P., Weichelt, H.K.: A non-conforming composite quadrilateral finite element pair for feedback stabilization of the Stokes equations. J. Numer. Math. **22**(3), 191–220 (2014). https://doi.org/10.1515/jnma-2014-0009
17. Benner, P., Saak, J., Uddin, M.M.: Balancing based model reduction for structured index-2 unstable descriptor systems with application to flow control. Numer. Algebr. Control Optim. **6**(1), 1–20 (2016). https://doi.org/10.3934/naco.2016.6.1
18. Benner, P., Werner, S.W.R.: MORLAB – the Model Order Reduction LABoratory (2020). arXiv preprint arXiv: 2002.12682 [Cs.MS]. https://doi.org/10.1007/978-3-030-72983-7_19
19. Breiten, T.: Interpolatory methods for model reduction of large-scale dynamical systems. Dissertation, Department of Mathematics, Otto-von-Guericke University, Magdeburg, Germany (2013). https://doi.org/10.25673/3917
20. Castagnotto, A., Cruz Varona, M., Jeschek, L., Lohmann, B.: sss & sssMOR: analysis and reduction of large-scale dynamic systems in MATLAB. at-Automatisierungstechnik **65**(2), 134–150 (2017). https://doi.org/10.1515/auto-2016-0137
21. Chebfun Developers: Chebfun — numerical computing with functions. https://www.chebfun.org/
22. Driscoll, T.A., Hale, N., Trefethen, L.N.: Chebfun Guide. Pafnuty Publications (2014). http://www.chebfun.org/docs/guide/
23. Fehr, J., Grunert, D., Holzwarth, P., Fröhlich, B., Walker, N., Eberhard, P.: Morembs—a model order reduction package for elastic multibody systems and beyond. In: Reduced-Order Modeling (ROM) for Simulation and Optimization: Powerful Algorithms as Key Enablers for Scientific Computing, pp. 141–166. Springer International Publishing, Cham (2018). https://doi.org/10.1007/978-3-319-75319-5_7
24. Freitas, F., Rommes, J., Martins, N.: Gramian-based reduction method applied to large sparse power system descriptor models. IEEE Trans. Power Syst. **23**(3), 1258–1270 (2008). https://doi.org/10.1109/TPWRS.2008.926693
25. Geuß, M., Butnaru, D., Peherstorfer, B., Bungartz, H., Lohmann, B.: Parametric model order reduction by sparse-grid-based interpolation on matrix manifolds for multidimensional parameter spaces. In: Proceedings of the European Control Conference, Strasbourg, France, pp. 2727–2732 (2014). https://doi.org/10.1109/ECC.2014.6862414

26. Geuß, M., Panzer, H., Wirtz, A., Lohmann, B.: A general framework for parametric model order reduction by matrix interpolation. In: Workshop on Model Reduction of Parametrized Systems II (MoRePaS II) (2012)
27. Golub, G.H., Van Loan, C.F.: Matrix Computations, 4th edn. Johns Hopkins Studies in the Mathematical Sciences. Johns Hopkins University Press, Baltimore (2013)
28. Gugercin, S., Antoulas, A.C., Beattie, C.: $\mathcal{H}_2$ model reduction for large-scale linear dynamical systems. SIAM J. Matrix Anal. Appl. **30**(2), 609–638 (2008). https://doi.org/10.1137/060666123
29. Gugercin, S., Li, J.R.: Smith-type methods for balanced truncation of large systems. In: Benner, P., Mehrmann, V., Sorensen, D. (eds.) Dimension Reduction of Large-Scale Systems. Lecture Notes in Computational Science and Engineering, vol. 45, pp. 49–82. Springer, Berlin/Heidelberg (2005)
30. Heinkenschloss, M., Sorensen, D.C., Sun, K.: Balanced truncation model reduction for a class of descriptor systems with application to the Oseen equations. SIAM J. Sci. Comput. **30**(2), 1038–1063 (2008). https://doi.org/10.1137/070681910
31. Himpe, C.: Comparing (empirical-Gramian-based) model order reduction algorithms (2020). arXiv preprint arXiv:2002.12226 [math.OC]. https://doi.org/10.1007/978-3-030-72983-7_7
32. Jbilou, K., Messaoudi, A.: A computational method for symmetric Stein matrix equations. In: Van Dooren, P., Bhattacharyya, S.P., Chan, R.H., Olshevsky, V., Routray, A. (eds.) Numerical Linear Algebra in Signals, Systems and Control. Lecture Notes in Electrical Engineering, vol. 80. Springer, New York (2011). https://doi.org/10.1007/978-94-007-0602-6_14
33. Kleinman, D.L.: On an iterative technique for Riccati equation computations. IEEE Trans. Autom. Control **13**(1), 114–115 (1968). https://doi.org/10.1109/TAC.1968.1098829
34. Kürschner, P.: Efficient low-rank solution of large-scale matrix equations. Dissertation, Otto-von-Guericke-Universität, Magdeburg, Germany (2016). http://hdl.handle.net/11858/00-001M-0000-0029-CE18-2. Shaker Verlag, ISBN 978-3-8440-4385-3
35. Lang, N.: Numerical methods for large-scale linear time-varying control systems and related differential matrix equations. Dissertation, Technische Universität Chemnitz, Germany (2017). https://www.logos-verlag.de/cgi-bin/buch/isbn/4700. Logos-Verlag, Berlin, ISBN 978-3-8325-4700-4
36. Lanzon, A., Feng, Y., Anderson, B.D.O.: An iterative algorithm to solve algebraic Riccati equations with an indefinite quadratic term. In: 2007 European Control Conference (ECC), pp. 3033–3039 (2007). https://doi.org/10.23919/ecc.2007.7068239
37. Laub, A.J., Heath, M.T., Paige, C.C., Ward, R.C.: Computation of system balancing transformations and other applications of simultaneous diagonalization algorithms. IEEE Trans. Autom. Control **32**(2), 115–122 (1987). https://doi.org/10.1109/TAC.1987.1104549
38. Li, T., Weng, P.C.Y., Chu, E.K.w., Lin, W.W.: Large-scale Stein and Lyapunov equations, Smith method, and applications. Numer. Algorithms **63**(4), 727–752 (2013). https://doi.org/10.1007/s11075-012-9650-2
39. Lin, Y., Simoncini, V.: A new subspace iteration method for the algebraic Riccati equation. Numer. Linear Algebr. Appl. **22**(1), 26–47 (2015). https://doi.org/10.1002/nla.1936
40. Mlinarić, P., Rave, S., Saak, J.: Parametric model order reduction using pyMOR (2020). arXiv preprint arXiv:2003.05825 [Cs.MS]. https://doi.org/10.1007/978-3-030-72983-7_17
41. Möckel, J., Reis, T., Stykel, T.: Linear-quadratic Gaussian balancing for model reduction of differential-algebraic systems. Int. J. Control **84**(10), 1627–1643 (2011). https://doi.org/10.1080/00207179.2011.622791
42. Moore, B.C.: Principal component analysis in linear systems: controllability, observability, and model reduction. IEEE Trans. Autom. Control **AC-26**(1), 17–32 (1981). https://doi.org/10.1109/TAC.1981.1102568
43. MORPACK (model order reduction package). https://tu-dresden.de/ing/maschinenwesen/ifkm/dmt/forschung/projekte/morpack
44. Mustafa, D., Glover, K.: Controller reduction by $\mathcal{H}_\infty$-balanced truncation. IEEE Trans. Autom. Control **36**(6), 668–682 (1991). https://doi.org/10.1109/9.86941

45. Oberwolfach Benchmark Collection: Steel profile. hosted at MORwiki – Model Order Reduction Wiki (2005). http://modelreduction.org/index.php/Steel_Profile
46. Panzer, H., Mohring, J., Eid, R., Lohmann, B.: Parametric model order reduction by matrix interpolation. at-Automatisierungstechnik **58**(8), 475–484 (2010)
47. Penzl, T.: A cyclic low rank Smith method for large sparse Lyapunov equations. SIAM J. Sci. Comput. **21**(4), 1401–1418 (2000). https://doi.org/10.1137/S1064827598347666
48. Penzl, T.: Eigenvalue decay bounds for solutions of Lyapunov equations: the symmetric case. Syst. Control Lett. **40**, 139–144 (2000). https://doi.org/10.1016/S0167-6911(00)00010-4
49. Penzl, T.: LYAPACK users guide. Technical Report SFB393/00-33, Sonderforschungsbereich 393, Numerische Simulation auf massiv parallelen Rechnern, TU Chemnitz, 09107 Chemnitz, Germany (2000). Available at http://www.tu-chemnitz.de/sfb393/sfb00pr.html
50. Poloni, F., Reis, T.: A deflation approach for large-scale Lur'e equations. SIAM J. Matrix Anal. Appl. **33**(4), 1339–1368 (2012). https://doi.org/10.1137/120861679
51. Pontes Duff, I., Kürschner, P.: Numerical computation and new output bounds for time-limited balanced truncation of discrete-time systems. Linear Algebra Appl. **623**, 367–397. https://doi.org/10.1016/j.laa.2020.09.029 (2019). arXiv preprint arXiv:1902.01652 [math.NA]
52. Rave, S., Saak, J.: A non-stationary thermal-block benchmark model for parametric model order reduction (2020). arXiv preprint arXiv:2003.00846 [math.NA]. https://doi.org/10.1007/978-3-030-72983-7_16
53. Reis, T., Stykel, T.: Balanced truncation model reduction of second-order systems. Math. Comput. Model. Dyn. Syst. **14**(5), 391–406 (2008). https://doi.org/10.1080/13873950701844170
54. Saak, J.: Efficient numerical solution of large scale algebraic matrix equations in PDE control and model order reduction. Dissertation, Technische Universität Chemnitz, Chemnitz, Germany (2009). http://nbn-resolving.de/urn:nbn:de:bsz:ch1-200901642
55. Saak, J., Köhler, M., Benner, P.: M-M.E.S.S. – the matrix equations sparse solvers library. https://doi.org/10.5281/zenodo.632897. See also: https://www.mpi-magdeburg.mpg.de/projects/mess
56. Saak, J., Voigt, M.: Model reduction of constrained mechanical systems in M-M.E.S.S. IFAC-PapersOnLine 9th Vienna International Conference on Mathematical Modelling MATHMOD 2018, Vienna, Austria, 21–23 Feb 2018 **51**(2), 661–666 (2018). https://doi.org/10.1016/j.ifacol.2018.03.112
57. Schmidt, M.: Systematic discretization of input/output maps and other contributions to the control of distributed parameter systems. Ph.D. Thesis, Technische Universität Berlin, Berlin (2007). https://doi.org/10.14279/depositonce-1600
58. Shank, S.D., Simoncini, V., Szyld, D.B.: Efficient low-rank solution of generalized Lyapunov equations. Numer. Math. **134**, 327–342 (2016). https://doi.org/10.1007/s00211-015-0777-7
59. Simoncini, V.: A new iterative method for solving large-scale Lyapunov matrix equations. SIAM J. Sci. Comput. **29**(3), 1268–1288 (2007). https://doi.org/10.1137/06066120X
60. Simoncini, V.: Analysis of the rational Krylov subspace projection method for large-scale algebraic Riccati equations. SIAM J. Matrix Anal. Appl. **37**(4), 1655–1674 (2016). https://doi.org/10.1137/16M1059382
61. Simoncini, V., Druskin, V.: Convergence analysis of projection methods for the numerical solution of large Lyapunov equations. SIAM J. Numer. Anal. **47**(2), 828–843 (2009). https://doi.org/10.1137/070699378
62. Simoncini, V., Szyld, D.B., Monsalve, M.: On two numerical methods for the solution of large-scale algebraic Riccati equations. IMA J. Numer. Anal. **34**(3), 904–920 (2014). https://doi.org/10.1093/imanum/drt015
63. Stillfjord, T.: Low-rank second-order splitting of large-scale differential Riccati equations. IEEE Trans. Autom. Control **60**(10), 2791–2796 (2015). https://doi.org/10.1109/TAC.2015.2398889
64. Stillfjord, T.: Adaptive high-order splitting schemes for large-scale differential Riccati equations. Numer. Algorithms **78**, 1129–1151 (2018). https://doi.org/10.1007/s11075-017-0416-8

65. Tombs, M.S., Postlethwaite, I.: Truncated balanced realization of a stable non-minimal state-space system. Int. J. Control **46**(4), 1319–1330 (1987). https://doi.org/10.1080/00207178708933971
66. Truhar, N., Veselić, K.: An efficient method for estimating the optimal dampers' viscosity for linear vibrating systems using Lyapunov equation. SIAM J. Matrix Anal. Appl. **31**(1), 18–39 (2009). https://doi.org/10.1137/070683052
67. Uddin, M.M.: Computational methods for model reduction of large-scale sparse structured descriptor systems. Dissertation, Department of Mathematics, Otto-von-Guericke University, Magdeburg, Germany (2015). http://nbn-resolving.de/urn:nbn:de:gbv:ma9:1-6535
68. Uddin, M.M.: Computational Methods for Approximation of Large-Scale Dynamical Systems. CRC Press, Boca Raton (2019). https://doi.org/10.1201/9781351028622
69. Weichelt, H.K.: Numerical aspects of flow stabilization by Riccati feedback. Dissertation, Otto-von-Guericke-Universität, Magdeburg, Germany (2016). http://nbn-resolving.de/urn:nbn:de:gbv:ma9:1-8693
70. Yue, Y., Feng, L., Benner, P.: An adaptive pole-matching method for interpolating reduced-order models (2019). arXiv preprint arXiv:1908.00820 [math.NA]

MORLAB—The Model Order Reduction LABoratory

Peter Benner and Steffen W. R. Werner

Abstract For an easy use of model reduction techniques in applications, software solutions are needed. In this paper, we describe the MORLAB, Model Order Reduction LABoratory, toolbox as an efficient implementation of model reduction techniques for dense, medium-scale linear time-invariant systems. Giving an introduction to the underlying programming principles of the toolbox, we show the basic idea of spectral splitting and present an overview about implemented model reduction techniques. Two numerical examples are used to illustrate different use cases of the MORLAB toolbox.

1 Introduction

For the modeling of natural processes as, e.g., fluid dynamics, chemical reactions, or the behavior of electronic circuits, power, or gas transportation networks, dynamical input-output systems are used

$$G : \begin{cases} 0 = f(x(t), Dx(t), \ldots, D^k x(t), u(t)), \\ y(t) = h(x(t), Dx(t), \ldots, D^k x(t), u(t)), \end{cases} \tag{1}$$

with states $x(t) \in \mathbb{R}^n$, inputs $u(t) \in \mathbb{R}^m$ and outputs $y(t) \in \mathbb{R}^p$. The operator D^j denotes the derivative or shift operator of order $j \in \mathbb{N}$ in case of underlying continuous- or discrete-time dynamics. Due to the demand for increasing the accu-

P. Benner · S. W. R. Werner (✉)
Max Planck Institute for Dynamics of Complex Technical Systems, Sandtorstr. 1, 39106 Magdeburg, Germany
e-mail: werner@mpi-magdeburg.mpg.de

P. Benner
e-mail: benner@mpi-magdeburg.mpg.de; peter.benner@ovgu.de

P. Benner
Faculty of Mathematics, Otto von Guericke University, Universitätsplatz 2, 39106 Magdeburg, Germany

P. Benner et al. (eds.), *Model Reduction of Complex Dynamical Systems*, International Series of Numerical Mathematics 171,
https://doi.org/10.1007/978-3-030-72983-7_19

racy of models, the number of states describing (1) is drastically increasing and, consequently, there is a high demand for computational resources (time and memory) when using (1) in simulations or controller design. A solution to this problem is given by model order reduction, which aims for the construction of a surrogate model $\widehat{G}$, with a much smaller number of internal states $\hat{x}(t) \in \mathbb{R}^r$, $r \ll n$, which approximates the input-to-output behavior of (1) such that

$$\|y - \hat{y}\| \leq \text{tol} \cdot \|u\|,$$

for an appropriately defined norm, a given tolerance tol and all admissible inputs u, where $\hat{y}$ is the output of the reduced-order system.

A software solution for model order reduction of dynamical systems is the **MORLAB**, **M**odel **O**rder **R**eduction **LAB**oratory, toolbox. Originating from [6], the toolbox is mainly developed as efficient open-source implementation of established matrix equation-based model reduction methods for dense, medium-scale, linear time-invariant systems, with its implementation compatible with MathWorks MATLAB and GNU Octave. In the latest version [12], MORLAB gives a large variety of balancing-based model reduction methods and also some non-projective methods. Most of those are not known to be implemented somewhere else. In contrast to other software solutions, the general philosophy of MORLAB is to work on invariant subspaces rather than with spectral decompositions, as the model reduction routines in the Control System Toolbox™ and Robust Control Toolbox™ in MATLAB, or projections on hidden manifolds, as, e.g., in the M-M.E.S.S. toolbox [36] and sssMOR toolbox [16]. An overview about different software packages for model order reduction can be found in the MORwiki.[1] Mainly the two spectral projection methods, the matrix sign function, and the right matrix pencil disk function, are used in the underlying implementations. Therefore, MORLAB is suited as backend source code for multi-step model reduction approaches, for example, using a pre-reduction step; see, e.g., [26, 37]. Additionally to model order reduction methods, the toolbox implements efficient matrix equation solvers, system-theoretic subroutines, and evaluation routines to examine original and reduced-order systems in the frequency and time domain. Due to the brevity of the paper, the additional main features are not further considered in detail.

In this paper, we will describe the underlying principles and structures of the MORLAB toolbox and give some applications of the software. The meta data of the latest MORLAB version [12] can be found in Table 1. In the following, Sect. 2 starts with an introduction of the programming principles that were used in MORLAB. Afterward, Sect. 3 gives the underlying ideas of the spectral splitting, on which the toolbox bases, followed by Sect. 4 with an overview about the implemented model reduction methods. In Sect. 5, two applications of using MORLAB as backend software are presented. The paper is concluded by Sect. 6.

[1] https://morwiki.mpi-magdeburg.mpg.de/morwiki/index.php/Comparison_of_Software

Table 1 Code meta data of the latest MORLAB version [12]

Name (shortname)	Model Order Reduction LABoratory (MORLAB)
Version (release-date)	5.0 (2019-08-23)
Identifier (type)	https://doi.org/10.5281/zenodo.3332716 (doi)
Authors	Peter Benner, Steffen W. R. Werner
Orcids	https://orcid.org/0000-0003-3362-4103, https://orcid.org/0000-0003-1667-4862
Topic (type)	Model Reduction (Toolbox)
License (type)	GNU Affero General Public License v3.0 (open)
Languages	MATLAB
Dependencies	MATLAB ($\geq$2012b), Octave ($\geq$4.0.0)
Systems	Linux, MacOS, Windows
Website	https://www.mpi-magdeburg.mpg.de/projects/morlab

2 Code Design Principles

The main aim of the MORLAB toolbox is to give efficient and comparable implementations of many different model reduction methods. Following certain design principles, which will be explained in more detail in the upcoming subsections, the following list of main features briefly summarizes the MORLAB toolbox.

Feature checklist

Open source and free	The toolbox is running under the GNU Affero General Public License v3.0 and is freely available on the project website and on Zenodo.
Fast and exact	Using spectral projection methods, the toolbox can outperform other established software in terms of accuracy and speed.
Unified framework	All model reduction routines share the same interface and allow for quick exchange and easy comparison between the methods.
Configurable	All subroutines can be configured separately using option structs.
Modular	Each subroutine can be called on its own by the user to be used and combined in various ways.
Portable	No binary extensions are required, which allows for running the toolbox with bare MATLAB or Octave installations.

In general, MORLAB uses spectral projection methods for all steps of the model reduction procedure. Figure 1 shows the different stages in MORLAB from the full-

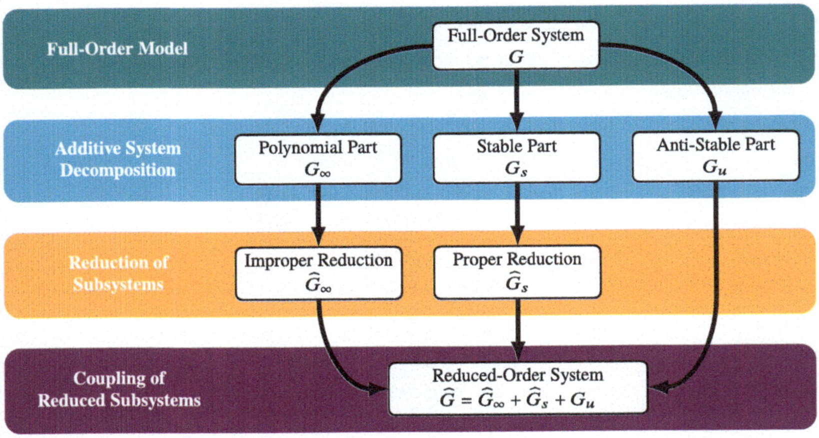

Fig. 1 General MORLAB workflow

order to the reduced-order model. First, the full-order model is decomposed into (at most) three subsystems that can usually be considered independently of each other for the application of model reduction techniques. This first main step, the additive system decomposition, is discussed in more detail in Sect. 3. Afterwards the model reduction methods are applied to the resulting subsystems. An overview of those can be found in Sect. 4. At the end, the reduced subsystems are coupled for the resulting reduced-order model. Based on this basic workflow, the different design principles applied in MORLAB are explained in the following. For the sake of brevity, mainly the model reduction routines are considered.

2.1 Toolbox Structure

The routines in MORLAB follow a strict structure and naming scheme to make them easy to find and interpret in terms of their objective. Describing first the general structure, the routines of the toolbox are divided by their purpose into the following subdirectories:

`checks/`	Contains subroutines that are used for internal checks of data, e.g., if the system structures fit to the model reduction methods.
`demos/`	Contains example scripts showing step-by-step explanations of the different main features of the toolbox.
`eqn_solvers/`	Contains the matrix equation solvers.
`evaluation/`	Contains functions to evaluate the full-order or reduced-order models in the time or frequency domain.

Table 2 Currently supported system classes

Class	System equations	Routine name
Continuous-time standard systems	$\dot{x}(t) = Ax(t) + Bu(t),$ $y(t) = Cx(t) + Du(t)$	`ct_ss`
Discrete-time standard systems	$x_{k+1} = Ax_k + Bu_k,$ $y_k = Cx_k + Du_k$	`dt_ss`
Continuous-time descriptor systems	$E\dot{x}(t) = Ax(t) + Bu(t),$ $y(t) = Cx(t) + Du(t)$	`ct_dss`
Discrete-time descriptor systems	$Ex_{k+1} = Ax_k + Bu_k,$ $y_k = Cx_k + Du_k$	`dt_dss`
Continuous-time second-order systems	$M\ddot{x}(t) = -Kx(t) - E\dot{x}(t) + B_u u(t),$ $y(t) = C_p x(t) + C_v \dot{x}(t) + Du(t)$	`ct_soss`

`mor/` Contains the model reduction routines.

`subroutines/` Contains auxiliary and system-theoretic routines that are used by the model reduction techniques, matrix equation solvers, or evaluation functions.

Considering the naming scheme of MORLAB, each function starts with `ml_` as assignment to the toolbox. This makes MORLAB routines easier to distinguish from other source codes and also allows for easy searching. Mainly, the model reduction routines, but some subroutines are also additionally named after the system classes they can be applied to. Currently, there are routines for continuous- (`ct`) and discrete-time (`dt`) dynamical systems with equations that describe standard (`ss`), descriptor (`dss`) or second-order state spaces (`soss`). The resulting different system classes, supported in the latest MORLAB version, are summarized in Table 2 with their names, system equations and the corresponding naming schemes.

2.2 Function Interfaces

A typical function call in MORLAB can be seen in Fig. 2. From before, we know that the called function is a MORLAB routine for continuous-time standard systems (see Table 2). The actual function name, `bt`, stands for the balanced truncation method. Figure 2 shows the principle idea in MORLAB to give an easy interface to the user. Here, `sys` contains the data of the original system, while `rom` gives the resulting reduced-order model in exactly the same format as the original model was given, indicating the purpose of using reduced-order models as surrogates for the original system. In general, MORLAB supports three different interfaces for model reduction methods. It is possible to pass directly the system matrices to the function (e.g.,

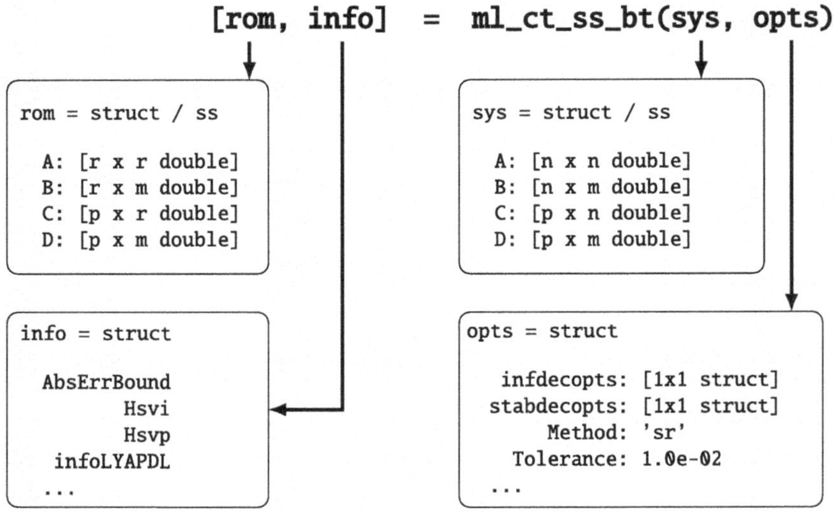

Fig. 2 Example function call of a model reduction routine in MORLAB

ml_ct_ss_bt(A, B, C, D, opts)) or to construct the system as an object by using the native data type struct, with appropriate naming of fields, or the state-space object (ss) introduced by the Control System Toolbox™ in MATLAB or the 'control' package in Octave. The latter format allows for easy interconnection to other model reduction software and also for using system-theoretic routines implemented in the two mentioned software libraries.

The second important part of the MORLAB interface for nearly all routines are the opts and info structs, as shown in Fig. 2. Supporting the feature of configurability, the opts struct allows the user the rearrangement of all computational parameters, which would be usually set by the function itself during runtime. In general, each MORLAB function that allows the user to change optional parameters for the computations has an opts struct for that purpose. As result, higher level routines can contain nested structs to change computational parameters of used subroutines. Figure 3 shows an example opts struct for the ml_ct_ss_bt. This struct again contains entries ending on opts denoting also opts structs for subroutines that are called by the main function. Beside changing computational parameters, a second aim of the opts struct is the a priori determination of system information. For example, if a system is known to be stable, the additive decomposition into the stable and anti-stable subsystems can be turned off using the opts struct to avoid unnecessary computations. For easy application, only entries, which the user wants to change, need to be existing in the struct. Also, the toolbox comes with an option constructor (ml_morlabopts), which creates a complete but empty opts struct for a given function name. The consistent naming of optional parameters between different routines allows the easy reuse of opts structs for different functions.

```
          lyapdlopts: [ 1x1 struct | {ml_morlabopts('ml_lyapdl_sgn_fac')} ]
              Method: [ 'bfsr' | {'sr'} ]
               Order: [ positive integer | {min(10,length(Hsv)) + Nu} ]
    OrderComputation: [ 'order' | {'tolerance'} ]
       stabsignmopts: [ 1x1 struct | {ml_morlabopts('ml_signm')} ]
        stabsylvopts: [ 1x1 struct | {ml_morlabopts('ml_sylv_sgn')} ]
     StoreProjection: [ 1 | {0} ]
           Tolerance: [ nonnegative scalar | {1.0e-02} ]
           UnstabDim: [ integer | {-1} ]

For more details see ml_ct_ss_bt.
```

Fig. 3 Example `opts` struct for the `ml_ct_ss_bt` function

The counterpart of the `opts` struct is the `info` struct. Here, information about the performance and results of the routine are collected. As for `opts`, the `info` struct can be nested as it contains structs starting with `info`, which give information about used subroutines. Also, this struct is used for optional outputs, e.g., projection matrices of a model reduction method can be stored here.

2.3 Documentation

MORLAB comes with an extensive documentation that is accessible in several ways. Each routine has a complete inline documentation, which can be displayed by the `help` command, containing the syntax, description, and literature references for background information. Besides, a complete overview about the existing MORLAB routines with short description can be generated by `help morlab`. As usual for MATLAB toolboxes, a full HTML documentation is provided in the toolbox and demo scripts can be used as a starting how-to to get into the main features of the toolbox.

3 Additive System Decomposition Approach

Most model order reduction methods are in a certain sense restricted with respect to the spectrum of the underlying system matrices, e.g., the classical balanced truncation method can only be applied to first-order systems with finite stable matrix pencils. Other software solutions use therefore either an eigendecomposition of the system matrices in the beginning or apply projections onto the hidden manifolds. In MORLAB, this problem is solved by working directly with the corresponding invariant subspaces of the matrix pencil. As shown in Fig. 1, this results in the additive decomposition of the full-order system into independent reducible subsystems, in the literature known as additive decomposition of the transfer function, which will be coupled at the end again. MORLAB has two different approaches for this additive decomposition based on either the solution of a Sylvester equation or on a block-wise projection approach. This gives MORLAB the advantage of handling unstructured systems, while staying efficient and accurate due to only computing the necessary deflating subspaces. For both approaches, the matrix sign and disk functions are used, as quickly defined below.

Let $Y \in \mathbb{R}^{n\times n}$ be a matrix with no purely imaginary eigenvalues, then the Jordan canonical form of Y can be written as

$$Y = S \begin{bmatrix} J_- & 0 \\ 0 & J_+ \end{bmatrix} S^{-1}, \tag{2}$$

where S is an invertible transformation matrix, J_- contains the k eigenvalues of Y with negative real parts and J_+ the $n-k$ eigenvalues with positive real parts. The *matrix sign function* is then defined as

$$\operatorname{sign}(Y) = S \begin{bmatrix} -I_k & 0 \\ 0 & I_{n-k} \end{bmatrix} S^{-1}, \tag{3}$$

with S the transformation matrix from (2); see, e.g., [35]. Efficient computations can be based on a Newton scheme.

Let $\lambda X - Y$, with $X, Y \in \mathbb{R}^{n\times n}$, be a regular matrix pencil with no eigenvalues on the unit circle and its Weierstrass canonical form be written as

$$\lambda X - Y = W \begin{bmatrix} \lambda I_k - J_0 & 0 \\ 0 & \lambda N - J_\infty \end{bmatrix} T^{-1}, \tag{4}$$

where W, T are invertible transformation matrices, $\lambda I_k - J_0$ contains the k eigenvalues inside the unit disk and $\lambda N - J_\infty$ the $n-k$ eigenvalues outside the unit disk. The *right matrix pencil disk function* is then defined by

$$\operatorname{disk}(Y, X) = T \left(\lambda \begin{bmatrix} 0 & 0 \\ 0 & I_{n-k} \end{bmatrix} - \begin{bmatrix} I_k & 0 \\ 0 & 0 \end{bmatrix} \right) T^{-1}, \tag{5}$$

with T, the right transformation matrix from (4). The computation follows the inverse-free iteration [1, 5] and a subspace extraction method [7, 38].

In the following subsections, the ideas of the additive decomposition for two general system classes are quickly summarized.

3.1 Standard System Case

Assume a continuous-time standard system

$$\begin{aligned} \dot{x}(t) &= Ax(t) + Bu(t), \\ y(t) &= Cx(t) + Du(t), \end{aligned} \tag{6}$$

with $A \in \mathbb{R}^{n\times n}$, $B \in \mathbb{R}^{n\times m}$, $C \in \mathbb{R}^{p\times n}$, $D \in \mathbb{R}^{p\times m}$, A having no eigenvalues on the imaginary axis and its representation in the frequency domain by the corresponding transfer function

$$G(s) = C(sI_n - A)^{-1}B + D,$$

for $s \in \mathbb{C}$. Most model reduction methods can only be applied to asymptotically stable systems, which means in case of (6) that A has only eigenvalues with negative real parts. Nevertheless, model reduction methods can be applied by decomposing the system (6) into two subsystems, where the system matrices contain either the stable or anti-stable system part, i.e., we search for a transformation matrix T such that

$$T^{-1}AT = \begin{bmatrix} A_s & 0 \\ 0 & A_u \end{bmatrix},$$

where A_s contains only the stable and A_u the anti-stable eigenvalues. Using T as state-space transformation and partitioning accordingly the input and output matrices yields the additive system decomposition of the system's transfer function

$$G(s) = G_s(s) + G_u(s).$$

Applying the matrix sign function (3) to A gives the appropriate spectral splitting, where the spectral projectors onto the deflating subspaces are given as

$$\mathcal{P}_s = \frac{1}{2}(I_n - \operatorname{sign}(A)) \quad \text{and} \quad \mathcal{P}_u = \frac{1}{2}(I_n + \operatorname{sign}(A)).$$

Let $QR\Pi^{\mathsf{T}} = I_n - \operatorname{sign}(A)$ be a pivoted QR decomposition, the dimension of the deflating subspace corresponding to the eigenvalues with negative real part is given by $0.5(n + \operatorname{tr}(\operatorname{sign}(A)))$, and we get

$$Q^{\mathsf{T}}AQ = \begin{bmatrix} A_s & W_A \\ 0 & A_u \end{bmatrix}.$$

By solving the standard Sylvester equation,

$$-A_uX + XA_s - W_A = 0, \tag{7}$$

the final transformation matrix and its inverse are given by

$$T = Q \begin{bmatrix} I_k & X \\ 0 & I_{n-k} \end{bmatrix} \quad \text{and} \quad T^{-1} = \begin{bmatrix} I_k & -X \\ 0 & I_{n-k} \end{bmatrix} Q^{\mathsf{T}}. \tag{8}$$

The MORLAB implementation uses the Newton iteration with Frobenius norm scaling for the computation of the matrix sign function as well as a matrix sign function-based solver for the Sylvester equation (7). Note that the actual transformation matrix (8) is never set up completely but only applied block-wise on the original system to avoid unnecessary computations.

Remark 1 (*Splitting of discrete-time standard systems*) In case of discrete-time standard systems, the implementation involves the matrix sign function of $(A + I_n)^{-1}(A - I_n)$ and the solution of the discrete-time Sylvester equation $A_u^{-1} X A_s - X - A_u^{-1} W_A = 0$ for doing the spectral splitting with respect to the unit circle.

3.2 Descriptor System Case

Now, we consider the case of continuous-time descriptor systems

$$\begin{aligned} E\dot{x}(t) &= Ax(t) + Bu(t), \\ y(t) &= Cx(t) + Du(t), \end{aligned} \tag{9}$$

with $E, A \in \mathbb{R}^{n \times n}$, $B \in \mathbb{R}^{n \times m}$, $C \in \mathbb{R}^{p \times n}$, $D \in \mathbb{R}^{p \times m}$, $\lambda E - A$ having no finite eigenvalues on the imaginary axis and its representation in the frequency domain by the corresponding transfer function

$$G(s) = C(sE - A)^{-1}B + D, \quad s \in \mathbb{C}.$$

In contrast to the previous section, an additional splitting for the algebraic part corresponding to the infinite eigenvalues of $\lambda E - A$ is necessary, i.e., we search for transformation matrices W, T such that

$$W(\lambda E - A)T = \lambda \begin{bmatrix} E_s & 0 & 0 \\ 0 & E_u & 0 \\ 0 & 0 & E_\infty \end{bmatrix} - \begin{bmatrix} A_s & 0 & 0 \\ 0 & A_u & 0 \\ 0 & 0 & A_\infty \end{bmatrix}, \tag{10}$$

where $\lambda E_s - A_s$ contains the finite stable eigenvalues, $\lambda E_u - A_u$ the finite anti-stable eigenvalues and $\lambda E_\infty - A_\infty$ only infinite eigenvalues. Then, the system and its transfer function accordingly decouple into the different parts

$$G(s) = G_s(s) + G_u(s) + G_\infty(s),$$

as shown in Fig. 1. For this purpose, the [14, Theorem 3] is used to construct block-wise orthogonal transformation matrices.

First, the splitting of the algebraic part is performed as $G = G_{su} + G_{\infty}$ by using the matrix disk function. In fact, the inverse-free iteration is applied to the matrix pencil $\lambda(\alpha A) - E$ for appropriate scaling parameter α to compute matrices $\widetilde{A}$ and $\widetilde{E}$, whose null spaces are the deflating subspaces of $\lambda(\alpha A) - E$ corresponding to the eigenvalues inside and outside the unit circle, respectively; see [1, 5]. Using a stabilized subspace extraction method [7, 38], the orthogonal projection matrices can be obtained and according to [14] combined into appropriate transformation matrices to get

$$\widetilde{W}(\lambda E - A)\widetilde{T} = \lambda \begin{bmatrix} E_{su} & 0 \\ 0 & E_{\infty} \end{bmatrix} - \begin{bmatrix} A_{su} & 0 \\ 0 & A_{\infty} \end{bmatrix},$$

where $\lambda E_{su} - A_{su}$ contains all the finite eigenvalues. Afterward, the generalized matrix sign function, working implicitly on the spectrum of $E_{su}^{-1} A_{su}$, is used such that the null spaces of $E_{su} - \operatorname{sign}(A_{su}, E_{su})$, and $E_{su} + \operatorname{sign}(A_{su}, E_{su})$ are the deflating subspaces corresponding to the eigenvalues left and right of the imaginary axis, respectively. Using the same subspace extraction method and block transformation, the block diagonalization (10) is accomplished.

Remark 2 (*Splitting of discrete-time descriptor systems*) In the discrete-time descriptor case, the second splitting with respect to the imaginary axis needs to be replaced by a splitting with respect to the unit disk. Although, this is the actual nature of the matrix disk function, for performance reasons, the generalized matrix sign function is used as $\operatorname{sign}(A_{su} - E_{su}, A_{su} + E_{su})$ replaces the sign functions above.

4 Model Reduction with the MORLAB Toolbox

Most of the model reduction methods in MORLAB belong to the class of projection-based model reduction, i.e., we are searching for truncation matrices $W, T \in \mathbb{R}^{n \times r}$, which are used to project the state space, $x \approx T\hat{x}$, and the corresponding equations. For example, given a continuous-time descriptor system (9), the reduced-order system is computed by

$$\begin{aligned} \underbrace{W^{\mathsf{T}} E T}_{\widehat{E}} \dot{\hat{x}}(t) &= \underbrace{W^{\mathsf{T}} A T}_{\widehat{A}} \hat{x}(t) + \underbrace{W^{\mathsf{T}} B}_{\widehat{B}} u(t), \\ \hat{y}(t) &= \underbrace{C T}_{\widehat{C}} \hat{x}(t) + \underbrace{D}_{\widehat{D}} u(t), \end{aligned} \tag{11}$$

with $\widehat{E}, \widehat{A} \in \mathbb{R}^{r \times r}$, $\widehat{B} \in \mathbb{R}^{r \times m}$, $\widehat{C} \in \mathbb{R}^{p \times r}$ and $\widehat{D} = D$. In the following, a very brief overview about the implemented model reduction methods in MORLAB is provided.

4.1 First-Order Methods

For the sake of generality in the MORLAB setting, only the method abbreviations are mentioned here. According to the naming scheme, see Sect. 2 and Fig. 2, the abbreviations have to be connected with the system classes to give the actual MORLAB function.

One of the oldest ideas for model reduction, and fitting with the spectral splitting approach from before, is modal truncation. While originally, a part of the eigenvector basis was used for the projection [18], the deflating subspaces from Sect. 3 are an appropriate choice when using shifting and scaling on the spectrum of the system matrices.

A large part of the model reduction methods in MORLAB are so-called balancing-related methods. In classical balanced truncation [29], the continuous-time Lyapunov equations

$$\begin{aligned} AP + PA^{\mathsf{T}} + BB^{\mathsf{T}} &= 0, \\ A^{\mathsf{T}}Q + QA + C^{\mathsf{T}}C &= 0, \end{aligned} \tag{12}$$

are solved for the system Gramians P and Q, which are then used by, e.g., the square root or balancing-free square root method to compute the reduced-order projection matrices; see, e.g., [27, 40]. The balancing-related methods are based on the idea of balanced truncation but replace the Lyapunov equations (12) by other matrix equations, which infuse different properties to the resulting methods. Some comments on the implementation of balancing-related methods in MORLAB are given for previous versions in [10] for the standard system case and the general idea of the implementation of model reduction for descriptor systems is given in [11].

Also, the Hankel-norm approximation is implemented. This method is non-projection-based, i.e., by construction, there are no W, T fulfilling (11) and also $\widehat{D} = D$ does not hold anymore. This method solves the optimal approximation problem in the Hankel semi-norm and is also a good guess for the $\mathcal{H}_\infty$ approximation problem [14, 21]. It can be seen as a refinement of the balanced truncation method since it is also based on the solution of (12).

As an overview for the current MORLAB version, Table 3 shows all the implemented model reduction methods for first-order continuous-time systems, with their routine abbreviation, comments on their properties and references for the standard and descriptor versions.

Remark 3 (*Discrete-time model reduction methods*) Currently, only the methods `mt`, `bt`, and `lqgbt` have discrete-time implementations for the standard and descriptor system cases. Discrete-time equivalents of the continuous-time matrix equations are solved for those methods.

Table 3 First-order model reduction methods

Method	Routine name	Comment	References
Balanced truncation	`bt`	preserves stability	[27, 29]
Balanced stochastic truncation	`bst`	preserves minimal phase	[9, 22]
Frequency-limited balanced truncation	`flbt`	local frequency approx.	[20, 24]
Time-limited balanced truncation	`tlbt`	local time approx.	[20, 23]
LQG balanced truncation	`lqgbt`	unstable system reduction	[9, 25]
$\mathcal{H}_\infty$ balanced truncation	`hinfbt`	unstable system reduction	[30]
Positive-real balanced truncation	`prbt`	preserves passivity	[19, 34]
Bounded-real balanced truncation	`brbt`	preserves contractivity	[32, 34]
Modal truncation	`mt`	preserves spectrum parts	[8, 18]
Hankel-norm approximation	`hna`	best approx. in Hankel-norm	[14, 21]

4.2 Second-Order Methods

In case of systems with second-order time derivatives, the toolbox implements different structure-preserving approaches. Given the system structure from Table 2, the reduced-order models will also have the form

$$\begin{aligned} \widehat{M}\ddot{\hat{x}}(t) &= -\widehat{K}\hat{x}(t) - \widehat{E}\dot{\hat{x}}(t) + \widehat{B}_u u(t), \\ \hat{y}(t) &= \widehat{C}_p\hat{x}(t) + \widehat{C}_v\dot{\hat{x}}(t) + \widehat{D}u(t), \end{aligned} \tag{13}$$

with $\widehat{M}\ \widehat{E}, \widehat{K} \in \mathbb{R}^{r\times r}$, $\widehat{B}_u \in \mathbb{R}^{r\times m}$, $\widehat{C}_p, \widehat{C}_v \in \mathbb{R}^{p\times r}$ and $\widehat{D} \in \mathbb{R}^{p\times m}$. MORLAB implements the second-order balanced truncation and balancing-related methods for this purpose. Originating in [17, 28, 33], the second-order balanced truncation approach uses a first-order realization of the original second-order system and then restricts to parts of the system Gramians to result in (13). In [13], a collection of the different construction formulas can be found that are all implemented in MORLAB, as well as the frequency- and time-limited second-order balanced truncation methods, which are also implemented in MORLAB. The naming of the methods follows the previous subsection.

5 Numerical Examples

In the following, two benchmark examples are shown to demonstrate possible applications of the MORLAB toolbox. The experiments reported here have been executed on a machine with 2 Intel(R) Xeon(R) Silver 4110 CPU processors running at 2.10 GHz and equipped with 192 GB total main memory. The computer is running on CentOS Linux release 7.5.1804 (Core) and using MATLAB 9.4.0.813654 (R2018a) with the MORLAB toolbox version 5.0 [12].

The source code of the implementations used to compute the presented results can be obtained from

https://doi.org/10.5281/zenodo.3865495

and is authored by Jens Saak and Steffen W. R. Werner.

5.1 Butterfly Gyroscope

As a first numerical example, we consider the butterfly gyroscope benchmark example from [31]; see [15] for the background. We will use the MORLAB toolbox as backend software for a two-step model reduction approach. Thereby, a fast pre-reduction step is used to create an accurate, medium-scale approximation of the original model and, afterward, more sophisticated model reduction methods are used to construct the final reduced-order model, see, e.g., [26, 37]. The model we consider now involves second-order time derivatives as it has the form

$$\begin{aligned} M\ddot{x}(t) + E\dot{x}(t) + Kx(t) &= B_{\mathrm{u}}u(t), \\ y(t) &= C_{\mathrm{p}}x(t), \end{aligned} \tag{14}$$

with a state-space dimension $n = 17\,361$ and $m = 1$, $p = 12$ inputs and outputs, respectively.

As in [37], we use the structure-preserving interpolation framework from [4] as an efficient pre-reduction method that preserves the system structure in an intermediate medium-scale approximation. Therefore, we compute the following matrices:

$$(\sigma_j^2 M + \sigma_j E + K)^{-1} B_{\mathrm{u}} = V_j \quad \text{and} \quad (\sigma_j^2 M + \sigma_j E + K)^{-\mathsf{H}} C_{\mathrm{p}}^{\mathsf{T}} = W_j,$$

with the interpolation point $\sigma_j \in \mathbb{C}$. For our experiments, we choose the interpolation points as $\pm$`logspace(0, 8, 100)`i and concatenate afterwards all V_j and W_j into a single truncation matrix, which is orthogonalized using the economy size QR decomposition. This approach leads to an intermediate reduced-order model with the structure as in (14) and state-space dimension 2 600. The frequency response of the original model and intermediate approximation can be seen in Fig. 4. The computa-

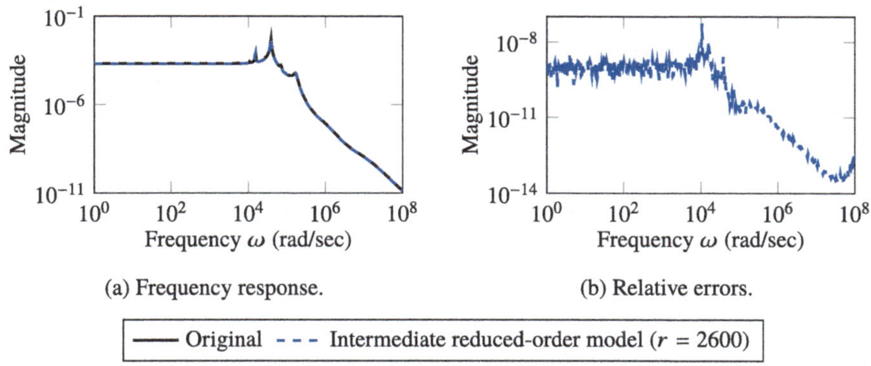

Fig. 4 Frequency-domain results of the intermediate reduced-order model for the butterfly gyroscope

tion of the frequency response of the original system took around 4.3 min, while the computation of the intermediate reduction took 2.4 and 3.4 min for the computation of the intermediate model's frequency response. The intermediate approximation is very accurate as it can be seen by the relative error in the right plot of Fig. 4, which was computed by

$$\frac{\|G(i\omega) - \widehat{G}(i\omega)\|_2}{\|G(i\omega)\|_2},$$

in the frequency range $\omega \in [10^0, 10^8]$.

The intermediate model is still too large for practical application, therefore, we apply now the second-order balanced truncation methods from MORLAB to it. The toolbox supports an all-at-once approach for balancing-related model reduction, i.e., the underlying Gramians are computed once and then used for several different reduced-order models. Therefore, we can compute all 8 different second-order balancing formulas from [13] at the same time and compare them afterward. Those computations took around 1.5 min. Figure 5 shows the resulting reduced-order models in the frequency domain with their relative approximation errors. Both plots were directly generated with the MORLAB routine `ml_sigmaplot`, which computes sigma and error plots for an arbitrary number of given models. The computation of all 8 frequency responses for the final reduced-order models took 0.34 s. The notation in the legend follows the formulas from [13]. Except for the pm and vpm models, all other reduced-order models are asymptotically stable. Clearly, the winners are p, v, pv, and so, which all have, in principle, the same size and error behavior.

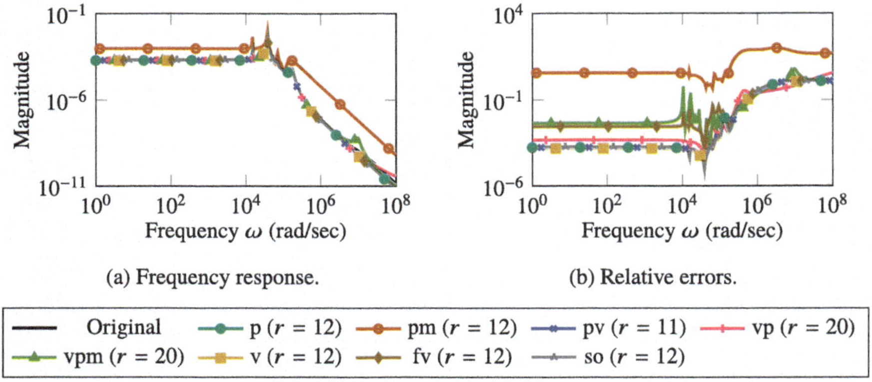

(a) Frequency response.

(b) Relative errors.

Fig. 5 Frequency-domain results of the final reduced-order models for the butterfly gyroscope

5.2 *Parametric Thermal Block Model*

As second example, we consider the parametric thermal block model as described in a Chap. 16 of this volume with the single parameter setup; see also [39]. Following this description, we consider the first-order generalized state-space system

$$\begin{aligned} E\dot{x}(t;\mu) &= A(\mu)x(t;\mu) + Bu(t), \\ y(t;\mu) &= Cx(t;\mu), \end{aligned} \tag{15}$$

where $A(\mu) = A_0 + \mu\,(0.2A_1 + 0.4A_2 + 0.6A_3 + 0.8A_4)$, with the parameter $\mu \in [10^{-6}, 10^2]$, the state-space dimension $n = 7\,488$ and $m = 1$, $p = 4$ inputs and outputs, respectively. The matrix pencil $\lambda E - A(\mu)$ is finite and stable for all parameter values μ in the range of interest. This example is also used by other reported software packages in this volume for easier comparison; see Chap. 7, 17, 18.

Although MORLAB does not implement parametric system classes yet, we will use the toolbox as model reduction backend for two-step parametric model reduction methods, i.e., we use MORLAB for the computation of nonparametric reduced-order models, which are afterward combined into a parametric reduced-order model. In the following, we will introduce some concepts of two-step parametric model order reduction, which are then applied to (15).

The first idea is taken from [3]. Given some nonparametric reduced-order models G_j computed for parameter samples μ_j, $j = 1, \ldots, k$, a global parameter interpolating system can be constructed in the frequency domain using Lagrange interpolation as

$$\widehat{G}(s,\mu) = \sum_{j=1}^{k} \ell_j(\mu) G_j(s), \tag{16}$$

with $\ell_j(\mu)$ Lagrange basis functions in the parameter μ with the knot vector $\mu_1, \ldots, \mu_k$. Rewriting the sum (16) gives a realization for the interpolating reduced-order model

$$\widehat{E} = \begin{bmatrix} \widehat{E}_1 & & \\ & \ddots & \\ & & \widehat{E}_k \end{bmatrix}, \quad \widehat{A} = \begin{bmatrix} \widehat{A}_1 & & \\ & \ddots & \\ & & \widehat{A}_k \end{bmatrix},$$

$$\widehat{B} = \begin{bmatrix} \widehat{B}_1 \\ \vdots \\ \widehat{B}_k \end{bmatrix}, \qquad \widehat{C} = \left[\ell_1(\mu)\widehat{C}_1, \ \ldots, \ \ell_k(\mu)\widehat{C}_k\right],$$

where $\widehat{E}_j$, $\widehat{A}_j$, $\widehat{B}_j$, $\widehat{C}_j$ are the matrices of the local reduced-order models. Thinking of other scalar function approximation methods, easy extensions of (16) come into mind. Replacing the Lagrange basis functions $\ell_j(\mu)$ by linear B-splines $b_{1,j}(\mu)$ over the knot vector $\mu_1, \ldots, \mu_k$, we can construct a piecewise linear interpolating reduced-order model. Another idea would be to use the variation diminishing B-spline approximation, which just needs some modifications of the knot vector used for the basis functions. In general, this transfer function interpolation-based approach comes with several advantages. First, it does not matter how the local reduced-order models were computed or which size they have. If all local reduced-order models are stable, the global interpolating one will be stable by construction, too. Also, instead of setting up the complete reduced-order model, it can be advantageous to use the local reduced-order models for simulations in parallel and combine the results at the end by the parametric output matrix.

A different approach is given by the piecewise approximation; see, e.g., [2]. For this method, let the local reduced-order models be computed by projection methods and the truncation matrices be collected as $W = [W_1, \ldots, W_k]$ and $T = [T_1, \ldots, T_k]$. The parametric reduced-order system is then computed using W, T as truncation matrices on the original system, as in (11). Concerning the parametric matrix $A(\mu)$ in (15), we note that

$$\begin{aligned} W^{\mathsf{T}} A(\mu) T &= W^{\mathsf{T}} A_0 T + \mu \left(0.2\, W^{\mathsf{T}} A_1 T + 0.4\, W^{\mathsf{T}} A_2 T + 0.6\, W^{\mathsf{T}} A_3 T + 0.8\, W^{\mathsf{T}} A_4 T\right) \\ &= \widehat{A}_0 + \mu \left(0.2\widehat{A}_1 + 0.4\widehat{A}_2 + 0.6\widehat{A}_3 + 0.8\widehat{A}_4\right). \end{aligned}$$

Using this method, we can preserve the exact parameter dependency in the reduced-order model. Variants of it, for example, use column compression of T and W to control the size of the resulting reduced-order model. Also, it needs to be noted that by concatenation of the projection matrices, original properties like stability preservation can be lost. Therefore, modifications like a one-sided projection by combining $[W, T]$ into a single basis can be used to handle most systems.

For our numerical example, we will use the following setup. For the parameter sampling points, we use 10 logarithmically distributed Chebyshev roots, i.e., let $\nu_1, \ldots, \nu_k$ be the Chebyshev roots in the interval $[-6, 2]$, the sampling points are given as $\mu_j = 10^{\nu_j}$. The local reduced-order models are computed by the balanced truncation routine from MORLAB (`ml_ct_dss_bt`) using 10^{-4} for the absolute error bound and we save the reduced-order models as well as the truncation matrices for the parametric approaches. The computation of the 10 nonparametric reduced-order models took 32.4 min. The following different parametric reduced-order models are then computed using the nonparametric results from MORLAB:

- two-sided piecewise approximation (TwoPW), where the final truncated projection matrices were compressed using singular value decompositions and a relative truncation tolerance of 10^{-4},
- one-sided piecewise approximation (OnePW), where the final truncated projection matrix was compressed using the basis concatenation and the singular value decomposition with relative truncation tolerance 10^{-4},
- transfer function interpolation using Lagrange basis functions (InterpLag),
- transfer function interpolation using linear B-splines (InterpBspline),
- transfer function approximation using the variation diminishing approximation with quadratic B-spline basis functions (VarDABspline).

Figure 6 shows the results in the frequency domain, where we computed the point wise relative errors as

$$\frac{\|G(i\omega, \mu) - \widehat{G}(i\omega, \mu)\|_2}{\|G(i\omega, \mu)\|_2},$$

in the ranges $\omega \in [10^{-4}, 10^{4}]$ and $\mu \in [10^{-6}, 10^{2}]$. The piecewise methods, TwoPW and OnePW, are the clear winners of the comparison. We note that TwoPW is unstable for all parameters, while OnePW is stable. Also, the interpolation approaches work nicely, where the interpolation property is clearly visible in the plots. The variation diminishing B-spline result, VarDABspline, seems to be a smoother version of InterpBspline.

In the time domain, we simulate the parametric systems using a pre-sampled white noise input signal. The relative errors shown in Fig. 7 are computed by

$$\sqrt{\sum_{j=1}^{4} \frac{|y_j(t; \mu) - \hat{y}_j(t; \mu)|^2}{|y_j(t; \mu)|^2}}$$

in the ranges $t \in [0, 1]$ and $\mu \in [10^{-6}, 10^{2}]$. The TwoPW is not shown in Fig. 7, since due to the instability in all parameters, no useful results were computed during the simulation. For the rest, we see that again OnePW performs overall very good. Also, we see that the B-spline approaches and classical Lagrange interpolation give more or less the same results. The computation times for the frequency responses as well as for the time simulations as shown in Figs. 7 and 6 for 100 logarithmically equidistant chosen test parameters are given in Table 4.

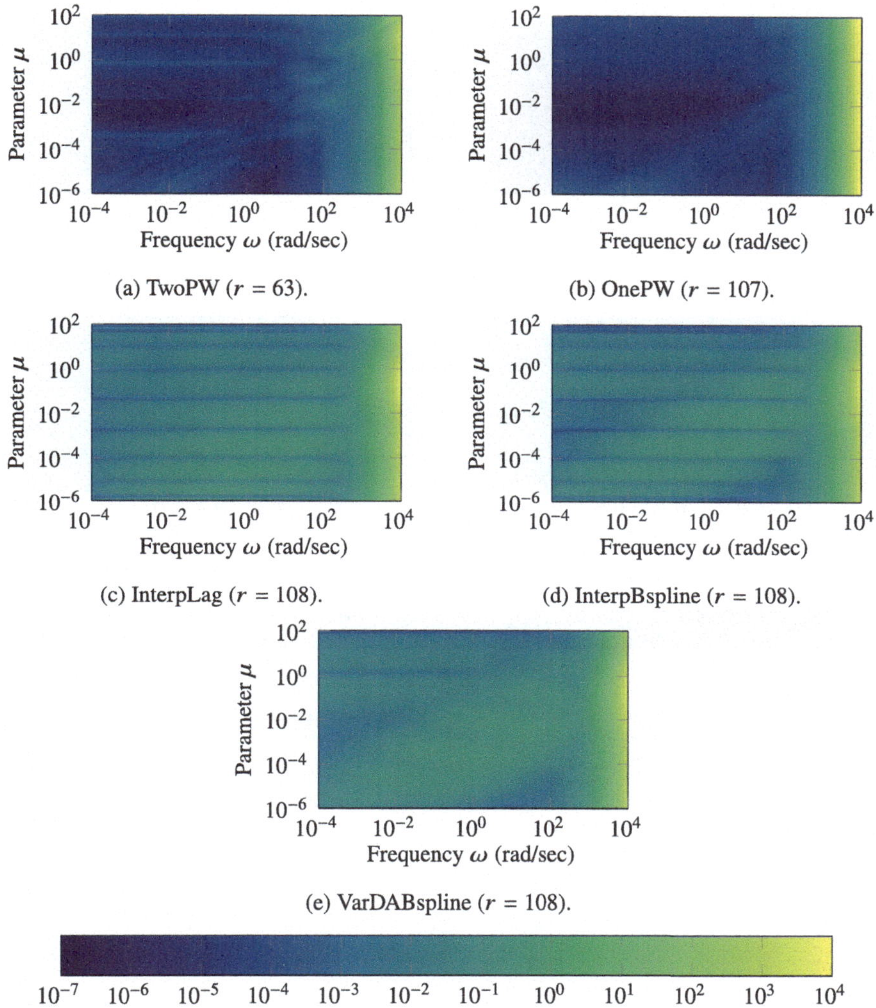

Fig. 6 Relative errors in the frequency domain of different parametric extensions for the thermal block model

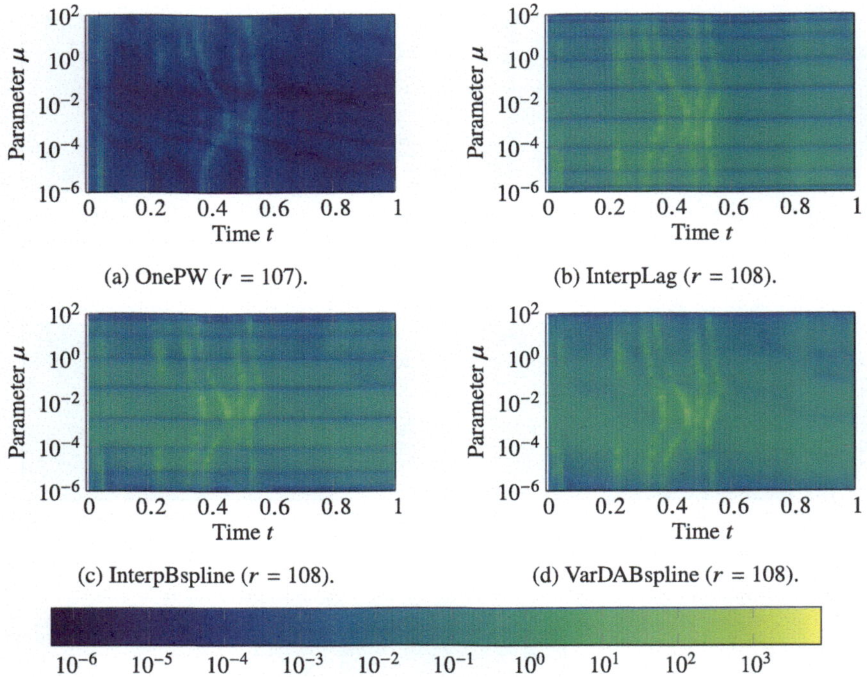

(a) OnePW ($r = 107$). (b) InterpLag ($r = 108$). (c) InterpBspline ($r = 108$). (d) VarDABspline ($r = 108$).

Fig. 7 Relative errors in the time simulation of different parametric extensions for the thermal block model

Table 4 Computation times of frequency responses and time simulations for the parametric thermal block model for 100 test parameters in seconds

System	Frequency response	Time simulation
Original model	647.51	31.65
TwoPW	1.32	—
OnePW	2.29	0.71
InterpLag	2.17	1.10
InterpBspline	2.23	0.69
VarDABspline	2.20	0.70

6 Conclusions

We presented the MORLAB toolbox as an efficient software solution for model reduction of dense, medium-scale linear time-invariant systems. We gave an overview of the main features and structure of the toolbox, as well as underlying programming principles. An important point when considering unstructured systems is the spectral splitting, which we showed in MORLAB to be based on spectral projection methods. Following the computational steps led to an overview of the implemented model reduction methods in MORLAB. We gave two numerical examples to illustrate how MORLAB can be used as backend software for different system types. In the first example, MORLAB provided the efficient, structure-preserving implementation of sophisticated model reduction methods that are used in two-step approaches. In the second example, we used MORLAB to generate local reduced-order models that were afterward combined by different techniques to construct parametric reduced-order systems.

Acknowledgements This work was supported by the German Research Foundation (DFG) Research Training Group 2297 "MathCoRe", Magdeburg.

References

1. Bai, Z., Demmel, J., Gu, M.: An inverse free parallel spectral divide and conquer algorithm for nonsymmetric eigenproblems. Numer. Math. **76**(3), 279–308 (1997). https://doi.org/10.1007/s002110050264
2. Baur, U., Beattie, C.A., Benner, P., Gugercin, S.: Interpolatory projection methods for parameterized model reduction. SIAM J. Sci. Comput. **33**(5), 2489–2518 (2011). https://doi.org/10.1137/090776925
3. Baur, U., Benner, P.: Modellreduktion für parametrisierte Systeme durch balanciertes Abschneiden und Interpolation (Model reduction for parametric systems using balanced truncation and interpolation). at-Automatisierungstechnik **57**(8), 411–420 (2009). https://doi.org/10.1524/auto.2009.0787
4. Beattie, C.A., Gugercin, S.: Interpolatory projection methods for structure-preserving model reduction. Syst. Control Lett. **58**(3), 225–232 (2009). https://doi.org/10.1016/j.sysconle.2008.10.016
5. Benner, P.: Contributions to the Numerical Solution of Algebraic Riccati Equations and Related Eigenvalue Problems. Logos-Verlag, Berlin, Germany, : Also: Dissertation. Fakultät für Mathematik, TU Chemnitz-Zwickau (1997).
6. Benner, P.: A MATLAB repository for model reduction based on spectral projection. In: 2006 IEEE Conference on Computer Aided Control System Design, 2006 IEEE International Conference on Control Applications, 2006 IEEE International Symposium on Intelligent Control, pp. 19–24 (2006). https://doi.org/10.1109/CACSD-CCA-ISIC.2006.4776618
7. Benner, P.: Partial stabilization of descriptor systems using spectral projectors. In: Van Dooren, P., Bhattacharyya, S.P., Chan, R.H., Olshevsky, V., Routray, A. (eds.) Numerical Linear Algebra in Signals, Systems and Control. Lecture Notes in Electrical Engineering, vol. 80, pp. 55–76. Springer, Netherlands (2011). https://doi.org/10.1007/978-94-007-0602-6_3
8. Benner, P., Quintana-Ortí, E.S.: Model reduction based on spectral projection methods. In: Benner, P., Mehrmann, V., Sorensen, D. (eds.) Dimension Reduction of Large-Scale Systems. Lecture Notes in Computational Science and Engineering, vol. 45, pp. 5–45. Springer, Berlin/Heidelberg (2005). https://doi.org/10.1007/3-540-27909-1_1
9. Benner, P., Stykel, T.: Model order reduction for differential-algebraic equations: a survey. In: Ilchmann, A., Reis, T. (eds.) Surveys in Differential-Algebraic Equations IV, Differential-Algebraic Equations Forum, pp. 107–160. Springer International Publishing, Cham (2017). https://doi.org/10.1007/978-3-319-46618-7_3
10. Benner, P., Werner, S.W.R.: Balancing related model reduction with the MORLAB toolbox. Proc. Appl. Math. Mech. **18**(1), e201800083 (2018). https://doi.org/10.1002/pamm.201800083
11. Benner, P., Werner, S.W.R.: Model reduction of descriptor systems with the MORLAB toolbox. In: IFAC-PapersOnLine 9th Vienna International Conference on Mathematical Modelling

MATHMOD 2018, Vienna, Austria, 21–23 Feb 2018 **51**(2), 547–552 (2018). https://doi.org/10.1016/j.ifacol.2018.03.092

12. Benner, P., Werner, S.W.R.: MORLAB – Model Order Reduction LABoratory (version 5.0) (2019). https://doi.org/10.5281/zenodo.3332716. See also: http://www.mpi-magdeburg.mpg.de/projects/morlab
13. Benner, P., Werner, S.W.R.: Frequency- and time-limited balanced truncation for large-scale second-order systems. Linear Algebra Appl. **623**, 68–103 (2021). Special issue in honor of P. Van Dooren, Edited by F. Dopico, D. Kressner, N. Mastronardi, V. Mehrmann, and R. Vandebril. https://doi.org/10.1016/j.laa.2020.06.024
14. Benner, P., Werner, S.W.R.: Hankel-norm approximation of large-scale descriptor systems. Adv. Comput. Math. **46**(3), 40 (2020). https://doi.org/10.1007/s10444-020-09750-w
15. Billger, D.: The butterfly gyro. In: Benner, P., Mehrmann, V., Sorensen, D.C. (eds.) Dimension Reduction of Large-Scale Systems. Lecture Notes in Computational Science and Engineering, vol. 45, pp. 349–352. Springer, Berlin/Heidelberg (2005). https://doi.org/10.1007/3-540-27909-1_18
16. Castagnotto, A., Cruz Varona, M., Jeschek, L., Lohmann, B.: sss & sssMOR: analysis and reduction of large-scale dynamic systems in MATLAB. at-Automatisierungstechnik **65**(2), 134–150 (2017). https://doi.org/10.1515/auto-2016-0137
17. Chahlaoui, Y., Lemonnier, D., Vandendorpe, A., Van Dooren, P.: Second-order balanced truncation. Linear Algebr. Appl. **415**(2–3), 373–384 (2006). https://doi.org/10.1016/j.laa.2004.03.032
18. Davison, E.J.: A method for simplifying linear dynamic systems. IEEE Trans. Autom. Control **AC-11**, 93–101 (1966). https://doi.org/10.1109/TAC.1966.1098264
19. Desai, U.B., Pal, D.: A transformation approach to stochastic model reduction. IEEE Trans. Autom. Control **29**(12), 1097–1100 (1984). https://doi.org/10.1109/TAC.1984.1103438
20. Gawronski, W., Juang, J.N.: Model reduction in limited time and frequency intervals. Int. J. Syst. Sci. **21**(2), 349–376 (1990). https://doi.org/10.1080/00207729008910366
21. Glover, K.: All optimal Hankel-norm approximations of linear multivariable systems and their L^{∞}-error norms. Int. J. Control **39**(6), 1115–1193 (1984). https://doi.org/10.1080/00207178408933239
22. Green, M.: A relative error bound for balanced stochastic truncation. IEEE Trans. Autom. Control **33**(10), 961–965 (1988). https://doi.org/10.1109/9.7255
23. Haider, K., Ghafoor, A., Imran, M., Malik, F.M.: Model reduction of large scale descriptor systems using time limited Gramians. Asian J. Control **19**(3), 1217–1227 (2017). https://doi.org/10.1002/asjc.1444
24. Imran, M., Ghafoor, A.: Model reduction of descriptor systems using frequency limited Gramians. J. Frankl. Inst. **352**(1), 33–51 (2015). https://doi.org/10.1016/j.jfranklin.2014.10.013
25. Jonckheere, E.A., Silverman, L.M.: A new set of invariants for linear systems - application to reduced order compensator design. IEEE Trans. Autom. Control **28**(10), 953–964 (1983). https://doi.org/10.1109/TAC.1983.1103159
26. Lehner, M., Eberhard, P.: A two-step approach for model reduction in flexible multibody dynamics. Multibody Syst. Dyn. **17**(2–3), 157–176 (2007). https://doi.org/10.1007/s11044-007-9039-5
27. Mehrmann, V., Stykel, T.: Balanced truncation model reduction for large-scale systems in descriptor form. In: Benner, P., Mehrmann, V., Sorensen, D.C. (eds.) Dimension Reduction of Large-Scale Systems. Lecture Notes in Computational Science and Engineering, vol. 45, pp. 83–115. Springer, Berlin/Heidelberg (2005). https://doi.org/10.1007/3-540-27909-1_3
28. Meyer, D.G., Srinivasan, S.: Balancing and model reduction for second-order form linear systems. IEEE Trans. Autom. Control **41**(11), 1632–1644 (1996). https://doi.org/10.1109/9.544000
29. Moore, B.C.: Principal component analysis in linear systems: controllability, observability, and model reduction. IEEE Trans. Autom. Control **AC-26**(1), 17–32 (1981). https://doi.org/10.1109/TAC.1981.1102568

30. Mustafa, D., Glover, K.: Controller reduction by $\mathcal{H}_\infty$-balanced truncation. IEEE Trans. Autom. Control **36**(6), 668–682 (1991). https://doi.org/10.1109/9.86941
31. Oberwolfach Benchmark Collection: Butterfly gyroscope. hosted at MORwiki – Model Order Reduction Wiki (2004). http://modelreduction.org/index.php/Butterfly_Gyroscope
32. Opdenacker, P.C., Jonckheere, E.A.: A contraction mapping preserving balanced reduction scheme and its infinity norm error bounds. IEEE Trans. Circuits Syst. **35**(2), 184–189 (1988). https://doi.org/10.1109/31.1720
33. Reis, T., Stykel, T.: Balanced truncation model reduction of second-order systems. Math. Comput. Model. Dyn. Syst. **14**(5), 391–406 (2008). https://doi.org/10.1080/13873950701844170
34. Reis, T., Stykel, T.: Positive real and bounded real balancing for model reduction of descriptor systems. Int. J. Control **83**(1), 74–88 (2010). https://doi.org/10.1080/00207170903100214
35. Roberts, J.D.: Linear model reduction and solution of the algebraic Riccati equation by use of the sign function. Int. J. Control **32**(4), 677–687 (1980). https://doi.org/10.1080/00207178008922881. (Reprint of Technical Report No. TR-13, CUED/B-Control, Cambridge University, Engineering Department, 1971)
36. Saak, J., Köhler, M., Benner, P.: M-M.E.S.S. – The Matrix Equations Sparse Solvers library. https://doi.org/10.5281/zenodo.632897. See also: https://www.mpi-magdeburg.mpg.de/projects/mess
37. Saak, J., Siebelts, D., Werner, S.W.R.: A comparison of second-order model order reduction methods for an artificial fishtail. at-Automatisierungstechnik **67**(8), 648–667 (2019). https://doi.org/10.1515/auto-2019-0027
38. Sun, X., Quintana-Ortí, E.S.: Spectral division methods for block generalized Schur decompositions. Math. Comput. **73**(248), 1827–1847 (2004). https://doi.org/10.1090/S0025-5718-04-01667-9
39. The MORwiki Community: Thermal block. hosted at MORwiki – Model Order Reduction Wiki (2020). https://morwiki.mpi-magdeburg.mpg.de/morwiki/index.php/Thermal_Block
40. Varga, A.: Efficient minimal realization procedure based on balancing. In: Prepr. of the IMACS Symposium on Modelling and Control of Technological Systems, vol. 2, pp. 42–47 (1991)

Correction to: Reduced-Order Methods in Medical Imaging

Saifon Chaturantabut, Thomas Freeze, Elias Salomão Helou, and Charles H. Lee

Correction to: Chapter "Reduced-Order Methods in Medical Imaging" in: P. Benner et al. (eds.), *Model Reduction of Complex Dynamical Systems*, International Series of Numerical Mathematics 171, https://doi.org/10.1007/978-3-030-72983-7_11

In the original version of this book, one of the author name "Nicole Hemming-Schroeder" has been removed from the Chapter "Reduced-Order Methods in Medical Imaging".

The chapter and book have been updated with the changes.

The updated version of this chapter can be found at
https://doi.org/10.1007/978-3-030-72983-7_11

P. Benner et al. (eds.), *Model Reduction of Complex Dynamical Systems*,
International Series of Numerical Mathematics 171,
https://doi.org/10.1007/978-3-030-72983-7_20

The manufacturer's authorised representative in the EU is Springer Nature Customer Service Centre GmbH, Europaplatz 3, 69115 Heidelberg, Germany. If you have any concerns regarding our products, please contact ProductSafety@springernature.com

Printed and bound by CPI Group (UK) Ltd, Croydon, CR0 4YY
27/07/2026
02177398-0010